Biofortification in Cereals

Rupesh Deshmukh • Altafhusain Nadaf •
Waquar Akhter Ansari • Kashmir Singh •
Humira Sonah
Editors

Biofortification in Cereals

Progress and Prospects

Springer

Editors
Rupesh Deshmukh
Central University of Haryana
Mahendragarh, Haryana, India

Altafhusain Nadaf
Department of Botany
Savitribai Phule Pune University
Pune, Maharashtra, India

Waquar Akhter Ansari
Department of Botany
Savitribai Phule Pune University
Pune, Maharashtra, India

Kashmir Singh
Department of Biotechnology
Panjab University
Chandigarh, India

Humira Sonah
National Agri-Food Biotechnology
Institute (NABI)
Mohali, Punjab, India

ISBN 978-981-19-4307-2 ISBN 978-981-19-4308-9 (eBook)
https://doi.org/10.1007/978-981-19-4308-9

This Springer imprint is published by the registered company Springer Nature Singapore Pte Ltd.
The registered company address is: 152 Beach Road, #21-01/04 Gateway East, Singapore 189721, Singapore

This book is dedicated to ***Prof. Henry T. Nguyen and Prof. Nagendra Kumar Singh,*** *two most eminent agriculture scientists. Their research contribution to the plant genomics field has helped in food security and inspired many scientists around the world.*

Prof. Henry T Nguyen
Curators' Distinguished Professor of Genetics, University of Missouri, USA

Prof. Nagendra Kumar Singh
National Professor, ICAR-National Institute for Plant Biotechnology, India

Contents

Editors and Contributors

About the Editors

Rupesh Deshmukh is working as Associate Professor, at Central University of Haryana, Mahendragarh, Haryana, India. Earlier he was visiting professor at Laval University in Quebec, Canada. He obtained his Ph.D. degree in Agriculture Biotechnology at SRTMU, Nanded, India. He went to Quebec, Canada, for his postdoc. He did another postdoc at Missouri University, USA. His interest spans diverse fields in plant science covering computational biology, gene identification through approaches like QTL mapping and GWAS, and elucidation of the molecular mechanism through the integration of omics approaches. His research group is exploring genomics and genome editing approaches to improve the nutritional quality of rice. He has published five books and more than 100 articles in reputed journals. He is a member of many academic societies and served as an Associate Editor for several reputed journals.

Altafhusain Nadaf is working as a Professor and Head at the Department of Botany, Savitribai Phule Pune University, Pune, in the area of biochemistry and molecular genetics of aromatic rice for the past 23 years. He was awarded Erasmus Mundus Action 2 India4EU II scholarship to visit the University of Bologna, Italy, as a visiting professor and DST-BOYSCAST Fellowship to work as a Visiting Fellow at the Centre for Plant Conservation Genetics, Southern Cross University, Lismore, Australia, for 1 year. He has to his credit four books and more than 60 research papers published in peer-reviewed national and international journals of repute. His research interests are biochemistry and molecular genetics of rice aroma volatiles. He has characterized scented rice varieties of India and other plant species containing rice aroma volatiles following the HS-SPME-GC-MS/FID approach. Moreover, the aroma enhancement and induction of aroma character have been achieved in his laboratory through transgenic approaches.

Waquar Akhter Ansari is working as DS Kothari Postdoctoral Fellow, at the Department of Botany, Savitribai Phule Pune University, Pune, Maharashtra, India. Earlier he worked as Research Associates in different national projects at ICAR-Indian Institute of Vegetable Research and ICAR-National Bureau of

Agriculturally Important Microorganism. He obtained his Ph.D. degree from the Department of Botany at Banaras Hindu University, Varanasi, Uttar Pradesh, India, in the subject of biotechnology. He has more than 10 years of research experience in the field of plant biotechnology. He has published more than 20 papers in reputed journals. He has received several awards for best oral and poster presentations at international and national conferences.

Kashmir Singh is presently working as a Professor at the Department at the Department of Biotechnology, Panjab University, India. He carried out his Ph.D. studies in India (CSIR-IHBT), where he got specialized in biotechnology and plant genomics. He did his Postdoctoral Research at Adam Mickiewicz University, Poland, Missouri State University, USA, and McGill University, Canada. His research is focused on the exploration of genomics for a better understanding of stress tolerance mechanisms in plants. He has published over 80 articles in reputed journals.

Humira Sonah is currently working as Ramalingaswami Fellow at National Agri-Food Biotechnology Institute (NABI), Mohali, India. Before joining NABI, Dr. Sonah was Visiting Professor at University Laval, Canada. She obtained her Ph.D. degree in Biotechnology at Banasthali University, India. Dr. Sonah went to Quebec, Canada, for her postdoc where she worked on the development of cost-efficient genotyping by sequencing (GBS) technology in plants. She and her research team developed GBS protocols for different crop species and subsequently used the SNP genotyping for genome-wide association studies (GWAS). She has developed genome-wide molecular markers for different crops. Her interest covers different fields in plant science including whole genome sequencing, computational biology, QTL mapping, GWAS, genomic selection, and integration of omics approaches. Dr. Sonah has published more than 90 articles in reputed journals. She is serving as a member of many academic societies and as a guest editor in reputed journals. Her name was featured in the "75 under 50: Scientists Shaping Today's India" list published by the Vigyan Prasar, Government of India.

Contributors

Avinash Ade Department of Botany, Savitribai Phule Pune University, Pune, India

Bharti Aggarwal Department of Biotechnology, Panjab University, Chandigarh, India
National Agri-Food Biotechnology Institute (NABI), Mohali, Punjab, India

Shahid Ahmed Division of Crop Improvement, ICAR-Indian Grassland and Fodder Research Institute, Jhansi, UP, India

Muhammad Waqas Ali School of Biosciences, University of Birmingham, Birmingham, UK

Sajad Ali Centre of Research for Development, University of Kashmir, Srinagar, Jammu and Kashmir, India

Waquar Akhter Ansari Department of Botany, Banaras Hindu University, Varanasi, Uttar Pradesh, India
Department of Botany, Savitribai Phule Pune University, Pune, India

Laxmipreeya Behera Department of Agricultural Biotechnology, Odisha University of Agriculture and Technology, Bhubaneswar, India

Ramesh Chauhan CSIR-Institute of Himalayan Bioresource Technology, Palampur, HP, India

Viswanathan Chinnusamy Division of Plant Physiology, ICAR-Indian Agricultural Research Institute, New Delhi, India

Ashish Kumar Dash Department of Soil Science and Agricultural Chemistry, Odisha University of Agriculture and Technology, Bhubaneswar, India

Pallavi Dhiman National Agri-Food Biotechnology Institute (NABI), Mohali, India

K. K. Dwivedi Division of Crop Improvement, ICAR-Indian Grassland and Fodder Research Institute, Jhansi, UP, India

Rahul Gajghate Division of Crop Improvement, ICAR-Indian Grassland and Fodder Research Institute, Jhansi, UP, India

Gunjan Guleria Department of Agronomy and Agrometeorology, Rani Lakshmi Bai Central Agricultural University, Jhansi, UP, India

S. J. Harishma Division of Crop Production, ICAR-CTCRI, Thiruvananthapuram, India

Rolly Kabange Department of Southern Area Crop Science, National Institute of Crop Science, Rural Development Administration, Miryang, Republic of Korea

Shaila Kadam University of Missouri, Columbia, MO, USA

Rahul Kaldate Department of Agricultural Biotechnology, Assam Agricultural University, Jorhat, India

Laveena Kaushal Department of Biotechnology, Panjab University, Chandigarh, India
National Agri-Food Biotechnology Institute (NABI), Mohali, Punjab, India

Kiran Khandagale Department of Botany, Savitribai Phule Pune University, Pune, India

Ram Krishna ICAR-Directorate of Onion and Garlic Research, Pune, India

Ashish Kumar Narayan Institute of Agricultural Sciences, Gopal Narayan Singh University, Rohtas, Bihar, India

Neeraj Kumar Division of Crop Improvement, ICAR-Indian Grassland and Fodder Research Institute, Jhansi, UP, India

Virender Kumar Department of Biotechnology, Panjab University, Chandigarh, India

Dharmendra Singh Lagoriya Department of Agricultural Biotechnology, Assam Agricultural University, Jorhat, India

Dongho Lee University of Missouri, Columbia, MO, USA

Devraj Lenka Department of Plant Breeding and Genetics, Odisha University of Agriculture and Technology, Bhubaneswar, India

Zahoor Ahmad Mir ICAR-National Institute for Plant Biotechnology, New Delhi, India

Ambika Prasad Mishra Faculty of Agriculture, Sri Sri University, Cuttack, India

Upasana Mohapatra Department of Plant Biotechnology, University of Agricultural Science, GKVK, Bengaluru, India

Ankit Moharana Department of Seed Science and Technology, Odisha University of Agriculture and Technology, Bhubaneswar, India

Muntazir Mushtaq ICAR-National Bureau of Plant Genetic Resources, New Delhi, India

Rizwana Begum Syed Nabi Department of Southern Area Crop Science, National Institute of Crop Science, Rural Development Administration, Miryang, Republic of Korea

Gunashri Padalkar Department of Biotechnology, Panjab University, Chandigarh, India
National Agri-Food Biotechnology Institute (NABI), Mohali, Punjab, India

Narayan Panda Department of Soil Science and Agricultural Chemistry, Odisha University of Agriculture and Technology, Bhubaneswar, India

Pranaya Pradhan Department of Nematology, Odisha University of Agriculture and Technology, Bhubaneswar, India

Nitish Ranjan Prakash ICAR-Central Soil Salinity Research Institute, Regional Research Station, Canning Town, West Bengal, India

Parichita Priyadarshini Division of Crop Improvement, ICAR-Indian Grassland and Fodder Research Institute, Jhansi, UP, India

Heresh Puren Department of Plant Breeding and Genetics, Assam Agricultural University, Jorhat, India

Maneet Rana Division of Crop Improvement, ICAR-Indian Grassland and Fodder Research Institute, Jhansi, UP, India

Rachna Rana Amity Institute of Organic Agriculture, Amity University, Noida, UP, India

Bodeddulla Jayasankar Reddy Department of Plant Breeding and Genetics, Assam Agricultural University, Jorhat, India

Jyoti Prakash Sahoo Department of Agricultural Biotechnology, Odisha University of Agriculture and Technology, Bhubaneswar, India

Sandhani Saikia Department of Agricultural Biotechnology, Assam Agricultural University, Jorhat, India

Kailash Chandra Samal Department of Agricultural Biotechnology, Odisha University of Agriculture and Technology, Bhubaneswar, India

Akashi Sarma Department of Plant Breeding and Genetics, Assam Agricultural University, Jorhat, India

Vinay Sharma International Crops Research Institute for the Semi-Arid Tropics, Hyderabad, India

Dhananjay Shirsat ICAR-Directorate of Onion and Garlic Research, Rajgurunagar, India

Aman Singh Department of Biotechnology, Panjab University, Chandigarh, India

Major Singh ICAR-Directorate of Onion and Garlic Research, Pune, India

Nisha Singh ICAR-National Institute for Plant Biotechnology, New Delhi, India

Sushil Kumar Singh DBT North East Centre for Agricultural Biotechnology, Assam Agricultural University, Jorhat, India

Sweta Sinha Bihar Agricultural University, Bhagalpur, Bihar, India

P. S. Soumia ICAR-Directorate of Onion and Garlic Research, Pune, India

Harsha Srivastava ICAR-National Institute for Plant Biotechnology, New Delhi, India

Rupesh Tayade Laboratory of Plant Breeding, School of Applied Biosciences, Kyungpook National University, Daegu, Republic of Korea

Anshika Tyagi ICAR-National Institute for Plant Biotechnology, New Delhi, India

Sunil M. Umate Wheat and Maize Research Unit, Agriculture Botany, Vasantrao Naik Marathwada Krishi Vidyapeeth, Parbhani, India

Sanskriti Vats Department of Biotechnology, Panjab University, Chandigarh, India
National Agri-Food Biotechnology Institute (NABI), Mohali, Punjab, India

Himanshu Yadav Department of Biotechnology, Panjab University, Chandigarh, India

Pragya Yadav Division of Plant Physiology, ICAR-Indian Agricultural Research Institute, New Delhi, India

Shashank Kumar Yadav Division of Plant Physiology, ICAR-Indian Agricultural Research Institute, New Delhi, India

Vinod Kumar Yadav University Department of Botany, Ranchi, India

Byung-Wook Yun Laboratory of Plant Functional Genomics, School of Applied Biosciences, Kyungpook National University, Daegu, Republic of Korea

---

The original version of the book has been revised. A correction to this book can be found at https://doi.org/10.1007/978-981-19-4308-9_15

Agronomical Approaches for Biofortification of Cereal Crops

1

Dharmendra Singh Lagoriya, S. J. Harishma, and Sushil Kumar Singh

Abstract

Cereals are the primary staple food crops based on a traditional diet in developing countries; sometimes, rice, wheat, or corn constitutes the entire diet. Half of the global community depends upon grains such as rice, wheat, and maize for consumption, which provide 30.4% of total energy and 20%–30% of protein for the average polish diet for day-to-day activities. More than a third of the world's population is deficient in micronutrients, vitamins, and minerals; notably, 60% suffer from iron and 30% from zinc deficiencies. The quantity of trace elements like iodine (I), selenium (Se), etc., are also found only in minute quantities in the cereal grains. These trace elements act as precursors of vitamins and minerals and are also necessary to fulfill dietary requirements. The WHO has predicted deficiencies of multinutrients to nearly 2 billion people worldwide, which gives birth to the global hidden hunger and malnutrition that affect children's mental and physical capabilities and development. The deficiencies of micronutrients can be alleviated by dietary diversification, extra input of mineral elements, food fortification, and crop biofortification. But this is probably not the case for poor people, especially in developing countries. Biofortification is an essential process of enriching crops with higher nutrients, vitamins, and minerals using agronomic methods, plant breeding, and biotechnological approaches. Biofortification aims to increase the nutritional content of the diet

D. S. Lagoriya (✉)
Department of Agricultural Biotechnology, Assam Agricultural University, Jorhat, India

S. J. Harishma
Division of Crop Production, ICAR-CTCRI, Thiruvananthapuram, India

S. K. Singh
DBT North East Centre for Agricultural Biotechnology, Assam Agricultural University, Jorhat, India

R. Deshmukh et al. (eds.), *Biofortification in Cereals*,
https://doi.org/10.1007/978-981-19-4308-9_1

by increasing the available content of micronutrients and vitamins. The agronomic approach to biofortification of food crops is reasonable and cost-effective, which depends on several factors such as management practices, soil factors, plant factors, etc. It is a practical solution to overcome micronutrient deficiencies in different cereals that optimize fertilizer application with different strategies to improve the nutrient quality of crops without scarification on yield with no objection to the acceptance of the product.

Keywords

Cereals · Biofortification · Agronomic biofortification · Hidden hunger · Micronutrients

1.1 Introduction

As the proverb "Health is wealth," health is the greatest asset for a human being. A good and nutritious diet is the secret of a healthy life, and it also depends on the intake of a balanced diet. To date, the global agriculture sector is focusing on the higher crop production to feed the growing population. The way of farming has changed after the green revolution for achieving higher production in all possible ways that have disturbed nature's harmony due to overapplication of fertilizer, high-yielding varieties, and overproduction from the soils. The use of an excessive and unbalanced amount of fertilizers in the field leads to a deficiency of essential nutrients in food crops which ultimately affects the health of human beings.

In these groups, micronutrients are the main precursors of amino acids, proteins, and vitamins. The deficiency of these micronutrients will be a significant cause of "hidden hunger" due to "micronutrient malnutrition." It is mainly seen in poor and developing countries, where staple food crops such as rice, wheat, ragi, millet, and maize dominate the diet (Khush et al. 2012). Current agricultural systems are trying to replace food crops with insufficient nutrients and focusing on nutrient-rich food crops to facilitate the fight against nutrient deficiencies. But in the present situation, cereals are the major food crops across the world. Its availability and a major component of the diet led to an imbalance in the nutrients. The replacement of traditional staple food crops is not possible in developing and nondeveloped countries due to the economic condition of the people. Therefore, the only possible way to fight such conditions is to enrich the micronutrient content in staple food crops to protect the peoples from micronutrient deficiency without changing their stable diet. Such enrichment is called biofortification.

Biofortification of crops is a promising, sustainable, and cost-effective technique of delivering micronutrients to populations with limited access to diverse diets and other micronutrients. In biofortification, the density of essential ingredients like vitamins, minerals, amino acids, etc., in important food crops is increased to upstand the quality of diets (Bouis and Saltzman 2017). It offers sustainable production of nutrient-rich food crops and assures its availability, especially to poor peoples in developing nations. Biofortified crops with better bioavailability of necessary micronutrients are the best way to create rich food availability through traditional farming and food trade activities. It can help to provide people with a wide variety of foods in the diet of malnourished and low-income groups in a viable way. Based on the examined economic perspective, the development of biofortified crops is a one-time investment that provides a cost-effective solution to combat micronutrient malnutrition. Once biofortified crops are generated, there is no further investment for supplementation, and they are incorporated into food resources at the time of processing (Pfeiffer and McClafferty 2007).

Agronomic biofortification helps in the better growth and development of plants. It helps improve the nutrient profile of the crops in which the required elements are present in less quantity or absent. It also helps develop better agronomic characters (yields, resistance to pests, tolerance to stress), increases food availability, and helps in the fight against poverty and starvation. Some research suggested that micronutrient quantity in the crops required more effort and information to increase the essential micronutrient content affected by different factors. These factors have an influential role, from crop sowing to crop harvest and postharvest storage. Among these factors, soil, pH, available nutrient, texture, organic matter content, soil water relationship, rainfall, and temperatures are essential. Future research for better management and agronomic practice needs to be identified to maintain or improve the crop yield and its nutritional content sustainably (Hornick 1992).

1.2 The Global Prevalence of Micronutrient Deficiencies

The source and sink relationship is well known to everyone. This relationship can also be compared with the host and guest. Similarly, food crops also depend on their host/sink for the nutrient's uptake. If the availability of nutrients is not sufficient to be utilized by the plants, it becomes deficient or limits minerals to plants, which ultimately affects the food synthesis system of a plant. However, a plant can survive or adjust to any environmental system, but it creates a severe issue for humans because humans eventually depend on the plant for food as a sink. The resulting deficiency of nutrients in staple food crops (wheat and rice) becomes a global health problem, affecting the physical and mental development of children and pregnant women due to malnutrition and ultimately resulting in increased susceptibility to the disease that caused illness to mental health, blindness, and loss of productivity. According to WHO-estimated reports globally, more than 2 billion people are affected by micronutrient deficiencies.

In contrast to macronutrient, micronutrient, vitamins, and minerals are required in lesser quantities, and among them, iron (Fe), zinc (Zn), calcium (Ca), iodine (I), vitamin A, B complex, and vitamin C help the cell to perform essential biological processes. Improper diet and imbalance in the nutrient were observed in the developing countries. Every year nearly 40% of preschool child and most pregnant women are suffering due to anemia. According to FAO, such types of malnutrition problems have been mostly observed in developing countries worldwide, whereas

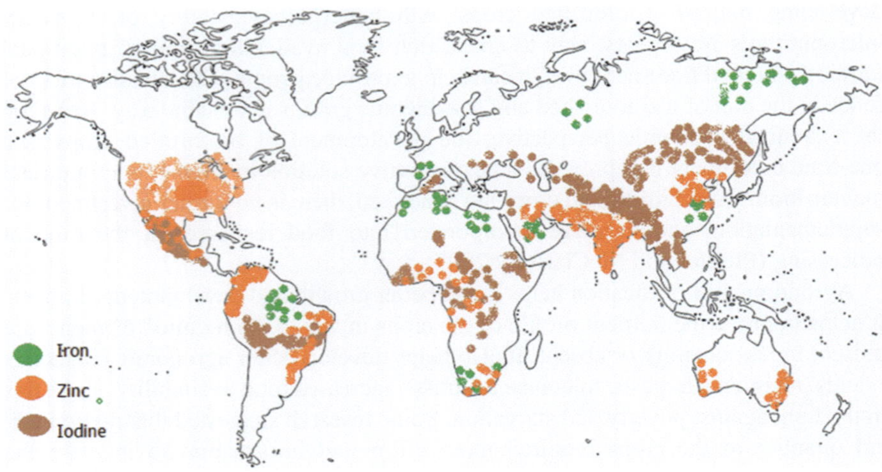

Fig. 1.1 Countrywide deficiency of mineral nutrient in the crop plants indicated in the world graph

780 million people out of 792.5 million world undernourished people belong to the developing countries (McGuire 2015).

These days, micronutrient deficiencies have become common among people due to change in eating habits toward fastfood that is unbalanced dietary habits. Most of the world's population is suffering from various nutritional deficiencies such as 60% iron (Fe), 30% zinc (Zn), and iodine deficiency, and 15% of the people are suffering from selenium deficiency other than calcium (Ca), copper (Cu), and magnesium (Mg) deficiency available communally (Kaur et al. 2020). According to a report published by WHO (World Health Organization), zinc deficiency is the fifth leading cause of disease and disorders in developing countries and 11th globally. Micronutrient deficiencies in soil vary from place to place and around the world. The soil present in almost all continents is the deficit in Zn. America and the southern parts of Asia mostly suffer from Zn deficiencies. Iron deficiency is predominant in southern parts of America, Africa, and northeast Asia. Iodine is deficient in soils of eastern regions of Asia and Central Africa (Fig. 1.1).

About half of the world's population are suffering from micronutrient malnutrition, including Se (selenium), Zn (zinc), and I (iodine), which are primarily associated with low dietary intake of micronutrients in a diet with a low variety of food (Mao et al. 2014). Crops are the primary source of essential nutrients, which provide food and energy to the living being. Among all the crops, three bowls of cereal crops are the stable and dominant food source to the four billion people and about 60% of plant-based energy intake by the human being (Frison 2016). But they do not get enough essential nutrients required in a human being's diet due to a deficit in soil and plants. The status of micronutrient deficiencies in the global soil system is displayed in the graph below (Fig. 1.2.).

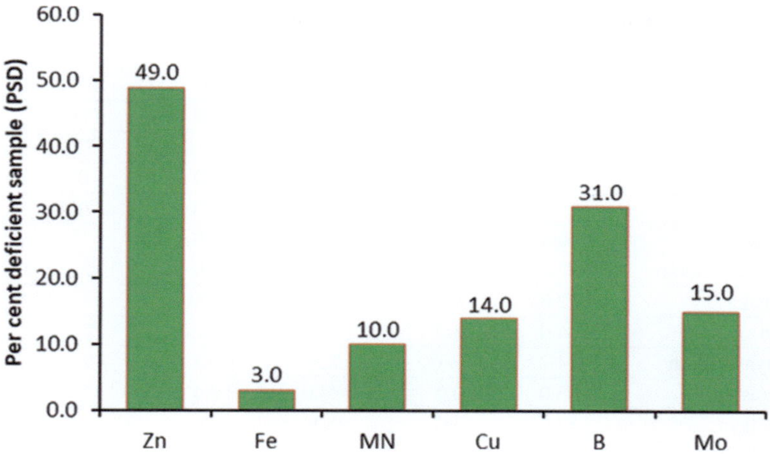

Fig. 1.2 Global status of micronutrient deficiency in a worldwide agricultural soil

1.3 Micronutrient Status in Indian Soil

India built up adequacy in food production in the mid-1960s after introducing high-yielding varieties that responded well to fertilizer application. The increased cropping intensity and exhausted uses of major fertilizers like NPK resulted in a lack of secondary nutrients and micronutrients. The shortcoming of nutrient or unavailability of essential nutrients from food crop becomes a critical hurdle for incrementing sustainable crop yield that has been commonly observed in crops such as cereal, oilseed, legumes, and vegetable crops. Crops grown in approximately half of the soils in the country suffer from several micronutrient deficiencies (Takkar et al. 1990). According to the study of more than two lakh samples of soil collected from 508 districts of the country, on average, 36.5%, 12.8%, 7.1%, 4.2%, and 23.2% of soils are deficient in Zn, Fe, Mn, Cu, and B, respectively. More than 50% of the samples are found inadequate in Zn and B in 110 and 63 districts, respectively (Table.1.1). The deficiency of nutrients in the Indian soil is represented in the graph in Fig. 1.3. Field-scale zinc (Zn) deficiency was first observed in the Tarai soils (molisols) of the Himalayan foothills, causing complete failure of the rice crop (Nene 1966). This type of Zn deficiencies has been observed strictly in alkaline soils where high yielding rice and wheat varieties are cultivated intensively. Initially, zinc (Zn) deficiency, and later iron (Fe) deficiency in rice, and manganese (Mn) deficiency have been observed in wheat, leading to reduced yield in crops.

Table. 1.1 The spectrum of micronutrient scarcity in soils of India (Shukla et al. 2021)

States	Percent deficient sample (PSD)					
	Zn	Cu	Fe	Mn	B	Mo
Andhra Pradesh	46.8	<1.0	2.8	1.2	53.0	49.0
Assam	34.0		2.0	20.0	17.0	
Bihar	54.0	3.0	6.0	2.0	1.0	1.0
Gujarat	23.9	4.0	8.0	4.0	57.0	8.0
Haryana	60.5	2.0	20.0	4.0		
Himachal Pradesh	42.0	0.0	27.0	5.0	32	5.0
Karnataka	72.8	5.0	35.0	17.0	53	49
Kerala	34.0	3.0	<0.1	0.2	17	
Madhya Pradesh	44.2	<1.0	7.3	0.2	49	30
Maharashtra	83.0	0.0	24.0	0.4		
Meghalaya	57.0	2.0	0.0	23.0		
Orissa	52.5	<1.0	0.0	0.6		
Pondicherry	8.0	4.0	2.0	3.0		
Punjab	46.1	1.1	14.0	2.3		
Rajasthan	21.0				14	
Tamil Nadu	58.6	6.0	17.0	6.0	68.0	
Uttar Pradesh	45.7	1.0	6.0	3.0	45	12
West Bengal	36.0	0.0	0.0	3.0	49	30

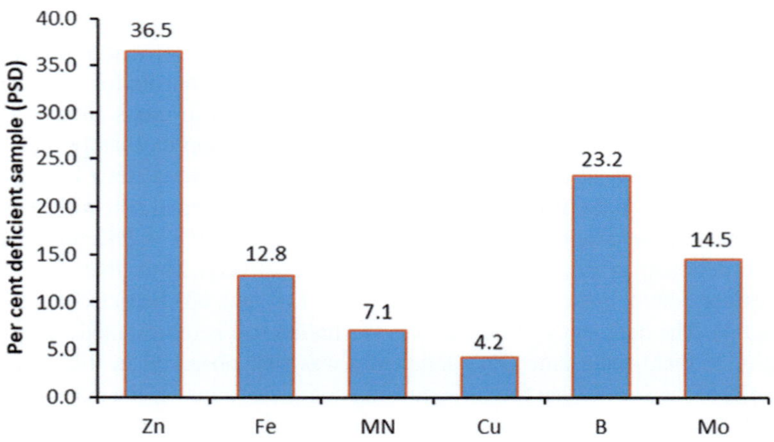

Fig. 1.3 Micronutrient deficiency in Indian soils (Shukla et al. 2019)

1.4 How to Fight Against Micronutrient Malnutrition?

The condition of malnutrition occurs due to the unavailability of essential supplements in diets such as vitamins, minerals, and micronutrients that contribute to the deaths of about 20% of children under the age of 5years. Some of the children go blind permanently every year due to vitamin A deficiency. Iron deficiencies cause anemia, mainly affecting children and women. In adults, it causes a reduction in the working and standing capacity, sometimes reproductive impairments. The strategies to fight hidden hunger globally are as follows:

1. Dietary diversification and food fortification.
2. Supplementation of specific micronutrients.
3. The intervention of horticulture crop to ensure regular consumption.
4. Prevention and control measure of public health and other diseases.

In India, a program is implemented to prevent deficiencies in preschool children and women, with special care of pregnant women. Under this program, there is a periodic distribution of vitamin A, iron, folic acid tablets, and iodized salts. However, these programmers have been running for decades. They do not have any biological impact on micronutrient malnutrition prevalence. The aim of fighting against malnutrition can be achieved successfully by utilizing nutrient-rich food in main diets instead of additional supplements. There is a way implemented to increase the nutrient of the crop, known as biofortification. Several biofortification techniques are used to enhance nutrients, such as agronomical biofortification, conventional plant breeding, genetic modification, etc. Agronomic biofortification benefits fast application and a simple process of biofortification of crops that can help develop crops in poor mineral soils—objectives and advantages of biofortification that help fight against micronutrient malnutrition.

The significant goals of biofortification are the following:

- To provide enriched food with the essential nutrients in the main diet rather than to take an additional supplement.
- To alleviate the micronutrient malnutrition from the developing countries.

Biofortification is a cost-effective approach to provide nutrient enhancement from the food source. It is a feasible and natural way of supplementing the nutrient. The process of biofortification improves soil's physical and chemical properties, and the plant gets benefitted after fortification in terms of production, vigor, quality, and stress tolerance. Soil health in cultivated land declines due to the depilation of various micronutrients in monoculture and improper fertilizer application (Bouis and Saltzman 2017). The process of agronomic biofortification helps in improving the physical and chemical properties of the soil, leading to higher crop yields.

1.4.1 Biofortification by Agronomic Practices

The process of biofortification requires a long time, but with agronomical practice, it can be achieved in a short time in a cost-effective manner (Hefferon 2015). The agronomic biofortification process is a simple, inexpensive method that requires a variety of physical methods to improve the nutritional and health status of crops and soils (Cakmak and Kutman, 2018). Agronomic strategies are effective in increasing crop yield and nutritional quality. These strategies include the use of appropriate amounts of NPK and S fertilizers along with other agrotechnical measures such as crop rotation, soil moisture management, tillage, and organic farming. The application of micronutrients, mineral, and nutrient-deficient soil is ultimately used to enhance the quality of crop grains and mineral and nutrient deficiencies used to supply through fertilizers. Fertilizer application to crops depends on the stage of their growth and development, which can be done through various methods such as seed treatment, soil application, and foliar application (Yang et al. 2011). Micronutrients can also be applied along with soil amendment substances to increase crop yield and nutritional quality (De Valença et al. 2017). Fertilizers, along with organic matter, significantly improved micronutrient content in soil and enhanced their bioavailability.

Additionally, the cropping system, intercropping, and crop rotations improve yield and quality of crops (Zuo et al. 2004). The application of zinc fertilizer and green manure has been seen as effective in the grain quality and quantity of Basmati rice in India (Pooniya and Shivay 2013). Foliar application is an effective and useful agronomic biofortification approach that provides mineral in the most appropriate ways known as phyto-available form (Lawson et al. 2015). However, it is not a feasible approach in windy and rainy areas. There is no single approach to get superior results in biofortification. Nevertheless, the incorporation of high Fe and Zn content into edible parts of plants requires integrated management of micronutrients.

Agronomic biofortification increases the targeted mineral's dietary intake directly from the plant to its edible portion of the crops. Such biofortified crops can reach the most vulnerable and poorest peoples, which impart a nutritionally significant impact on farmers and consumers' lives. Today more than one million people in Asia, Africa, and Latin America benefit from using biofortified foods. There are different agencies and organization who have been playing a prominent and outstanding role such as Harvest Plus and Biofort; grand challenges on global health, India biofortification program, and organization also play an essential role in food crops such as wheat and maize from the International Maize and Wheat Improvement Center (CIMMYT), vitamin-A- and zinc-enriched maize varieties from the International Institute of Tropical Agriculture (IITA). Biofortification aims to ensure the availability of essential micronutrients, vitamins, and minerals required in amounts of less than 1 mg/day. Agronomic biofortification strategies are conducive and virtuously effective in improving food quality. This approach entirely depends on the application of fertilizer and micronutrients. The application of micronutrients depends on crops' requirements without wastage of these supplements to give a fair and cost-effective way to produce biofortified crops. There are different ways to provide these supplements to the plant; the most common and conventional fertilizer supply method to the plant is the soil's application; another one is a foliar application (Cakmak 2010). An integrated approach of micronutrient application emerged as the most significant and critical factor in agronomic biofortification. It becomes more effective by combining NPK with organic fertilizer and high yielding varieties (De Valença et al. 2017). Eventually, it will result in the increment of nutritional quality and yield of the crops that directly benefit human health by making the availability of vital microelements.

1.4.1.1 The Agronomic Measure of Biofortification

- Site-specific application of fertilizers.
- Application of fertilizer with coating, e.g., zinc-coated fertilizer.
- Use of soil amendments in the problematic soils, e.g., lime in acidic soil and gypsum in salt-affected soils.
- Application of biofertilizer and green manures, e.g., *Rhizobium, Azotobacter,* and *Azospirillum.*
- Balanced application of N, P, K, and S fertilizers.
- Crop rotation.
- Soil moisture management and tillage practice.

1.5 Agronomical Practices for Biofortification in Cereals

Cereals *are a*n essential part of the dietary requirement in most people. Therefore, the biofortification of cereals is more imperious (Cassman 1999). In poor and low-income countries, people do not have access to the right quality of foods, health care, and living condition. In such places, especially women are more vulnerable to chronic diseases. To date, we are trying to focus on the biofortification of significant and stable food crops of cereals such as wheat, rice, maize, barley, sorghum, millets, oats, and rye. It can provide most of the essential mineral elements required for humans' well-being (Graham et al. 2007). The flow of nutrients always occurs in a pathway from producer to consumer or from source to sink; in the case of essential micronutrients, they flow from soil to plant and then to humans, which is the final consumer in most cases. The micronutrients are absorbed by the plant and deposited in their edible form for storage such as Fe, Zn, Co, Mn, I, Se, Mo, Co, and Ni, with varying amounts. Several factors are involved in agronomic biofortification's success to gratify micronutrients deficiencies from the population. These factors affect the bioavailability of nutrients from food to humans such as the uptake of nutrients from the soil to crop. These factors include the allocation of nutrients within the plant, the retransfer of the part of the plant consumed by humans (food from the crop), the process of preparing food (from food to humans), and the physiological state of the human body that determines its ability to absorb and use nutrients (Fig.1.4.) (Mayer et al. 2011). The soil has various complex systems that influence the availability of micronutrient uptake from soil to crops, such as physical and

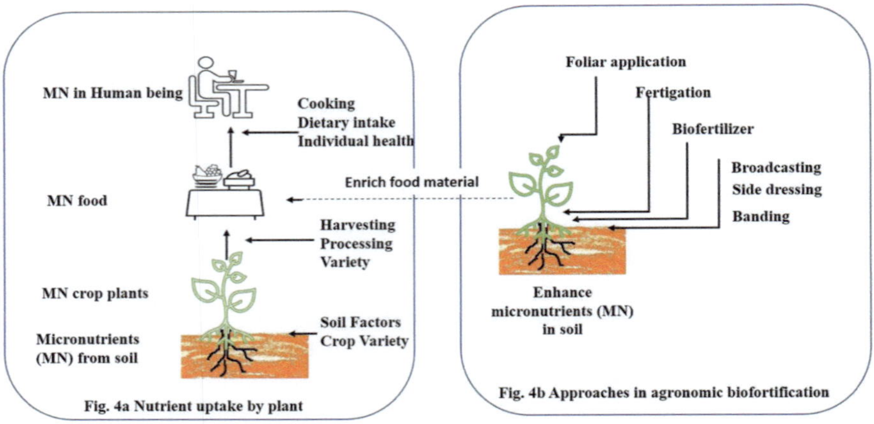

Fig. 1.4 The flow of micronutrients from plant to human

chemical properties, including pH, organic carbon, soil moisture, aeration, and interactions of other elements. It is also influenced by the farmers' crop variety, as the absorptions depend upon the structure and functioning of the root systems. The diversity of microorganisms in the rhizosphere region promotes plant growth and enhances nutrient mobility, thereby improving the nutritional status of the edible parts.

1.5.1 Biofortification Through Fertilization

The deficiency of micronutrients in the soil can be improved by adding mineral fertilizers to the soil. Vigilant use of chemical fertilizers can increase the soil nutrient status and absorption by the crops, leading to the biofortification of crops and enhancing the yield and nutritional quality sustainably with the balanced combination of primary and secondary micronutrients in the small amounts (Voortman and Bindraban 2015). For example, adequate application of phosphate fertilizers can stimulate root growth and, thus, micronutrient uptake. It has been observed that the presence of P fertilizer or the addition of small amounts of P fertilizer in the soil does affect the availability of Zn that lead to deficiencies of Zn in the soil, resulting in the reduction of Zn uptake from soil to plants (Zingore 2011; Singh et al. 1988).

1.5.1.1 Fertilizers Application to Enhance Micronutrient Elements in Cereal Crops

The green revolution helps to sustain and enhance the productivity of agricultural products. But this was achieved by utilization of higher dose of chemical fertilizer and high yielding varieties. Fertilizer application is an important step in maintaining soil fertility and increasing the yield of crops. It has been observed that the application of NPK fertilizer improves the yield and vigor of the crop (Brady and Weil

2008). Fertilizers provide essential nutrients to the crop plants for germination, flower development, and grain formation. Many fertilizers used to work with soil and help in maintaining soil moistures, airflow, and better root developments. The proper supplies of nutrients in the crop plants can be achieved by exploring micronutrients' genetic variability in the crop plants. Various external and internal factors must be considered to determine the inadequate supply of the micronutrients Zn, Fe, Mn, Cu, and Se. Because of high environmental factors, epically the soil variability required higher knowledge and advanced technology to determine the exact demand and nutrient variability among the crop plants. That can be succeeded by agronomic intervention and the external supplies of the fertilizer to the crop plant. To supply these essential nutrients to the crop plants there are various commercial fertilizers are available, which apply to the plant using different mangement practices, cropping systems, and equipments.

1. Broadcasting: Applied in larger fields, incorporated with the help of cultivator, left in the soil to filter the nutrients.
2. Banding: Used to apply for a small number of fertilizer requirements in small areas, in the no-tillage cropping system, involved in the furrows as the in-band pattern, e.g., anhydrous ammonia and liquid nitrogen.
3. Side-dressing: Fertilizer used to apply in the soil after the plant growth (early to mid-growth). Fertilizer can be broad using a hand near the plant or banded down in the middle row, e.g., nitrogen fertilizer is more often used in this way.
4. Fertigation: Application of fertilizer with irrigation of water known as fertigation. For example, nitrogen and potassium are most used in this method, but the application of phosphorous required extra care during application because it is used to form clog due to the variation in the pH.
5. Foliar application: Plants have more advantages in using the foliar application because of the plants' rapid absorption and utilization. At the plant's critical requirement, this method can also be applied in the soil, but it is not suitable for soil application.

There are different criteria for the classification of fertilizer. The most common standards are defined by Arnon 1954, for the requirement of nutrients to the crop plants, and are categorized in the following manner.

1. Based upon nutrient composition:
   (a) Straight—e.g., urea, ammonium sulphate, potassium sulphate.
   (b) Complex—e.g., DAP.
   (c) Mix fertilizer.
      - Open formulated fertilizer mixture—e.g., quality of ingredients are disclosed.
      - Closed formulated fertilizer mixture—e.g., quality of ingredients are not disclosed.
2. Based on physical forms:

Table 1.2 Percentage nutrient content in micronutrient fertilizers

S. No	Source	% Nutrient
1	Magnesium sulfate (Epsom salts)	9% Mg
2	Magnesium-potassium sulfate	11% Mg
3	Dolomitic limestone	9% Mg
4	Magnesium oxysulfate (granular)	36% Mg
5	Calcium sulfate (gypsum)	15%–18% S
6	Ammonium sulfate	24% S
8	Borax	10%–15% B
9	Solubor	20.5% B
10	Calcitic limestone	35% Ca
11	Calcium sulfate (gypsum)	22.5% Ca
12	Iron sulfate	40% Fe
13	Manganese oxysulfate	28% Mn
14	Manganese chelates (soluble powder)	20% Mn
15	Zinc oxysulfate	36% Zn
16	Zinc chelates (soluble powder)	25% Zn

(a) Solid fertilizer—e.g., in the form of crystals, prill, and super granules.
(b) Liquid fertilizer—e.g., N, P, and K entirely dissolved in water.

3. Secondary and micronutrient fertilizer; required in small quantity (PB1637 2017) (Table 1.2).

1.5.2 Application of Micronutrient Fertilizers

The uptake of nutrients in plants can be increased by the use of fertilizers containing micronutrients. This can be achieved either by applying fertilizer to the soil or directly applying foliar to the plant. Plants can now absorb nutrients and enhance the nutritional quality of crop production. However, it is a crucial practice used to strengthen zinc concentrations in cereal grains. The practical methods are applying combined micronutrient compounds and N, P, K fertilizers or formulation of compound fertilizers by coating complexes with micronutrients or by a bulk blending of micronutrients with granular fertilizers (Singh and Singh 2013).

1.5.3 Application of Other Soil Amendments

The bioavailability of micronutrients can also be improved by using soil additives such as lime, organic matter, etc. This will enhance soil conditions and accelerate the absorption of micronutrients in plants.

1.5.4 Inoculation of Biofertilizers to the Soil

Biofertilizers are natural fertilizers that contain microbial strains. These microbes keep the soil in a healthy condition and increase the availability of nutrients to the plants. Various microorganisms such as *Bacillus, Pseudomonas, Rhizobium, Azotobacter,* etc., are generously available in the soil, which can be used to increase the mineral phyto-availability in the cropping system (Smith and Read 2010). The addition of various microorganisms increases the bioavailability of multiple micronutrients by converting them from complex organic forms to more available inorganic forms through the solubilization process. The addition of nitrogen-fixing bacteria in nitrogen-limited soils increases productivity by fixing atmospheric nitrogen (Sprent et al. 2004). The high availability of nitrogen stimulates plant growth and also helps in the absorption of other micronutrients. Mycorrhizal fungi, found attached to crops, can release organic acids, siderophores, and enzymes, which can degrade organic compounds and increase mineral concentrations in the edible part of plants (Cavagnaro 2008). There are various successful agronomic biofortification examples in cereal crops, such as rice, wheat, maize, barley, and sorghum.

1.6 Rice

Worldwide, cereal crops are grown over a large area and provide more energy. Rice is one of the most consumed cereal crops of the rural household. More than half of the world's population depend on rice for their one-time meals. The Asian continent has more than 90% of rice consumption and production worldwide (Cavagnaro 2008). India is the second-largest producer and the largest consumer of rice (USDA 2019). In rice, zinc deficiency is a significant problem that occurs in rice grains. Foliar supply of zinc fertilizers is the standard method to overcome zinc deficiency of rice crops. It increases the concentration of zinc in rice grains (Singh and Singh 2013). However, foliar sprays serve as a valuable method of fertilizer applications into the various crops. But in rice, soil application of Zn fertilizers works well in soils deficient in Zn in rice (Guo et al. 2016). The application of zinc fertilizer for the efficient improvement of Zn content in grain depends on the soil and climatic factors of the regions. In general, zinc sulfate heptahydrate ($ZnSO_4.7H_2O$) at 10–20 kg/ha can be applied in zinc-deficient soil.

To maximize the uptake of micronutrients and enrichments of crop produce, the combined application of fertilizers in the soil and foliar application substantiates a practical approach (Singh and Singh 2013). However, iron's foliar application has a strong effect in promoting iron concentration in the rice grain. Also, the fortification of ferrous sulfate in germinating rice plantlets increases the iron concentration in brown rice (Yuan et al. 2013). Similarly, foliar application of selenium in rice works very well. Selenium is an essential trace element for humans that helps enhance immunity by acting as a potential antioxidant (Ram et al. 2016).

1.7 Wheat

Agronomic biofortification can be effectively used to improve the quality of wheat grain. The incorporation of iron and foliar application of nitrogenous fertilizers have been correlated with higher iron accumulation (Zou et al. 2012). In all the cases, foliar application of zinc worked well and helped reduce Zn deficiency disease, particularly in zinc-deficient soil and conjointly improved its bioavailability and reduced antinutrient factors like phytate (Yang et al. 2011). Zn deficiency in wheat is primarily due to the shortage of soil moisture because of irregular and scanty rainfalls, so proper irrigation management also triggers the Zn content in wheat grains. Zinc fertilizer has significant effects on the increment of wheat grain. The wheat-growing area has also been observed with the application of NPKs in the field from 1994 in Turkey with the record yield improvement of 400,000 tons per annum in 10–15 years. In rural areas of Turkey, the improved quality of wheat with Zn has directly benefited human health because the majority of Turkey's population depends on wheat for more than 50% of calorie uptakes on wheat (Cakmak 2008). Since 1984, compound fertilizers supplemented with Se have been used, which results in increased selenium in humans. Researchers have also investigated the role of biofertilizers and chemical and organic fertilizers, in improving grain yields. The addition of organic manure was found to be effective positively with the essential element in wheat grains. Conjugate application of chemical and organic fertilizers was found to strengthen the micronutrient uptake in wheat crops (White and Broadley 2005). Integrated use of AMF (arbuscular mycorrhizal fungi) and fertilizers was also found to have a synergistic effect on micronutrient uptake in wheat. We cannot neglect the role of microorganisms' in modern agriculture with various advantages supporting micronutrients to the plants. The Zn solubilizing bacteria *Bacillus aryabhattai* has a significant role in plants' growth by nutrient mobilization from the vertisols in Central India (Ramesh et al. 2014). However, iron enrichment in wheat grain has been achieved successfully by integrating organic and chemical fertilizers (Ramzani et al. 2016).

1.8 Maize

Zinc is the major micronutrient affecting the yield and nutrient quality of maize grains. To enhance Zn concentration, various treatments for zinc fertilizer and foliar applications were performed in maize crops (Alvarez and Rico 2003). Agronomic approaches with the integration of PGPR (plant growth-promoting rhizobacteria) helps in the uptake of nutrients from the soil to plants which has become an effective strategy for biofortification of staple crops. The use of biological agents with ash can efficiently reduce the Cd content in maize grains (Fahad et al. 2015). Selenium content can enrich food grains with biofertilizers as part of agro-based biofortification that ultimately impact the health of humans and animals (Ros et al. 2016).

1.9 Barley

Barley is one of the most critical cereal grains known from ancient times for its nutritional quality, health benefits, and delicious flavor worldwide. Barley is used as animal feed, fodder, and for malting purposes. In the ancient literature of the world, barley is also known as "king of grains" due to its health benefits. Barley is a storehouse of dietary fibers ranging from 11% to 34% for nutritional fiber, while soluble dietary fiber is present in 3% to 20%. Soluble fibers contain beta-glucan, pectin, and hemicellulose, which have been associated with many health benefits.

Even though being known as the king of cereal, barley requires improvement in the nutrient profiles that can be achieved by applying various organic and inorganic biofertilizers. The application of biofertilizer and inorganic manure and compost has worked well in barley to improve zinc and iron concentrations (Maleki et al. 2011). Some of the barley's antinutritional factors reduce the bioavailability of nutrients in barley that can be reduced by increasing nutritional enhancer levels. Biofortification barely enhances the concentration of Se in its grain, which helps in the adequate supply in the food of humans and animals with a lot of health benefits such as prevention of cancers, immunity boost up, and prevention of several cardiovascular diseases. Se-biofortified grains help in the development of good beers. The organic form of selenomethionine is more efficiently absorbed in the human body compared to the inorganic forms, which is also not affected by the antinutritional factors (Rodrigo et al. 2014).

1.10 Sorghum

Sorghum is also known as "great millet" with a nutty flavor and chewy texture, which were initially domesticated in Africa. This crop often suffers the challenge of growing in nutrient-poor and contaminated soil. Cultivation of sorghum is often faced with the challenge of growing it in nutrient-depleted soil with little care. But the biofortification strategy in sorghum has a positive impact on the quality and quantity of its yield. In addition, the use of Azospirillum alone or in combination with other bacteria that can provide soluble phosphate content in soil results in increased grain yield and nutrient content such as proteins, amino acids, etc. (Patidar and Mali 2004).

1.11 Application of Prebiotics as Micronutrient Promoters

Application of prebiotics to the crop plant works as a booster for crop plants' growth and development. The use of prebiotic works as a stimulator to the plant and helps absorb several minerals to improve the update of micronutrients by increasing the availability of Fe, Zn, Se, etc. (Choudhari et al. 2008). These micronutrients also work as a biostimulant by enhancing crop productivity, nutritional quality, and plant capacity to withstand environmental stress. These micronutrients become available in the soil due to the process of mineralization. Still, they are not readily available to the plants due to adsorption by the soil colloids (microscopic particles of soil) or becoming a saline soil solution. The best alternative source is for the easy uptake of micronutrients by the plants' decomposition of organic matter. There are different factors to consider for the availability and uptake of micronutrients in crops (Sims 1986: Laurent et al. 2020).

- Low (less than 2.0%) and higher (over 30% to a depth of 30 cm) percentage of organic matter in the soil.
- Cool or wet region soil.
- Higher pH of the soil (molybdenum is an exception).

Adsorption reaction, precipitation reaction, nutrient cation form, variable charge in minerals (e.g., Fe oxide), and environmental conditions play a major role in the availability of micronutrients to plants. Among the micronutrients present in the soil, the plant used to uptake most of them in their cationic form (Fe, Mn, Cu, Zn, Ni, and B), whereas Mo and Cl took in their anionic form (Welch and Shuman 1995). Unfortunately, there is a lack of knowledge about the uses of prebiotic in the crop plant for biofortification of crops. The benefits of plant biostimulants or prebiotics increase the interest of farmers and agrochemical industries as they increase nutrient use efficiency. Using this prebiotics with the arbuscular mycorrhizal fungi in crop plants' roots helps uptake water and nutrient from the soil, especially in the low fertile soils (Rouphael et al. 2015). Similarly, foliar and root application of protein hydrolysates helps in the C and N metabolism and uptake of micronutrients that can increase the yield and quality of produce (Colla et al. 2015). The quality and quantity of micronutrients have been observed to effectively increase in the olive tree after the foliar application of Se fertilization (Mattioli et al. 2020).

1.12 Limitations of Agronomic Biofortification

The efficiency of agronomic biofortification depends only on the method of application of fertilizers. There are different methods of application of fertilizers that have differential effects on the concentration of micronutrients of the grain. A high dose of micronutrient fertilizers may sometimes reduce crop yield, affecting the plant's physiology. Sometimes the use of micronutrients as fertilizers can increase grain yield rather than the micronutrient content of the grain. Foliar application of fertilizers is most effective compared to the other methods. But the foliar application requires knowledge of the crop development stage to get better results with the enriched quantities in the grains. In other stages, the foliar application will not be much helpful in the cereal crops.

1.13 Conclusion of the Agronomic Biofortification

The agronomic biofortification approach depends on micronutrient application using various sources and agronomic practices to make them available to the crop plant to utilize the soil nutrients in their growth and development efficiently. Application of fertilizer with the enriched content of micronutrients improves the health of the soil and crop plant. It reduces the negative effect on the environment when it is used at inappropriate rates. The agronomic biofortification approaches are working very effectively with the combination of micronutrients, resulting in an increment of plant yield and nutrition of the crop that ultimately benefited humans and animal health. It works faster than the other approaches for cereal crops and many more in an economical way.

References

Alvarez JM, Rico MI (2003) Effects of zinc complexes on the distribution of zinc in calcareous soil and zinc uptake by maize. J Agric Food Chem 51(19):5760–5767

Bouis HE, Saltzman A (2017) Improving nutrition through biofortification: a review of evidence from HarvestPlus, 2003 through 2016. Glob Food Sec 12:49–58

Brady NC, Weil RR (2008) Soil colloids: seat of soil chemical and physical acidity. Nat Propert Soils 5:311–358

Cakmak I (2008) Enrichment of cereal grains with zinc: agronomic or genetic biofortification? Plant Soil 302(1–2):1–17

Cakmak I (2010) Biofortification of cereals with zinc and iron through fertilization strategy. In: 19th World Congress of Soil Science, Brisbane

Cassman KG (1999) Ecological intensification of cereal production systems: yield potential, soil quality, and precision agriculture. Proc Natl Acad Sci 96(11):5952–5959

Cavagnaro TR (2008) The role of arbuscular mycorrhizas in improving plant zinc nutrition under low soil zinc concentrations: a review. Plant Soil 304(1–2):315–325

Choudhari A, Shinde S, Ramteke BN (2008) Prebiotics and probiotics as health promoter. Veterinary world 1(2):59–61

Colla G, Nardi S, Cardarelli M, Ertani A, Lucini L, Canaguier R, Rouphael Y (2015) Protein hydrolysates as biostimulants in horticulture. Sci Horticul 196:28–38

De Valença AW, Bake A, Brouwer ID, Giller KE (2017) Agronomic biofortification of crops to fight hidden hunger in sub-Saharan Africa. Glob Food Sec 12:8–14

Fahad S, Hussain S, Saud S, Hassan S, Shan D, Chen Y et al (2015) Grain cadmium and zinc concentrations in maize influenced by genotypic variations and zinc fertilization. CLEAN–Soil, Air, Water 43(10):1433–1440

Frison EA (2016) From uniformity to diversity: a paradigm shift from industrial agriculture to diversified agroecological systems, vol 29, p 237

Graham RD, Welch RM, Saunders DA, Ortiz-Monasterio I, Bouis HE, Bonierbale M et al (2007) Nutritious subsistence food systems. Adv Agron 92:1–74

Guo JX, Feng XM, Hu XY, Tian GL, Ling N, Wang JH et al (2016) Effects of soil zinc availability, nitrogen fertilizer rate and zinc fertilizer application method on zinc biofortification of rice. J Agric Sci 154(4):584–597

Hefferon KL (2015) Nutritionally enhanced food crops; progress and perspectives. Int J Mol Sci 16(2):3895–3914

Hornick SB (1992) Factors affecting the nutritional quality of crops. Am J Altern Agric 7:63–68

Kaur T, Rana KL, Kour D, Sheikh I, Yadav N, Yadav AN et al (2020) Microbe-mediated biofortification for micronutrients: present status and future challenges. In: *Trends of microbial biotechnology for sustainable agriculture and biomedicine systems: perspectives for human health*. Elsevier, Amsterdam, pp 1–17

Khush GS, Lee S, Cho JI, Jeon JS (2012) Biofortification of crops for reducing malnutrition. Plant Biotechnology Reports 6(3):195–202

Laurent C, Bravin MN, Crouzet O, Pelosi C, Tillard E, Lecomte P, Lamy I (2020) Increased soil pH and dissolved organic matter after a decade of organic fertilizer application mitigates copper and zinc availability despite contamination. Sci Total Environ 709:135927

Lawson PG, Daum D, Czauderna R, Meuser H, Härtling JW (2015) Soil versus foliar iodine fertilization as a biofortification strategy for field-grown vegetables. Front Plant Sci 6:450

Maleki, F. S., Chaichi, M. R., Mazaheri, D., Tavakkol, A. R., & Savaghebi, G. (2011). Barley grain mineral analysis as affected by different fertilizing systems and by drought stress

Mao H, Wang J, Wang Z, Zan Y, Lyons G, Zou C (2014) Using agronomic biofortification to boost zinc, selenium, and iodine concentrations of food crops grown on the loess plateau in China. J Soil Sci Plant Nutr 14(2):459–470

Mattioli S, Rosignoli P, D'Amato R, Fontanella MC, Regni L, Castellini C et al (2020) Effect of feed supplemented with selenium-enriched olive leaves on plasma oxidative status, mineral profile, and leukocyte dna damage in growing rabbits. Animals 10(2):274

Mayer AB, Latham MC, Duxbury JM, Hassan N, Frongillo EA (2011) A food systems approach to increase dietary zinc intake in Bangladesh based on an analysis of diet, rice production and processing. In: Combating micronutrient deficiencies: food-based approaches, pp 254–267

McGuire S (2015) FAO, IFAD, and WFP. The state of food insecurity in the world 2015: meeting the 2015 international hunger targets: taking stock of uneven progress. Rome: FAO, 2015. Adv Nutr 6(5):623–624

Nene YL (1966) Symptoms, cause and control of Khaira disease of paddy. Bull Indian Phytopathol Soc 3:97–191

Patidar M, Mali AL (2004) Effect of farmyard manure, fertility levels and bio-fertilizers on growth, yield and quality of sorghum (Sorghum bicolor). Indian J Agron 49(2):117–120

PB1637 (2017) PB1637 fertilizers and their use. The University of Tennessee Agricultural Extension Service. PB1637-10M-11/99 E12-2015-00-117-00, https://trace.tennessee.edu/utk_agexcrop/52

Pfeiffer WH, McClafferty B (2007) HarvestPlus: breeding crops for better nutrition. Crop Sci 47: S-88

Pooniya V, Shivay YS (2013) Enrichment of Basmati rice grain and straw with zinc and nitrogen through ferti-fortification and summer green manuring under Indo-gangetic plains of India. J Plant Nutr 36(1):91–117

Ram H, Rashid A, Zhang W, Duarte AÁ, Phattarakul N, Simunji S et al (2016) Biofortification of wheat, rice and common bean by applying foliar zinc fertilizer along with pesticides in seven countries. Plant Soil 403(1):389–401

Ramesh A, Sharma SK, Sharma MP, Yadav N, Joshi OP (2014) Inoculation of zinc solubilizing Bacillus aryabhattai strains for improved growth, mobilization and biofortification of zinc in soybean and wheat cultivated in Vertisols of central India. Appl Soil Ecol 73:87–96

Ramzani PMA, Khalid M, Naveed M, Ahmad R, Shahid M (2016) Iron biofortification of wheat grains through integrated use of organic and chemical fertilizers in pH affected calcareous soil. Plant Physiol Biochem 104:284–293

Rodrigo S, Santamaria O, Chen Y, McGrath SP, Poblaciones MJ (2014) Selenium speciation in malt, wort, and beer made from selenium-biofortified two-rowed barley grain. J Agric Food Chem 62(25):5948–5953

Ros GH, Van Rotterdam AMD, Bussink DW, Bindraban PS (2016) Selenium fertilization strategies for bio-fortification of food: an agro-ecosystem approach. Plant Soil 404(1–2):99–112

Rouphael Y, Franken P, Schneider C, Schwarz D, Giovannetti M, Agnolucci M et al (2015) Arbuscular mycorrhizal fungi act as biostimulants in horticultural crops. Sci Hortic 196:91–108

Shukla AK, Behera SK, Satyanarayana T, Majumdar K (2019) Importance of micronutrients in Indian agriculture. Better Crops South Asia 11:6–10
Shukla AK, Behera SK, Prakash C, Tripathi A, Patra AK, Dwivedi BS, Trivedi V, Rao C, Chaudhari SK, Das S, Singh AK (2021) Deficiency of phyto-available sulphur, zinc, boron, iron, copper, and manganese in soils of India. Sci Rep 11(1):1–13
Sims JT (1986) Soil pH effects on the distribution and plant availability of manganese, copper, and zinc. Soil Sci Soc Am J 50(2):367–373
Singh A, Singh RK (2013) Agronomic biofortification technique for enhancing the zinc content in rice.National Conference on Systematic approach in the implementation of informational and resource-saving technologies in food-crop production. Prerequisite for Eco-Balancing
Singh JÁ, Karamanos RE, Stewart JWB (1988) The mechanism of phosphorus-induced zinc deficiency in bean (Phaseolus vulgaris L.). Can J Soil Sci 68(2):345–358
Smith SE, Read DJ (2010) Mycorrhizal symbiosis. Academic press
Sprent JI, Hardarson G, Broughton WJ (2004) Maximising the use of biological nitrogen fixation in agriculture. Springer Science & Business Media
Takkar PN, Mehta SK, Chhibba IM (1990) A decade of micronutrient research. Res. Bull., AICRP Micronutr IISS Bhopal 1:1–136
United States Department of Agriculture (USDA). (2019). World Agricultural Production. Available online at: https://apps.fas.usda.gov/psdonline/circulars/production.pdf
Voortman, R. L., & Bindraban, P. S. VFRC Report 2015
Welch RM, Shuman L (1995) Micronutrient nutrition of plants. Crit Rev Plant Sci 14(1):49–82
White PJ, Broadley MR (2005) Biofortifying crops with essential mineral elements. Trends Plant Sci 10(12):586–593
Yang XW, Tian XH, Lu XC, Cao YX, Chen ZH (2011) Impacts of phosphorus and zinc levels on phosphorus and zinc nutrition and phytic acid concentration in wheat (Triticum aestivum L.). J Sci Food Agric 91(13):2322–2328
Yuan L, Wu L, Yang C, Lv Q (2013) Effects of iron and zinc foliar applications on rice plants and their grain accumulation and grain nutritional quality. J Sci Food Agric 93(2):254–261
Zingore S (2011) Maize productivity and response to fertilizer use as affected by soil fertility variability, manure application, and cropping system. Better crops 95(1):4–6
Zou CQ, Zhang YQ, Rashid A, Ram H, Savasli E, Arisoy RZ et al (2012) Biofortification of wheat with zinc through zinc fertilization in seven countries. Plant Soil 361(1–2):119–130
Zuo Y, Liu Y, Zhang F, Christie P (2004) A study on the improvement of iron nutrition of peanut intercropping with maize on nitrogen fixation at early stages of growth of peanut on calcareous soil. Soil Sci Plant Nutr 50(7):1071–1078

Molecular Approaches for Biofortification of Cereal Crops

2

Heresh Puren, Bodeddulla Jayasankar Reddy, Akashi Sarma, Sushil Kumar Singh, and Waquar Akhter Ansari

Abstract

Cereals are edible grains and the primary staple food crops globally. They are the major source of energy, nutrients, and bioactive substances, which provide potential health benefits in humans and livestock animals. The green revolution enhanced the crop yield potential but did not enhance the nutritional requirements, which can now be overcome by genetic revolution. There is a colossal shortage of micronutrient consumption by a preponderance of the population globally for hidden hunger. Biofortification is a new paradigm, an instrumental intervention for providing nutrients to the people deprived of access to micronutrients while keeping the promise of least cost and sustainability. It can provide enough calories and essential nutrients to meet the requirement for the sound health of the target population.

Meanwhile, there is a continuous increase in population in developing nations, and climate change is a serious challenge to fulfill the food and nutritional requirements. Thus, a more rapid and efficient strategy for biofortification is the need of the hour. Molecular breeding approaches are proven suitable for biofortification by transferring mapped QTLs/genes into nutrient-deficient crops such as QPM maize, high Fe, and Zn-rich rice lines, golden rice enriched with vitamin A, etc. However, molecular breeding advises for limited employment, citing insufficient genetic variation among crops. Furthermore, genetic

H. Puren · B. J. Reddy · A. Sarma
Department of Plant Breeding and Genetics, Assam Agricultural University, Jorhat, India

S. K. Singh (✉)
DBT North East Centre for Agricultural Biotechnology, Assam Agricultural University, Jorhat, India

W. A. Ansari
Department of Botany, Banaras Hindu University, Varanasi, Uttar Pradesh, India

R. Deshmukh et al. (eds.), *Biofortification in Cereals*,
https://doi.org/10.1007/978-981-19-4308-9_2

engineering technologies go beyond the agricultural gene pool to increase the concentration of micronutrients. Similarly, recent genome sequencing technologies have paved the way for identifying new regulatory genes and miRNA within the crops, which can be further modified by genome editing approaches using CRISPR-Cas technology. This chapter highlights the strategies of molecular methods for the biofortification of crops substantiated to be precise and effective strategies for potential economic enrichment of nutritional status to overcome the hidden hunger worldwide.

Keywords

Malnutrition · Cereals · Biofortification · Molecular approaches · CRISPR-Cas

2.1 Introduction

Cereals originated from the Latin word "Cerealis" or "Ceres," which is worshiped as "Goddess of Grain." These are the staple foods globally since ancient times and reserve historical significance in human civilization. Cereals with the grains as the economic part belong to the family Gramineae. They pose a massive impact across the globe with the production of 2789.8 metric tons (mt) comprising crops like wheat (734 mt), rice (782 mt), maize (1147.6 mt), millets (31 mt), barley (141.4 mt), sorghum (59.34 mt), oats (23.05 mt), and rye (11.27 mt) (Anonymous 2020). Current estimates of annual cereal consumption are 175–190 and 155 kg/capita, both in developing and developed countries, respectively (OECD/FAO 2019). They are the primary source of various nutrients in both developed and developing countries, viz., energy, carbohydrate (nearly 75%), starch (25–27%), fiber (13%), fat (2%), and protein (6–15%) for humans and livestock (Ram and Mishra 2010). Cereals also contain a range of micronutrients, vitamin E, vitamin B complex, non-starch polysaccharides, and bioactive substances that provide potential health benefits. In contrast, micronutrients are required for proper functioning and essential metabolic activities in humans. Whole grain cereals are considered a healthy food for their nutritional value and worldwide accessibility to a significant proportion of the weak and underprivileged populations.

There is no uniformity in nutritional enrichment globally, which is a worrying concern identified by WHO, especially in children and women concerning essential micronutrients. Micronutrient deficiencies were observed, with an average of 11% in Asia and Africa. The recent reports published as *State of food security and nutrition in the world* (Anonymous 2020) suggested a rise of graph for hungry people since 2014 all across the world. The data shows the trends for 2018 as nearly 700 million or more people that account for 9.2% of the global population facing food insecurity, while ~1.3 billion people, i.e., 17.2% of the whole human population, did not have access to nutritious food at moderately severe levels. Altogether, this scenario prevailed at 26.4% of the population, i.e., 2 billion people in 2018 and 25.9% in 2019. The new report of FAO (Food and Agriculture Organization) suggests that in the last 5 years, 60 million people have joined the line of hunger. Subsequently, this data would cross the 840 million mark by 2030 with the current trends. The report also estimated that before the COVID-19 pandemic, 8.9% of the total population (~690 million) was undernourished. Global economic outlook predicts that post to the pandemic, there may be the addition of around 83–132 million malnourished people in the existing list by the end of the year 2020. This is a horrifying figure adding to the existing burden of malnutrition and thus making the challenge much more complicated. Another report of FAO reveals the facts and figures for the status of various forms of malnutrition which says that in the year 2019, children below 5 years of age were vulnerable to deprived growth as stunted, wasted, and overweight with an estimation of 144 million (21.3%), 47 million (6.9%), and 38.3 million (5.6%), respectively. This figure clears the picture that instead of all the efforts, the world is nowhere close to the SDG target of ending hunger by a stipulated timeline, i.e., 2030. Besides, the rise in obesity among adults goes from developed countries like America to developing countries like India, concisely every part of the world. The rapid rise in the growth of the human population, climatic catastrophe, and unprecedented pandemic like COVID-19 and desert locust attacks are causing setbacks to sustain food security. Consequently, the quality of diet continues to be deteriorating, posing a risk for undernutrition and worsening the food insecurity for the world altogether. In a nutshell, facts about malnutrition are the thumbnails describing the comprehensive efforts taken against malnutrition that are not concrete enough to achieve the target of "zero hunger" and "end malnutrition" to sustain food security by 2030.

We have not lost the battle yet; however, to procure the target, there has to be the assurance of proper accessibility of diets rich with balanced nutrients which are capable of cutting down the expenses of health by 97% and the greenhouse gas emission by 41–74% till 2030 if the healthy diets enriched with balanced nutrition become accessible to the whole world (Anonymous 2020). But the problems lie here, as more than three billion people who dwell in poverty simply cannot afford the expensive healthy food. Low-cost starchy staple food provides only dietary energy needs. However, it is not balanced in nutrients. Simultaneously, most of these people merely possess the expense to have a healthy balanced meal every day, which is roughly five times costlier than any starchy staple food (Anonymous 2020). In a similar context, the FAO annual report mentioned that in Africa and South Asia, nearly 57% of the population cannot make expenses to afford the healthy balanced diets and thus stated the projection of health costs to exceed 1.3 trillion USD per year by 2030 regarding mortality and no-communicable diseases due to deficient diet intake. Cereal crops are the primary food for the people of many developing nations. The recommendation of WHO/FAO for the consumption of vegetables and fruits at the rate of 400 g per person daily is not feasible for countries with a low monetary economy. Therefore, they rely on cereals as the main staple diet, viz., maize, wheat, rice, barley, sorghum, oats, millets, and rye to have wholesome nutrition. Cereals are more accessible and provide most nutrition requirements in every region globally, particularly in Africa and Asia, where cereals are the national staple diet. As more population increases in these regions, cereals can ideally become economically feasible and accessible healthy diets. However, some problems are associated with these crops, such as deficiencies of some micronutrients and nutritional components like inadequate bioavailability of Fe and Zn; lack of vitamin A and essential amino acid, viz., lysine, and tryptophan; and gluten content in wheat cause severe allergic coeliac disease. Enhancing the bioavailability of deficient nutritional content in these cereal crops can efficiently overcome these problems. Biofortification is an efficient way that offers a feasible solution to the nutritional improvement of cereal as a staple food crop and ensures the accessibility of a healthy balanced diet to the target population in the whole world. This is one of the major concerns in the current global scenario to mitigate the prevailing malnutrition status among children and adults belonging to all the world regions. Biofortification of cereal crops can restore food security in the whole world by reducing hunger and malnutrition.

Therefore, a potential economic approach is required to enhance the nutritional components of cereal crops by biofortification. Thus, a molecular approach like molecular marker-assisted breeding, genetic engineering, agrobacterium-mediated transformation, tissue culture techniques, transgenic techniques, and gene editing techniques like CRISPERcas9 presents the efficient opportunity to foster the nutritional component of crops with more certainty and lesser time stipulation than the conventional approaches. Molecular approaches have so far been proven efficient for biofortification in earlier efforts, for example, high vitamin A content-rich rice genotypes, Golden Rice-1 and Golden Rice-2 (IARI-ICAR); high Fe and Zn content-rich rice genotype, BR-29 (IARI-ICAR); high lysine and tryptophan content-rich maize genotypes, HQPM-1 (CCS HAU, against pellagra), HQPM-5, HQPM-7, Protina (IARI), Shaktiman, and Vivek Hybrid-9 (VPKAS Almora), etc. Hence, this chapter imposes the molecular approaches for enhancing the nutrient quotient of cereal crops. Its significant potential is to boost the accessibility of nutrient-enriched diets for the targeted population of all the world regions and to reduce malnutrition to ascertain food security altogether. It is a crucial tool to help us achieve the sustainable goal of zero hunger and end malnutrition by 2030 (Fig. 2.1).

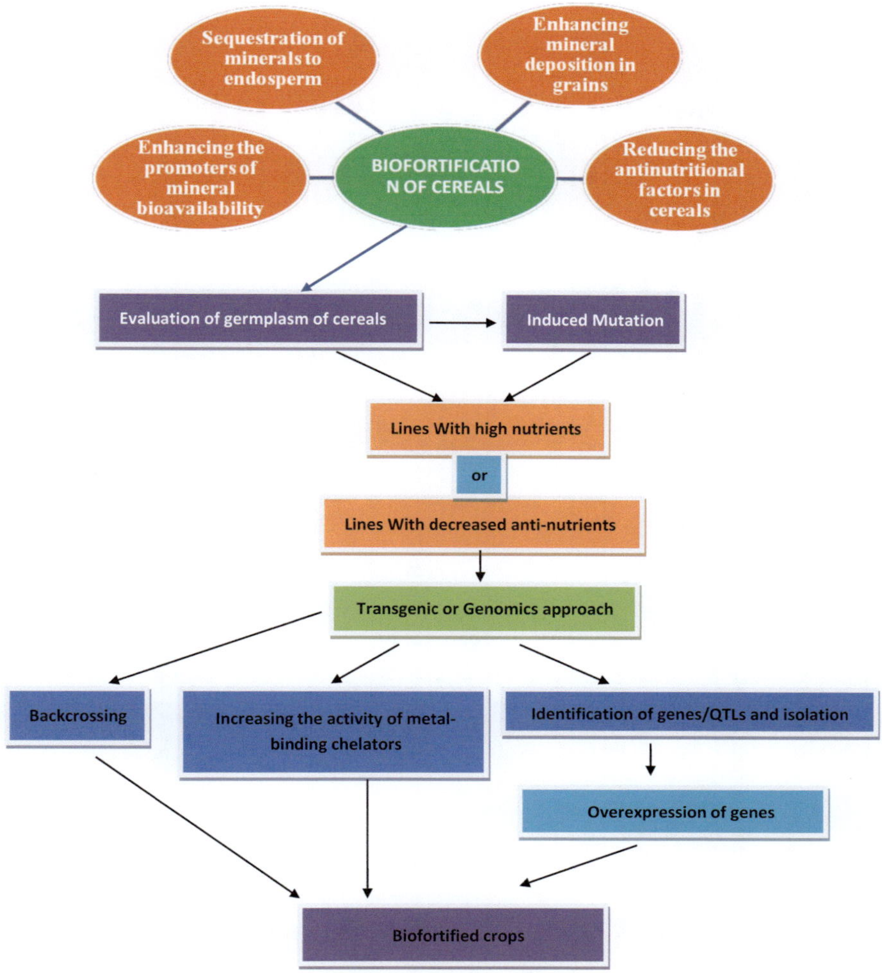

Fig. 2.1 Molecular approaches for quality improvement in cereal crops

2.2 Rice (Table 2.1)

Table 2.1 QTLs identified for different traits for biofortification in cereal crops

Traits/ QTL	No. of QTL	Parentage/cross	Type of population	Marker intervals/closest marker for major QTL	PVE	Chromosome/ chromosome arms/ chromosome position	References
Grain protein content							
qGPC	2	Zhenshan 97 (Indica rice) × Minghui 63 (Indica rice)	RIL—238 lines	C952-Wx-2	6.0–13.0	6, 7	Tan et al. (2001)
qGPC	4	Caiapo (Indica rice) × *Oryza glaberrima* (IRGC103544)	DH—312 lines	RM226–RM297	4.8–15.0	1, 2, 6, 11	Aluko et al. (2004)
qGPC	5	Gui630 (Indica rice) × 02428 (Japonica rice)	DH—81 lines	C22-RG449d, ZG34B-G20, RG435-RG172a	6.9–35.0	1, 4, 5, 6, 7	Hu et al. (2004)
qGPC	1	*Oryza sativa* (V20A) × *Oryza glaberrima* (accession 103544)	BC_3(TC) F_1 families—308 lines		9.0–10.0	8	Li et al. (2004)
qGPC	3	Moritawase (Japonica rice) × Koshihikari (Japonica rice)	RIL—92 lines		2.3–16.3	2, 6, 9	Wada et al. (2006)
qGPC	2	Koshihikari/Kasalath (Indica rice)//Koshihikari (Japonica rice)	BIL—92 lines	R1952 and R2447	14.3–14.8	6, 10	Takeuchi et al. (2007)
qGPC	2	Chuan (Indica rice) × Nanyangzhan (Japonica rice)	RIL—286 lines		2.69–4.50	6, 7	Zhang et al. (2011)

qGPC	5	Xieqingzao (Indica rice) × Milyang 46 (Indica rice)	RIL—209 lines	RM251–RM282, RM190–RZ516	3.9–19.3	3, 4, 5, 6, 10	Lou et al. (2009)
qGPC	9	Zhenshan 97 (Indica rice) × Minghui 63 (Indica rice)	RIL—241 lines		1.60–9.26	2, 3, 5, 6, 7, 10,11, 12	Yu et al. (2009)
qGPC	3	Samgang (Tongil variety) × Nagdong (Japonica variety)	DH—120 lines	RM287-RM26755, 11025-RM287	6.92–22.98	1, 11	Shi et al. (2009)
qGPC	9	Asominori (Japonica rice) × IR24 (Indica rice)	CSSL—66 lines	R1982, XNpb113, C1350, G1149	3.0–53.7	1, 2, 3, 6, 8, 11	Qin et al. (2009)
qGPC	10	Japonica rice (Asominori) × Indica rice (IR24)	RIL—71 lines	R265B-XNpb36, C1003-C688, XNpb212-G1318, C606-XNpb238, R1854-R2373, XNpb24-C562, XNpb338-C796, R758-XNpb15, XNpb268-R411	8.53–23.70	1, 3, 4, 6, 7, 8, 9, 10, 12	Zhang et al. (2011)
qGPC	2	Zhenshan 97B (Indica rice) × Delong 208 (Indica rice)	RIL—188 lines	RM445–RM418	7.2–25.9	1, 7	Zhong et al. (2011)
qGPC	4	Koshihikari/Kasalath (Indica rice)//Koshihikari (Japonica rice)	BIL—182 lines	R250-C746, C16-C809, C847-C596	6.26–12.11	2, 3, 7, 10	Zheng et al. (2011)
qGPC	1	Cheongcheong (Indica rice) × Nagdong (Indica rice)	DH—133 lines	RM12532–RM555	39–41	2	Lee et al. (2014)
qGPC	1	CJ06 (Japonica cultivar) × TN1 (Indica rice cultivar)	DH—116 lines	RM216-RM467	12.3–15.8	10	Leng et al. (2014)
qGPC	3	Cheongcheong (Indica rice) × Nagdong (Indica rice)	DH—133 lines	RM506-RM1235, RM24934-RM25128, RM219-RM23914	39–40	8, 9, 10	Yun et al. (2014)

(continued)

Table 2.1 (continued)

Traits/ QTL	No. of QTL	Parentage/cross	Type of population	Marker intervals/closest marker for major QTL	PVE	Chromosome/ chromosome arms/ chromosome position	References
qGPC	5	M201 (*Oryza sativa*) × JY293 (*Oryza sativa*)	RIL—234 lines	RM423-RM6375, GS3-SLAF13430	6.74–13.50	1, 2, 3, 4	Xu et al. (2015)
qGPC	1	Sasanishiki (Japonica variety) × Habataki (Indica variety)	CSSL—39 lines	RM7124	10.38–15.43	1	Yang et al. (2015)
qGPC	1	Cheongcheong (Indica rice) × Nagdong (Indica rice)	DH—120 lines	RM8261	14	7	Bruno et al. (2017)
qGPC	3	Naveen/*Oryza sativa* (ARC10075)//Naveen (*Oryza sativa*)	BC_3F_5—200 lines	CSCWR_Os01g02590—61041, CSCWR_Os02g10740—65058	6.70–17.35	1, 2, 7	Chattopadhyay et al. (2019)
Phytate concentration							
IP6	2	Azucena (Upland Japonica var) × IR64 (Indica var)	DH—129 lines	RM305–RM178; RM247–RM179	24.3; 15.4	5, 12	Stangoulis et al. (2007)
Iron and zinc							
GZn; GFe	2; 3	IR64 (Indica variety) × Azucena (Japonica variety)	DH—129 lines	**GZn**: RM235–RM17, RM34–RM237; **GFe**: RM270–RM17, RM53–RM300, RM137–RM325A	12–15; 13.8–18	1, 12; 2, 8, 12	Stangoulis et al. (2007)
GZn; GFe	3; 2	Zhengshan97 (Indica cultivar) × Minghui 63 (Indica cultivar)	RIL—241 lines	**GZn:** R3166-RG360, C794-RG118; **GFe**: C472-R2638, RG236-C112	5.3–18.61; 11.11–25.81	5, 7, 11; 1,9	Lu et al. (2008)
GZn; GFe	2; 1	*Oryza sativa* ssp. Indica (Teqing) × *Oryza rufipogon* Griff	IL—85 lines	**Zn**: RM152	5–11; 7	5, 8; 2	Garcia-Oliveira et al. (2009)

GZn; GFe	4; 4	Bala (Indica rice) × Azucena (Japonica rice)	RIL—79 lines	**GZn**: G1082, G20, AB0601, C223; **GFe**: R1440, C949, R1618	11.2–14.8; 9.7–21.4	6, 7, 10; 1, 3, 4, 7	Norton et al. (2010)
GZn	2	ZYQ8 (Indica cultivar) × JX17 (Japonica cultivar)	DH—127 lines	**GZn**: CT206-G177, RZ516-G30	10.83–12.38	4, 6	Zhang et al. (2011)
GZn; GFe	6; 7	Madhukar (Indica rice) × Swarna (Indica rice)	RIL—168 lines	**GZn**: RM501–OsZip2, RM7–RM517, RM260–RM7102, RM234–RM248, RM248–RM8007, RM17–RM260; **GFe**: RM243–RM488, RM488–RM490, RM574–RM122, RM234–RM248 RM248–RM8007, RM17–RM260, RM 260–RM7102	29–35; 69–71	3, 7, 12; 1, 5, 7, 12	Agarwal et al. (2014)
GZn; GFe	3; 8	PAU201 (Indica rice) × Palman 579 (Indica rice)	F_2—247 lines	**Zn**: 8RM474–RM184; RM491–RM519, RM228–RM496, RM53–RM521, RM221–RM208	4.7–19.1; 2.4–26.8	2, 10; 2, 3, 7,10, 12	Kumar et al. (2014)
GZn-4; GFe-1	4; 1	Ce258 (Indica cultivar) × IR75862 (Japonica breeding line)/ZGX1 (Indica cultivar) × IR75862 (Japonica breeding line)	BIL—200 and 201 lines	**GZn**: RM293–RM85, RM407–RM152; **GFe**: RM3–RM340	2–24.4; 10.2–18.3	3, 6, 7, 8; 6, 11	Xu et al. (2012)
GFe	1	Swarna (Indica cultivar) × Moroberekan (Japonica rice)	RIL—60 lines	RM490-RM5	39	1	Indurkar et al. (2015)

(continued)

Table 2.1 (continued)

Traits/ QTL	No. of QTL	Parentage/cross	Type of population	Marker intervals/closest marker for major QTL	PVE	Chromosome/ chromosome arms/ chromosome position	References
GZn; GFe	6; 3	*Oryza sativa* (XB) × *Oryza rufipogon* (DWR accession)	BIL—202 lines	**GZn**: RG172-RM340; RG123-RG172, RG510-RZ251	5.3–11.8; 6.1–28.2	3, 4, 6, 7, 10, 12; 3, 6, 9	Hu et al. (2016)
GZn	4	Nipponbare (*Oryza sativa*)/ W1627 (*Oryza meridionalis*)// Nipponbare	BRIL—151 lines	RM171-RM590, RM573, RM6, RM24085-RM566	15.0–21.9	2, 9, 10	Ishikawa et al. (2017)
GZn; GFe	8; 1	PSBRc82 (Indica cultivar) × Korean rice (Joryeongbyeo & PSBRc82) × IR69428 (Indica breeding line)	DH—130 and 97 lines	**GZn**: 2140834-2147095, 13048465-13057679, 8803052-8832534, 6025827-6047367, 606341-id6006214, 2110566-id2009463, 2783884-2785595, 10858811-id11000778	7.5–22.8; 9.4	2, 3, 6, 8, 11, 12; 4	Swamy et al. (2018)
GZn	8	IR64 (Indica cultivar) × IR69428 (Breeding line) and BR29 (Indica cultivar) × IR75862 (Breeding line)	DH—111 and 146 lines	**GZn**: wd9002310-9831169, 5645339-5648872, 2048774-2054640, 3538410-3548096, 7062019-7089136, 5027770-5077125, 10907196-id11001107	8.6–27.7	2, 3, 5, 7, 8, 9,11	Descalsota-Empleo et al. (2019)
GZn; GFe	1; 5	PAU201 (Indica rice) × Palman (Indica rice)	F_4—579 lines	**GZn**: RM585-RM3; **GFe**: RM2488-RM440, RM440-RM31, RM440-RM31, RM432-RM429, RM566-RM434	25; 34.6–95.2	6; 5, 7, 9	Kumar et al. (2019)

2 QTL clusters	2	Zhenshan 97 (Indica rice) × Nanyangzhan (Indica rice)	RIL—190 lines	**Tyr/Pro/Thr/Gly/Ala/Glu/Asp/Val/Met/Lys/Ser/Asp/Ile/Leu/Phe/His/Arg/Cys:** RM472-RM104	4.05–33.3	1, 7	D'Mello et al. (2003)
10 (His) + 8 (Arg)	10; 8	Zhenshan 97 (Indica rice) × Minghui 63 (Indica rice)	RIL—241 lines	**His:** R321–RM55, RZ398–RM204, RG101–G393, C1003B–RG103, RG118–C794, RM53–RZ599, RM258–RG561, RG424–R2549, RG528–RG128, RM20b–C732; **Arg:** C734b–RZ649, R321–RM55, RG424–R2549, RM258–RG561, R3203–RM20A, RM53–RZ599, RG528–RG128, RM20b–C732	12–35 (His); 16–33 (Arg)	1, 2, 3, 6, 7, 10, 11, 12 (His); 2, 3, 5, 6, 7, 10, 11, 12	Zheng et al. (2011)
12	12	Zhenshan 97 (Indica rice) × Minghui 63 (Indica rice)	RIL—241 lines	**Ser**: R2632–C39, Val: RG173–RM81A, **Met**: RZ536–TEL3	3.4–48.8	1, 11	Lu et al. (2008)
3 QTL clusters	3	Zhenshan 97B (Indica rice) × Delong 208 (Indica rice)	RIL—188 lines	**Asp/Thr/Ser/Gly/Val/Ile/Phe/Lys/Taa:** RM328–RM107, **Asp/Thr/Ser/Glu/Gly/Ala/Cys/Val/Met/Ile/Phe/Arg/Pro/Taa:** MRG186–MRG4499, **Asp/Thr/Glu/Gly/Ala/Val/Leu/Phe/Arg/Pro/Taa:** RM493–RM562	4.2–31.7	1, 7, 9	Zhong et al. (2011)
6	6	Dasanbyeo (*Oryza sativa*) × TR22183 (*Oryza sativa*)	RIL—172 lines	id3015453-id3016090, id3001422 fd10	10.2–12.4		Yoo et al. (2017)

(continued)

Table 2.1 (continued)

Traits/QTL	No. of QTL	Parentage/cross	Type of population	Marker intervals/closest marker for major QTL	PVE	Chromosome/chromosome arms/chromosome position	References
Other micro-minerals							
As: qAs1.1; qAs10.1	1; 1	PSBRc82 × Joryeongbyeo/PSBRc82 × IR69428	DH (130 and 97 lines)	RM206639-RM214137 (11.2); RM10721590-RM10722207 (7)	27; 25.8	1 (11 cM), 10 (124 cM)	Swamy et al. (2018)
Ca: qCa2.1	1			RM2094246-id2009186 (40.4)	37	2 (5 cM)	
Co: qCo1.1; qCo4.1	1; 1			id1013855-RM827062 (15.8); id4008544-RM4572241 (10.8)	29.2; 29.2	1 (67 cM), 4 (114 cM)	
Cu: qCu4.1	1			RM4285667-RM4314701 (31.8)	30.4	4 (17 cM)	
Mn: qMn2.1	1			2094246-id2009186 (59.5)	43	2 (5 cM)	
Mo: qMo1.1; qMo1.2	2; 1			id1014853-854218 (20.1); id1014853-854218 (29.2); 815155-id1014853(24.3)	51.2; 42.8; 31.5	1 (80 cM); 1 (88 cM)	
Na: qNa1.1; qNa10.1	1; 1			id1006604-267954 (10.7); 10594456-10635878 (8.4)	30.7; 26.7	1 (10 cM); 10 (105 cM)	

PVE phenotypic variance effect

2.3 Wheat

No. of QTLs	Parentage	Population type and their size	Chromosomes position/ chromosome arms	PVE range for the major QTLs	References
Grain protein content (Gpc-b1)					
6	Messapia (Durum wheat) × *Triticum turgidum* L. var. dicoccoides (MG4343)	RIL—65 lines	4BS, 5AL, 6AS, 6BS, and 7BS	6.0–23.5	Blanco et al. (2012)
1	*Triticum turgidum* (L.) var. dicoccoides	RICL—85 lines	6BS	66	Joppa et al. (1997)
1	*Triticum aestivum* (PH132) × *Triticum aestivum* (WL711)	RIL—100 lines	2DL	18.73	Prasad et al. (1999)
2	*Triticum aestivum* (Courtot) × *Triticum aestivum* (Chinese Spring)	DH—187 lines	1B, 6A	7.0–17.0	Perretant et al. (2000)
9	*Triticum aestivum* (PH132) × *Triticum aestivum* (WL711)	RIL—106 lines	2BL, 7AS	2.9–7.2	Dholakia et al. (2001)
1	*Triticum aestivum* (PH132) × *Triticum aestivum* (WL711)	RIL—100 lines; NIL—10 lines	5AL	6.2	Singh et al. (2001)
7	Messapia (Durum wheat) × *Triticum turgidum* L. var. dicoccoides (MG4343)	RIL—65 lines	4BS, 6AS, 5AL, 7AS, 7BS, 6BS	6.5–31.7	Blanco et al. (2002)
2	*Triticum aestivum* (Opata 85) × synthetic hexaploid wheat (W7984)	RIL—114 lines	2DS, 7AS		Börner et al. (2002)
10	*Triticum aestivum* (Renan) × *Triticum aestivum* (Récital)	RIL—194 lines	1A, 2AS, 3AL, 3BS, 4AS, 4DL, 5BL, 6AL, 7AS, 7DL	4.1–10.4	Groos et al. (2004)
13	*Triticum aestivum* (WL711) × *Triticum aestivum* (PH132)	RIL—100 lines	7AS, 2AS, 2DL, 2BL, 3DS, 4AL, 6BS, 7DS	2.95–32.44	Prasad et al. (2003)
3	*Triticum turgidum* (L.) var dicoccoides (LDNDic-5B) × LDN	RICL—133 lines	5B	10.0–33.0	Gonzalez-Hernandez et al. (2004)
3	*Triticum aestivum* (Renan) × *Triticum aestivum* (Récital)	RIL—194 lines	3A, 4D, 7D	6.2–9.6	Groos et al. (2003)

(continued)

No. of QTLs	Parentage	Population type and their size	Chromosomes position/ chromosome arms	PVE range for the major QTLs	References
7	*Triticum aestivum* (WL711) × *Triticum aestivum* (PH132)	RIL—110 lines	2DS, 3AL, 2AS, 1DL, 5AL, 7DL	8.38–16.58	Kulwal et al. (2005)
2	Canadian Spring wheat (AC Karma) × *Triticum aestivum* (87E03-S2B1)	DH—185 lines	4D, 7B	12.6–32.7	Huang et al. (2006)
4	*Triticum aestivum* (Opata85) × Synthetic hexaploid wheat (W7984)	RIL—114 lines	2DS, 5AL, 6DS	15.0–32.0	Nelson et al. (2006)
13	*Triticum aestivum* (Arche) × *Triticum aestivum* (Recital)	DH—222 lines	2D, 4B, 2A, 1B, 3B, 3D, 5A, 5B, 7D	5.5–24.7	Laperche et al. (2007)
3	*Triticum aestivum* (Chuan 35050) × *Triticum aestivum* (Shannong 483)	RIL—131 lines	5AL, 3BL, 6AS	8.64–21.23	Sun et al. (2008)
16	*Triticum aestivum* (Neixiang188) × *Triticum aestivum* (Yanzhan)	RIL—198 lines	3B, 2B, 1B, 2A, 2B, 3A, 4D, 5B, 5D, 7B, 7D	3.2–14.5	Li et al. (2009)
13	*Triticum aestivum* (kukri) × *Triticum aestivum* (Janz)	DH—160 lines	1B, 2A, 3AS, 3B, 4B, 4D, 5A, 5B, 7AL, 7D		Mann et al. (2009)
1	Indian durum wheat (PDW 233) × Bhalegaon 4 (a landrace)	RIL—140 lines	7B	9.64	Patil et al. (2009)
10	Durum wheat (Langdon) × Wild emmer accession (G18–16)	RIL—152 lines	2AL, 2BL, 3BL, 4AL, 5AS, 5BL, 6AS, 6BL, 7AL, 7BS	2.8–9.7	Peleg et al. (2009)
1	*Triticum aestivum* (Chara) × an advanced breeding line (WW2449)	DH—190 lines	4A	20	Raman et al. (2009)
9	Durum breeding line (DT695) × Durum wheat cultivar (Strongfield)	DH—185 lines	2B, 7A, 1A, 1B, 2A, 5B, 6B, 7A	16–46	Suprayogi et al. (2009)
2	Chinese hard wheat line (Ning7840) × Soft wheat cultivar (Clark)	RIL—132 lines	3AS, 4B	11.2–16.8	Sun et al. (2010)
3	*Triticum aestivum* (MN98550) × *Triticum aestivum* (MN99394)	RIL—139 lines	5AL, 2BS, 6DL	4.5–16.8	Tsilo et al. (2010)
4	*Triticum aestivum* (Huapei 3) × *Triticum aestivum* (Yumai 57)	DH—168 lines	3A, 3B, 5D, 6D	3.09–8.40	Zhao et al. (2010)

(continued)

No. of QTLs	Parentage	Population type and their size	Chromosomes position/ chromosome arms	PVE range for the major QTLs	References
10	Durum breeding line (C1113) × Durum cultivar (Kofa)	RIL—93 lines	3BS, 7BL, 5AS, 2BS, 4AL, 5BL, 2AL, 1BS, 7AS, 3BL	9.3–21.6	Conti et al. (2011)
10	Svevo × Ciccio (both elite durum wheat cultivars)	RIL—120 lines	3BS, 2BL, 1AL, 4AL, 2AS, 4BL, 1AS, 6BS, 5AL, 7BL	7.8–40.2	Blanco et al. (2012)
2	Oste-Gata × Massara-1 (durum wheat genotypes)	F_2 derived F_3 and F_4—151 lines	1A, 5BL	5.31–9.44	Golabadi et al. (2012)
9	*Triticum aestivum* (Weimai 8) × *Triticum aestivum* (Jimai 20)	RIL—485 lines	2B, 3A, 4A, 4D, 5B, 7A, 7B	3.06–9.79	Li et al. (2012a)
10	*Triticum aestivum* (Weimai 8) × *Triticum aestivum* (Yannong 19)	RIL—229 lines	5A, 1A, 2D, 1B, 4B, 2A, 3A, 5D, 6B, 7D	6.29–53.04	Li et al. (2012a)
9	Synthetic wheat (Am3) × Synthetic wheat (Laizhou953)	BC_5F_2:F_6 families—82 lines	6A, 1A, 2D, 3A, 4B, 5D, 6B, 6D, 7B	2.2–11.5	Li et al. (2012b)
1	*Triticum aestivum* (BR34) × *Triticum aestivum* (Grandin)	RIL—118 lines	5BL	16.3	Simons et al. (2012)
7	*Triticum aestivum* (Weimai 8) × *Triticum aestivum* (Luohan 2)	RIL—302 lines		4.15–9.73	Wang et al. (2012)
5	*Triticum aestivum* (Xiaoyan 54) × *Triticum aestivum* (Jing 411)	RIL—182 lines		1.14–9.25	Xu et al. (2012)
5	*Triticum aestivum* (CO940610) × *Triticum aestivum* (Platte)	DH—185 lines	5BS, 6AL, 6BS, 7BS, 7DL	5.6–12.3	El-Feki et al. (2013)
2	*Triticum aestivum* (Choteau) × *Triticum aestivum* (Yellowstone)	RIL—97 lines	3B, 5B	17–19	Heo and Sherman (2013)
13	*Triticum aestivum* (Huapei 3 × Yumai 57; Nuomai 1 × Gaocheng 890; 1Shannong 01–35 Gaocheng 9411);	DH—68 lines, RIL—256 lines, RIL—182 lines	2A, 1B, 1D, 2B, 2D, 3B, 4B, 5B, 6D, 7A	0.84–10.51	Deng et al. (2015)

(continued)

No. of QTLs	Parentage	Population type and their size	Chromosomes position/ chromosome arms	PVE range for the major QTLs	References
4	*Triticum aestivum* (Drysdale) × *Triticum aestivum* (gladius)	RILs (155)	2B, 2D, 3D, 5A	–	Maphosa et al. (2015)
12	*Triticum aestivum* (CD87) × *Triticum aestivum* (Katepwa)	DH lines (180)	1D, 2A, 2B, 2D, 4A, 4B, 5A, 5B, 5D, 6A, 6B, 6D, 7A	–	Moore et al. (2015)
11	*Triticum aestivum* (WCB414) × *Triticum aestivum* (WCB617)	RIL—163 lines	6B, 1A, 5B, 2B. 7B, 1B, 2D, 3D, 4B	4.7–16.5	Echeverry-Solarte et al. (2015)
4	*Triticum aestivum* (Ning7840) × *Triticum aestivum* (Clark)	RIL—127 lines	4BS, 5AL, 5BL, 3A	11.5–22	Li et al. (2016)
12	*Triticum aestivum* (RAC875) × *Triticum aestivum* (Kukri)	DH—156 lines	7A, 6A, 3D,1B, 2D, 4B, 5A, 5B, 5D	7.00–17.00	Mahjourimajd et al. (2016)
1	*Triticum aestivum* (Kitami 81) × *Triticum aestivum* (Kachikei 63)	DH—94 lines	2BS	32.1	Terasawa et al. (2016)
1	*Triticum aestivum* (Berkut) × *Triticum aestivum* (Krichauff)	DH—138 lines	1A	17.7	Tiwari et al. (2016)
9	*Triticum aestivum* (Chuan 35050) × *Triticum aestivum* (Shannong 483)	RIL—131 lines	7D, 4A, 4B, 5D, 1A, 1D, 2A, 2D,	4.1–32.7	Sun et al. (2016)
8	*Triticum turgidum* (Duilio) × *Triticum turgidum* (Avonlea)	RIL—134 lines	5A, 3B, 4A, 7B, 1B, 2B, 7A	10–14	Marcotuli et al. (2017)
12	*Triticum durum* var. Svevo × *Triticum turgidum* ssp. *dicoccoides* accession no. Y12-3	RIL—208 lines	4B, 5A, 6B, 1A, 1B, 2A, 3A, 4A, 6A, 7B	2.6–26.6	Fatiukha et al. (2019)
Grain zinc (GZn), grain iron (GFe), and grain selenium (Se) contents					
Zinc conc.-4 and Zn content-7	*Triticum aestivum* Hanxuan10 × *Triticum aestivum* Lumai 14	DH—119 lines	4D, 5A, 4A, 7A and 7A, 2D, 1A, 3A, 4A, 4D, 5A	5.3–11.9; 4.6–14.6	Shi et al. (2008)
GZn-4; GFe-1	*Triticum aestivum* (RAC875-2) × *Triticum aestivum* (cascades)	DH—90 lines	3D, 4B, 6B, 7A, 3D		Genc et al. (2009)
GZn-2; GFe-3	*Triticum boeoticum* (Tb5088) × *Triticum monococcum* (Tm14087)	RIL—93 lines	7A, 2A, 7A	7.0–12.6; 9.0–18.8	Tiwari et al. (2009)

(continued)

No. of QTLs	Parentage	Population type and their size	Chromosomes position/ chromosome arms	PVE range for the major QTLs	References
GZn-6; GFe-11	Durum wheat (cv. Langdon) × wild emmer (accession G18-16)	RIL—152 lines	2A, 7A, 5A, 6B, 7B, 5A, 7A, 2A, 2B, 3A, 3B, 4B, 5A, 6A, 6B, 7B	1.3–23.5; 0.8–17.8	Peleg et al. (2009)
GZn-2; GFe-2	*Triticum aestivum* (Xiaoyan 54) × *Triticum aestivum* (Jing 411)	RIL—182 lines	4B, 5A, 5A	4.23–6.88; 3.27–3.43	Xu et al. (2012)
GFe-4	*Triticum aestivum* (Hanxuan 10) × *Triticum aestivum* (Lumai 14)	DH—120 lines	5A, 4D, 7A, 7B	6.1–14.6	Shi et al. (2013)
GZn-2; GFe-6	*Triticum aestivum* (Tabassi) × *Triticum aestivum* (Taifun)	RIL—118 lines	4A, 1A, 7B, 3D, 4D, 2A, 7D	40.22–50.79; 8.94–47	Roshanzamir et al. (2013)
GZn-3	*Triticum aestivum* (PBW343) × *Triticum aestivum* (Kenya Swara)	RIL—177 lines	1BS, 2B, 3AL	10–15	Hao et al. (2014)
GZn-4; GFe-4	Synthetic hexaploid (SHW-L1) × *Triticum aestivum* (Chuanmai 32)	RIL—171 lines	2D, 3D, 4D, 5D, 2B, 5B, 5D, 7D	5.5–8.6; 5.4–9.5	Crespo-Herrera et al. (2016)
GZn-3; GFe-4	*Triticum aestivum* (Chuanmai 42) × *Triticum aestivum* (Chuannong 16)	RIL—127 lines	5B, 3D, 4D, 4A, 5A, 4D, 5B	13.8–15.9; 9.2–19.1	Crespo-Herrera et al. (2017)
GZn-5; GFe-5	*Triticum spelta* (PI348449) × *Triticum aestivum* (HUW 234)	RIL—185 lines	2B, 2A, 3D, 6A, 6B, 1A, 3B, 2A	4.25–16.46; 5.6–25.95	Srinivasa et al. (2014)
GZn-2; GFe-1	*Triticum aestivum* (Berkut) × *Triticum aestivum* (Krichau)	DH—138 lines	1B, 2B, 2B	23.1–35.9; 22.2	Tiwari et al. (2016)
GZn-3; GFe-5	*Triticum aestivum* (SeriM82) × *Triticum dicoccoides/Aegilops Tauschii* (SHW CWI76364)	RIL—140 lines	4BS, 6AL, 6BL, 4BS, 7DS, 2BL, 2DS, 6AL,1D, 6B, 7B, 7A, 3A, 1B, 2B, 3D, 6A, 6B, 7B, 6B, 2B, 7B, 1B, 2A	8.3–17.3; 7.5–14.5	Velu et al. (2017)
GZn-10; GFe-7	*Triticum aestivum* (Adana99) × *Triticum sphaerococum* (70711)	RIL—127 lines	2B, 3A	9–31; 9–18	Pu et al. (2018)

(continued)

No. of QTLs	Parentage	Population type and their size	Chromosomes position/ chromosome arms	PVE range for the major QTLs	References
GZn-4; GFe-3	*Triticum spelta* (Bubo) × resynthesized hexaploid wheat (Turtur)	RIL—188 lines	1B, 7B, 6A, 3A, 4B, 5B	2.86–16.75; 5.49–10.35	Krishnappa et al. (2017)
GZn-12; GFe-7	Synthetic hexaploid wheat (Louries) × *Triticum spelta* (Bateleur)	RIL—188 lines	1A, 1B, 3B, 7B, 3D, 4A, 5B, 6A, 7D, 5B, 2A, 4D, 4A, 2B, 3B, 5B	3.30–32.79; 5.79–21.14	Krishnappa et al. (2017)
GZn-5; GFe-4	*Triticum aestivum* (WH542) × synthetic derivative (*Triticum dicoccon*) PI94624/ *Aegilops sqarrosa* (409/ BCN)	RIL—286 lines	2A, 4A, 5A, 7A, 7B; 2A, 5A, 7A, 7B	3.2–14.4; 2.3–6.8	Crespo-Herrera et al. (2017)
4	*Triticum aestivum* (SHW-L1) × *Triticum aestivum* (Chuanmai 32)	RIL—171 lines	5B, 3D, 7D	6.4–28.5	
1	*Triticum aestivum* (Chuanmai 42) × *Triticum aestivum* (Chuannong 16)	RIL—127 lines	4D	35.1	
7	*Triticum aestivum* (TN18) × *Triticum aestivum* (LM6)	RIL—184 lines	2B, 5B	7.44–15.57	Wang et al. (2017a)
6	Synthetic wheat (SHW-L1) × *Triticum aestivum* (Chuanmai 32)	RIL—171 lines	3D, 5A	8.17–28.38	Pu et al. (2014)
7	*Triticum dicoccoides* (Landon) × Wild emmer wheat (accession no. G18-16)	RIL—152 lines	7A, 5A, 7B, 1A, 1B, 3A	1.4–18.6	Yan et al. (2018)
QFe. pau-2A	*Triticum boeoticum* accession pau 5088 × *Triticum monococcum* accession pau14087	RIL—93 lines	2A , Xwmc382-Xbarc124	23.6 •	Tiwari et al. (2009)
QFe. pau-7A			7A, Xgwm473-Xbarc29	153.8	
QFe. pau-7A			7A, Xcfd31-Xcfa2049	72.6	
QZn. pau-7A			7A, Xcfd31-Xcfa2049	72.6	
QZn. pau-7A			7A, Xgwm473-Xbarc29	153.8	
Grain yellow pigment content					
2	*Triticum aestivum* (Schomburgk) × *Triticum aestivum* (Yarralinka)	SSD—150 lines	3A, 7A	13–41	Parker et al. (1998)

(continued)

No. of QTLs	Parentage	Population type and their size	Chromosomes position/ chromosome arms	PVE range for the major QTLs	References
3	*Triticum turgidum* L. var. *durum* (Omrabi5) × *Triticum dicoccoides* (acc.600545)	RIL—114 lines	7AL, 7BL	6.0–53.0	Elouafi et al. (2001)
1	*Triticum aestivum* (Trident) × *Triticum aestivum* (Molineux)	DH—182 lines	7B	48–77	Kuchel et al. (2006)
5	*Triticum turgidum* L. var *durum* (PDW 233) × *Triticum turgidum* var *durum* (Bhalegaon 4)L	RIL—140 lines	7A, 1A, 3B, 5B, 7B	5–55.22	Patil et al. (2008)
1	*Triticum turgidum* L. var *durum* (UC1113) × *Triticum turgidum* L.var *durum* (Kofa)	RIL—93 lines	7A		Zhang and Dubcovsky (2008)
4	*Triticum aestivum* (PH82-2) × *Triticum aestivum* (Neixiang 188)	RIL—240 lines	1B, 7A, 1A, 4A,	1.5–33.9	Zhang et al. (2009)
5	*Triticum turgidum* L. var *durum* (Latino) × *Triticum turgidm* L. var *durum* (Primadur)	F_2:F_3 families—121 lines	2A, 3B, 5A, 7A	9.4–53.2	Blanco et al. (2011)
6	Ajana × WAWHT2074; Carmah × WAWHT2046; Ajanax WAWHT204 (*Triticum aestivum*)	DH—179, 121 and 127 lines	7B, 2D, 3A, 7A, 4D, 5B, 7B	4.0–36.0	Crawford et al. (2011)
15	*Triticum turgidum* L. var *durum* (UC1113) × *Triticum turgidum* L. var *durum* (Kofa)	RIL—93 lines	1BL, 4AL, 7BL, 6AL, 2AS, 5AS, 5AL, 5BL, 7AS, 7AL, 7BS	6–42.7	Roncallo et al. (2012)
13	*Triticum aestivum* (Chuan 35050) × *Triticum aestivum* (Shannong 483)	RIL—131 lines	5B, 6A, 1A, 1B, 2D, 4A, 4D, 5D, 6D, 7B	4.1–16.5	Zhang and Dubcovsky (2008)
7	*Triticum turgidum* L. var *durum* (Svevo) × *Triticum turgidum* L. var *durum* (Ciccio)	RIL	1B, 5B, 7A, 2A, 2B, 5A, 7B	19.3–51.6	Colasuonno et al. (2014)
16	*Triticum aestivum* (Gaocheng 8901) × *Triticum aestivum* (Zhoumai 16)	RIL—176 lines	5AL, 2DL, 5BS, 1B. 1R, 2AL, 2B-1,5AS, 5BL, 6BL, 7AS, 7BL	5.7–30.8	Zhai et al. (2016)

PVE phenotypic variance effect

2.4 Maize

S. No	Trait	Parentage	Population type with size	No. of QTL	Chromosome/ chromosome bin	chromosome position (cM)	Closest marker to the QTL	LOD	Reference
Micromineral content									
1	**P**	B84 (BSSL line) × Os6-2 (relative line of C103)	F_4 families (294 lines)	8	3-05; 3-09; 4-08; 6-03; 6-05; 8-05; 9-02; 10-07	10; 48; 48; 20; 34; 26; 16; 32	umc59e; bnlg1257; ZM0819; umc1887; ZM1367; ZM0353; bnlg0244; bnlg1839	5.98; 6.48; 6.69; 4.89; 5.42; 4.50; 8.39; 8.67	Šimić et al. (2012)
2	**Fe**			3	2-05; 6-03; 8-06	40; 24; 32	ZM1368; ZM0960; ZM0825	4.44; 4.93; 4.44	
3	**Fe/ P**			7	1-05; 2-05; 3-05; 4-07; 6-01; 6-03; 10-06	34; 40; 16; 32; 10; 24; 26	ZM0845; ZM1368; bnlg1456; bnlg1784; bnlg0426; ZM0960; ZM1315	4.55; 7.82; 6.54; 4.73; 4.73; 6.75; 5.95	
4	**Zn**			1	4-08	44	ZM1362	5.19	
5	**Zn/ P**			1	3-08	16	bnlg1456	5.21	
6	**Mg**			4	5-03; 6-01; 8-05; 9-07	18; 16; 24; 44	bnlg1046; bnlg426; bnlg1782; bnlg0128	4.44; 4.65; 5.03; 7.96	
7	**Mg/ P**			8	1-01; 1-03; 3-05; 4-08; 5-03; 9-02; 9-07; 10-04	0; 20; 16; 48; 12; 16; 34; 12	bnlg1014; phi109275; bnlg1456; ZM0819; ZM0215; bnlg244; umc1675; ZM0363	6.00; 5.96; 5.65; 5.34; 14.00; 5.28; 5.02; 12.77	

Chr chromosome

2.5 Oats

Character	Position of QTL	Marker interval	Closest marker to the QTL	LOD score	R^2 (%)	Additive effects	Reference
β-glucan	QBg1. jki-A1	17–21	E36M55_1–E36M52_3	7.9	31.27	−0.14	Matthias et al. (2014)
	QBg2. jki-A10	21–83	E33M48_2–E34M49_1	3.51	16.27	0.17	
	QBg1. jki-B1	6–9	E32M55_1–E36M56_a	3.39	24.98	−0.16	
Groat protein	QPr1. jki-A1	20–24	E36M52_3–E35M62_3	14.11	48.84	−0.76	
	QPr1. jki-B1	16–21	E37M47_c–E37M47_b	9.88	45.62	−0.65	

2.6 Pearl Millet

Trait	Position of QTL	Support interval	Marker interval	LOD	R^2 (%)	Additive effects	References
Iron (Fe)	1/54	52–56	pgpb10531-pgpb9130	25.36	31.9	9.7	Kumar et al. (2018)
	1/30	26–36	Xipes017-pgpb12900	6.22	9	4	
	3/20	8–26	Xipes188-pgpb6069	6.59	9.5	0.4	
	4B/8	04–08	pgpb9502-pgpb6039	6.87	10.4	−0.6	
	7/86	84–90	pgpb8427-pgpb13221	8.58	12.2	5.3	
	7/120	118–122	pgpb6825-Xipes195	9.7	14	0.1	
	7/16	0–30	pgpb11956-pgpb9273	7.25	12.5	−1.9	
	7/108	106–110	pgpb11938-pgpb8987	8.83	12.5	4.9	
	E/14	8–14	pgpb10727-Xipes179	9.36	14.3	3.1	
	A/0	0–2	pgpb8445-pgpb11206	7.67	12.4	4	
	D/20	14–24	pgpb10660-pgpb8626	7	11.6	1.2	
	3/110	106–116	*Xpsmp2214-Xipes142*	4.68	19.4	4.5	Kumar et al. (2016)
Fe_OP	1/30	10–38	*Xpsmp322-Xipes181*	4.34	18.1	0.7	
	5/118	94–124	*pgpb11029-pgpb8456*	4.39	18.3	2.6	

(continued)

Trait	Position of QTL	Support interval	Marker interval	LOD	R^2 (%)	Additive effects	References
Zinc (Zn)	1/182	178–186	pgpb10397-pgpb10394	6.5	9.4	1.7	Kumar et al. (2018)
	1/54	52–56	pgpb10531-pgpb9130	23.93	30.4	6.7	
	1/22	14–26	pgpb10483-pgpb11463	6.68	11.6	−2.2	
	4B/8	4–10	pgpb9502-pgpb6039	7.33	11.1	−0.6	
	5/112	108–118	pgpb13229-pgpb12681	8.17	11.6	2.7	
	7/112	110–116	pgpb12329-pgpb9721	7.58	10.9	2.8	
	7/82	74–88	Xipes198-pgpb8427	7.16	10.2	2.7	
	H/16	10–22	pgpb8779-pgpb12691	6.68	11.6	2.1	
	3/110	106–117	*Xpsmp2214-Xipes142*	9.66	35.9	6.8	
Zn_OP	3/110	106–116	*Xpsmp2214-Xipes142*	14.96	50.1	3.7	Kumar et al. (2016)
	7/96	90–98	*Xpsmp2040-pgpb10727*	4.77	197	−1.8	

2.6.1 Molecular Approaches for Biofortification in Cereal Crops

Biofortification strategies for the target nutrient trait improvement among cereal crops and taking time constraints and population growth rates into account may involve one or more molecular approaches. These approaches deal with the genome sequencing of the crop plant and subsequent modification at the genomic level, including alteration or editing of the target sequence and study of expression analysis of the variant. The desirable variants with the transgressive segregation for the particular target trait are selected, and their expression stability is analyzed. Further, the validation at both the molecular and phenotypic levels must be done to ascertain the target trait expression. The molecular approaches for the biofortification of target nutrient traits in cereal crops are as follows:

1. **Genomic Approaches**
   (a) Genome sequencing/resequencing
   (b) RNA sequencing
   (c) QTL mapping

    (d) Micro RNA discovery
    (e) Epigenomics and SNPs
2. **Genome Engineering**
    (a) Molecular breeding
    (b) Mutation breeding
    (c) Cis and transgenics
    (d) Genome editing (TALEN, ZFN, CRISPERcas)

2.6.1.1 Genomic Approaches

(a) *Genome Sequencing/Resequencing*

Sequencing of several crops and non-crop plants has led to the identification of novel alleles for the nutrient trait, which can be incorporated into the crop by various approaches like marker-assisted breeding. Rice is the first milestone achieved in sequencing under the collaborative project on the "International Rice Genome Sequencing Project" in 2005. Since then, over 3000 rice accessions have been resequenced worldwide, resulting in the discovery of the most recent unique genes and QTLs, and thus impeccably moulding the wealthy status of micronutrient quotient in distinct genotypes.

(b) *QTL Mapping*

Liu (1997) had described the QTL as the loci consisting of many genes with a small and cumulative effect that governs the quantitative traits. Hence, the variation so produced is relatively continuous than the discrete, unlike qualitative characteristics. Many other pieces of researches support his description. Gande (2014) and his colleagues emphasized the importance of QTLs regarding the functional expressions like uptake, transport, and packing up of micronutrients into the grain and the coordinated terms of these specific genes/QTLs at the time of deprivation or surplus micronutrient element conditions for employing the layout to improve upon the respective micronutrient concentration. These QTLs can be sequenced and analyzed further using the techniques like *knock out gene* or *overexpression analysis*, which provides valuable information in determining the "gene or sequence of interest." Once the loci get sequenced, the piece of information to design the functional marker can be obtained. In such a way, the QTL can be mapped using fine mapping or used in breeding programs like MAS (marker-assisted selection) to improve the trait.

(c) *Micro RNA Discovery*

MicroRNAs or miRNAs were discovered later, and ever since, various molecular biologists worldwide have conducted research on their functions, expressions, structure, etc., in plant biology. Their role has been clarified after intensive investigations in the various metabolical and gene expression regulation processes. Unlike DNA, it does not carry exons or codes for any functional genes. Instead, its counterpart functions are primarily mediators and helpers, such as meditating in mRNA degradation posttranscriptionally to help regulate gene expression. Recently, some researchers reported their role in the uptake and transport of some nutrients (Kehr 2013; Fischer et al. 2013). This report is supported by some other studies that provide evidence about their potential role in regulating zinc and iron metabolic regulation and signal transduction pathway during the deficiency of any of the essential microminerals in rice crops (Agarwal et al. 2015).

(d) *Epigenetics*

The term epigenetic summarizes the sum of expression caused due to regulation control by the cellular components such as chromatin structure, histone modification, DNA methylation, etc. In a study of stress signaling and pathway, Chinnusamy and Zhu (2009) reported the regulation of gene expression by epigenetic modification in the sequence of DNA upon the arrival of stress signal in the response of stress occurrence as the tolerance or resistance mechanism in crop plants. Duan et al. (2015) reported that Fe plays a determining role in DNA methylation patterns as an act according to epigenetic regulation. The basis of epigenetic regulation for stress resistance/tolerance has been revealed with the advent of tools and techniques of molecular biology, which has opened a new horizon for stress breeding against various abiotic and biotic stresses. The success against the stress through conventional breeding is minimal; however, the concept of epigenetic regulation is being exploited against stress like drought, cold, salinity, pest, etc., which is very useful in reducing crop loss and subsequently increasing yield potential.

2.6.1.2 Genome Engineering

(a) *Mutagenesis or Mutation Breeding*

The term "mutation" refers to the heritable changes in the genetic coding instructions of the DNA. These changes can either be detrimental or beneficial. The beneficial mutants are subjected to mutation breeding. These mutants may be originated spontaneously, naturally, or artificially induced by physical mutagen like ionizing or nonionizing radiation and chemical mutagen like base analogs, alkylating agents, acridine dyes, etc. The mutant so produced causes a new genetic variation that did not exist naturally; thus, it introduces the functional variability that can be exploited in the breeding program to improve any specific traits. Numerous mutants have been reported related to various characteristics, viz., high yield, early maturity, disease resistance, grain quality characters, abiotic stress resistance, etc., among different crops, viz., soybean, cotton, rice, etc. Henceforth, it is capable of shaping nutritional breeding like other traits. In one such study, Wang et al. (2017a, b) reported that "Lilizhi," a mutant rice variety, showed up to a 35% increase in the zinc content than the original rice genotypes by increasing the rate of transport of storage Zn from roots to the grain.

(b) *Molecular Breeding*

This is the most widely used method for biofortification in cereals such as Fe, Zn, provitamin A, grain protein content, in wheat, rice, maize, millets, etc. It requires the gene or allele of interest possessing a specific target trait and set of polymorphic DNA markers such as SSR, SNPs, etc., for the background selection. It reduces the time and effort required in the conventional approach. It selects various stages like foreground selection with the marker at flanking zone, recombinant selection, and background selection to recover the recurrent genome as maximum as possible. On the other hand, genomic selection (GS) with more advantages is a modern approach. It estimates the genotypic breeding value of the individual. GS is more suitable for the transfer of complex QTLs than MAS. The fundamental requisition for this method is the nearest or linked markers to the trait of interest, polymorphic DNA markers, and specific QTL or specific genes.

(c) *Transgenics*
Since the achievement of nutritional targets through conventional breeding is marginal and quite complicated, it draws attention to other alternatives that are target-specific and ascertain the success. One such option is "transgenics"; however, it delineates the turmoil while it comes to the ethical perception, yet significant achievements have been garlanded with transgenic technology. Many researchers have reported the enhancement of storage Fe in rice seeds with a range of 1.5- to 3.7-fold due to ferritin through introgressing the gene *OsGluB1 proSoya-ferH1* using transgenic technology in various researches (Goto et al. 1999; Lucca et al. 2002; Vasconcelos et al. 2003; Qu et al. 2005; Paul et al. 2012). Various researchers have reported two- to fourfold enhancement of Fe translocation due to overexpression of NAS genes by introgressing gene *OsActin1 Pro-HvNAS1* and *35S Pro-HvNAS1* (Masuda et al. 2009; Lee et al. 2012; Ishimaru et al. 2010; Johnson et al. 2011). Similarly, enhancement of Fe and Zn has been reported by overexpression of "OsNAS2" and "OsZIP4" genes by various researchers such as Ishimaru et al. (2007). Wirth et al. (2009) reported enhancing transgenes, NAS, and ferritin by 6-fold and 1.3–1.5-fold, respectively. In such a way, transgene portrays the efficient method to improve complex traits in less time frame and with high precision. There is an implicit need for a legitimate framework to address the use of transgene for dire complications regarding nutritional improvement programs.

(d) *Genome Editing*
A genome represents the complete set of DNA, all the genes, and information needed to build and maintain an organism. The function of these genes can be elucidated through genome editing by manipulating the gene sequences. Genome editing requires a "molecular scissor," i.e., an engineered nucleases enzyme that cuts any specific line at specific points. These cleaved sequences were then inserted, replaced, or removed accordingly. The nuclease cleaving follows different specificity such as DSB (double-strand break), homologous recombination, and NHEJ (nonhomologous end-joining). Genome editing, however, engrosses three different forms of nucleases, viz., TALEN (transcription activator-like effector nucleases), and ZFN (zinc finger nuclease), and the recent one are CRISPER (clustered regularly interspaced short palindromic repeats). ZFN consists of two protein domains, viz., DNA-binding domain for a transcription factor and the nuclease domain comprises FokI restriction enzyme for specific cleavage. TALEN contains an NLS, N-terminal translocation signal, and transcriptional activator domain. CRISPER has three components, viz., protospacer adjacent motif (PAM), CRISPER RNA

(Cr RNA), and trans-activating Cr RNA (trscr RNA). It is associated with "cas" protein. Together, it produces precise incisions at a specific sequence. Precision, cleavage, and ligation specificity facilitate the genome editing process to create mutation or substitutions in a cell. It is helpful in gene transfer, transcriptional modifications, resistance against abiotic and biotic stresses, multiple gene editing, gene silencing, etc., and thus, a convenient tool in creating controlled desirable gene expressions like promoter sequence modification or regulatory gene sequence modification. Such a powerful concept is, therefore, being applied broadly for various traits by many researchers.

2.6.2 Molecular Strategies Achieved for Biofortification in the Following Cereals

2.6.2.1 Rice

Rice is the foremost energy package add-on for almost the hemisphere's population, predominantly in Asian countries. Being staple food should essentially supply wholesome nutrition to the consumers; however, it is not so. It lacks a few essential nutritional components such as Fe, Zn, provitamin A, and protein, which stands for the growth and development of young children and infants. Hence, it poses a worrying census regarding malnutrition status at a considerable scale. To surmount this vulnerable problem, various molecular strategies have been employed, which served promising results. The development of "golden rice" using three genes (from "daffodil" and bacteria *Erwinia caratovora*) for the biosynthesis of β-carotenoids in the edible grains with the help of transgenic technology earned praises all over the world. Many of the novel laureates worldwide recommend using golden rice in developing countries to overcome malnutrition.

Similarly, introgression of genes and QTLs for "NAS" and "ferritin" using transgenic technology in several programs produced a noticeable increase in Fe and Zn content by two- to sixfolds. The bioavailability of the nutrients has been an issue that is circumvented by using transgenic assays by many researchers. Mutagenesis has been very useful in producing numerous "mutants" shown to enhance the higher protein and mineral content by increasing the transport of storage minerals from roots to grains. Genome editing added the magical tool in the bag of molecular breeding, with the advent of "CRISPERcas9" editing for the complex genome trait expression, and regulation is pretty handy. Being the first fully sequenced cereal crop, most of the genes and QTLs in rice have been studied, and most molecular markers have been designed. The sequence data is available in various database banks such as NCBI (National Center for Biotechnological Information), DDBJ (DNA Database Bank of Japan), etc.; apart from this, software like SSRit/Gramene is also available to provide complete information about SSR markers. It paves an easy way for marker-assisted breeding to address the improvement of rice breeding programs. Several QTLs/genes are available for nutritional traits. The use of marker-assisted breeding like MAS, MARS, MABB, and genomic selection has reduced the time and enhanced the precision with which even the complex traits can be improved.

2.6.2.2 Wheat

Wheat has an impetus position in the global market. The *hexaploid* wheat and *diploid* wheat are widely consumed wheat varieties. They show significant variation for the Fe and Zn content. Grain protein content in the wheat grain is a concern for wheat breeders. Deficiencies in micro minerals content, viz., Fe, Zn, and Se (selenium), grain protein, and provitamin A are significant nutritional gaps in wheat. For the biofortification of wheat, the most successful and used method is marker-aided selection (MAS). Grain protein content (GPC) is the most concerning nutrient in wheat. Several QTLs have been mapped by various researchers worldwide. *Gpc-B1* is the most accountable QTL identified for the grain protein and mineral content that has been transferred from parent *dicoccoides* species to cultivar *hexaploid* species. However, the nature of QTLs depends upon the uptake, transport, and packing of the nutrient in the grain.

Due to such complexity of these QTLs, conventional breeding could not be part of the winning side. Yet, a variety like "Zinc Shakti" with up to 40% high Zn content has been developed by conventional back cross-breeding; still, no such variety could be developed for Fe content. Mapping study reveals that the Fe and Zn are associated together in the same loci. The QTLs have been mapped for Fe, Zn, and selenium. Marker-assisted selection can unarguably improve these traits by introgressing these QTLs. However, the complex polygenic nature of the QTLs and their interaction with the environment poses problems for MAS. Recently, the announced method of "genomic selection" or genome-wide selection (GWS) can counteract such issues as the genetic breeding value is estimated in this method.

2.6.2.3 Maize

Maize withholds a virtuous place among the cereal crops. Primarily, maize is used as a staple food in corn. Grains of maize critically are deprived of amino acids *lysine* and *tryptophan*. The deficiency of such amino acid causes severe diseases like pellagra upon consumption. Lack of these amino acids subsequently leads to insufficient protein in the endosperm. Another concern is the poor bioavailability of micro minerals like Fe and Zn content and the lack of provitamin A and *phytate* in kernels. Phytate content is responsible for seedling growth and germination. Besides, it also inhibits various types of cancer upon consumption. However, it inhibits the bioavailability of micro minerals such as Fe, Zn, Mg, etc. However, several QTLs have been reported, which govern the high accumulation of micro minerals, viz., Fe and Zn. The endosperm protein content has been improved with the intervention of mutant *opaque-2*. The first QPM maize "Vivek QPM-9" has been developed by the introgression of the opaque-2 mutant using MAS. Similarly, low phytase mutants, viz., *lpa1, lpa2,* and *lpa3,* with a significant reduction in *phytase* have been reported and introgress in cultivars using MAS.

2.6.2.4 Sorghum

Sorghum is an essential hardy cereal crop after rice, wheat, and maize and is the cheap source of energy and nutrients, primarily in semiarid tropical areas. The main obstacle in sorghum is the presence of antinutritional factors like polyphenols and phytate, which inhibit the bioavailability of Fe and Zn. Wide variability is reported for the micro mineral content in sorghum. However, lines with high Fe and Zn content have been developed under the HarvestPlus programs. Much of the molecular strategies have not been exploited in this crop. Yet, the prospect holds the scope for molecular approaches like MAS, GS, genome engineering, etc., for the precise and immediate development of sorghum cultivars.

2.6.2.5 Millets

Millets are the chief source of calories after cereals in Asia and Africa's arid and semiarid parts. They are a staple food, nutrient-rich, and resistant to extreme ecological conditions. Millets play a crucial role in nutritional security as they provide an appreciable amount of proteins, dietary fiber, vitamins, and minerals. To combat micronutrient malnutrition, the biofortification of significant food crops was shown to be an economical method. The HarvestPlus group noticed the need for biofortification in millets. So, to tackle iron deficiency, the HarvestPlus group included pearl millet as one of the staple crops under the Biofortification Challenge Program (BCP) (Fig. 2.2).

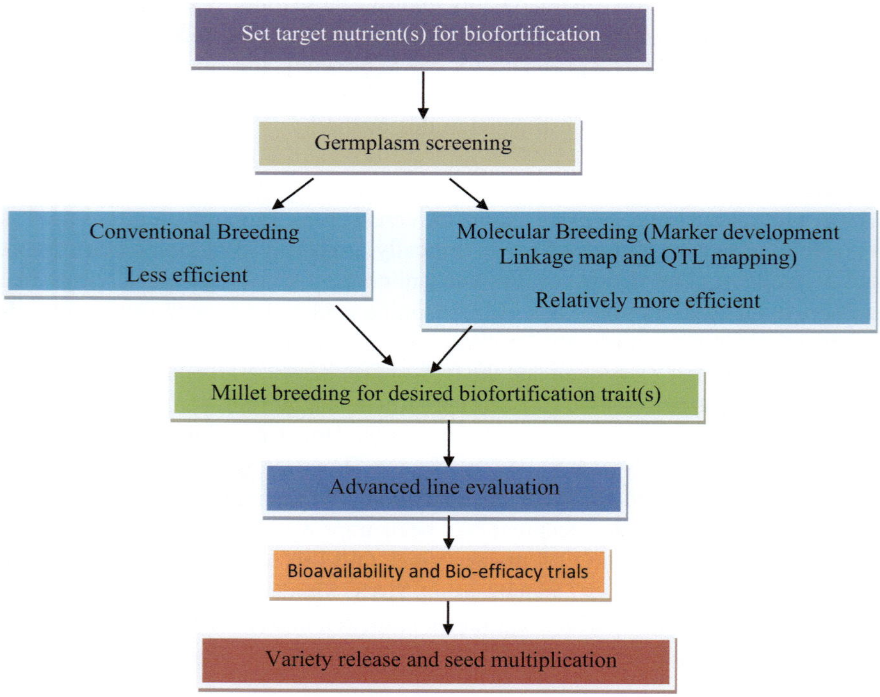

Fig. 2.2 Flow chart for the general breeding procedure of biofortification of millets

(a) **Pearl Millet**

Genetic diversity and QTL mapping studies were conducted for pearl millet genome analysis using RFLP and SSR markers. RFLP and SSR markers were also utilized to construct the first consensus linkage map with four different mapping populations of pearl millet. Accessibility of a high-density consensus molecular map permits us to locate the position of QTLs/genes strictly in the genetic background of several mapping populations. Different marker techniques, viz., single-strand conformation polymorphism–single nucleotide polymorphisms (SSCP-SNPs), conserved intron scanning primers (CISP), and diversity arrays technology (DArT) were also tried in genome studies of pearl millet. However, SNP-based assays and genotyping by sequencing (GBS) technologies were yet to be utilized widely in pearl millet. A severe constrain for genomic studies in pearl millet is the unavailability of high-density genetic maps. The discovery of genes/QTLs responsible for various nutrients (essential amino acids, Fe, and Zn) present in grains of pearl millet can be accelerated by applying high-throughput technologies. The draft genome sequence of the pearl millet is almost ready and expected to be released soon in the public domain. These genomic tools will be handy for analyzing quantitative as well as quality traits.

(b) **Foxtail Millet**

Genomic studies of foxtail millet were mainly done with RFLP and RAPD markers until the generation of SSRs. However, these markers are considered essential genomic tools in foxtail millet to investigate economically important traits at the molecular level. Accelerated creation of structural genomics resources was achieved through the utilization of NGS technologies in foxtail millet. Two research groups have sequenced the foxtail millet genome independently. One was at the Beijing Genomics Institute (BGI), China, and the other at the Joint Genome Institute (JGI) of the Department of Energy, USA. The Beijing Genomics Institute (BGI) research group, China, has done the genome sequencing of foxtail millet cultivar Zhang Gu and variety A-2. Then they identified many SNPs, InDels, and structural variations (for marker development) through comparison of genome sequences of both, i.e., the genome of cultivar Zhang Gu and variety A-2. The other group at the JGI (Joint Genome Institute) of the Department of Energy, USA, has sequenced foxtail millet genotypes Yagu1, A-10, and also a recombinant inbred line (RIL) population derived by crossing Yagu1 and A-10, which leads to the construction of high-density SNP map from resequencing information of the RIL population. Recently, with the help of resequencing information of 916 diverse genotypes of foxtail millet with 0.8 million SNP and SNP haplotype map was constructed. Such a haplotype map may be utilized for the discovery of genes governing grain quality traits employing association mapping-based techniques.

(c) **Other Millets**

Limited research was done in other minor millets at the genomic level. Genetic diversity analysis was done in proso millet with the help of RAPD and AFLP markers. However, SSR markers have been developed in proso millet from microsatellite enriched genomic DNA libraries and through the transferability of microsatellites from related species of proso millet (wheat, oat, rice, barley, and switchgrass). Some species of millets have only a few shreds of evidence on the utilization of DNA markers, and their genomes are yet to be characterized. Consequently, the immediate focus should be on generating more genomic resources for minor millets, which could be employed to analyze and improve qualitative traits.

2.6.2.6 Oats

The oat (*Avena sativa*) is an annual grass commonly grown in temperate regions requiring lower summer temperatures. Oats are used for human consumption as well as livestock feed. Despite the richness of various essential nutrients, the use of human consumption is limited due to their blend organoleptic taste and spoilage. Though oats are highly nutritious, yet it is not cultivated on a large scale mainly due to reduced market demand and less economic revenue generation compared to other crops. As a result, investments in oat breeding are consistently decreasing. However, the release of biofortified varieties with higher and stable yields can put them in demand.

Mapping studies with RFLP markers revealed that the genomic regions governing β-glucan content in oats are present on chromosomes number 7 and 11. QTLs for traits β-glucan and protein content were mapped with the help of an integrated molecular linkage map constructed by several DNA markers like RFLP, RAPD, SCAR, and AFLP. Though rarely used, CAPS and SCAR were designed for the genomic regions controlling β-glucan and protein content. Like other cereal crops, huge numbers of SSR markers were developed in oat too. High-density linkage maps are needed in oats for high-resolution mapping of quantitative traits due to their large genome size. Genes governing qualitative characteristics could be discovered by oat genome analysis with the constructed SNP consensus map framework.

2.6.3 Limitations in Biofortification Through Molecular Approaches

1. ***Presence of Antinutrient***
   The digestibility of nutrients present in millets is an issue due to the presence of antinutrients. Phytic acid (present in most of the cereals) is one of the essential antinutrients that have a solid ability to inhibit the absorption of crucial minerals, viz., Zn, Ca, and Fe. Antinutritional factors present in crops like phenols, phytase, etc., though they have beneficial functions yet inhibit the expression of other microelements such as Fe, Zn, Mg, Se, etc.
2. ***Yield Reduction with an Improvement of Micronutrient***
   A negative and nonsignificant association was found between Fe and Zn with the grain yield in the breeding lines and cultivars of pearl millet due to their genetic architecture. They were developed through selections for improvement of yield

Table 2.2 Antinutrient profile of millets

Crop	Antinutrient
Pearl millet	• Phytate and goitrogens
Finger millet	• Tannins
	• Oxalates
	• Phytate
	• Non-starch polysaccharides—glucan
	• Protease inhibitors

and not for the micronutrients. But there is a positive correlation between Zn and Fe; therefore, breeding for high-iron pearl millet leads to enhancement of Zn associated with the iron.

3. ***Floral Biology and Mode of Pollination***
   The development of mapping population requires artificial hybridization. Some millets (foxtail millet, finger millet, Kodo millet, etc.) are self-pollinated, which makes artificial hybridization among the selected parental lines tedious. Emasculation and pollination are difficult due to the small flower size.
4. ***Bioavailability of the Micronutrients***
   The bioavailability of micro minerals and nutrients needs to be taken into consideration. The gene/QTLs for deficient traits may enhance the expression; yet, it does not determine the bioaccessibility of the nutrient.
5. ***The Complex Interaction of QTLs with the Environment***
   The complex interaction of QTLs with the climate and additive gene action of multiple loci poses problems in marker-assisted selection (Table 2.2).

2.7 Conclusion

Molecular approaches have reduced the complexity of biofortification of micronutrients in cereal crops, which otherwise was not much virtuous through solemn dependence upon conventional breeding. Mapping studies for the variability among the micro mineral contents like Fe and Zn, grain protein content, vitamin A, and antinutritional factors have revealed several QTLs. Upon introgression, several cultivars yield manifold higher nutrient values. With the advent of molecular markers like SSR, SNPs, etc., marker-assisted breeding has simplified the biofortification of crops, viz., high Fe and Zn, protein, and β-carotenoids. The deficient amino acid content in maize has been improved by developing QPM maize cultivars using MAS. Genomic regions with the coding sequence for target expression have been revealed with the advancement of genome sequencing. Genetic engineering techniques like genome editing via CRISPERcas9, TALEN, and ZFN modify traits' expression of the undesirable genetic sequence. Transgenic technology has been instrumental in resolving complex difficulties to produce genetically novel characteristics such as "Golden Rice-1" and "Golden rice-2" with higher provitamin A concentrations. The expression of Fe and Zn in rice by enhancing expression of NAS genes and introgression of *ferritin* could be possible due to transgenics.

Further, mutagenic breeding added benefits by producing novel mutants for improving traits like high Fe content, low phytase content, etc. Conclusively, the sum of genomic and genetic engineering approaches have enhanced the biofortification in multiple cereal crops, viz., rice, wheat, maize, sorghum, millets, oats, etc. It dealt with all the genetic complexities and turned the roller coaster of biofortification difficulties into a smooth road wide open for easy passage. It also saves time under urgency. Altogether, a slide shifting of paradigm from worrying "hidden hunger" status to sufficient healthy society. In this way, we can achieve the goal of "end malnutrition" and "zero hunger" by 2030 and sustain food security.

2.8 Future Prospects

Genomic selection can be a potential approach in the introgression of complex polygenic traits/QTLs. GWAS (genome-wide association mapping) can map-wide variability and identify many more potential trait-specific QTLs. The various assays to verify the bioavailability of micronutrients using transgenic technologies, *overexpression analysis*, *knockout gene*, etc., can be done in the future to assess the stable expression of introgressing gene/QTL. More strategies for biofortification in crops like sorghum, oats, barley, rye, and millets need to be implemented for an environmentally friendly way of enhancing nutrients.

References

Agarwal S, Tripura Venkata VGN, Kotla A, Mangrauthia SK, Neelamraju S (2014) Expression patterns of QTL-based and other candidate genes in Madhukar × Swarna RILs with contrasting levels of iron and zinc in unpolished rice grains. Gene 546:430–436

Agarwal S, Mangrauthia SK, Sarla N (2015) Expression profiling of iron deficiency responsive micro RNAs and gene targets in rice seedlings of Madhukar × Swarna recombinant inbred lines with contrasting levels of iron in seeds. Plant Soil 396:1–14

Aluko G, Martinez C, Tohme J, Castano C, Bergman C, Oard JH (2004) QTL mapping of grain quality traits from the interspecific cross Oryza sativa × O. glaberrima. Theor Appl Genet 109:630

Anonymous (2020) Statistics of Food and Agriculture Organization of the United Nations—FAOSTAT. Update June 15, 2020. www. fao.org

Blanco A, Pasqualone A, Troccoli A, Di Fonzo N, Simeone R (2002) Detection of grain protein content QTLs across environments in tetraploid wheats. Plant Mol Biol 48:615–623

Blanco A, Colasuonno P, Gadaleta A, Mangini G, Schiavulli A, Simeone R, Digesù AM, De Vita P, Mastrangelo AM, Cattivelli L (2011) Quantitative trait loci for yellow pigment concentration and individual carotenoids compounds in durum wheat. J Cereal Sci 54:255–264

Blanco A, Mangini G, Giancaspro A, Giove S, Colasuonno P, Simeone R, Signorile A, De Vita P, Mastrangelo AM, Cattivelli L et al (2012) Relationships between grain protein content and grain yield components through quantitative trait locus analyses in a recombinant inbred line population derived from two elite durum wheat cultivars. Mol Breed 30:79–92

Börner A, Schumann E, Fürste A, Cöster H, Leithold B, Röder M, Weber W (2002) Mapping of quantitative trait loci determining agronomic important characters in hexaploid wheat (Triticum aestivum L.). Theor Appl Genet 105:921–936

Bruno E, Choi YS, Chung IK, Kim KM (2017) QTLs and analysis of the candidate gene for amylose, protein, and moisture content in rice (*Oryza sativa* L.). 3 Biotech 7:40

Chattopadhyay K, Behera L, Bagchi TB, Sardar SS, Moharana N, Patra NR, Chakraborti M, Das A, Marndi BC, Sarkar A, Ngangkham U (2019) Detection of stable QTLs for grain protein content in rice (*Oryza sativa* L.) employing high throughput phenotyping and genotyping platforms. Sci Rep 9:3196

Chinnusamy V, Zhu JK (2009) Epigenetic regulation of stress responses in plants. Curr Opin Plant Biol 12(2):133–139

Colasuonno P, Gadaleta A, Giancaspro A, Nigro D, Giove S, Incerti O, Mangini G, Signorile A, Simeone R, Blanco A (2014) Development of a high-density SNP-based linkage map and detection of yellow pigment content QTLs in durum wheat. Mol Breed 34:1563–1578

Conti V, Roncallo PF, Beaufort V, Cervigni GL, Miranda R, Jensen CA, Echenique VC (2011) Mapping of main and epistatic effect QTLs associated to grain protein and gluten strength using a RIL population of durum wheat. J Appl Genet 52:287–298

Crawford AC, Stefanova K, Lambe W, McLean R, Wilson R, Barclay I, Francki MG (2011) Functional relationships of phytoene synthase 1 allele on chromosome 7A controlling flour colour variation in selected Australian wheat genotypes. Theor Appl Genet 123:95–108

Crespo-Herrera LA, Velu G, Singh RP (2016) Quantitative trait loci mapping reveals pleiotropic effect for grain iron and zinc concentrations in wheat. Ann Appl Biol 169:27–35

Crespo-Herrera LA, Govindan V, Stangoulis J, Hao Y, Singh RP (2017) QTL mapping of grain Zn and Fe concentrations in two hexaploid wheat RIL populations with ample transgressive segregation. Front Plant Sci 8:1800

Deng Z, Hu S, Chen F, Li W, Chen J, Sun C, Zhang Y, Wang S, Song X, Tian J (2015) Genetic dissection of interaction between wheat protein and starch using three mapping populations. Mol Breed 35:12

Descalsota-Empleo GI, Amparado A, Inabangan-Asilo MA, Tesoro F, Stangoulis J, Reinke R, Swamy BM (2019) Genetic mapping of QTL for agronomic traits and grain mineral elements in rice. Crop J 7:560

Dholakia BB, Ammiraju JSS, Santra DK, Singh H, Katti MV, Lagu MD, Tamhankar SA, Rao VS, Gupta VS, Dhaliwal HS et al (2001) Molecular marker analysis of protein content using PCR-based markers in wheat. Biochem Genet 39:325–338

D'Mello JPF (2003) Amino acids as multifunctional molecules. In: D'Mello JPF (ed) Amino acids in animal nutrition, 2nd edn. CABI Publishing, Cambridge, MA, pp 1–14

Duan G, Wang X, Tang K (2015) MET18 connects the cytosolic iron-sulfur cluster assembly pathway to active dna demethylation in *Arabidopsis*. PLoS Genet 11(10):e1005559

Echeverry-Solarte M, Kumar A, Kianian S, Simsek S, Alamri MS, Mantovani EE, McClean PE, Deckard EL, Elias E, Schatz B et al (2015) New QTL alleles for quality-related traits in spring wheat revealed by RIL population derived from supernumerary × non-supernumerary spikelet genotypes. Theor Appl Genet 128:893–912

El-Feki WM, Byrne PF, Reid SD, Lapitan NL, Haley SD (2013) Quantitative trait locus mapping for end-use quality traits in hard winter wheat under contrasting soil moisture levels. Crop Sci 53:1953–1967

Elouafi I, Nachit MM, Martin LM (2001) Identification of a microsatellite on chromosome 7B showing a strong linkage with yellow pigment in durum wheat (Triticum turgidum L. var. durum). Hereditas 135:255–261

Fatiukha A, Lupo I, Lidzbarsky G, Klymiuk V, Korol AB, Pozniak C, Fahima T, Krugman T (2019) Grain protein content QTLs identified in a durum × wild emmer wheat mapping population tested in five environments. bioRxiv

Fischer JJ, Beatty PH, Good AG (2013) Manipulation of microRNA expression to improve nitrogen use efficiency. Plant Sci 210:70–81

Gande NK, Kundur PJ, Soman R (2014) Identification of putative candidate gene markers for grain zinc content using recombinant inbred lines (RIL) population of IRRI38 X Jeerigesanna. Afr J Biotechnol 13(5):657–663

Garcia-Oliveira AL, Tan L, Fu Y, Sun C (2009) Genetic identification of quantitative trait loci for contents of mineral nutrients in rice grain. J Integr Plant Biol 51:84

Genc Y, Verbyla AP, Torun AA, Cakmak I, Willsmore K, Wallwork H, McDonald GK (2009) Quantitative trait loci analysis of zinc efficiency and grain zinc concentration in wheat using whole-genome average interval mapping. Plant Soil 314:49

Golabadi M, Arzani A, Mirmohammadi Maibody S (2012) Identification of microsatellite markers associated with grain protein content in durum wheat grown under drought stress at terminal growth stages. Cereal Res Commun 40:215–224

Gonzalez-Hernandez JL, Elias EM, Kianian SF (2004) Mapping genes for grain protein concentration and grain yield on chromosome 5B of Triticum turgidum (L.) var. Dicoccoides. Euphytica 139:217–225

Goto F, Yoshihara T, Shigemoto N (1999) Iron fortification of rice seed by the soybean ferritin gene. Nat Biotechnol 17:282–286

Groos C, Robert N, Bervas E, Charmet G (2003) Genetic analysis of grain protein-content, grain yield and thousand-kernel weight in bread wheat. Theor Appl Genet 106:1032–1040

Groos C, Bervas E, Charmet G (2004) Genetic analysis of grain protein content, grain hardness and dough rheology in a hard × hard bread wheat progeny. J Cereal Sci 40:93–100

Hao Y, Velu G, Peña RJ, Singh S, Singh RP (2014) Genetic loci associated with high grain zinc concentration and pleiotropic effect on kernel weight in wheat (Triticum aestivum L.). Mol Breed 34:1893–1902

Heo H, Sherman J (2013) Identification of QTL for grain protein content and grain hardness from winter wheat for genetic improvement of spring wheat. Plant Breed Biotechnol 1:347–353

Hu ZL, Li P, Zhou MQ, Zhang ZH, Wang LX, Zhu LH, Zhu YG (2004) Mapping of quantitative trait loci (QTLs) for rice protein and fat content using doubled haploid lines. Euphytica 135:47

Hu BL, Huang DR, Xiao YQ, Fan YY, Chen DZ, Zhuang JY (2016) Mapping QTLs for mineral element contents in brown and milled rice using an *Oryza sativa* × *O. rufipogon* backcross inbred line population. Cereal Res Commun 44:57

Huang XQ, Cloutier S, Lycar L, Radovanovic N, Humphreys DG, Noll JS, Somers DJ, Brown PD (2006) Molecular detection of QTLs for agronomic and quality traits in a doubled haploid population derived from two Canadian wheats (Triticum aestivum L.). Theor Appl Genet 113: 753–766

Indurkar AB, Majgahe SK, Sahu VK, Vishwakarma A, Premi V, Shrivastatva P (2015) Identification, characterization and mapping of QTLs related grain Fe, Zn and protein contents in rice (*Oryza sativa* L.). Electron J Plant Breed 6:1059

Ishimaru Y, Masuda H, Bashir K, Inoue H, Tsukamoto T, Takahashi M et al (2010) Ricemetal-nicotianamine transporter, OsYSL2, is required for the long-distance transport of iron and manganese. Plant J 62(3):379–390

Ishikawa R, Iwata M, Taniko K, Monden G, Miyazaki N, Orn C, Tsujimura Y, Yoshida S, Ma JF, Ishii T (2017) Detection of quantitative trait loci controlling grain zinc concentration using Australian wild rice, Oryza meridionalis, a potential genetic resource for biofortification of rice. PLoS One 12:pe0187224

Ishimaru Y, Masuda H, Suzuki M, Bashir K, Takahashi M, Nakanishi H, Mori S, Nishizawa NK (2007) Overexpression of the *OsZIP4* zinc transporter confers disarrangement of zinc distribution in rice plants. J Exp Bot 58:2909–2915

Johnson AAT, Kyriacou B, Callahan DL (2011) Constitutive overexpression of the *OsNAS* gene family reveals single-gene strategies for effective iron- and zinc-biofortification of rice endosperm. PLoS One 6:e24476

Joppa LR, Du C, Hart GE, Hareland GA (1997) Mapping gene(s) for grain protein in tetraploid wheat (Triticum turgidum L.) using a population of recombinant inbred chromosome lines. Crop Sci 37:1586–1589

Kehr J (2013) Systemic regulation of mineral homeostasis by microRNAs. Front Plant Sci 4:145

Krishnappa G, Singh AM, Chaudhary S, Ahlawat AK, Singh SK, Shukla RB, Jaiswal JP, Singh GP, Solanki IS (2017) Molecular mapping of the grain iron and zinc concentration, protein content and thousand kernel weight in wheat (Triticum aestivum L.). PLoS One 12:e0174972

Kuchel H, Langridge P, Mosionek L, Williams K, Jefferies SP (2006) The genetic control of milling yield, dough rheology and baking quality of wheat. Theor Appl Genet 112:1487

Kulwal P, Kumar N, Kumar A, Gupta RK, Balyan HS, Gupta PK (2005) Gene networks in hexaploid wheat: interacting quantitative trait loci for grain protein content. Funct Integr Genom 5:254–259

Kumar J, Jain S, Jain RK (2014) Linkage mapping for grain iron and zinc content in F2 population derived from the cross between PAU201 and Palman 579 in rice (*Oryza sativa* L.). Cereal Res Commun 42:389

Kumar S, Hash CT, Thirunavukkarasu N, Singh G, Rajaram V, Rathore A, Senapathy S, Mahendrakar MD, Yadav RS, Srivastava RK (2016) Mapping quantitative trait loci controlling high iron and zinc content in self and open-pollinated grains of pearl millet [Pennisetum glaucum (L.) R. Br.]. Front Plant Sci 7:1636

Kumar S, Hash CT, Nepolean T, Mahendrakar MD, Satyavathi CT, Singh G, Rathore A, Yadav RS, Gupta R, Srivastava RK (2018) Mapping grain iron and zinc content quantitative trait loci in an Iniadi-derived immortal population of pearl millet. Genes 9:248

Kumar N, Jain RRK, Chowdhury VK (2019) Linkage mapping of QTLs for grain minerals (iron and zinc) and physio morphological traits for development of mineral rich rice (Oryza sativa L.). Indian J Biotechnol 18:69–80

Laperche A, Brancourt-Hulmel M, Heumez E, Gardet O, Hanocq E, Devienne-Barret F, Le Gouis J (2007) Using genotype × nitrogen interaction variables to evaluate the QTL involved in wheat tolerance to nitrogen constraints. Theor Appl Genet 115:399–415

Lee S, Kim YS, Jeon US, Kim YK, Schjoerring JK, An G (2012) Activation of rice nicotianamine synthase 2 (OsNAS2) enhances iron availability for biofortification. Mol Cells 33(3):269–275

Lee GH, Yun BW, Kim KM (2014) Analysis of QTLs associated with the rice quality related gene by double haploid populations. Int J Genomics 2014:781832

Leng Y, Xue D, Yang Y, Hu S, Su Y, Huang L, Wang L, Zheng T, Zhang G, Hu J, Gao Z (2014) Mapping of QTLs for eating and cooking quality-related traits in rice (*Oryza sativa* L.). Euphytica 197:99

Li J, Xiao J, Grandillo S, Jiang L, Wan Y, Deng Q, Yuan L, McCouch SR (2004) QTL detection for rice grain quality traits using an interspecific backcross population derived from cultivated Asian (*O. sativa* L.) and African (*O. glaberrima* S.) rice. Genome 47:697

Li Y, Song Y, Zhou R, Branlard G, Jia J (2009) Detection of QTLs for bread-making quality in wheat using a recombinant inbred line population. Plant Breed 128:235–243

Li J, Cui F, Ding AM, Zhao CH, Wang XQ, Wang L, Bao YG, Qi XL, Li XF, Gao JR et al (2012a) QTL detection of seven quality traits in wheat using two related recombinant inbred line populations. Euphytica 183:207–226

Li Y, Zhou R, Wang J, Liao X, Branlard G, Jia J (2012b) Novel and favorable QTL allele clusters for end-use quality revealed by introgression lines derived from synthetic wheat. Mol Breed 29: 627–643

Li C, Bai G, Chao S, Carver B, Wang Z (2016) Single nucleotide polymorphisms linked to quantitative trait loci for grain quality traits in wheat. Crop J 4:1–11

Liu BH (1997) Statistical genomics: linkage mapping and QTL analysis. CRC Press, Boca Raton

Lou J, Chen L, Yue G, Lou Q, Mei H, Xiong L, Luo L (2009) QTL mapping of grain quality traits in rice. J Cereal Sci 50:145

Lu K, Li L, Zheng X, Zhang Z, Mou T, Hu Z (2008) Quantitative trait loci controlling Cu, Ca, Zn, Mn and Fe content in rice grains. J Genet 87:305

Lucca P, Hurrell R, Potrykus I (2002) Fighting iron deficiency anemia with iron-rich rice. J Am Coll Nutr 21(Suppl 3):184S–190S

Mahjourimajd S, Taylor J, Rengel Z, Khabaz-Saberi H, Kuchel H, Okamoto M, Langridge P (2016) The genetic control of grain protein content under variable nitrogen supply in an Australian wheat mapping population. PLoS One 11:e0159371

Mann G, Diffey S, Cullis B, Azanza F, Martin D, Kelly A, McIntyre L, Schmidt A, Ma W, Nath Z et al (2009) Genetic control of wheat quality: interactions between chromosomal regions determining protein content and composition, dough rheology, and sponge and dough baking properties. Theor Appl Genet 118:1519–1537

Maphosa L, Langridge P, Taylor H, Emebiri LC, Mather DE (2015) Genetic control of grain protein, dough rheology traits, and loaf traits in a bread wheat population grown in three environments. J Cereal Sci 64:147–152

Marcotuli I, Gadaleta A, Mangini G, Signorile A, Zacheo S, Blanco A, Simeone R, Colasuonno P (2017) Development of a high-density SNP-based linkage map and detection of QTL for β-glucan, protein content, grain yield per spike and heading time in durum wheat. Int J Mol Sci 18:1329

Masuda H, Usuda K, Kobayashi T, Ishimaru Y, Kakei Y, Takahashi M et al (2009) Overexpression of the barley nicotianamine synthase gene HvNAS1 increases iron and zinc concentrations in rice grains. Rice 2(4):155–166

Matthias HH, Yu J, Beuch S, Weber WE (2014) Quantitative trait loci for quality and agronomic traits in two advanced backcross populations in oat (Avena sativa L.). Plant Breed 133(5):588–601

Moore CM, Richards RA, Rebetzke GJ (2015) Phenotypic variation and QTL analysis for oil content and protein concentration in bread wheat (Triticum aestivum L.). Euphytica 204:371–382

Nelson JC, Andreescu C, Breseghello F, Finney PL, Gualberto DG, Bergman CJ, Pena RJ, Perretant MR, Leroy P, Qualset CO et al (2006) Quantitative trait locus analysis of wheat quality traits. Euphytica 149:145–159

Norton GJ, Deacon CM, Xiong L, Huang S, Meharg AA, Price AH (2010) Genetic mapping of the rice ionome in leaves and grain: identification of QTLs for 17 elements including arsenic, cadmium, iron, and selenium. Plant Soil 329:139

OECD/FAO (2019) OECD-FAO agricultural outlook 2019-2028, OECD Agriculture statistics (database). https://doi.org/10.1787/agr-outl-data-en

Parker GD, Chalmers KJ, Rathjen AJ, Langridge P (1998) Mapping loci associated with flour color in wheat (Triticum aestivum L.). Theor Appl Genet 97:238–245

Patil RM, Oak MD, Tamhankar SA, Sourdille P, Rao VS (2008) Mapping and validation of a major QTL for yellow pigment content on 7AL in durum wheat (Triticum turgidum L. ssp. durum). Mol Breed 21:485–496

Patil RM, Oak MD, Tamhankar SA, Rao VS (2009) Molecular mapping of QTLs for gluten strength as measured by sedimentation volume and mixograph in durum wheat (Triticum turgidum L. ssp durum). J Cereal Sci 49:378–386

Paul S, Ali N, Gayen D, Datta SK, Datta K (2012) Molecular breeding of Osfer2 gene to increase iron nutrition in rice grain. GM Crops Food 3(4):310–316

Peleg Z, Cakmak I, Ozturk L, Yazici A, Jun Y, Budak H, Korol AB, Fahima T, Saranga Y (2009) Quantitative trait loci conferring grain mineral nutrient concentrations in durum wheat × wild emmer wheat RIL population. Theor Appl Genet 119:353–369

Perretant MR, Cadalen T, Charmet G, Sourdille P, Nicolas P, Boeuf C, Tixier MH, Branlard G, Bernard S (2000) QTL analysis of bread-making quality in wheat using a doubled haploid population. Theor Appl Genet 100:1167–1175

Prasad M, Varshney RK, Kumar A, Balyan HS, Sharma PC, Edwards KJ, Dhaliwal HS, Roy JK, Gupta PK (1999) A microsatellite marker associated with a QTL for grain protein content on chromosome arm 2DL of bread wheat. Theor Appl Genet 99:341–345

Prasad M, Kumar N, Kulwal P, Röder M, Balyan H, Dhaliwal H, Gupta P (2003) QTL analysis for grain protein content using SSR markers and validation studies using NILs in bread wheat. Theor Appl Genet 106:659–667

Pu ZE, Ma YU, He QY, Chen GY, Wang JR, Liu YX, Jiang QT, Wei LI, Dai SF, Wei YM et al (2014) Quantitative trait loci associated with micronutrient concentrations in two recombinant inbred wheat lines. J Integr Agric 13:2322–2329

Pu Z, Pei Y, Yang J, Ma J, Li W, Liu D, Wang J, Wei Y, Zheng Y (2018) A QTL located on chromosome 3D enhances the selenium concentration of wheat grain by improving phytoavailability and root structure. Plant Soil 425:287–296

Qin Y, Kim SM, Sohn JK (2009) QTL analysis of protein content in double haploid lines of rice. Korean J Crop Sci 54:165

Qu LQ, Yoshihara T, Ooyama A, Goto F, Takaiwa F (2005) Iron accumulation does not parallel the high expression level of ferritin in transgenic rice seeds. Planta 222(2):225–233

Ram S, Mishra B (2010) Cereals processing and nutritional quality. New India Publishing Agency, New Delhi, p 188

Raman R, Allen H, Diffey S, Raman H, Martin P, McKelvie K (2009) Localisation of quantitative trait loci for quality attributes in a doubled haploid population of wheat (Triticum aestivum L.). Genome 52:701–715

Roncallo PF, Cervigni GL, Jensen C, Miranda R, Carrera AD, Helguera M, Echenique V (2012) QTL analysis of main and epistatic effects for flour color traits in durum wheat. Euphytica 185: 77–92

Roshanzamir H, Kordenaeej A, Bostani A (2013) Mapping QTLs related to Zn and Fe concentrations in bread wheat (Triticum aestivum) grain using microsatellite markers. Iran J Genet Plant Breed 2:10–17

Shi R, Li H, Tong Y, Jing R, Zhang F, Zou C (2008) Identification of quantitative trait locus of zinc and phosphorus density in wheat (Triticum aestivum L.) grain. Plant Soil 306:95–104

Shi CH, Shi Y, Lou XY, Xu HM, Zheng X, Wu JG (2009) Identification of endosperm and maternal plant QTLs for protein and lysine contents of rice across different environments. Crop Pasture Sci 60:295

Shi RL, Tong YP, Jing RL, Zhang FS, Zou CQ (2013) Characterization of quantitative trait loci for grain minerals in hexaploid wheat (Triticum aestivum L.). J Integr Agric 12:1512–1521

Šimić D, Mladenović Drinić S, Zdunić Z, Jambrović A, Ledenčan T, Brkić J et al (2012) Quantitative trait loci for biofortification traits in maize grain. J Hered 103(1):47–54

Simons K, Anderson JA, Mergoum M, Faris JD, Klindworth DL, Xu SS, Sneller C, Ohm JB, Hareland GA, Edwards MC et al (2012) Genetic mapping analysis of bread-making quality traits in spring wheat. Crop Sci 52:2182–2197

Singh H, Prasad M, Varshney RK, Roy JK, Balyan HS, Dhaliwal HS, Gupta PK (2001) STMS markers for grain protein content and their validation using near-isogenic lines in bread wheat. Plant Breed 120:273–278

Srinivasa J, Arun B, Mishra VK, Singh GP, Velu G, Babu R, Vasistha NK, Joshi AK (2014) Zinc and iron concentration QTL mapped in a Triticum spelta × T. aestivum cross. Theor Appl Genet 127:1643–1651

Stangoulis JC, Huynh BL, Welch RM, Choi EY, Graham RD (2007) Quantitative trait loci for phytate in rice grain and their relationship with grain micronutrient content. Euphytica 154:289

Sun H, Lü J, Fan Y, Zhao Y, Kong F, Li R, Wang H, Li S (2008) Quantitative trait loci (QTLs) for quality traits related to protein and starch in wheat. Prog Nat Sci 18:825–831

Sun X, Marza F, Ma H, Carver BF, Bai G (2010) Mapping quantitative trait loci for quality factors in an inter-class cross of the US and Chinese wheat. Theor Appl Genet 120:1041–1051

Sun X, Wu K, Zhao Y, Qian Z, Kong F, Guo Y, Wang Y, Li S (2016) Molecular genetic analysis of grain protein content and flour whiteness degree using RILs in common wheat. J Genet 95:317–324

Suprayogi Y, Pozniak CJ, Clarke FR, Clarke JM, Knox RE, Singh AK (2009) Identification and validation of quantitative trait loci for grain protein concentration in adapted Canadian durum wheat populations. Theor Appl Genet 119:437–448

Swamy BM, Descalsota GIL, Nha CT, Amparado A, Inabangan-Asilo MA, Manito C, Tesoro F, Reinke R (2018) Identification of genomic regions associated with agronomic and biofortification traits in DH populations of rice. PLoS One 13:e0201756

Takeuchi Y, Nonoue Y, Ebitani T, Suzuki K, Aoki N, Sato H, Ideta O, Hirabayashi H, Hirayama M, Ohta H, Nemoto H (2007) QTL detection for eating quality including glossiness, stickiness, taste and hardness of cooked rice. Breed Sci 57:231

Tan YF, Sun M, Xing YZ, Hua JP, Sun XL, Zhang QF, Corke H (2001) Mapping quantitative trait loci for milling quality, protein content and color characteristics of rice using a recombinant inbred line population derived from an elite rice hybrid. Theor Appl Genet 103:1037

Terasawa Y, Ito M, Tabiki T, Nagasawa K, Hatta K, Nishio Z (2016) Mapping of a major QTL associated with protein content on chromosome 2B in hard red winter wheat (Triticum aestivum L.). Breed Sci 66:471–480

Tiwari VK, Rawat N, Chhuneja P, Neelam K, Aggarwal R, Randhawa GS, Dhaliwal HS, Keller B, Singh K (2009) Mapping of quantitative trait loci for grain iron and zinc concentration in diploid A genome wheat. J Hered 100:771–776

Tiwari C, Wallwork H, Arun B, Mishra VK, Velu G, Stangoulis J, Kumar U, Joshi AK (2016) Molecular mapping of quantitative trait loci for zinc, iron and protein content in the grains of hexaploid wheat. Euphytica 207:563–570

Tsilo TJ, Hareland GA, Simsek S, Chao S, Anderson JA (2010) Genome mapping of kernel characteristics in hard red spring wheat breeding lines. Theor Appl Genet 121:717–730

Vasconcelos M, Datta K, Oliva N, Khalekuzzaman M, Torrizo L, Krishnan S et al (2003) Enhanced iron and zinc accumulation in transgenic rice with the ferritin gene. Plant Sci 164(3):371–378

Velu G, Tutus Y, Gomez-Becerra HF, Hao Y, Demir L, Kara R, Crespo-Herrera LA, Orhan S, Yazici A, Singh RP et al (2017) QTL mapping for grain zinc and iron concentrations and zinc efficiency in a tetraploid and hexaploid wheat mapping populations. Plant Soil 411:81–99

Wada T, Uchimura Y, Ogata T, Tsubone M, Matsue Y (2006) Mapping of QTLs for physicochemical properties in japonica rice. Breed Sci 56:253

Wang LIN, Cui FA, Wang J, Jun LI, Ding A, Zhao C, Li X, Feng D, Gao J, Wang H (2012) Conditional QTL mapping of protein content in wheat with respect to grain yield and its components. J Genet 91:303–312

Wang P, Wang H, Liu Q, Tian X, Shi Y, Zhang X (2017a) QTL mapping of selenium content using a RIL population in wheat. PLoS One 12:e0184351

Wang Y, Mei S, Wang Z (2017b) Metabolite profiling of a zinc-accumulating rice mutant. J Agric Food Chem 65:3775. https://doi.org/10.1021/acs.jafc.7b00105

Wirth J, Poletti S, Aeschlimann B (2009) Rice endosperm iron biofortification by targeted and synergistic action of nicotianamine synthase and ferritin. Plant Biotechnol J 7:631–644

Xu Y, An D, Liu D, Zhang A, Xu H, Li B (2012) Molecular mapping of QTLs for grain zinc, iron and protein concentration of wheat across two environments. Field Crops Res 138:57–62

Xu F, Sun C, Huang Y, Chen Y, Tong C, Bao J (2015) QTL mapping for rice grain quality: a strategy to detect more QTLs within sub-populations. Mol Breed 35:105

Yan J, Xue WT, Yang RZ, Qin HB, Zhao G, Tzion F, Cheng JP (2018) Quantitative trait loci conferring grain selenium nutrient in durum wheat × wild emmer wheat RIL population. Czech J Genet Plant Breed 54:52–58

Yang Y, Guo M, Li R, Shen L, Wang W, Liu M, Zhu Q, Hu Z, He Q, Xue Y, Tang S (2015) Identification of quantitative trait loci responsible for rice grain protein content using chromosome segment substitution lines and fine mapping of qPC-1 in rice (*Oryza sativa* L.). Mol Breed 35:130

Yoo SC (2017) Quantitative trait loci controlling the amino acid content in rice (*Oryza sativa* L.). J Plant Biotechnol 44:349

Yu YH, Li G, Fan YY, Zhang KQ, Min J, Zhu ZW, Zhuang JY (2009) Genetic relationship between grain yield and the contents of protein and fat in a recombinant inbred population of rice. J Cereal Sci 50:121

Yun BW, Kim MG, Handoyo T, Kim KM (2014) Analysis of rice grain quality associated quantitative trait loci by using genetic mapping. Am J Plant Sci 5(9):1125

Zhai S, He Z, Wen W, Jin H, Liu J, Zhang Y, Liu Z, Xia X (2016) Genome-wide linkage mapping of flour color-related traits and polyphenol oxidase activity in common wheat. Theor Appl Genet 129:377–394

Zhang W, Dubcovsky J (2008) Association between allelic variation at the phytoene synthase 1 gene and yellow pigment content in the wheat grain. Theor Appl Genet 116:635–645

Zhang Y, Wu Y, Xiao Y, He Z, Zhang Y, Yan J, Zhang Y, Xia X, Ma C (2009) QTL mapping for flour and noodle colour components and yellow pigment content in common wheat. Euphytica 165:435

Zhang X, Zhang G, Guo L, Wang H, Zeng D, Dong G, Qian Q, Xue D (2011) Identification of quantitative trait loci for Cd and Zn concentrations of brown rice grown in Cd-polluted soils. Euphytica 180:173

Zhao L, Zhang KP, Liu B, Deng ZY, Qu HL, Tian JC (2010) A comparison of grain protein content QTLs and flour protein content QTLs across environments in cultivated wheat. Euphytica 174: 325–335

Zheng L, Zhang W, Chen X, Ma J, Chen W, Zhao Z, Zhai H, Wan J (2011) Dynamic QTL analysis of rice protein content and protein index using recombinant inbred lines. J Plant Biol 54:321

Zhong M, Wang LQ, Yuan DJ, Luo LJ, Xu CG, He YQ (2011) Identification of QTL affecting protein and amino acid contents in rice. Rice Sci 18:187

Molecular Breeding Approaches for Biofortification of Cereal Crops

3

Vinay Sharma, Nitish Ranjan Prakash, and Ashish Kumar

Abstract

Over three billion population in this world is suffering from micronutrient malnutrition; among them women and children are more sufferers, mainly in developing countries. The effective and sustainable way to enhance the food value is by breeding cereal crops having a good amount of micronutrient content. Several reports of quantitative trait locus (QTL)/loci for iron (Fe) and zinc (Zn) content in rice, wheat, and millets; provitamin A, vitamin E, and quality protein in maize; starch content in rye; and malting quality in barley have been discussed in detail. A detailed discussion was also made on molecular approaches for biofortification, viz., dissection of genome-wide genomic regions linked with biofortification traits, marker-assisted selection (MAS), and genomic selection. This chapter emphasizes a different breeding approach to develop micronutrient-rich cereal cultivars to overcome malnutrition and ultimately eradicate hidden hunger.

The original version of the chapter has been revised. A correction to this chapter can be found at https://doi.org/10.1007/978-981-19-4308-9_15

V. Sharma
International Crops Research Institute for the Semi-Arid Tropics, Hyderabad, India

N. R. Prakash
ICAR-Central Soil Salinity Research Institute, Regional Research Station, Canning Town, West Bengal, India

A. Kumar (✉)
Narayan Institute of Agricultural Sciences, Gopal Narayan Singh University, Rohtas, Bihar, India

R. Deshmukh et al. (eds.), *Biofortification in Cereals*,
https://doi.org/10.1007/978-981-19-4308-9_3

3.1 Introduction

The world population is increasing and, with the limited natural resources we have, is creating a great burden on it. Dwindling resources, overexploitation of land and water, climate change, and other associated problems have created an impact of concern on food and its nutritional quality over the years. The quantum jump gained through the green revolution in cereal production has been neutralized now by the demands of the increasing population. Because of the severe impact of improper distribution of foods arising out of faulty economic policy and income disparity, a mass of people are still in poverty and are not able to get enough nutritious food (Pfeiffer and McClafferty 2007). Today we are on a verge of food surplus in terms of cereals and pulses, but a large section of the poor population is not able to get a balanced diet. As per the UN-FAO report, almost 11% of the population of the world is malnourished (FAOSTAT 2013). With further deepening crisis, almost 151 million children aged below 5 years are stunted, and 94% of them live in Asia and Africa. India is burdened with the largest population of undernourished children in the world. With all these in mind, several organizational efforts have been focusing on alleviating hunger, malnutrition, and poverty from this planet and thus helping in attaining United Nations Sustainable Development Goals (UN-SDG) of "Zero Hunger," "No Poverty," "Good Health and Well-Being." As poor people are not even able to get enough food, a balanced diet is a distant dream for them. International and national organizational efforts are to provide a healthy dietary supplement to people through fortified food items or biofortified crop produce. The biofortified crops are alternative to the biofortified food supplements which are easy to produce, less costly, and easy to distribute (Tripathi et al. 2018).

Breeding for crops rich in nutrients is called biofortification, which is a feasible, long-lasting, and frugal way to address the problem of hidden hunger by delivering more micronutrients to people through a normal diet (Saltzman et al. 2013). Rather than mixing micronutrients with food supplements, biofortification is depending on utilizing plant's biosynthetic machinery to produce a nutrient-enriched crop. The availability of genetic diversity for several nutrient-enriched germplasm is a prime requirement of crop breeding for biofortification (Mayer et al. 2008). The discovery of quality protein maize (QPM) mutant having protein enriched with lysine and tryptophan by Mertz et al. (1964) was a milestone in crop biofortification. Several such plants have also been identified in the germplasm of cereals such as sorghum, pearl millet, rice, etc. These germplasm resources are useful for defining the breeding strategy for biofortification and unearthing the genetics of traits for biofortification (Neeraja et al. 2017). In India, several biofortified maize hybrids rich in lysine, tryptophan, provitamin A, and vitamin E have been released for cultivation (Hossain et al. 2019). Apart from this, iron and zinc enhancement in rice, pearl millet, sorghum, and other millets have been targeted (Yadava et al. 2018). High-protein crop varieties have also been developed in rice and wheat.

Studying the genetics of traits affecting the nutrient quantity, nutrient bioavailability, storability, digestibility, food palatability, etc., are important points to be considered. Molecular studies regarding physiology and biochemistry of such traits are being assisted nowadays by precision and high-throughput phenotyping (Francis et al. 2017). The phenotyping of traits for breeding goals is important to improve genetic analysis and accuracy. Despite the remarkable progress in crop genomics, the lack of cost-effective high-performance phenotyping assays prevents the comprehensive usage of genetic mechanisms that underlies the biofortification traits (Hossain et al. 2019). High-throughput platforms of phenotyping will be needed for measuring the genetic effects that contribute to phenotypic variation. The use of high-throughput platforms such as near-infrared spectral, ultrasound-assisted assays, and mass spectrophotometry (Belalcazar et al. 2016), for phenotyping traits such as oil content, protein content, iron and zinc concentration, vitamin content, etc., with reliable statistical models recently paved the way for phenotyping in a more efficient and precise manner. The discovery of molecular markers associated with biofortification traits will be an area of interest as these traits are complex and far-reaching to phenotype (Francis et al. 2017). QTL mapping and genome-wide association studies (GWAS) are trending techniques for finding linked molecular markers for use in breeding activities, especially in today's post-genomics era where the cost of genotyping is getting lower day by day and the use of molecular marker in assisting the cereal breeding have been increased (Tripathi et al. 2018; Mishra et al. 2022).

Breeding for biofortified crop varieties has been tremendously augmented by the advancement in genetic and genomic tools. Genomics has helped in unearthing the genetics of micronutrient enhancement, fine mapping of target genes, marker-assisted selection (MAS), and transfer of genes (Pixley et al. 2013a, b). Micronutrients such as iron and zinc are required in small quantities for humans and plants, excess accumulation of which is toxic. Physiological studies determining the optimum concentration of these micronutrients are required for the successful transfer of genes governing these traits (Shahzad et al. 2014). Apart from this, the development of easily accessible phenotyping tools is required for efficient breeding. Enzyme-based detection of micronutrient content, cheap amino acid analysis from leaf and grain samples, imaging-based phenotyping, etc., are nowadays available for genetic analysis of traits governing nutrient availability in food (Fahlgren et al. 2015). Marker-assisted backcross breeding (MABB) has been utilized in maize for the transfer of genes such as *crt-RB1* (governing provitamin A), *opaque-2* and *opaque-16* (for high lysine and tryptophan), and *lpa1* and *lpa2* (for high iron and zinc) to develop biofortified hybrids (Gupta et al. 2015; Yadava et al. 2018; Hossain et al. 2019). With the advent of transgenic methods, it has become easy to transfer the genes from any species to our crop for improvement of micronutrient availability. In this case, golden rice was a breakthrough in utilizing a transgenic approach to develop high provitamin A rice by utilizing a gene from bacteria (Beyer et al. 2002). With all products in hand, popularization of the biofortified crop is also difficult if the crop quality in terms of taste and preferences is not considered and improved nutrient availability is not at the cost of yield.

Molecular analysis of any trait is giving us a complete understanding of its function, physiology, metabolic pathway, associated genes and enzyme, and a network of associated regulatory functional elements, which in turn assist in precise breeding (Francis et al. 2017, Kumar et al. 2021). Targeting a trait that is generally

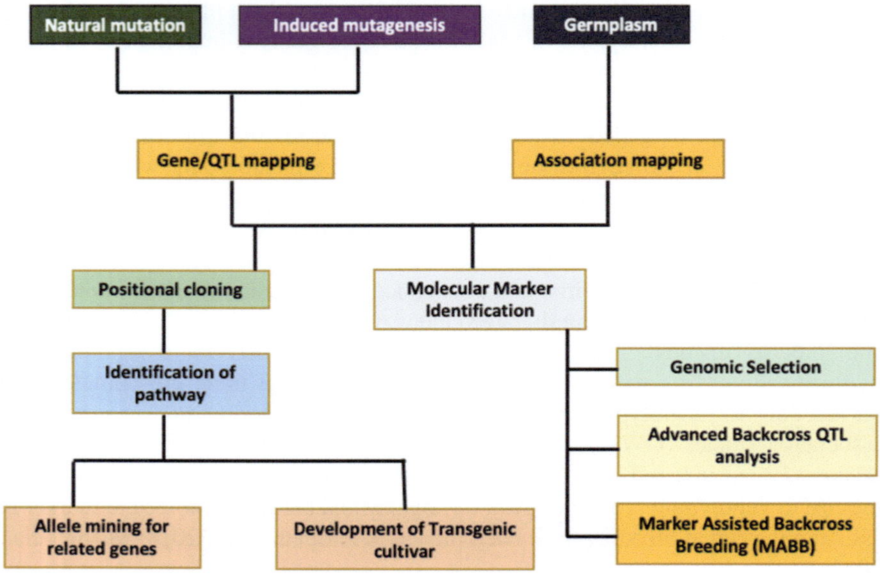

Fig. 3.1 Schematic diagram representing the flow of molecular breeding process in the development of biofortified varieties

governed by several genes results in severe background effects. This effect from the background dilutes the impact of the gene in the selection and thus limits breeding efficiency. The best example of this was the quality protein maize (QPM) mutant. This maize mutant has been used only when we were able to select a line with the best modifier genes in the background (Sofi et al. 2009). Therefore, unraveling the molecular mechanism of a trait will help in precisely targeting a trait and allele mining for a desirable phenotype and ultimately selecting the best germplasm. Apart from these, gene-based markers will help in more precisely genotyping the breeding population. Genomic selection for target traits along with the other desirable attributes is enhancing the breeding program of today's era. Now we are on the verge of predicting the phenotype of plants with more accuracy because of the use of vast computational tools, big data analysis, high-throughput genotyping, high-throughput phenotyping, and phenomics facilities (Prakash et al. 2017). Certainly, the breeding for biofortification will next be based on breeding-by-design. The schematic diagram represents different approaches of molecular breeding for the development of biofortified varieties, which is shown in Fig. 3.1.

3.2 Molecular Approaches for Biofortification

Different traits governing nutrient concentration in cereals are governed by one or more genes. This makes breeding for biofortification an interesting and difficult task. Generally, most of the traits for biofortification are polygenic with varying levels of background effect depending on the genotype (Saltzman et al. 2013). Several large effect QTLs or major genes have been used in biofortification breeding. On the basis of physiological and molecular, breeding for biofortification must target the following approach:

- **Increasing the concentration/bioaccumulation**: It includes the increase in more accumulation of micronutrients such as Fe or Zn in seed or economically important plant parts. It may be done by improved absorption of iron and zinc via the roots and their efficient transportation into economically important plant parts such as the seeds in rice (Masuda et al. 2012). Several metal transporter proteins, metal chelators, and metal chaperones, metal-associated storage proteins are responsible for efficient uptake, transportation, accumulation, and bioavailability of these nutrients (Palmgren et al. 2008).
- **Upregulating the enzyme involved in the biosynthesis of target nutrient**: Upregulation of genes governing biosynthesis of provitamin A in crops will automatically enhance the accumulation of provitamin A in seeds. An increase in the activity of *phytoene synthase 1* (*psy1*) will enhance the accumulation of phytoene (a precursor in β-carotene biosynthetic pathway, in turn, increasing the accumulation of β-carotene in maize kernel) (Buckner et al. 1990).
- **Downregulating the genes involved in the degradation of nutrient biomolecule**: In provitamin A, the biosynthetic pathway is consistently being converted into β-cryptoxanthin, thus reducing the availability of β-carotene (which has provitamin A activity). This is governed by a gene called *β-carotene hydroxylase* (*crtRB1*). A leaky mutation in this gene has resulted in higher provitamin A content in maize kernel and has been used in breeding programs (Muthusamy et al. 2014).
- **Downregulation of genes involved in branching pathways**: Accumulation of a particular nutrient in seed or any other plant part requires more diversion of substrate and energy into its biosynthesis. Similarly in maize, a gene called *lycopene epsilon cyclase* (*lcyE*) gene changes more *lycopene* to *α-carotene* and *lutein* and thus reduces *β-carotene* content. A leaky mutation in this gene has been successfully used by breeders to enhance provitamin A content in maize (Harjes et al. 2008).
- **Enhancing the bioavailability**: Fe and Zn are required in plants in very small quantities and cereals store them in terms of phytates. Unfortunately, these metals stored in terms of phytates are least available for getting absorbed in the human digestive system. Therefore, breeding for reducing the phytate accumulation of iron and zinc will certainly enhance their bioavailability in humans (Sperotto et al. 2012).

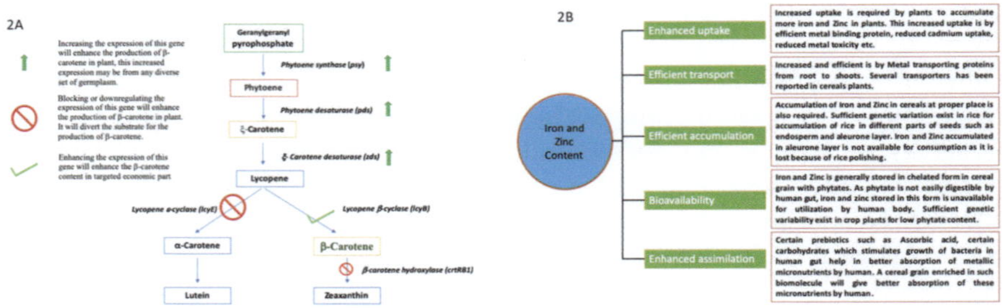

Fig. 3.2 (**a**) Diagrammatic representation showing a different approach for enhancing provitamin A content in cereals. A similar approach can also be applied for enhancing bioaccumulation, absorption, and bioavailability of iron and zinc in cereals. (**b**) Criteria used for selecting target traits for breeding for metallic micronutrients in plants for biofortification

- **Transgenic approach**: If the genes responsible for a particular nutrient are not available in that crop, it can be transferred from other organisms. This scheme is generally used when there is a nonavailability of variation for that particular trait in all the available germplasm. This may be used to transfer a new gene (*psy* gene from maize to rice), to increase the expression of desired gene to manifold by changing promoter and regulatory sequences (increasing Fe and Zn accumulation in rice using 2′-deoxymugenic acid (DMA) and nicotianamine (NA) as metal-chelating ligands) and by decreasing the expression of genes responsible for degradation of nutrient biomolecule (Beyer et al. 2002; Banakar et al. 2017).

Utilization of any approach to combat hidden hunger via biofortification will rely on the availability of trait variations, genetics of traits, mode of transfer, easy and efficient phenotyping assay. Most of the time, breeding for biofortification has not been started because of the nonavailability of source germplasm, high cost involved in genotyping and phenotyping, and lack of sufficient information about the genetics of trait (Neeraja et al. 2017). A detailed pictorial representation regarding the above approaches is given in Fig. 3.2a, b.

3.3 Studying Genetics Using Molecular Approaches

With the advancement in genetic analysis, genomic tools, biotechnology, and transgenics, crop breeding for biofortification has become easy. Several reports have come describing the mapping of loci governing biofortification traits in crop plants.

3.3.1 Quantitative Trait Loci (QTLs) for Biofortification Traits

Identification of genomic regions that contributes to target traits of biofortification will aid the transferring of chromosomal segments (called QTL) to elite/popular varieties using MAS. QTL is a genomic region controlling continuous traits which is regulated by many genes of smaller effects (Liu 1997). Understanding the genetic basis of biofortification traits has made slow progress due to the complex nature of quantitative traits and influence of genotype × environment (G × E) interactions (Mahender et al. 2016). Low to moderate heritability estimates of the contents of nutrients (especially, minor elements) in crops may limit the understanding of its genetics (Manickavelu et al. 2017). In this context, the identification of genomic regions that account for significant phenotypic variation is a key to clarifying the genetic basis of quantitative traits controlled by multiple genes (Sharma et al. 2017). Extensive research efforts have been done and are underway for the improvement of grain quality traits in cereal crops such as rice (Li et al. 2004; Verma et al. 2022), wheat (Huang 2006; McCartney et al. 2006), maize (Cook 2012), and sorghum (Sukumaran et al. 2012), and large phenotypic variation for quality traits, amino acid composition, micronutrient density, and protein content has been reported. Identifying genes that influence traits of interest will help to exploit grain quality and texture which facilitate the development of new cultivars (Bean et al. 2016). Table 3.1 summarizes the QTL identified in recent studies in key cereal crops for various biofortification traits.

Rice Polished grains of rice typically contain low Zn and Fe levels which do not fulfill the average daily human requirement and adversely affect human health. Several studies of QTL mapping for high Zn and Fe were reported in rice using diverse populations/backgrounds and intraspecific/interspecific crosses (Garcia-Oliveira et al. 2009; Norton et al. 2010; Anuradha et al. 2012; Swamy et al. 2016). Several major effects of QTLs in wild species and deep water rice for Zn level in grain with ≥30% phenotypic variation have been reported (Neelamraju et al. 2012). QTLs for Zn and Fe homeostasis have been identified on the different chromosomes, and their expression has been detected in recently released Zn rice cultivars (Swamy et al. 2016). Another study Anuradha et al. (2012) conducted in RIL population (Madhukar × Swarna) identified 14 QTLs on linkage group (L.G.) 7 and 12 associated with Zn and Fe concentration in grains. Similarly, genomic regions for Zn and Fe content have been found in RIL derived from Jalmagna × Swarna, and 2 SSR markers (RM3322 and RM7488) were found to be linked with Fe and Zn (Kiranmayi and Manorama 2014). Swamy et al. (2018a, b) have identified varied candidate genes in the vicinity of the QTLs region for grain Zn concentration (*OsNRAMP*, *OsNAS*, *OsZIP*, *OsYSL*, *OsFER*, and *OsZIFL* family). Such genomic regions and candidate genes help to further understand the genetic basis for the concentration of grain micronutrients and can help for marker-assisted breeding. QTL mapping using a back-cross recombinant inbred population (*O. sativa* "*Nipponbare*" × *O. meridionalis* W1627) for grain Zn concentration has identified four QTLs on L.G., 2, 9, and 10, respectively (Ishikawa et al. 2017a, b). Other studies in rice for Mn content identified (Liu et al. 2017) a major locus *qGMN7.1* in

Table 3.1 Description of cereal crops and their corresponding details of trait of study, mapping population, and QTLs

Crops	Population name	QTLs /chromosome	Target traits	PVE %	References
Maize	Ye478 × Wu312 (RIL)	*qMnCC1-1, qMnCC1-2, MnCC2-1, qMnCC2-1, MnCC4-1, qMnCC4-2, qZnCT4-1, qZnCT4-2, qZnCC5-1, qZnCC5-1*	Mn, Zn	6.22–27.7	Gu et al. (2015)
	By804 × B73 (RIL)	*qbc6-1, qbc/bcry10-1, qbc/tc10-1, qbcry6-1, qlut6-1, qtc6-1, qtva6-1, qzea10-1, qzea/tc10-1*	Carotenoid related traits	16.91–43.71	Jittham et al. (2017)
	Mu6 × SDM (MuS)	*CQZnK1, CQZnK9-1, YNZnK2, YNZnK6*	Zn	9.87–12.58%	Qin et al. (2012)
		CQZnC1, CQZnC7-1, YNZnC2, YNZnC7-1	Zn	7.67–15.34%	
		CQFeK2-1, YNFeK2-1, YNFeK9	Fe	10.02–17.01%	
		CQFeC4, CQFeC10-1, YNFeC2-1, YNFeC10-1	Fe	6.36–10.06%	
	Mo17 × SDM (MoS)	*CQZnK7, CQZnK9-2, CQZnK10, YNZnK9, YNZnK10*	Zn	6.34–21.33%	
		CQZnC5, CQZnC7-2, YNZnC7-2, YNZnC10	Zn	7.23–19.88%	
		CQ FeK2-2, YNFeK2-2	Fe	17.31, 21.12%	
		CQFeC2, CQFeC10-2, YNFeC2-2	Fe	10.75–16.15%	
	178 × P53 (RIL)	*qZn6-3, qZn6-5*	Zn	16.8, 11.7%	
		qMn1-1, qMn1-2, qMn1-5, qMn3-2, qMn3-3, qMn3-4, qMn4-3	Mn	10.49–15.35%	Zhang et al. (2017)
	B84 × Os6–2 (Ils)	*32 significant QTLs*	Biofortification traits	6.7–19.9%	Šimić et al. (2012)

	B73 × Zheng58 (F2:4)	*qOIL06-01, qOIL08-01, qPRO02-01, qPRO06-01*	Oil and protein concentration	10.2, 11.1, 11.4, 10.5%	Yang et al. (2016)
	By804 × B73 (F2:3)	*oilc1, oilc4-1*	Oil concentration	13.13, 10.14%	Zhang et al. (2008)
	8984 × GY220 (RIL)	*qzPRO1-5-1, qPRO1-8-1*	Grain protein content	11.3, 13.4%	Yang et al. (2014)
	GY220 × 8984, GY220 × 8622 (RIL)	*qOIL1-3-1, qOIL1-4-1, qxOIL2-6-1*	Oil content	11.1, 11, 14.47%	Yang et al. (2012)
	W64a × A632	*3 major QTLs*	Carotenoids (β-carotene, β-cryptoxanthin)	11.8–25.4%	Wong et al. (2003)
	K22 × CI7, K22 × Dan340 (F2:4)	*qd1-1, qc5-1/qd5-1, and qc5-2)*	Tocopherols content (αT)	53, 30, 25%	Shutu et al. (2012)
	N6 × NC296	*ZmVTE2*	Vitamin E content	21.67%	Fenton et al. (2018)
Wheat	PBW343 × Kenya Swara	*QGzncpk.cimmyt-1BS, QGzncpk.cimmyt-3AL, QGzncpk.cimmyt-2Bc, QGzncpk.cimmyt-2D*	Zinc content	10–26%	Velu and Singh (2019)
	(Saricanak98 × MM5/4) (tetraploid)	*Qzneff.sar-6A, Qzneff.sar-6B*	Zn efficiency	N. A.	Velu et al. (2016a)
	(Berkut × Krichauff) doubled haploid	*QZn.bhu-1B, QZn.bhu-2*	Zn	22.2–35.9%	Tiwari et al. (2016), Velu et al. (2016a)
	Tetraploid (Saricanak98 × MM5/4)	*QGzn.sar-1B, QGzn.sar-6B, QGZn.sar-1B*	Zn	N. A.	Velu et al. (2016a)
	Hexaploid (Adana99 × 70711)	*QGzn.ada-6B, QGzn.ada-1D QGzn.ada-7B*	Zn	N. A.	Velu et al. (2016a)
	Doubled haploid (Berkut × Krichauff)	*IA QGPC.bhu-1A*	Protein contents	N. A.	Tiwari et al. (2016)

(continued)

Table 3.1 (continued)

Crops	Population name	QTLs /chromosome	Target traits	PVE %	References
	Doubled haploid, (Berkut × Krichauff), Hexaploid (Adana99 × 70711)	*QGfe.ada-2B, QGZn.ada-2B, QFe.bhu-2B*	Co-localized for Zn and Fe	N. A.	Tiwari et al. (2016), Velu et al. (2016a)
	Triticum dicoccoides	*GPC-B1 (6 7 BS)*	Fe, Zn, and protein	N. A.	Uauy et al. (2006), Distelfeld et al. (2007)
	Triticum dicoccoides	*TtNAM-B1*	Fe and Zn, protein	N. A.	Distelfeld et al. (2007)
	Triticum boeoticum × *Triticum monococcum* (RIL)	*QFe.pau-7A, QFe.pau-2A*	Fe	N. A.	Tiwari et al. (2009)
	Tetraploid (Saricanak98 × MM5/4)	*QGfe.sar-5B*	Fe	N. A.	Velu et al. (2016a)
	Doubled haploid	*3D, 4B, 6B, and 7A*	Zn	N. A.	Genc et al. (2009)
	Tetraploid Chinese (Langdon × accession G18-16)	*4B and 5A*	Zn and Fe	N. A.	Peleg et al. (2009)
	Triticum spelta L. × synthetic hexaploid RIL	*QGZn.cimmyt-7B_1P2, QGZn.cimmyt-7B_P1, QGFe.cimmyt-4A_P2, QFe.cimmyt-3A_P1*	Zn and Fe	10.3–32.7%	Crespo-Herrera et al. (2017)
	WH542 × synthetic derivative	*QGFe.iari-2A, QGFe.iari-5A, QGFe.iari-7A, QGFe.iari-7B, QGZn.iari-2A, QGZn.iari-4A, QGZn.iari-7B*	Fe, Zn	20–32%	Krishnappa et al. (2017)
	Seri M82 × SHW CWI76364	*One major QTL (4BS)*	Fe, Zn	19.60%	Crespo-Herrera et al. (2016)
	Yumechikara × Kitahonami (DH)	*QGpc.2B-yume*	Protein	32.10%	Terasawa et al. (2016)
	Tainong18 × Linmai6 (RIL)	*QSsece-1B, QSsece-7D.1, QSsece-7D.2, QRsec-6A, QSsec-4B*	Se	10.99–20.22%	Wang et al. (2017)

Rice	1689 × IR64	*OsNAS2, SferH-1*	Fe, Zn	N. A.	Trijatmiko et al. (2016)
	Ce258 and Zhongguangxiang1 × IR75862	*qFe1, qFe2 and qZn12, 4B and 5A*	Fe, Zn and protein	N. A.	Xu et al. (2015)
	Zhengshan 97 × Minghui 63 (RIL)	*qZn-5, qZn-7, qZn-11*	Zn	N. A.	Lu et al. (2008)
	Teqing × *O. rufipogon* (IL)	*qZn5-1, qZn8-1, qZn12-1*	Zn	5–19%	Garcia-Oliveira et al. (2009)
		qMn10-1	Mn	5%	
	IR64 × Azucena (DH)	*qZn1-1, qZn12-1*	Zn	N. A.	Stangoulis et al. (2007)
	ZYQ8 × JX17 (DH)	*qZn4, qZn6*	Zn	N. A.	Zhang et al. (2011)
	Madhukar × Swarna (RIL)	*qZn3.1, qZn7.1, qZn7.2, qZn7.3, qZn12.1, qZn12.2*	Zn	29–35%	Anuradha et al. (2012)
	Bala × Azucena	*qZn7*	Zn	12%	Norton et al. (2010)
	Sasanishiki × Habataki (BIL)	*qSZn2, qSZn12*	Zn	16.7, 21.3%	Ishikawa et al. (2010)
	TeQing × Lemont (IL)	*qZn2, qZn4, qZn5, qZn6, qZn7, qZn9*	Zn	8.10%	Zhang et al. (2014)
	Lemont × TeQing (RIL)	*qZn10*	Zn	4.40%	Zhang et al. (2014)
	Goami 2 × Hwaseonchal (DH)	*qFe1.1, qFe1.2, qFe1.3, qFe4, qFe6, qFe7, qFe11*	Fe	11–40.50%	Jeong et al. (2020)
		qZn7	Zn	15.36%	Jeong et al. (2020)
	IR05F102 × IR69428 (DH)	*qFe9.1, qFe12.1*	Fe	11.79–13.34%	Calayugan et al. (2020)
		qZn5.1, qZn9.1, qZn12.1	Zn	12.15–15.26%	Calayugan et al. (2020)
	IR64 × IR69428 and BR29 × IR75862 (DH)	*qZn2.1, qZn3.1, qZn5.1, qZn5.2, qZn7.1, qZn8.1, qZn9.1, qZn11.1*	Zn	8.6–27.7%	Descalsota-Empleo et al. (2019)

(continued)

Table 3.1 (continued)

Crops	Population name	QTLs /chromosome	Target traits	PVE %	References
	PSBRc82 × Joryeongbyeo and PSBRc82 × IR69428 (DH)	*qZn2.1, qZn3.1, qZn6.1, qZn6.2, qZn8.1, qZn11.1, qZn12.1, qZn12.2*	Zn	7.5–22.8%	Swamy et al. (2018a, b)
	RILs (Liang-You-Pei-Jiu-advanced self-fertilization population)	*GZC4, GZC5, GZC6*	Zn	2.74–4.64%	Liu et al. (2020)
	MAGIC (16 founders of indica and japonica pools)	*QZn, QZn, QZn7*	Zn	17.57–20.10%	Zaw et al. (2019)
	O. sativa "Nipponbare" × *O. meridionalis* W1627 (BIL)	*qGZn2-1 and qGZn2-2, qGZn9, qGZn10*	Zn	15–21.9%	Ishikawa et al. (2017a, b)
	93-11 × PA64s (RIL)	*qGMN7.1*	Mn	15.6–22.8%	Liu et al. (2017)
Sorghum	BTx642 × BTxARG-1, BTxARG-1/P850029 (RIL)	*DGAT1 gene*	Protein	N. A.	Boyles (2017)
	390 sorghum accessions from the Sorghum Association Panel	*Alpha-amylase 3 gene*	Protein	N. A.	Rhodes et al. (2017)
Pearl millet	130 accessions	*16 significant MTAs*	Zn and Fe	11.3–13.4%	Anuradha et al. (2017)
	ICMB 841-P3 × 863B-P2(RIL)	*3/110# (QTL position)*	Fe and Zn	18.9, 30.9%	
		2/30, 5/118 (QTL position)	Fe_OP	16.10%	Kumar et al. (2016)
		3/110#, 7/96 (QTL position)	Zn_OP	41.70%	
	ICMS 8511-S1-17-2-1-1-B-P03 × AIMP 92901-S1-183-2-2-B-08 (RIL)	*11 QTL for Fe and 8 QTL for Zn*	Fe and Zn	9.0–31.9% and 9.4–30.4%	Kumar et al. (2018)
Rye	S120 × S76 (RIL)	*QAa1R-M1, QAa1R-M2, QAa1R-M3, QAa2R-M1, QAa3R-M1, QAa3R-M2, QAa4R-M1, QAa4R-M2, QAa4R-M3,*	Alpha-amylase activity	6.09–23.32%	Myśków et al. (2012)

		QAa5R-M1, QAa6R-M1, QAa6R-M2, QAa6R-M3, QAa7R-M1			
	541 × Ot1-3 (RIL)	*AMY_WPUT/13_1, AMY_WPUT/13_2, AMY_WPUT/13_3, AMY_DANKO/13_2, AMY_DANKO/14_2, AMY_DANKO/14_3, AMY_WPUT/14_1, AMY_WPUT/14_2*	Alpha-amylase activity	10.3–38.8%	Milczarski et al. (2017)
Barley	TX9425 × Naso Nijo	*QMe.NaTx-2H/MLOC_60943.2*	Malt extract	48.40%	Wang et al. (2015)
	Steptoe × Morex	*QTL2 (HvTLP8)*	Malt extract	37.60%	Singh et al. (2017)
	Nure × Tremois	*Qme1.2*	Malt extract	11.30%	Laidò et al. (2009)
	Amazone × St. 2730e × Kym	*QAa.S42-2H.a, QAa.S42-4H.a, QAa.S42-4H.c*	Diastatic power	11.2–15.8%	von Korff et al. (2008)
	TX9425 × Naso Nijo	*QDp.NaTx-5H, QDp.NaTx-7H*	Diastatic power	13–20.9%	Wang et al. (2015)
	Admiral × Navigator	N.A.	Soluble protein	13.90–14.95%	Cu et al. (2016)
	Victoriana × Sofiara	*QSNI-3-1*	Soluble protein	27.31%	Kochevenko et al. (2018)
	Admiral × Navigator	N.A.	Soluble protein	12.41–14.37%	Cu et al. (2016)
	Nure × Tremois	*Qpc1.1, Qpc6.1*	Grain protein content	18.7, 29.1%	Laidò et al. (2009)
	Triumph × Morex	N.A.	Grain protein content	13.4–62.3%	Elía et al. (2010)
	Brenda × HS213	*Qpc2.1*	Malt protein content	14.72%	Li et al. (2005)
	Brenda × HS213	*Qpc7.1*	Malt protein content	12.02%	Li et al. (2005)

the RIL population (93-11 and PA64s) on chromosome 7 explaining 23% phenotypic variance. Further, fine mapping of this genomic region detected 5 candidate genes (*LOC_Os07g15350, LOC_Os07g15360, LOC_Os07g15390, LOC_Os07g15400,* and *LOC_Os07g15370)* within the 49.3 kb target region. Later on, *LOC_Os07g15370* (*OsNRAMP5*) was reported as a potential gene for the accumulation of high grain Mn. With the advancement in genomic technologies and availability of rice genome annotation, it is easy to detect genes within QTL regions and way forward the development of functional markers which can be used for MAS breeding for biofortification trait programs. Further, literature for detailed rice biofortification QTL studies are available in a recent publication (Sharma et al. 2020).

Wheat Genetic variability for grain micronutrients concentration has enabled the progress of biofortified wheat via conventional breeding methods (Velu et al. 2014). Genetic diversity is being explored at CIMMYT under the HarvestPlus program, which has shown that there is desirable variability for grain quality traits that exist in the crop wild relatives of wheat (Cakmak et al. 2010; Velu et al. 2016a, b). In this case, a major effect of QTL *Gpc-B1* has been identified for protein level in wild emmer wheat on chromosome 6 (Distelfeld et al. 2007). In another study, Crespo-Herrera et al. (2016) identified a major loci *QGZn.cimmyt-7B_1P2* on chromosome 7B for Zn concentration in grain with 32.7% of phenotypic variation explained (PVE), while QTL (*QGFe.cimmyt-4A_P2*) for grain Fe content was detected on chromosome 4A, explaining 20% PVE (Crespo-Herrera et al. 2017). However, Velu et al. (2016a) reported that two major QTLs in two RIL populations for grain Zn concentration co-localized on chromosome 1B and 6B. In another study, Krishnappa et al. (2017) reported 4 loci for Fe content in grain, explaining up to 20% PVE and 5 loci for Zn content with 32% PVE, respectively. Similarly, 16 loci for Se content at various stages, viz., seedling, shoot, and grain were reported (Wang et al. 2017). Previous, QTL studies on Se content also reported five QTL involvement (Pu et al. 2014). Two wheat databases contain information for wheat-specific SSR markers and mapped SSR markers (https://wheat.pw.usda.gov/ggpages/SSRclub/) and Integrated Plant Breeding Platform (https://www.integratedbreeding.net/104/communities/genomics-crop-info/agricultural-genomics/markers/ssrmarkers/wheat).

Maize Several mapping studies have been conducted to discover the genomic regions associated with biofortification traits in maize. Recently, a major QTL (ZmVTE2) for vitamin E content was identified in cross N6 × NC296 with 21.67% PVE (Fenton et al. 2018). Another study (Shutu et al. 2012) conducted in ($F_{2:4}$)-derived population reported three major QTLs (*qd1-1, qc5-1/qd5-1, and qc5-2*) for tocopherol content (αT) explaining PVE ranging from 25 to 53%. Similarly two QTLs (*qZn6-3, qZn6-5*) for Zn content with 11.7–16.8% PVE and 7 QTLs (*qMn1-1, qMn1-2, qMn1-5, qMn3-2, qMn3-3, qMn3-4, qMn4-3*) for Mn content explaining PVE range from 10.49 to 15.35% have been reported in 178 × P53 (RIL) population (Zhang et al. 2017). Another study (Jittham et al.

2017) reported loci for carotenoid-related traits explaining PVE ranging from 16.91 to 43.71% in the RIL population (By804 × B73). Gu et al. (2015) identified loci for Mn and Zn content in the RIL population (Ye478 × Wu312) with 6.22–27.7% PVE.

Pearl millet and sorghum Pearl millet and sorghum were considered to be crops with minimal genetic and genomic resources. Tremendous progress of genomic technologies in recent years made it possible for the generation of genetic stocks. Recently, various $F_{2:3}$ and $F_{2:4}$ mapping populations (H77/833-2 × PRLT 2/89-33) (early maturing inbred line, 150 F_8 population) and ICMB 841 × 863B (agronomically elite inbred seed parent, 106 F_6 population) were developed in pearl millet for grain Zn and Fe concentration from a diverse genetic background panel of Asian, American, and African origin (Kumar et al. 2016). In ICRISAT, progress has been made for mapping loci for grain Fe and Zn content in the RIL population (ICMB 841-P3 × 863-P2) (Kumar et al. 2016). Two major loci were detected for Fe and Zn content on chromosome 3 with 19% PVE for Fe QTL and 36% PVE for Zn QTL, respectively. Similarly, for open-pollinated seeds, 2 loci each for Fe and Zn concentration have been detected on L.G. 3, 5, and 7 with 16 and 42% PVE (Kumar et al. 2016). In sorghum, interesting findings which is associated with the genetic control of grain composition through mutation include QTL for amylose content which is 12 kb from waxy locus (a gene that regulates the production of amylose and improved starch digestibility) (Lichtenwalner et al. 1978; Rooney and Pflugfelder 1986), sugary which enhanced the content of sucrose (Boyer and Liu 1983), and high lysine, which increase lysine content and digestibility of protein (Singh and Axtell 1973). QTL mapping and genome-wide association mapping studies have reported many QTLs for grain composition, and additional study should be conducted for the detection of genes linked to biofortification-related traits variation in sorghum. Recently, a major QTL was identified in chromosome SBI-10 for crude fat with 28.1% PVE (Boyles 2017). Earlier, a study in sorghum has identified two major QTLs for high protein a-kafirin digestibility on chromosome 1, which can be used for introgression in popular/elite cultivars for agronomic advantages. Another study Kotla et al. (2016) reported loci for Fe and Zn content in sorghum.

Rye In rye, the main determinants of quality are starch content, alpha-amylase activity, and pentosan concentration. In comparison to wheat, protein quality and content play a key role. Previous, mapping studies target single traits based on the line per se performance, e.g., α-amylase activity (Masojć and Milczarski 2009). In a recent study, QTL mapping in two biparental populations (Petkus gene pool) for important traits has been done (Miedaner et al. 2012). In another study, QTL mapping in the RIL population (S120 × S76) for alpha-amylase activity has been reported with 14 QTL with 6.09–23.32% PVE (Myśków et al. 2012). More recently, 5 major QTL (in 2013) and 4 major QTLs (in 2014) were identified for alpha-amylase activity with 10.3–38.8% PVE in the RIL population (541 × Ot1-3) (Milczarski et al. 2017).

Barley In barley, the main objective for breeding programs is to enhance malting quality. Approximately 30% of the worldwide produced barley is used for malting, therefore breeding for the development of high-quality malt barley varieties is a key objective (Walker and Panozzo 2016; Kochevenko et al. 2018). A previous study reported *QTL2*, for malting quality on chromosome 4H using Steptoe × Morex' doubled haploid population. To date, more than 200 QTLs/genes for malting quality have been identified. However, using modern technologies numbers of QTLs have been successfully applied for MAS (Igartua et al. 2000; Rae et al. 2007; Li et al. 2010; Xu et al. 2018). In a recent study, a major QTL, viz., *QMe.NaTx-2H*, was reported for malt extract on chromosome 2 (2H) with 48.4% PVE (Wang et al. 2015). In other studies, 2 major loci for alpha-amylase have been identified on chromosome 5H with 12.4 and 25.6% PVE (Mohammadi et al. 2015; Zhou et al. 2016). Protein is one of the essential constituents of malt products. To date, only a few QTLs have been reported for malt protein content and grain protein content. QTL mapping using a double haploid population reported 2 major QTLs (*qPC2.1* and *qPC7.1*) on chromosome 2H and 7H with 14.72% and 12.2% PVE, respectively (Li et al. 2005). A similar study by Laidò et al. (2009) identified two major loci (qPC1.1 and qPC6.1) for protein content in chromosomes 1H and 6H explaining 29.1% and 18.7% PVE, respectively. In another study, 3 major QTLs for soluble protein was reported on chromosomes 3H, 5H, and 7H with 27.31%, 33.6%, and 29% PVE, respectively.

3.3.1.1 Dissection of Genome-Wide Genomic Regions Associated with Biofortification Traits

The potential of the genome-wide association studies/mapping (GWAS) to underpinning traits of interest in diverse genetic background germplasm with utmost resolution makes this approach suitable for associating nutrient content variation with genetic variants in crops. Recently performed GWAS study using 378 brown rice accessions for Fe, Zn, Se, Cd, and Pb mineral elements reported 20 QTLs with the variation in concentration (Huang et al. 2015). However, QTL identified on chromosomes 5, 7, and 11 has significant potential concerning rice breeding. Similarly, another study using USDA mini core collection identified 37 genomic regions associated with the amount of minerals like Zn, Fe, Mn, Mg, and K (Nawaz et al. 2015).

In wheat, Gorafi et al. (2018) reported 6 QTLs for Zn and Fe concentration using 47 synthetic hexaploid germplasm. Another study reported 24 candidate genes, linked with the biosynthesis of carotenoid using 233 tetraploid wheat accessions (Colasuonno et al. 2017). In the case of maize, GWAS analysis using 923 lines for kernel Fe and Zn concentration reported 20 SNPs linked with Zn content in the kernel and 26 marker-trait association for Fe concentration. These results are useful for maize mineral biofortification, and in the future, it may enable the cloning of identified genomic regions in the background of target traits (Hindu et al. 2018). Other studies, by using the 380 lines identified 7 significant marker-trait associations (MTAs) on chromosomes 1, 2, 8, and 10 explaining up to 16% PVE for β-carotene (*βC*) (Suwarno et al. 2015). Association mapping studies in sorghum detected loci for mineral elements, polyphenol, protein, fat, and starch content (Shakoor et al. 2016; Rhodes et al. 2017). De Alencar Figueiredo et al. (2010) reported an MTA for genes associated with the starch synthesis pathway (*Sh2, Bt2, SssI, Ae1*, and *Wx*). Another study using a diverse set of 265 lines detected a significant MTA on chromosomes 2 and 4 for protein content in grain (Rhodes et al. 2017). In the case of pearl millet, Anuradha et al. (2017) detected 3 SSR markers; Xpsmp2261 with 13.34% PVE, Xipes 0180 with 11.40% PVE, and Xipes 0096 with 11.38% PVE were found to be associated with Fe and Zn concentration of grain across three environments.

3.3.2 Molecular Breeding Approaches for Biofortification

The advancement of genomic tools and their continuous usage in crop breeding has very fruitful results in the past two decades. Several MAS-derived varieties have been released. Nowadays, a decrease in sequencing cost and advancement in efficient phenotyping platforms has paved the way to use the genomic selection approach in breeding for biofortification traits.

3.3.2.1 Marker-Assisted Selection (MAS)

QTL mapping and introgression of a gene through MAS offer options to develop new higher quality cultivars with potential yields (Varshney et al. 2012). In India, Vivek Hybrid-9 has been improved for quality protein maize (QPM) and released by the name Vivek QPM 9 (Gupta et al. 2013). Improved version Vivek QPM-9 has the same yield as the original hybrid; besides, it has 41% higher tryptophan and 30% higher lysine than its original hybrid, respectively. In India, it was the first commercially available MAS-based maize variety. Recently, using MAS, three commercial hybrids HM-4, HM-8, HM-9 have been developed in QPM version (Hossain et al. 2018). These versions have significantly improved lysine content in the endosperm (48–74%) and tryptophan (55–100%). These three hybrids of the QPM "Pusa HM-4 Improved," "Pusa HM-8 Improved," and "Pusa HM-9 Improved" for commercial cultivation are now available (Yadava et al. 2017). In another study, Zunjare et al. (2018) have combined two *crtRB1* and *lcyE* genes to the genetic background of QPM hybrids. i.e., HQPM-1, HQPM-4, HQPM-5, and HQPM-7, and a substantial increase in proA (9.0–12.9 ppm) has been recorded. Further, Das et al. (2018) proA-enriched hybrids QPM parental lines were also targeted by introgression of *VTE4* favorable allele for the improvement of vitamin E using MAS. Several studies have been reported for the enhancement of Fe and Zn concentration in QPM genetic background (Pandey et al. 2015; Mallikarjuna et al. 2015). Such high-yielding multinutrient QPM hybrids could be effectively used in maize biofortification all over the world and deliver a holistic promise of nutrition security.

In the case of rice, MAS approach has been widely employed to improve the quality traits in rice (Luo et al. 2014, 2016). For instance, II-32B (a maintainer line used in hybrid breeding) was used as the recurrent parent, and Yixiang B was used as donor parent for a quality trait (which contained three quality genes *Wx*, *SSIIa,* and *fgr* involved in controlling amylose content, gelatinization temperature, and fragrance). The functional markers used in this study are Wx-(CT) 17 (SSR marker) for amylose content, SSIIa-TT (SNP marker) for gelatinization temperature, and fgr-E7 for aroma. The improved version of II-32B variety was developed by two backcrosses and three selfings. On the other hand, the traditional approach will require six to eight backcrosses to achieve the same results (Boopathi 2012). Therefore, by reducing breeding time and cycle, MAS can speed up breeding and help in developing biofortified rice varieties. So far, MAS has achieved several positive results in terms of rice grain quality improvement (Yi et al. 2009; Wang et al. 2010; Ni et al. 2011; Lau et al. 2017).

Similarly, through the MABC approach in wheat, a QTL region for enhancing GZnC (grain zinc concentration) was introgressed from PBW 343 to Kenya Swara. In wheat, rich genetic diversity for Fe and Zn is present in CWR and landraces which provide novel combinations of alleles for genetic improvement of such traits in wheat. In the past, several novel alleles were introduced from elite germplasm resulting in the development of biofortified wheat varieties such as "Zinc Shakti (Chitra)," WB-02, HPBW 01 (PBW 1 Zn), Zincol-2016, and BARI-Gom 33 (Singh and Velu 2017). These have been evolved from the crosses between elite lines and *T. dicoccum* derived synthetic hexaploids or *T. spelta* accessions which showed nearly 40% more Zn concentration in grain over local checks (Velu et al. 2015). These studies revealed that the novel alleles from wild relatives/elite germplasm contribute significantly to improve the genetic enhancement of traits in wheat. At CIMMYT, a biofortification breeding program using MAS has shown that the early generation selection of targeted crosses aids the identification and introgression of several novel alleles for biofortification traits.

In a similar study in pearl millet through marker-assisted breeding, strategy-validated loci associated with Fe and Zn content on L.G. 3 along with downy mildew resistance QTLs have been moved into the genetic background of hybrid HHB 67 Improved (pollen parent). These two QTLs' introgression lines (ILs) have been crossed with the HHB 67 (seed parent) to develop HHB 67 improved hybrids (Kumar et al. 2016). In the national testing system of India, these QTL ILs were tested together with improved test-cross hybrids.

3.3.2.2 Genomic Selection for Biofortification Traits in Cereals

MAS strategy is worthwhile only for those traits which are regulated by a few major loci. Genomic selection (GS) has been considered to be more effective than MAS for complex quantitative traits which are regulated by small effect/minor QTLs. To avoid bias or loss of information, GS uses a massive number of markers with whole-genome coverage, ensuring that all the genes will be in linkage disequilibrium (LD) with at least some markers to generate genotypic data for all the individuals, while in MAS, the individual is chosen from a population based on markers linked to a trait of interest (Spindel et al. 2015). GS makes possible the prediction of target biofortification traits at the early stages of breeding, even before multilocations or several years of evaluation by field trials. It is also able to predict the phenotypic performance of individuals without comprehensive phenotyping evaluation and thus help in achieving genetic gain to improve biofortification traits (Battenfield et al. 2016). GS builds a prediction model based on selected training population's phenotypic and marker genotypic data (Meuwissen et al. 2001). The developed prediction model is then implemented in a genotyped test population for estimating the genomic-estimated breeding values (GEBVs) for each line/individual in the test population (Bhat et al. 2016). Based on GEBVs, promising individuals are chosen as parents from the test population without phenotypic evaluation or progeny screening for future crossing or next-generation breeding programs (Meuwissen et al. 2001). For instance, in the case of rice, micronutrient concentration is low and further decreased after milling (Mayer et al. 2008). Consumers who only depend on rice for micronutrients mostly suffered from malnutrition. A study conducted by Lu et al. (2008) identified 10 loci for Mn, Ca, Zn, Fe, and Cu concentration in grain, and those were detected on L.G. 1, 2, 4, 5, 7, 9, and 11. In another study, Garcia-Oliveira et al. (2009) reported 26 QTLs on chromosomes 1, 9, and 12 for Zn, Fe, Mn, Cu, Ca, Mg, P, and K traits, constituting 45% PVE. Similarly, four QTLs were detected for Zn and Fe found on different L.G. (Norton et al. 2010). These studies conclude that the QTLs for biofortification-related traits are distributed throughout the genome and introgression of a favorable trait with several minor QTLs via MAS is not possible. In this case, GS has the potential to improve the nutrition quality of rice grain, especially those mineral elements which are regulated by minor QTL effect. GS approach provides aid for the selection of potential individuals in the breeding population by the ease and which results in achieving higher genetic gain by reducing the selection period for the improvement of crops.

Battenfield et al. (2016) applied the GS approach for the enhancement of biofortification traits in wheat using 5520 lines. In this study, phenotyping data of 5220 lines from the year 2010 to 2015 and 3075 SNPs have been utilized to compute the parameters of GS models. As their study progressed, the prediction precision of their GS model was increased over time ranging from 0.32 for grain hardness to 0.62 for mixing time. Similarly, 330 diverse genetic background line panels were used to estimate genome-wide prediction for Zn and Fe concentration in grain (Velu et al. 2016b). While the accuracy levels differed between locations, moderate to high prediction precision (up to 0.69 for Zn and 0.73 for Fe) emphasized GS importance in the improvement of Fe and Zn content in wheat. Another study by Manickavelu et al. (2017) identified moderate to high prediction accuracies for Mg, K, P, Mn, Fe, and Zn content in wheat. In maize, a prediction precision of up to 0.71 with an average of 0.43 for carotenoid traits of grain was detected (Owens et al. 2014). Manickavelu et al. (2017) advocate the combination of GS with speed breeding which enhances remote breeding for the improvement of nutrient density in crops based on whole-genome predictions.

3.4 Status of Biofortification in Cereal Crops

3.4.1 Iron- and Zinc-Biofortified Rice

Fe and Zn are crucial mineral elements required for the human body. The International Rice Research Institute and CIAT along with NARES developed Zn rice. In polished rice, the breeding target of 28 ppm Zn reported an increase of 12 ppm beyond the baseline. In India and Bangladesh, three varieties with Zn level of up to 75% have been released, and other varieties are under testing evaluation in nine more countries (Saltzman et al. 2016). Several QTL studies were undertaken aimed at the detection of high Zn and Fe controlling genomic region, while two meta-QTLs for Zn and three for Fe content in grain were detected. Therefore, selected five rice backcross ILs of recurrent parent Bio-226/Sampada showed higher Zn ($>$21 μg/g) and Fe ($>$13 μg/g) grain content (Dixit et al. 2019). In Bangladesh "BRRI dhan 62" was released as the first high-level Zn content variety with a target Zn level up to 92%.

3.4.2 Zinc-Biofortified Wheat

Breeding for Zn-enriched wheat in India and Pakistan, led by CIMMYT along with national partners. In India, 4 varieties with a 40–80% Zn level were released (Saltzman et al. 2016). The newly developed varieties are 20–40% superior in the Zn content level (+8 to +12 ppm additional level) in grain (Velu et al. 2015). Recently, the Punjab Agricultural University has developed the first high-zinc-rich variety (PBW1Zn) and released it for cultivation in North Western Plain Zone. Another, high Zn and Fe (WB2) variety was developed by the Indian Institute of Wheat and Barley Research, India. The next varieties' wave is in pipeline, combining high Zn target levels with different maturity types. The advancement of potential breeding lines is under evaluation across multiple environments both in Asia and Africa.

3.4.3 Provitamin A Orange Maize

CIMMYT and IITA in collaboration with NARES led the orange maize breeding. The released varieties are both open-pollinated and hybrids with 40–50% of the 15 ppm target level of provitamin A. Recent advances in genomic technologies open a new area for the discovery of QTL and important genes associated with provitamin A carotenoid which resulted in acceleration of genetic gain for increasing the content of provitamin A in maize (Babu et al. 2013). In Zambia, participatory farmers' meeting confirmed the strong preference for orange maize varieties (Chibwe et al. 2013). *Lycopene epsilon cyclase (lcyE)* and *beta-carotene hydroxylase 1 (crtRB1)* are identified as the most important alleles for enhancing provitamin A (Suwarno et al. 2014). Till now, in Africa, approximately 40 provitamin A maize single-cross hybrids, three-way hybrids, and synthetics have been released. The Zn concentration in maize grain ranged from 17 to 42 ppm (Pixley et al. 2013a, b), while maize hybrids have been identified with a target increase of more than 80% (+12 ppm additional Zn level) in breeding programs. Often high-content Zn lines also have a high level of protein content (known as Quality Protein Maize). Further research efforts are ongoing to develop provitamin A maize with improved stability of carotenoid (Ortiz et al. 2016).

3.4.4 Iron-Enriched Pearl Millet

Pearl millet breeding research is led by the Indian Institute of Millets Research, ICRISAT along with ICAR's All India Coordinated Pearl Millet Improvement Project and the private sector. The breeding target of 77 ppm Fe in pearl millet, one hybrid and one open-pollinated variety with about 85% target level, has been released in India. Private sector partners commercialized the hybrid varieties with a 65–85% target level. ICRISAT has developed an improved version of the popular variety ICTP 8203 and is officially released for Maharashtra (India). In 2014, ICTP 8203-Fe has been released and recommended under the name of "Dhanashakti" for cultivation with more than 80% of Fe target goal (Rai et al. 2014). Recently, this variety participated in "Nutrifarm" a pilot project of the Indian government to overcome Fe deficiency in the country (Singh and Uddeen 2016). The first hybrid ICMH-1201 has developed at ICRISAT for high Fe and yield and has been tested for 3 consecutive years under 48 field trials. This hybrid has an additional Fe content +28 ppm (>90% of target level) and grain yield of 38% higher than ICTP 8203. This hybrid was released under its brand name Shakti-1201 (Govindaraj et al. 2016). In Western and Central Africa, more than 90% of the region is under pearl millet cultivation. In Niger and Sudan, studies on landraces and other locally adapted pearl millet material have shown variation in mineral density (Bashir et al. 2014). Two open-pollinated varieties (GB 8735 and ICTP 8203) were chosen as rapid tracking candidates in Senegal, Niger, and Ghana.

3.4.5 Zinc- and Iron-Rich Sorghum

In 2013, multilocation testing and on-farm adaptation trials were carried out for Zn- and Fe-rich sorghum hybrids. ICRISAT screened and analyzed approximately 2200 lines to detect Zn concentration ranging from 13 to 47 ppm and Fe concentration of 20–70 ppm (Kumar et al. 2009). Potential donor parents and hybrids have been identified after initial screening. Guinea landrace was detected as a source for enhancing micronutrient levels in sorghum (Kumar et al. 2015). At ICRISAT, breeders developed biofortified sorghums with a Zn and Fe grain concentration of up to 50–60% higher than the popular Indian sorghum varieties (30 ppm for Fe level and 20 ppm for Zn level). Now they are being evaluated under multilocation trials in the AICSIP for their release. A list of the released varieties taken from the recent studies in major cereal crops for various biofortification traits is presented in Table 3.2.

3.5 Conclusion and Future Prospects

Biofortification of cereals is a primary concern for researchers, industries, and policymakers. The only solution is for reducing malnutrition issues in Africa and Asia by developing nutrient-rich quality cereals. Global program partners, such as CGIAR's HarvestPlus and national initiatives, led the foundation for achieving these targets. While the HarvestPlus consortium has done great efforts, it is still required to assess the success of biofortification programs with defined priorities. Improving micronutrient levels in cereals requires a clear understanding of complex traits, bioavailability process, uptake, transport, and sequestration mechanism of target trait. Mutual efforts of plant breeders, biotechnologists, molecular biologists, and nutritionists are required to achieve target levels of nutrient-rich cereals. The lack of natural variation in source germplasm selected for quality enhancement limits the conventional breeding approach. In this scenario, application of biotechnology such as genetic engineering, genome editing, etc., complement breeding approach using single overexpression of gene cassette and conferred multiple nutrient traits' enhancement. In order to concurrently increase the level of several micronutrients, it's also important to consider the stability of quality up to the long term. Advances in genomic technologies have been significantly complementing crop improvement for enhancing the nutritional quality status of cereals. Various approaches such as QTL mapping, GWAS, and genomic selection have led to the identification of novel genes/alleles which are useful for targeting important quality traits and selection of potential individuals from a diverse population. These strategies provide the benefit of reducing several years of phenotyping laborious and time-consuming work. Besides, omics approaches also help in understanding the gene network which is involved in the enhancement of quality traits involved in different pathways. Usage of these high-throughput technologies in breeding for biofortification in cereals may enhance the productivity and quality level with essential nutrients. We expect that in the future, these multidisciplinary efforts using advanced technologies will help the biofortification breeding efficiently for the development of nutrient-rich cereal crops to solve the emergent secret of hidden hunger.

Table 3.2 Cereal crops varieties/hybrids released through breeding approach for biofortification

Cereal	Target trait	Status	Variety/country	Reference/source
Maize				
Orange Maize	Vitamin A	Released	Ghana: CSIR-CRI Honampa (OPV)	CIMMYT, International Institute of Tropical Agriculture (IITA), HarvestPlus
			Nigeria:Sammaz 39 (OPV), Ife maizehyb-3, Ife maizehyb-4, Sammaz 38 (OPV),	
			Zambia: GV662A, GV664A, GV665A	
Quality protein maize	Lysine and Tryptophan	Released	India: CML176, CML176 × CML186, HQPM-1, HQPM4, HQPM-5, HQPM-7, VivekQPM-9, FQH-4567	Surinder Vasal and Evangelina Villegas, CIMMYT
			China: CML140, CML194, P70	
			South Africa: QS-7705	
			Ghana: GH-132-28	
			Brazil: BR-451, BR-473	
			Peru: INIA	
			Colombia: ICA	
			Honduras: HQ-31	
			El Salvador: HQ-61	
			Guatemala: HB-Proticta	
Wheat				
	Zinc	Released	India: PBW1Zn	Punjab Agricultural University, India
	Zinc and iron	Released	India: WB2	Indian Institute of Wheat and Barley Research, India
		Released	Pakistan: NR 419, 42, 421, Zincol	CIAT, CIMMYT, HarvestPlus
	Carotene	Released	India: HI 8627	IARI
	Anthocyanins (colored wheat)	Released	China: Black-grained wheat	Havrlentova et al. (2014)
		Released	Austria: Indigo	Havrlentova et al. (2014)
		Registered/ Research	India: NABIMG-9, NABIMG-10, NABIMG-11	Garg et al. (2016)

(continued)

Table 3.2 (continued)

Cereal	Target trait	Status	Variety/country	Reference/source
Rice				
	Zinc and iron	Released	Bangladesh: BRRIdhan 62, BRRIdhan 72, BRRIdhan 64	CIAT, HarvestPlus
	Iron	Research Traditional Variety	India, Philippines: IR68144-3B-2-2-3 (improved line) Jalmagna	IRRI, Gregorio et al. (2000)
	Zinc	Research Traditional Variety	Jalmagna	Gregorio et al. (2000)
Sorghum	Iron	Released	India: ICSR 14001, ICSH 14002, **Hybrids:** ICSA 661 × ICSR 196, ICSA 318 × ICSR 94, ICSA 336 × IS 3760	ICRISAT, HarvestPlus
	Iron	Released	Nigeria: 12KNICSV (Deko)-188 12KNICSV-22 (Zabuwa)	ICRISAT, HarvestPlus
Pearl Millet	Zinc and iron		India: Dhanashakti, **Hybrid**: ICMH 1201 (Shakti-1201)	ICRISAT, HarvestPlus

References

Anuradha K, Agarwal S, Rao YV (2012) Mapping QTLs and candidate genes for iron and zinc concentrations in unpolished rice of Madhukar × Swarna RILs. Gene 508:233–240

Anuradha N, Satyavathi CT, Bharadwaj C, Nepolean T, Sankar SM, Singh SP, Meena MC, Singhal T, Srivastava RK (2017) Deciphering genomic regions for high grain iron and zinc content using association mapping in pearl millet. Front Plant Sci 8:412

Babu R, Rojas R, Gao S, Yan J, Pixley K (2013) Validation of the effects of molecular marker polymorphisms in LcyE and CrtRB1 on provitamin A concentrations for 26 tropical maize populations. Theor Appl Genet 126(2):389–399

Banakar R, Alvarez Fernandez A, Díaz-Benito P, Abadia J, Capell T, Christou P (2017) Phytosiderophores determine thresholds for iron and zinc accumulation in biofortified rice endosperm while inhibiting the accumulation of cadmium. J Exp Bot 68(17):4983–4995

Bashir EMA, Ali AM, Ismail MI, Parzies HK, Haussmann BIG (2014) Patterns of pearl millet genotype-by-environment interaction for yield performance and grain iron (Fe) and zinc (Zn) concentrations in Sudan. Field Crops Res 166:82–91

Battenfield SD, Guzmán C, Gaynor RC, Singh RP, Peña RJ, Dreisigacker S, Fritz AK, Poland JA (2016) Genomic selection for processing and end-use quality traits in the CIMMYT spring bread wheat breeding program. Plant Genome 9(2):1–12

Bean SR, Wilson JD, Moreau RA, Galant A, Awika JW, Kaufman RC, Adrianos SL, Ioerger BP (2016) Structure and composition of the sorghum grain. In: Ciampitti I, Prasad V (eds) Sorghum: state of the art and future perspectives. American Society of Agronomy and Crop Science Society of America, Madison

Belalcazar J, Dufour D, Andersson MS, Pizarro M, Luna J, Londoño L, Ceballos H (2016) High-throughput phenotyping and improvements in breeding cassava for increased carotenoids in the roots. Crop Sci 56(6):2916–2925

Beyer P, Al-Babili S, Ye X, Lucca P, Schaub P, Welsch R, Potrykus I (2002) Golden rice: introducing the β-carotene biosynthesis pathway into rice endosperm by genetic engineering to defeat vitamin A deficiency. J Nutr 132(3):506S–510S

Bhat JA, Ali S, Salgotra RK, Mir ZA, Dutta S, Jadon V, Tyagi A, Mushtaq M, Jain N, Singh PK, Singh GP (2016) Genomic selection in the era of next generation sequencing for complex traits in plant breeding. Front Genet 7:221

Boopathi NM (2012) Genetic mapping and marker assisted selection: basics, practice and benefits. Springer, New Delhi

Boyer CD, Liu KC (1983) Starch and water-soluble polysaccharides from sugary endosperm of sorghum. Phytochemistry 22:2513–2515

Boyles RE (2017) Genetic dissection of sorghum grain quality traits using diverse and segregating populations. Theor Appl Genet 130(697):716

Buckner B, Kelson TL, Robertson DS (1990) Cloning of the y1 locus of maize, a gene involved in the biosynthesis of carotenoids. Plant Cell 2(9):867–876

Cakmak I, Pfeiffer WH, McClafferty B (2010) Biofortification of durum wheat with zinc and iron. Cereal Chem 87:10–20

Calayugan MIC, Formantes AK, Amparado A, Descalsota-Empleo GI, Nha CT, Inabangan-Asilo MA, Swe ZM, Hernandez JE, Borromeo TH, Lalusin AG, Mendioro MS (2020) Genetic analysis of agronomic traits and grain iron and zinc concentrations in a doubled haploid population of rice (*Oryza sativa* L.). Sci Rep 10(1):1–14

Chibwe E, De Groote H, Gitonga Z, Birol E (2013) A participatory evaluation of vitamin A biofortified maize in Zambia. HarvestPlus, Washington, DC. Unpublished

Colasuonno P, Lozito ML, Marcotuli I, Nigro D, Giancaspro A, Mangini G, De Vita P, Mastrangelo AM, Pecchioni N, Houston K, Simeone R, Gadaleta A, Blanco A (2017) The carotenoid biosynthetic and catabolic genes in wheat and their association with yellow pigments. BMC Genomics 18:122

Cook JP (2012) Genetic architecture of maize kernel composition in the nested association mapping and inbred association panels. Plant Physiol 158:824–834

Crespo-Herrera LA, Velu G, Singh RP (2016) QTL mapping reveals pleiotropic effect for grain iron and zinc concentrations in wheat. Ann Appl Biol 169(1):27–35

Crespo-Herrera LA, Velu G, Singh RP, Hao Y, Stangoulis J (2017) QTL mapping of grain Zn and Fe concentrations in two hexaploid wheat RIL populations with ample transgressive segregation in wheat. Front Plant Sci 8:1800

Cu ST, March TJ, Stewart S, Degner S, Coventry S, Box A, Stewart D, Skadhauge B, Burton RA, Fincher GB, Eglinton J (2016) Genetic analysis of grain and malt quality in an elite barley population. Mol Breed 36:129

Das AK, Hossain F, Muthusamy V, Zunjare RU, Chauhan HS, Baveja A, Jaiswal SK, Bhata JS, Guleria SK, Saha S, Singh AK et al (2018) Genetic analyses of kernel tocopherols in maize possessing novel allele of *γ-tocopherol methyl transferase* (*ZmVTE4*). In: International symposium on biodiversity and biobanking—biodiverse 2018, 27–29 Jan 2018, IIT Guwahati, Guwahati, pp 112–113

De Alencar Figueiredo LF, Sine B, Chantereau J, Mestres C, Fliedel G, Rami JF, Glaszmann JC, Deu M, Courtois B (2010) Variability of grain quality in sorghum: association with polymorphism in Sh2, Bt2, SssI, Ae1, Wx and O2. Theor Appl Genet 121(6):1171–1185

Descalsota-Empleo GI, Amparado A, Inabangan-Asilo MA, Tesoro F, Stangoulis J, Reinke R, Swamy BM (2019) Genetic mapping of QTL for agronomic traits and grain mineral elements in rice. Crop J 7(4):560–572

Distelfeld A, Cakmak I, Peleg Z, Ozturk L, Yazici AM, Budak H, Saranga Y, Fahima T (2007) Multiple QTL effects of wheat Gpc-B1 locus on grain protein and micronutrient concentrations. Physiol Plant 129:635–643

Dixit S, Singh UM, Abbai R, Ram T, Singh VK, Paul A, Virk PS, Kumar A (2019) Identification of genomic region(s) responsible for high iron and zinc content in rice. Sci Rep 9(1):1–8

Elía M, Swanston JS, Moralejo M, Casas A, Pérez-Vendrell AM, Ciudad FJ et al (2010) A model of the genetic differences in malting quality between European and north American barley cultivars based on a QTL study of the cross Triumph × Morex. Plant Breed 129:280–290

Fahlgren N, Gehan MA, Baxter I (2015) Lights, camera, action: high-throughput plant phenotyping is ready for a close-up. Curr Opin Plant Biol 24:93–99

FAOSTAT (2013) Food and agricultural organization of the United Nations. http://www.fao.org

Fenton ME, Owens BF, Lipka AE, Ortiz D, Tiede T, Mateos-Hernandez M, Ferruzzi M, Rocheford T (2018) High-density linkage mapping of vitamin E content in maize grain. Mol Breed 38:31

Francis D, Finer JJ, Grotewold E (2017) Challenges and opportunities for improving food quality and nutrition through plant biotechnology. Curr Opin Biotechnol 44:124–129

Garcia-Oliveira AL, Tan L, Fu Y, Sun C (2009) Genetic identification of quantitative trait loci for contents of mineral nutrients in rice grain. J Integr Plant Biol 51:84–92

Garg M, Chawla M, Chunduri V, Kumar R, Sharma S, Sharma NK et al (2016) Transfer of grain colors to elite wheat cultivars and their characterization. J Cereal Sci 71:138–144

Genc Y, Verbyla AP, Torun AA, Cakmak I, Willsmore K, Wallwork H, McDonald GK (2009) Quantitative trait loci analysis of zinc efficiency and grain zinc concentration in wheat using whole genome average interval mapping. Plant Soil 314:49–66

Gorafi YSA, Ishii T, Kim JS, Elbashir AAE, Tsujimoto H (2018) Genetic variation and association mapping of grain iron and zinc contents in synthetic hexaploid wheat germplasm. Plant Genet Resour 16(1):9–17

Govindaraj M, Rai KN, Shanmugasundaram P (2016) Intra-population genetic variance for grain iron and zinc contents and agronomic traits in pearl millet. Crop J 4(1):48–54

Gregorio GB, Senadhira D, Htut H, Graham RD (2000) Breeding for trace mineral density in rice. Food Nutr Bull 21:382–386

Gu R, Chen F, Liu B, Wang X, Liu J, Li P, Pan Q, Pace J, Soomro AA, Lübberstedt T, Mi G, Yuan L (2015) Comprehensive phenotypic analysis and quantitative trait locus identification for grain mineral concentration, content, and yield in maize (*Zea mays* L.). Theor Appl Genet 128:1777–1789

Gupta HS, Raman B, Agrawal PK, Mahajan V, Hossain F, Thirunavukkarasu N (2013) Accelerated development of quality protein maize hybrid through marker-assisted introgression of opaque-2 allele. Plant Breed 13:277–282

Gupta HS, Hossain F, Muthusamy V (2015) Biofortification of maize: an Indian perspective. Indian J Genet 75(1):1–22

Harjes CE, Rocheford TR, Bai L, Brutnell TP, Kandianis CB, Sowinski SG et al (2008) Natural genetic variation in lycopene epsilon cyclase tapped for maize biofortification. Science 319(5861):330–333

Havrlentova M, Psenakova I, Zofajova A, Ruckschloss L, Kraic J (2014) Anthocyanins in wheat seed—a mini review. Nova Biotechnol Chim 13(1):1–12

Hindu V, Palacios-Rojas N, Babu R, Suwarno WB, Rashid Z, Usha R, Saykhedkar GR, Nair SK (2018) Identification and validation of genomic regions influencing kernel zinc and iron in maize. Theor Appl Genet 131(7):1443–1457

Hossain F, Muthusamy V, Pandey N, Vishwakarma AK, Baveja A, Zunjare RU, Gupta HS (2018) Marker-assisted introgression of opaque2 allele for rapid conversion of elite hybrids into quality protein maize. J Genet 97(1):287

Hossain F, Muthusamy V, Zunjare RU, Gupta HS (2019) Biofortification of maize for protein quality and provitamin—a content. In: Nutritional quality improvement in plants. Springer, Cham, pp 115–136

Huang XQ (2006) Molecular detection of QTLs for agronomic and quality traits in a doubled haploid population derived from two Canadian wheats (*Triticum aestivum* L.). Theor Appl Genet 113(4):753–766

Huang Y, Sun C, Min J, Chen Y, Tong C, Bao J (2015) Association mapping of quantitative trait loci for mineral element contents in whole grain rice (*Oryza sativa* L.). J Agric Food Chem 63: 10885–10892

Igartua E, Edney M, Rossnagel BG, Spaner D, Legge WG, Scoles GJ, Eckstein PE, Penner GA, Tinker NA, Briggs KG, Falk DE (2000) Marker-based selection of QTL affecting grain and malt quality in two-row barley. Crop Sci 40:1426–1433

Ishikawa S, Abe T, Kuramata M, Yamaguchi M, Ando T, Yamamoto T, Yano M (2010) A major quantitative trait locus for increasing cadmium-specific concentration in rice grain is located on the short arm of chromosome 7. J Exp Bot 61(3):923–934

Ishikawa R, Iwata M, Taniko K, Monden G, Miyazaki N, Orn C, Tsujimura Y, Yoshida S, Ma JF, Ishii T (2017a) Detection of quantitative trait loci controlling grain zinc concentration using Australian wild rice, *Oryza meridionalis*, a potential genetic resource for biofortification of rice. PLoS One 12(10):e0187224

Ishikawa R, Iwata M, Taniko K, Monden G, Miyazaki N, Orn C et al (2017b) Detection of quantitative trait loci controlling grain zinc concentration using Australian wild rice, Oryza meridionalis, a potential genetic resource for biofortification of rice. PLoS One 12(10): e0187224

Jeong OY, Lee JH, Jeong EG, Chun A, Bombay M, Banzon Ancheta M, Ahn SN (2020) Analysis of QTL responsible for grain iron and zinc content in doubled haploid lines of rice (*Oryza sativa*) derived from an intra-japonica cross. Plant Breed 139(2):344–355

Jittham O, Fu X, Xu J, Chander S, Li J, Yang X (2017) Genetic dissection of carotenoids in maize kernels using high-density single nucleotide polymorphism markers in a recombinant inbred line population. Crop J 5(1):63–72

Kiranmayi SL, Manorama K (2014) Identification of markers associated with iron and zinc concentration in recombinant inbred lines of brown rice. Indian J Genet Plant Breed 74(4): 423–429

Kochevenko A, Jiang Y, Seiler C, Surdonja K, Kollers S, Reif JC, Korzun V, Graner A (2018) Identification of QTL hot spots for malting quality in two elite breeding lines with distinct tolerance to abiotic stress. BMC Plant Biol 18:106

Kotla A, Phuke R, Hariprasanna K, Mehtre SP, Rathore A, Radhika K, Hash CT, Reddy BVS, Patil JV, Jabeen F, Shashikanth D, Pagde LS, Deshpande SP, Kumar A (2016) Identification of QTLs and underlying candidate genes controlling grain Fe and Zn concentration in sorghum [Sorghum bicolor (L). Moench]. In: Plant and animal genome conference XXIV, January 08-13, 2016, San Diego, CA

Krishnappa G, Singh AM, Chaudhary S, Ahlawat AK, Singh SK, Shukla RB, Jaiswal JP, Singh GP, Solanki IS (2017) Molecular mapping of the grain iron and zinc concentration, protein content and thousand kernel weight in wheat (*Triticum aestivum* L.). PLoS One 12(4):e0174972

Kumar AA, Reddy BV, Ramaiah B, Reddy PS, Sahrawat KL, Upadhyaya HD (2009) Genetic variability and plant character association of grain Fe and Zn in selected core collection accessions of sorghum germplasm and breeding lines. J SAT Agric Res 7:1–4

Kumar AA, Anuradha K, Ramaiah B, Grando S, Frederick H, Rattunde W, Virk P, Pfeiffer WH (2015) Recent advances in sorghum biofortification research. Plant Breed Rev 39:89–124

Kumar R, Sharma V, Suresh S, Ramrao DP, Veershetty A, Kumar S, Priscilla K, Hangargi B, Narasanna R, Pandey MK, Naik GR (2021) Understanding omics driven plant improvement and de novo crop domestication: some examples. Front Genet 12:637141

Kumar S, Hash CT, Thirunavukkarasu N, Singh G, Rajaram V, Rathore A, Senapathy S, Mahendrakar MD, Yadav RS, Srivastava RK (2016) Mapping quantitative trait loci controlling high iron and zinc content in self and open pollinated grains of pearl millet [*Pennisetum glaucum* (L.) R. Br.]. Front Plant Sci 7:1636

Kumar S, Hash CT, Nepolean T, Mahendrakar MD, Satyavathi CT, Singh G, Rathore A, Yadav RS, Gupta R, Srivastava RK (2018) Mapping grain iron and zinc content quantitative trait loci in an iniadi-derived immortal population of pearl millet. Genes 9(5):248

Laidò G, Barabaschi D, Tondelli A, Gianinetti A, Stanca AM, Li Destri Nicosia O, Di Fonzo N, Francia E, Pecchioni N (2009) QTL alleles from a winter feed type can improve malting quality in barley. Plant Breed 128:598–605

Lau WC, Rafii MY, Ismail MR, Puteh A, Latif MA, Asfaliza R, Miah G (2017) Development of advanced fragrant rice lines from MR269 × basmati 370 through marker-assisted backcrossing. Euphytica 213:11

Li J, Xiao J, Grandillo S, Jiang L, Wan Y, Deng Q, Yuan L, McCouch SR (2004) QTL detection for rice grain quality traits using an interspecific backcross population derived from cultivated Asian (*O. sativa* L.) and African (*O. glaberrima* S.) rice. Genome 47(4):697–704

Li JZ, Huang XQ, Heinrichs F, Ganal MW, Röder MS (2005) Analysis of QTLs for yield, yield components, and malting quality in a BC_3-DH population of spring barley. Theor Appl Genet 110:356–363

Li C, Cakir M, Lance R (2010) Genetic improvement of malting quality through conventional breeding and marker-assisted selection. In: Zhang G, Li C (eds) Genetics and improvement of barley malt quality. Springer, Berlin, pp 260–292

Lichtenwalner RE, Ellis EB, Rooney LW (1978) Effect of incremental dosages of the waxy gene of sorghum on digestibility. J Anim Sci 46:1113–1119

Liu BH (1997) Statistical genomics: linkage, mapping, and QTL analysis. CRC Press, Boca Raton

Liu C, Chen G, Li Y, Peng Y, Zhang A, Hong K, Jiang H, Ruan B, Zhang B, Yang S, Gao Z (2017) Characterization of a major QTL for manganese accumulation in rice grain. Sci Rep 7(1):17704

Liu C, Ding S, Zhang A, Hong K, Jiang H, Yang S, Ruan B, Zhang B, Dong G, Guo L, Zeng D (2020) Development of nutritious rice with high zinc/selenium and low cadmium in grains through QTL pyramiding. J Integr Plant Biol 62(3):349–359

Lu K, Li L, Zheng X, Zhang Z, Mou T, Hu Z (2008) Quantitative trait loci controlling Cu, Ca, Zn, Mn and Fe content in rice grains. J Genet 87:305–310

Luo Y, Zakaria S, Basyah B, Ma T, Li Z, Yang J, Yin Z (2014) Marker-assisted breeding of Indonesia local rice variety Siputeh for semi-dwarf phenotype, good grain quality and disease resistance to bacterial blight. Rice 7(1):33

Luo Y, Ma T, Zhang A, Ong KH, Li Z, Yang J, Yin Z (2016) Marker-assisted breeding of the rice restorer line Wanhui 6725 for disease resistance, submergence tolerance and aromatic fragrance. Rice 9:66

Mahender A, Anandan A, Pradhan SK, Pandit E (2016) Rice grain nutritional traits and their enhancement using relevant genes and QTLs through advanced approaches. Springerplus 5(1): 2086

Mallikarjuna MG, Thirunavukkarasu N, Hossain F, Bhat JS, Jha SK, Rathore A, Agrawal PK, Pattanayak A, Reddy SS, Gularia SK, Singh AM, Manjaiah KM, Gupta HS et al (2015) Stability performance of inductively coupled plasma mass spectrometry-phenotyped kernel minerals concentration and grain yield in maize in different agro-climatic zones. PLoS One 10(9): e0139067

Manickavelu A, Hattori T, Yamaoka S, Yoshimura K, Kondou Y, Onogi A, Matsui M, Iwata H, Ban T (2017) Genetic nature of elemental contents in wheat grains and its genomic prediction: toward the effective use of wheat landraces from Afghanistan. PLoS One 12:e0169416

Masojć P, Milczarski P (2009) Relationship between QTLs for preharvest sprouting and alpha-amylase activity in rye grain. Mol Breed 23(1):75–84

Masuda H, Ishimaru Y, Aung MS, Kobayashi T, Kakei Y, Takahashi M, Higuchi K, Nakanishi H, Nishizawa NK (2012) Iron biofortification in rice by the introduction of multiple genes involved in iron nutrition. Sci Rep 2:543

Mayer JE, Pfeiffer WH, Beyer P (2008) Biofortified crops to alleviate micronutrient malnutrition. Curr Opin Plant Biol 11:166–170

McCartney CA, Somers DJ, Lukow O, Ames N, Noll J, Cloutier S, Humphreys DG, McCallum BD (2006) QTL analysis of quality traits in the spring wheat cross RL4452 × 'AC Domain'. Plant Breed 125:565–575

Mertz ET, Bates LS, Nelson OE (1964) Mutant gene that changes the protein composition and increases the lysine content of maize endosperm. Science 145:279–280

Meuwissen THE, Hayes BJ, Goddard ME (2001) Prediction of total genetic value using genome-wide dense marker maps. Genetics 157:1819–1829

Miedaner T, Hübner M, Korzun V, Schmiedchen B, Bauer E, Haseneyer G, Wilde P, Reif JC (2012) Genetic architecture of complex agronomic traits examined in two testcross populations of rye (*Secale cereale* L.). BMC Genomics 13(1):706

Milczarski P, Masojć P, Krajewski P, Stochmal A, Kowalczyk M, Angelov M, Ivanova V, Schollenberger M, Wakuliński W, Banaszak Z, Banaszak K (2017) QTL mapping for benzoxazinoid content, preharvest sprouting, α-amylase activity, and leaf rust resistance in rye (*Secale cereale* L.). PLoS One 12(12):e0189912

Mishra A, Singh PK, Bhandawat A, Sharma V, Sharma V, Singh P, Roy J, Sharma H (2022) Analysis of SSR and SNP markers. InBioinformatics, Academic Press, pp 131–144

Mohammadi M, Blake TK, Budde AD, Chao S, Hayes PM, Horsley RD, Obert DE, Ullrich SE, Smith KP (2015) A genome-wide association study of malting quality across eight US barley breeding programs. Theor Appl Genet 128:705–721

Muthusamy V, Hossain F, Thirunavukkarasu N, Choudhary M, Saha S, Bhat JS, Prasanna BM, Gupta HS (2014) Development of β-carotene rich maize hybrids through marker-assisted introgression of β-carotene hydroxylase allele. PLoS One 9(12):e113583

Myśków B, Stojałowski S, Łań A, Bolibok-Brągoszewska H, Rakoczy-Trojanowska M, Kilian A (2012) Detection of the quantitative trait loci for α-amylase activity on a high-density genetic map of rye and comparison of their localization to loci controlling preharvest sprouting and earliness. Mol Breed 30(1):367–376

Nawaz Z, Kakar KU, Li XB, Li S, Zhang B, Shou HX, Shu QY (2015) Genome-wide association mapping of quantitative trait loci (QTLs) for contents of eight elements in brown rice (*Oryza sativa* L.). J Agric Food Chem 63:8008–8016

Neelamraju S, Mallikarjuna Swamy BP, Kaladhar K, Anuradha K, Venkateshwar Rao Y, Batchu AK, Agarwal S, Babu AP, Sudhakar T, Sreenu K, Longvah T, Surekha K, Rao KV et al (2012) Increasing iron and zinc in rice grains using deep water rices and wild species—identifying genomic segments and candidate genes. Qual Assur Safety Crops Foods 4:138

Neeraja CN, Babu VR, Ram S, Hossain F, Hariprasanna K, Rajpurohit BS, Prabhakar TL, Prasad KS, Sandhu JS, Datta SK (2017) Biofortification in cereals: progress and prospects. Curr Sci 113(6):1050–1057

Ni D, Zhang S, Chen S, Xu Y, Li L, Li H, Wang Z, Cai X, Li Z, Yang J (2011) Improving cooking and eating quality of Xieyou57, an elite indica hybrid rice, by marker-assisted selection of the *Wx* locus. Euphytica 179:355–362

Norton GJ, Deacon CM, Xiong L, Huang S, Meharg AA, Price AH (2010) Genetic mapping of the rice ionome in leaves and grain: identification of QTLs for 17 elements including arsenic, cadmium, iron and selenium. Plant Soil 329:139–153

Ortiz D, Rocheford T, Ferruzzi MG (2016) Influence of temperature and humidity on the stability of carotenoids in biofortified maize (*Zea mays* L.) genotypes during controlled post-harvest storage. J Agric Food Chem 64(13):2727–2736

Owens BF, Lipka AE, Magallanes-Lundback M, Tiede T, Diepenbrock CH, Kandianis CB, Kim E, Cepela J, Mateos-Hernandez M, Buell CR, Buckler ES (2014) A foundation for provitamin A biofortification of maize: genome-wide association and genomic prediction models of carotenoid levels. Genetics 198(4):1699–1716

Palmgren MG, Clemens S, Williams LE, Krämer U, Borg S, Schjørring JK, Sanders D (2008) Zinc biofortification of cereals: problems and solutions. Trends Plant Sci 13(9):464–473

Pandey N, Hossain F, Kumar K, Vishwakarma AK, Nepolean T, Vignesh M, Manjaiah KM, Agrawal PK, Guleria SK, Reddy SS, Gupta HS et al (2015) Microsatellite marker-based genetic diversity among quality protein maize (QPM) inbred lines differing for kernel iron and zinc. Mol Plant Breed 6:1–10

Peleg Z, Cakmak I, Ozturk L, Yazici A, Jun Y, Budak H, Korol AB, Fahima T, Saranga Y (2009) Quantitative trait loci conferring grain mineral nutrient concentrations in durum wheat × wild emmer wheat RIL population. Theor Appl Genet 119:353–369

Pfeiffer WH, McClafferty B (2007) HarvestPlus: breeding crops for better nutrition. Crop Sci 47: S-88

Pixley K, Palacios N, Babu R, Mutale R, Surles R, Simpungwe E (2013a) Biofortification of maize with provitamin A carotenoids. In: Tanumihardjo SA (ed) Carotenoids and human health. Humana Press, Totowa, pp 271–292

Pixley K, Rojas NP, Babu R, Mutale R, Surles R, Simpungwe E (2013b) Biofortification of maize with provitamin A carotenoids. In: Carotenoids and human health. Humana Press, Totowa, pp 271–292

Prakash K, Kumar R, Prakash NR, Singh S, Singh L (2017) Pre-breeding in the omics era: a review in vegetable crops. Chem Sci Rev Lett 6(22):752–762

Pu ZE, Yu M, He QY, Chen GY, Wang JR, Liu YX et al (2014) Quantitative trait loci associated with micronutrient concentrations in two recombinant inbred wheat lines. J Integr Agric 13: 2322–2329

Qin H, Cai Y, Liu Z, Wang G, Wang J, Guo Y, Wang H (2012) Identification of QTL for zinc and iron concentration in maize kernel and cob. Euphytica 187:345–358

Rae SJ, Macaulay M, Ramsay L, Leigh F, Matthews D, O'Sullivan DM, Donini P, Morris PC, Powell W, Marshall DF, Waugh R (2007) Molecular barley breeding. Euphytica 158:295–303

Rai KN, Patil HT, Yadav OP, Govindaraj M, Khairwal IS, Cherian B, Rajpurohit BS, Rao AS, Shivade H, Kulkarni MP (2014) Notification of crop varieties and registration of germplasm: pearl millet variety 'Dhanashakti'. Indian J Genet Plant Breed 74(3):405–406

Rhodes DH, Hoffmann L Jr, Rooney WL, Herald TJ, Bean S, Boyles R, Brenton ZW, Kresovich S (2017) Genetic architecture of kernel composition in global sorghum germplasm. BMC Genomics 18:15

Rooney LW, Pflugfelder RL (1986) Factors affecting starch digestibility with special emphasis on sorghum and corn. J Anim Sci 63:1607–1623

Saltzman A, Birol E, Bouis HE, Boy E, De Moura FF, Islam Y, Pfeiffer WH (2013) Biofortification: progress toward a more nourishing future. Glob Food Secur 2(1):9–17

Saltzman A, Andersson MS, Asare-Marfo D, Lividini K, De Moura FF, Moursi M, Oparinde A, Taleon V (2016) Biofortification techniques to improve food security

Shahzad Z, Rouached H, Rakha A (2014) Combating mineral malnutrition through iron and zinc biofortification of cereals. CRFSFS 13(3):329–346

Shakoor N, Ziegler G, Dilkes BP, Brenton Z, Boyles R, Connolly EL (2016) Integration of experiments across diverse environments identifies the genetic determinants of variation in Sorghum bicolor seed element composition. Plant Physiol 170:1989–1998

Sharma V, Verma RK, Dey PC, Chetia SK, Baruah AR, Modi MK (2017) QTLs associated with yield attributing traits under drought stress in upland rice cultivar of Assam. Oryza 54:253–257

Sharma V, Saini DK, Kumar A, Kesh H, Kaushik P (2020) Breeding for biofortification traits in rice: means to eradicate hidden hunger. In: Agronomy. IntechOpen, London

Shutu X, Dalong Z, Ye C, Yi Z, Shah T, Ali F, Qing L, Zhigang L, Weidong W, Jiansheng L, Xiaohong Y (2012) Dissecting tocopherols content in maize (*Zea mays* L.), using two segregating populations and high-density single nucleotide polymorphism markers. BMC Plant Biol 12(201):1–14

Šimić D, Mladenović Drinić S, Zdunić Z, Jambrović A, Ledenčan T, Brkić J, Brkić A, Brkić I (2012) Quantitative trait loci for biofortification traits in maize grain. J Hered 103(1):47–54

Singh R, Axtell JD (1973) High lysine mutant gene (hl that improves protein quality and biological value of grain sorghum 1. Crop Sci 13(5):535–559

Singh SK, Uddeen R (2016) Nutri-farms for mitigating malnutrition in India. In: Biofortification of food crops. Springer, New Delhi, pp 461–477

Singh RP, Velu G (2017) Zinc-biofortified wheat: harnessing genetic diversity for improved nutritional quality. Science brief: biofortification no. 1–4, Crop Trust online portal

Singh S, Tripathi RK, Lemaux PG, Buchanan BB, Singh J (2017) Redox-dependent interaction between thaumatin-like protein and b-glucan influences malting quality of barley. Proc Natl Acad Sci U S A 114:7725–7730

Sofi PA, Wani SA, Rather AG, Wani SH (2009) Quality protein maize (QPM): genetic manipulation for the nutritional fortification of maize. IJPBCS 1(6):244–253

Sperotto RA, Ricachenevsky FK, de Abreu WV, Fett JP (2012) Iron biofortification in rice: it's a long way to the top. Plant Sci 190:24–39

Spindel J, Begum H, Akdemir D et al (2015) Genomic selection and association mapping in rice (*Oryza sativa*): effect of trait genetic architecture, training population composition, marker number and statistical model on accuracy of rice genomic selection in elite, tropical rice breeding lines. PLoS Genet 11:e1004982

Stangoulis JC, Huynh BL, Welch RM, Choi EY, Graham RD (2007) Quantitative trait loci for phytate in rice grain and their relationship with grain micronutrient content. Euphytica 154:289–294

Sukumaran S, Xiang W, Bean SR, Pedersen JF, Kresovich S, Tuinstra MR (2012) Association mapping for grain quality in a diverse sorghum collection. Plant Genome J 5:126–135

Suwarno WB, Pixley KV, Palacios-Rojas N, Kaeppler SM, Babu R (2014) Formation of heterotic groups and understanding genetic effects in a provitamin A biofortified maize breeding program. Crop Sci 54(1):14–24

Suwarno WB, Pixley KV, Palacios-Rojas N, Kaeppler SM, Babu R (2015) Genome-wide association analysis reveals new targets for carotenoid biofortification in maize. Theor Appl Genet 128: 851–864

Swamy BP, Rahman MA, Inabangan-Asilo MA, Amparado A, Manito C, Chadha-Mohanty P, Reinke R, Slamet-Loedin IH (2016) Advances in breeding for high grain zinc in rice. Rice 9:49

Swamy BPM, Descalsota GIL, Nha CT, Amparado A, Inabangan-Asilo MA, Manito C, Tesoro F, Reinke R (2018a) Identification of genomic regions associated with agronomic and biofortification traits in DH populations of rice. PLoS One 13(8):1–20

Swamy BPM, Kaladhar K, Anuradha K, Batchu AK, Longvah T, Sarla N (2018b) QTL analysis for grain iron and zinc concentrations in two *O. nivara* derived backcross populations. Rice Sci 25(4):197–207

Terasawa Y, Ito M, Tabiki T, Nagasawa K, Hatta K, Nishio Z (2016) Mapping of a major QTL associated with protein content on chromosome 2B in hard red winter wheat (*Triticum aestivum* L.). Breed Sci 66(4):471–480

Tiwari VK, Rawat N, Chhuneja P, Neelam K, Aggarwal R, Randhawa GS, Dhaliwal HS, Keller B, Singh K (2009) Mapping of quantitative trait loci for grain iron and zinc concentration in diploid a genome wheat. J Hered 100(6):771–776

Tiwari C, Wallwork H, Arun B, Mishra VK, Velu G, Stangoulis J, Kumar U, Joshi AK (2016) Molecular mapping of quantitative trait loci for zinc, iron and protein content in the grains of hexaploid wheat. Euphytica 207:563–570

Trijatmiko KR, Duenas C, Tsakirpaloglou N, Torrizo L, Arines FM, Adeva C, Balindong J, Oliva N, Sapasap MV, Borrero J, Rey J, Francisco P, Nelson A, Nakanishi H, Lombi E, Tako E et al (2016) Biofortified indica rice attains iron and zinc nutrition dietary targets in the field. Sci Rep 6:13

Tripathi A, Choudhary JR, Prakash NR (2018) Bio fortification-breeding for nutritional security. J Pharmacogn Phytochem 7(3):1266–1270

Uauy C, Distelfeld A, Fahima T, Blechl A, Dubcovsky J (2006) A NAC gene regulating senescence improves grain protein, zinc, and iron content in wheat. Science 314:1298–1301

Varshney RK, Ribaut JM, Buckler ES, Tuberosa R, Rafalski JA, Langridge P (2012) Can genomics boost productivity of orphan crops? Nat Biotechnol 30(12):1172–1176

Velu G, Singh RP (2019) Genomic approaches for biofortification of grain zinc and iron in wheat. In: Quality breeding in field crops. Springer, Cham, pp 193–198

Velu G, Ortiz-Monasterio I, Cakmak I, Hao Y, Singh RP (2014) Biofortification strategies to increase grain zinc and iron concentrations in wheat. J Cereal Sci 59:365–372

Velu G, Singh RG, Balasubramaniam A, Mishra VK, Chand R, Tiwari C, Joshi AK, Virk P, Cherian B, Pfeiffer WH (2015) Reaching out to farmers with high zinc wheat varieties through public-private partnerships: an experience from eastern-Gangetic plains of India. Adv Food Technol Nutr Sci 1(3):73–75

Velu G, Tutus Y, Gomez-Becerra HF, Hao Y, Demir L, Kara R et al (2016a) QTL mapping for grain zinc and iron concentrations and zinc efficiency in a tetraploid and hexaploid wheat mapping populations. Plant Soil 411:81–99

Velu G, Crossa J, Singh RP, Hao Y, Dreisigacker S, Perez-Rodriguez P, Joshi AK, Chatrath R, Gupta V, Balasubramaniam A, Tiwari C (2016b) Genomic prediction for grain zinc and iron concentrations in spring wheat. Theor Appl Genet 129(8):1595–1605

Verma RK, Chetia SK, Sharma V, Baishya S, Sharma H, Modi MK (2022) GWAS to spot candidate genes associated with grain quality traits in diverse rice accessions of North East India. Mol Biol Rep 49:5365–5377

von Korff M, Wang H, Léon J, Pillen K (2008) AB-QTL analysis in spring barley: III. Identification of exotic alleles for the improvement of malting quality in spring barley (H. vulgare ssp. spontaneum). Mol Breed 21:81–93

Walker CK, Panozzo JF (2016) Genetic characterization, expression and association of quality traits and grain texture in barley (*Hordeum vulgare* L.). Euphytica 212:1–15

Wang C, Zhang Y, Zhu Z, Chen T, Zhao L, Lin J, Zhou L (2010) Development of a new japonica rice variety Nanjing 46 with good eating quality by marker assisted selection. Rice Genomics Genet 1

Wang J, Yang J, Zhang Q, Zhu J, Jia Q, Hua W, Shang Y, Li C, Zhou M (2015) Mapping a major QTL for malt extract of barley from a cross between Tx9425 × Naso Nijo. Theor Appl Genet 128:943–952

Wang P, Wang H, Liu Q, Tian X, Shi Y, Zhang X (2017) QTL mapping of selenium content using a RIL population in wheat. PLoS One 12(9):e0184351

Wong JC, Lambert RJ, Tadmor Y, Rocheford TR (2003) QTL associated with accumulation of tocopherols in maize. Crop Sci 43(6):2257–2266

Xu Q, Zheng TQ, Hu X, Cheng LR, Xu JL, Shi YM, Li ZK (2015) Examining two sets of introgression lines in rice (*Oryza sativa* L.) reveals favorable alleles that improve grain Zn and Fe concentrations. PLoS One 10:e0131846

Xu Y, Zhang XQ, Harasymow S, Westcott S, Zhang W, Li C (2018) Molecular marker-assisted backcrossing breeding: an example to transfer a thermostable b-amylase gene from wild barley. Mol Breed 38:63

Yadava DK, Choudhury PR, Hossain F, Kumar D (2017) Biofortified varieties: sustainable way to alleviate malnutrition. Indian Council of Agricultural Research, New Delhi

Yadava DK, Hossain F, Mohapatra T (2018) Nutritional security through crop biofortification in India: status & future prospects. Indian J Med Res 148(5):621

Yang G, Li Y, Wang Q, Zhou Y, Zhou Q, Shen B, Zhang F, Liang X (2012) Detection and integration of quantitative trait loci for grain yield components and oil content in two connected recombinant inbred line populations of high-oil maize. Mol Breed 29(2):313–333

Yang GH, Dong YB, Li YL, Wang QL, Shi QL, Zhou Q (2014) QTL verification of grain protein content and its correlation with oil content by using connected RIL populations of high-oil maize. Genet Mol Res 13:881–894

Yang Z, Li X, Zhang N, Zhang YN, Jiang HW, Gao J, Kuai BK, Ding YL, Huang XQ (2016) Detection of quantitative trait loci for kernel oil and protein concentration in a B73 and Zheng58 maize cross. Genet Mol Res 15(3):10–4238

Yi M, Nwe KT, Vanavichit A, Chai-arree W, Toojinda T (2009) Marker assisted backcross breeding to improve cooking quality traits in Myanmar rice cultivar Manawthukha. Field Crops Res 113(2):178–186

Zaw H, Raghavan C, Pocsedio A, Swamy BM, Jubay ML, Singh RK, Bonifacio J, Mauleon R, Hernandez JE, Mendioro MS, Gregorio GB (2019) Exploring genetic architecture of grain yield and quality traits in a 16-way indica by japonica rice MAGIC global population. Sci Rep 9(1): 1–11

Zhang J, Lu XQ, Song XF, Yan JB, Song TM, Dai JR, Rocheford T, Li JS (2008) Mapping quantitative trait loci for oil, starch, and protein concentrations in grain with high-oil maize by SSR markers. Euphytica 162(3):335–344

Zhang X, Zhang G, Guo L, Wang H, Zeng D, Dong G, Qian Q, Xue D (2011) Identification of quantitative trait loci for Cd and Zn concentrations of brown rice grown in Cd-polluted soils. Euphytica 180:173–179

Zhang M, Pinson SR, Tarpley L, Huang XY, Lahner B, Yakubova E, Baxter I, Guerinot ML, Salt DE (2014) Mapping and validation of quantitative trait loci associated with concentrations of 16 elements in unmilled rice grain. Theor Appl Genet 127:137–165

Zhang H, Liu J, Jin T, Huang Y, Chen J, Zhu L, Zhao Y, Guo J (2017) Identification of quantitative trait locus and prediction of candidate genes for grain mineral concentration in maize across multiple environments. Euphytica 213(4):90

Zhou G, Panozzo J, Zhang XQ, Cakir M, Harasymow S, Li C (2016) QTL mapping reveals genetic architectures of malting quality between Australian and Canadian malting barley (*Hordeum vulgare* L.). Mol Breed 36:70

Zunjare RU, Hossain F, Muthusamy V, Baveja A, Chauhan HS, Bhat JS, Thirunavukkarasu N, Saha S, Gupta HS (2018) Development of biofortified maize hybrids through marker-assisted stacking of β-carotene hydroxylase, lycopene-ε-cyclase and Opaque2 genes. Front Plant Sci 9: 178

Genome-Editing Approaches for Biofortification of Cereal Crops

4

Shaila Kadam, Dongho Lee, and Pallavi Dhiman

Abstract

Hidden hunger is an overshadowed form of hunger resulting from inadequate micronutrients (Fe, Zn, I, Mg, vitamin A, etc.) supply and especially occurred in countries where cereal crops are the major source of nutrition. To ensure guaranteed food security and combat hidden hunger, recent technological advancement techniques such as clustered regularly interspaced short palindromic repeats/CRISPR-associated protein (CRISPR/Cas) look promising. Over the past decades, CRISPR/Cas has transformed the genome-editing field. The exploration of this tool is a possible solution to increase the intrinsic nutritional quality of crops, i.e., through biofortification by targeting key genes. This genome-editing tool confers a precise range of gene editing such as altering a few nucleotides to deleting or inserting a gene segment in a targeted genome. In this chapter, we have highlighted the impacts of hidden hunger and possible solutions to alleviate this problem. This chapter mainly emphasizes on increasing the micronutrient content (Fe, and provitamin A, low phytate) by targeting key genes for each trait through genome-editing tools in cereal crops. We also have addressed recent advancements in CRISPR-mediated precise gene editing, and base editing, multiplex genome-editing approaches like polycistronic tRNA processing, Csy4 endoribonuclease, and Drosha-based multiplex genome editing has been deliberated. The challenges to implement genome editing in polyploid crops and the future prospectus of CRISPR/Cas approaches in cereal crops for

Shaila Kadam, Dongho Lee and Pallavi Dhiman contributed equally.

S. Kadam (✉) · D. Lee
University of Missouri, Columbia, MO, USA

P. Dhiman
National Agri-Food Biotechnology Institute (NABI), Mohali, India

R. Deshmukh et al. (eds.), *Biofortification in Cereals*,
https://doi.org/10.1007/978-981-19-4308-9_4

biofortification have been discussed. This chapter provides a comprehensive review suggesting the exploration of these genome-editing tools in crop improvement.

Keywords

Hidden hunger · Micronutrients · ZFNs · TALENs · CRISPR · Biofortification · Multiplex genome editing

4.1 Introduction

Plant-derived food makes up a major part of the human diet and ensures the nutritional health and well-being of humans. There are at least 17 nutrients that humans require including Fe, Zn, Cu, I, Se, and vitamin A in varying amounts that are essential and integral to human health (Linder 1991; O'Dell and Sunde 1997; Grusak and DellaPenna 1999). Not all plant foods contain all the essential nutrients needed, and not a single serving can measure up all the dietary needs (Grusak and DellaPenna 1999). A well-balanced and ideal diet should include grains, fruits, vegetables, and animal-food sources for daily nutrient needs (Dwyer 1994). Cereal crops make up a large proportion of the daily food intake; in consequence, mineral intake mainly in the developing countries is less than the daily requirement (Welch and Graham 2002). Cereals are edible grains belonging to the Gramineae family grown worldwide. According to Food and Agriculture Organization (FAO), cereal crops, including wheat, rice, rye, millet, sorghum, barley, oats, buckwheat, triticale, quinoa, and fonio, are harvested for dry grains. Cereals are major sources of nutrition in developed as well as developing countries as they hold a good amount of carbohydrates, proteins, lipids, and vitamins. According to FAO's report in March 2020, world cereal production elevated almost 62 million tons (2.3%) since 2018 making it to 2719 million tons in 2019 which is higher than the hike from 2017 to 2018 that was 1.77% (48 million tons) (Table 4.1).

However, these crop-based diets (rice, wheat, maize, etc.) unfortunately fall far short in providing all essential micro and macronutrients. Current estimates reveal that around 2 billion people (33% of the world's population) can suffer from iron deficiency. Similarly, 1.5 billion (26% of world's population) are at a risk of iodine deficiency (Gilani and Nasim 2007; Ahmed et al. 2012) (Table 4.2). The risk of zinc deficiency prevails among about 30% of the world's population (NairK and Choudhury 2013). Iron deficiency anemia is globally affecting one-fourth of women and children which causes impairment of cognitive skills, physical activity, perinatal mortality, mild mental retardation, and maternal mortality (Stoltzfus and Dreyfuss 1998; Stoltzfus et al. 2004). Zinc deficiency causes cognitive impairment, retarded growth, and immune system impairment (Gibson 2006; Cakmak 2009) (Fig. 4.1).

Table 4.1 World cereal production (in million tons)

Continents	2017	2018	2019	% Change (2018 over 2017)	% Change (2019 over 2018)
Asia	1201.20	1198.50	1214.40	−0.2	1.3
Europe	523.7	497.4	540.1	−5	8.6
Central America and Caribbean	44.1	42.4	41	−3.8	−3.3
South America	215.9	197.4	227.6	−8.5	15.3
North America	494.6	495.2	480.6	0.1	−2.9
Africa	191.4	196.7	187.6	2.7	−4.6
Oceanic	34.6	29.9	28	13.5	−6.5
Total	*2705.5*	*2657.5*	*2719.4*	*1.8*	*2.3*

Source: FAO Cereal Supply and Demand Brief, 5 March 2020

Table 4.2 Micronutrients function, deficiency, and affected population

Micronutrient	Functions	Deficiency effects	Deficiency prevalence
Iron	Oxygen carrier, component of hemoglobin and myoglobin and various enzymes, role in myelin formation in brain	Anemia, low energy, risk of maternal mortality, premature births.	~1.62 billion
Zinc	Key component of enzymes and proteins, regulates gene expression, coding for zinc finger proteins, supports synaptic vesicles, neurotransmitter	Retard growth, reduced appetite, hair loss, diarrhea, impaired immune system, impotency	~1.2 billion
Iodine	Component of thyroid hormones, regulates cell activity, growth, essential for embryonic and postnatal development	Goiter, hypothyroidism, impaired neurological development of the fetus, cretinism	~1.8 billion
Vitamin A	Antioxidant, vision, immune system	Visual impairment, night blindness, impaired body defense system, risk of infectious diseases like diarrhea and measles	~200 million

Sources: (Maberly et al. 1994, Biesalski et al. 2016, Zimmermann 2016, Camaschella 2019, and Von Grebmer et al. 2014

Anti-nutritional contents like phytic acid (in grains), tannin, and lignins reduce the bioavailability of micronutrients (Gupta et al. 2015). Phytate content in cereals is concentrated in the aleurone layer (Lásztity and Lásztity 1990). Phytate has the ability to bind with essential minerals including iron, calcium, and zinc causing decreased absorption of these minerals. Tannins can bind to proteins causing their precipitation, decreasing their digestibility (Rawat et al. 2013).

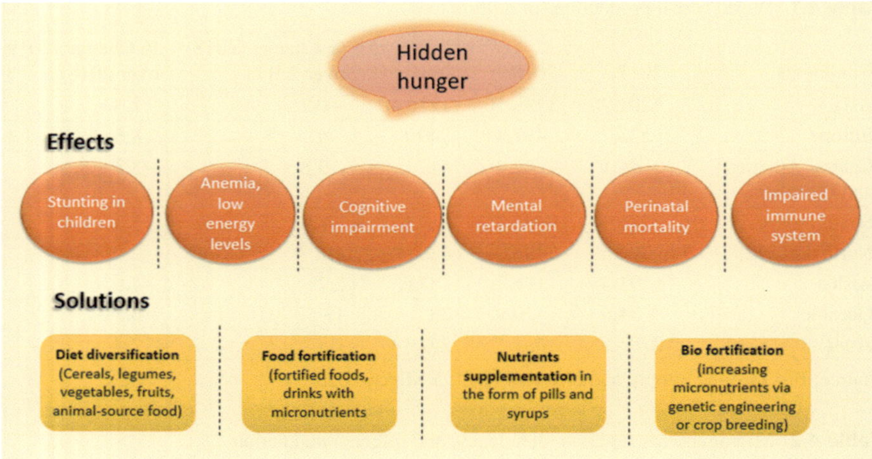

Fig. 4.1 Effects of hidden hunger and the different solutions

Unlike vitamins that are synthesized in plants, mineral nutrients are obtained from the soil. Hence, pathways of mineral uptake in plants must be understood such as micronutrients localization, transport proteins, xylem transport system, etc. The phloem pathway primarily confers nutrient supply in the grains. Hence, phloem loading system can be considered for the improvement of nutrient content in plants (Grusak and DellaPenna 1999).

Low uptake of micronutrients from the soil limits plants' growth and productivity which in turn contributes to malnutrition in the human population (Sanchez and Swaminathan 2005; Nubé and Voortman 2011). Genetic modifications in crop plants such as overexpression of transporters and overexpression of the metal chelator nicotinamide synthase are evident to elevate root uptake for the nutrients. In addition, overexpression of the metal storage protein in grains and reduction of anti-nutritional components of grains have been proffered to improve the grain nutritional quality (Singer et al. 2019). Biofortification via modern biotechnology techniques will provide invigorating cereal crops and combat the nutrient deficiency in the human population (Bhullar and Gruissem 2013; Mottiar 2013) which will thereby improve food affluence and standard of living. Therefore, strategies including conventional breeding methods and genome-editing approaches are required to facilitate the dietetic value of crop plants to ensure adequate dietary supply.

Conventional breeding methods receive less acceptance as:

- Some unwanted traits like low yield and poor agronomic performance can be introduced by breeding.
- Micronutrients enhanced by the conventional breeding method may be lost during processing such as polishing, milling, cooking, etc.
- Market prices of fortified crops are not satisfactory for the farmers.

Among several approaches to solve hidden hunger (Fig. 4.1), the genome-editing approach is the most promising strategy for improving nutrient contents in cereal crops.

4.1.1 Global Issue of Hidden Hunger

Definition: "Micronutrient deficiency (also referred to as hidden hunger) is outlined as a sort of undernutrition caused by low intake and the absorption of vitamins and minerals that ensures proper health and development of children and well-being of adults" (von Grebmer et al. 2014). Though its consequences don't seem to be directly visible at the incipient stage, over the long term these nutrient deficiencies cause serious diseases.

Even though in populations where food supply is adequate, around 2 billion people suffer from micronutrient deficiency globally because the food consumed is insufficient in providing all required nutrients (WHO 2006). Based on a recent report in FAO, out of about 840 million people where the diet did not serve enough energy and nutrition, around 799 million were accounted from developing countries (Kennedy et al. 2003; Amoroso 2016). This is known as "hidden hunger" or micronutrient deficiency which is a crucial public health condition among poor populations and emergent nations (Gödecke et al. 2018). Around seven million children aged <5 years die annually as a consequence of malnutrition. Lack of necessary micronutrients including zinc, iron, iodine, and vitamin A in body is characterized by various health problems especially among women and children (Hunger 2012; Von Grebmer et al. 2014). Iron and zinc deficiencies are typical micronutrient deficiencies (Bouis 2003; Cakmak 2008). Deficiency of vitamin A incriminates to night blindness, corneal diseases, and even deaths of one million children annually (West Jr 2002; Ortiz-Monasterio et al. 2007) (Table 4.2). Food diversification, supplementation, fortification, and biofortification are possible approaches to address these micronutrient deficiencies. Supplementation is an approach where nutrients are given in a form of syrups or pills (Kennedy et al. 2003). The procedure of adding micronutrients to products during food processing, which includes iodized salt, vitamin D-rich milk, etc., is referred to as fortification (Kennedy et al. 2003; Dary and Hurrell 2006). Biofortification refers to altering crops biologically through breeding or genetic engineering. Biofortification ensures nutrient supply without changing the traditional diet patterns of people (Murgia et al. 2012).

4.1.1.1 Iron Deficiency

Iron deficiency being a widespread single nutrient deficiency can be related to severe pathologies (Stoltzfus and Dreyfuss 1998). Out of around two billion global anemia cases, "iron deficiency anemia" accounted for around 50% of cases mostly among infants and pregnant women (Oski 1993; Killip et al. 2007). Iron is an integral part of many redox reactions, activation and decomposition of peroxides, and cofactor in enzymatic pathways (Ilbert and Bonnefoy 2013). About 20% of perinatal and 10%

of maternal mortality is accountable to iron deficiency. The average per day iron requirement in children aged 2–11 years is approximately 13,000–15,000 μg/day and 16,000 μg/day in 12–19 years old children and teenagers. In men and women (>19 years), the daily requirement is 20,000 μg and 18,000 μg, respectively, and 15,000 μg/day in pregnant women (DRI 2005).

4.1.1.2 Zinc Deficiency

Zinc is a crucial microelement for modulating cell differentiation, expression of various genes, and integrity of protein structure and cofactor of many enzymatic reactions (Black 1998; NairK and Choudhury 2013). It is regarded as the second most abundant micronutrient in the human body. Zinc deficiency especially among infants in economically developing nations consequently leads to illness and death from infectious diseases (Black 2003). About one-third of the world's population is affected by zinc deficiency, which is estimated to be 4% to 73% across subregions. The first case of Zn deficiency was reported in the 1960s, and in 2004 the World Health Organization (WHO) and the United Nations International Children's Emergency Fund (UNICEF) included Zn supplements in the treatment of acute diarrhea (Gibson 2006). However, severe zinc deficiency is rare, and mild-to-moderate zinc deficiency is quite common across the world. Severe zinc deficiency results in disorders like short stature, hypogonadism, impaired immune function, skin disorders, cognitive dysfunction, anorexia etc. Zinc deficiency is attributable for approximately 16% of respiratory tract infections and 10% of diarrheal disease (Gibson 2006). The recommended daily Zn intake in infants is 2 mg (0–6 months) and 3 mg (7–12 months); in children, 3 mg (1–3 years) and 5 mg (4–8 years); in adults, 8–12 mg (Shukla et al. 2009), 8–9 mg (women); and in pregnant and lactating women, 12–13 mg n (Hyman and Times).

4.1.1.3 Vitamin a Deficiency

Vitamin A deficiency (VAD) is a major nutritional concern in sub-Saharan Africa which is assessed by biochemical and clinical indications. Vitamin A is crucial for visual system functioning, cell growth, and the immune system. Xerophthalmia caused by VAD is the main cause of blindness among children. VAD prevails among 21.1% (254 million) preschool-age children and 5.6% of pregnant women (Rice et al. 2004). Vitamin deficiency is determined by the level of serum retinol (<70 μmol/l) in children and women.

4.1.2 The Aim of Biofortification

Biofortification refers to a process of enrichment of vitamin and mineral contents in the edible parts of crops through traditional breeding, agronomic practices, genetic engineering, etc. (Bouis 2018). Biofortification ensures the delivery of micronutrients to populations that lack access to diverse diets. Biofortification aims to bring down the occurrence of diseases and mortality related to inadequate micronutrient supply through staple diet. It facilitates better food productivity as well as security and quality of life, especially among impoverished populations (Wakeel et al. 2018).

4.2 Genome-Editing Advancement

4.2.1 Underlying Concepts of Genome Editing

Genome editing is a concept of using a set of tools and techniques that virtually target any genomic sequences of interest. Genome editing is built on nuclease-based engineering for elucidating a gene function. They are based on the induction of double-stranded breaks (DSBs) in DNA. These DSBs are eventually repaired via the nonhomologous end-joining (NHEJ) or homology-directed repair (HDR), which are cell's inherent mechanisms (Petolino 2015). NHEJ repair mechanism causes frameshift mutations if random deletions and insertions (indels) in the gene occur in the coding region, thus resulting in a gene knockout. NHEJ mechanism is also exploited to introduce desired DNA sequence in a homology-independent manner resulting in gene stacking. The HDR-mediated pathway is more precise than NHEJ for inserting or replacing gene sequences and introducing point mutations (Fig. 4.2). As whole-genome sequences of various crops are available, plant species are easy to transform and regenerate. Advancement in genome-editing techniques has made it accessible to get desired traits. The DSBs in DNA can be induced by three widely exploited enzyme-based systems, namely, zinc finger nucleases (ZFN), transcription activator-like effector nucleases (TALEN), and clustered regularly interspaced short palindromic repeats/CRISPR-associated nucleases (CRISPR/Cas) (Ma and Liu 2015) (Table 4.3). These site-specific nucleases allow reverse genetics, targeted transgene introduction, and genome engineering in an efficient, precise, and optimized way (Bortesi and Fischer 2015). The presence of antinutrients like polyphenols, phytic acid, and tannins are the major restrictions to the biofortification of cereal crops. The tools like ZFNs, TALENs, and CRISPR have been employed to reduce antinutrient as well as to elevate the nutritional content in order to combat hidden hunger and increase food security.

4.2.2 Zinc-Finger Nucleases (ZFNs)

The ZFN advent began in 2002 as the first target-specific engineered nucleases that transfigured the area of genome editing (Vats et al. 2019). ZFN technology has facilitated efficacious, site-specific gene modifications. ZFNs are the first DNA-binding proteins which in combination with nonspecific type IIS restriction enzyme nuclease domain *FokI* and zinc-finger DNA-binding domains introduces a DSB at predetermined sites (Gaj et al. 2013). NHEJ being an error-prone pathway repairs DSBs in DNA leading to insertion and deletions (indels). Structurally ZFN includes two domains: DNA-binding domain consisting of 3–6 individual Cys_2-His_2 zinc-finger repeats that recognizes between 9–18 base pairs and a DNA-cleavage

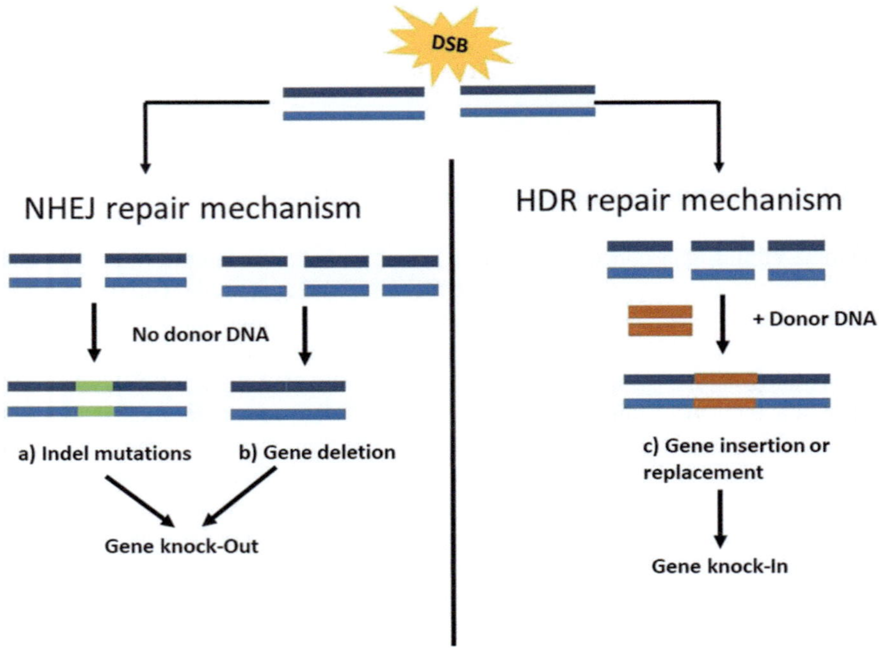

Fig. 4.2 Generalized scheme showing non-homologous end-joining (NHEJ) and homologous recombinant DNA (HRD) repair mechanisms (**a**, **b**) NHEJ repair leading to indel mutations and gene deletion causing gene knock-out, (**c**) HDR repair by donor DNA insertion leading to gene insertion or replacement causing gene knock-in. Modified from Bortesi and Fischer (2015) and Chen et al. (2019)

domain. ZFN monomer pair is required for targeting a genome sequence by attaching in reverse arrangement flanking a 5–6 bp region (Urnov et al. 2010) (Fig. 4.3). In addition to site-specific mutagenesis, ZFN expression can also be used for gene stacking or gene replacement (Weinthal et al. 2010). Numerous studies of genome editing using ZFNs have been reported in many plants including arabidopsis (Lloyd et al. 2005; Zhang et al. 2010), tobacco (Cai et al. 2009; Petolino et al. 2010), maize (Shukla et al. 2009), petunia (Marton et al. 2010), soybean (Curtin et al. 2011), rice (Cantos et al. 2014; Jung et al. 2018), fig, and apple (Peer et al. 2015). A key gene *IPK1* encoding inositol-1,3,4,5,6-pentakisphosphate 2-kinase that catalyzes the final step in phytate biosynthesis in maize seeds has been targeted by ZFN-mediated insertional mutagenesis to bring down the phytate content (Shukla et al. 2009).

Table 4.3 Comparison among ZFNs, TALENs, and CRISPR/Cas9 genome editing and improving tools

Comparison	ZFNs	TALENs	CRISPR/Cas9
Mechanism	FokI introduces DSBs using zinc-finger protein	FokI introduces DSBs using TALEs	Produces DSBs by wtCas9 and single-stranded nicks by Cas9
Module	Protein-based DNA targeting	Protein-based DNA targeting	RNA-based DNA targeting
Mode of action	Double-stranded breaks in target DNA	Double-stranded breaks in target DNA	Double-stranded breaks or single-stranded nicks in target
Catalytic domain	Restriction endonuclease fokI domain	Restriction endonuclease fokI domain	RuvC and HNH
Recognition site	9–18 bp per ZFN monomer, 18–36 bp per ZFN pair	14–20 bp per monomer, 20–40 bp per pair	20 bp guide sequence +2 bp PAM up to 44 for double nicking
Nature	Gene knockout	Gene knockout	Gene knockout/in
Off-target effects	High	Limited	Negligible in plants, high in other systems
Specificity	Only a few positional mismatches are acceptable	Only a few positional mismatches are acceptable	Positional and multiple mismatches are tolerated
Targeting	Targeting non-G-rich sequences is problematic	5′ targeted base must be at T for each TALEN monomer	Target sequence preceeding a PAM
Programmable/ ease of use	Difficult, require more protein engineering	Difficult, require more protein engineering	Easier than TALEN, require more time and effort than RNAi
Structure	Dimer	Dimer	Monomer
Multiplex genome editing	Difficult	Difficult	Possible
Mutation rate	High	Medium	Low

Sources: Khandagale and Nadaf (2016), Abdallah et al. (2015)

4.2.3 Transcription Activator-like Effector Nucleases (TALENs)

Like ZFNs, TALENs are also engineered nucleases having TALE repeats (targets the DNA) and *FokI* restriction enzyme (cleaves the DNA). A single tale repeat targets a single nucleotide base (instead of three as in ZFNs) conferring malleable target design and more number of probable targets (Zhang et al. 2018a, b). Repeat variable di-residues (RVD) are polymorphic amino acids that determine the specificity of TALEN and are present at 12 and 13 positions (Ansari et al. 2020). The DNA-binding domain is acquired from proteins produced by plant bacterial pathogens *Xanthomonas* via type II restriction system. Transcription activator-like (TAL) effectors are delivered to the plant cell during infection which specifically

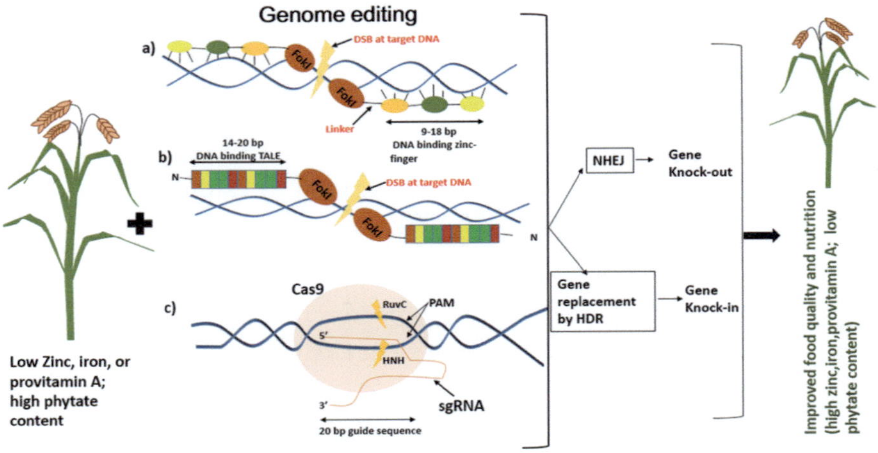

Fig. 4.3 Schematic representation of genome engineering by ZFNs, TALENs, and CRISPR/Cas9 genome engineering tools: a) ZFNs: Zinc fingers fused with FokI endonuclease inducing DSBs in desired DNA. Zinc finger domains recognize one triplet of consecutive nucleotides in DNA, FokI joins to the C-terminal of zinc-finger domains to form zinc-finger nuclease monomer. The two ZFN monomers bind to both the DNA strands positioning their fokI nuclease in such a way to form dimer causing DSBs in the DNA. b) TALENs: just like ZFNs, TALENs induce DSBs following dimerization of their fokI nucleases. The target site is recognized by two monomers of TALE repeats. N-terminal contains nuclear localization signal, and C-terminal is fused with fokI, c) CRISPR/Cas9: Cas9 is guided by sgRNA to the target DNA sequence. Cas9 cleavage domains RuvC and HNH are guided by PAM thus inducing DSBs at the target site. DSBs are repaired by NHEJ or HDR leading to the modifications in DNA sequences (insertion, deletions [indels], substitution) and conclusively in the encoded proteins. Images sources: (Carroll 2011; LaFountaine et al. 2015)

bind to plant gene promoters and upregulates the expression leading to accelerated pathogen virulence (Voytas 2013) (Fig. 4.3). TALEN-mediated disruption of bacterial blight susceptible gene OsSWEET14 was the first report of genome editing via TALENs (Li et al. 2012). TALENs have been employed to alter the nutritional content in plants. The high oleic acid content in soybean by disrupting fatty acid desaturase genes (Haun et al. 2014), purple tomatoes with high anthocyanin by TALEN-mediated gene insertion (Čermák et al. 2015), and fragrant rice by betaine aldehyde dehydrogenase (OsBADH2) gene knockout have been obtained (Shan et al. 2015).

Both ZFNs and TALENs limit their application due to laborious assembly, complicated designs, difficulty in protein synthesis, and the requirement of validation. Although the literature contains many studies in which ZFNs and TALENs are applied to target different traits like disease resistance, herbicide tolerance, minimizing sugars, high oleic acid content, early flowering, increasing anthocyanin content, etc. (reviewed by Zhang et al. 2018a, b). However, we have found only a few studies showing their use in increasing micronutrient profile of crop plants.

4.2.4 Clustered Regularly Interspaced Short Palindromic Repeats/CRISPR-Associated Protein (CRISPR/Cas)

However, TALENs and ZFNs were evidently used for genome editing, but CRISPR/Cas system was utilized more providently in cereal crops due to its design, specificity, multiplexing, cost-effectiveness, short time, and focused targeting (Ansari et al. 2020). CRISPR/Cas9 system consists of a single endonuclease, CRISPR repeats spacer array associated with a Cas9 protein which is an RNA-mediated type II prokaryotic defense system against phages and other invading genetic material (Shan et al. 2013; Chen et al. 2019). CRISPR/Cas9 was established in 2013 for its use in rice, wheat, *Nicotiana benthamiana,* and arabidopsis (Nekrasov et al. 2013; Shan et al. 2013; Li et al. 2016). There are three CRISPR/Cas9 systems present in bacteria that are markedly different from each other in their molecular mechanism. Out of those three, *Streptococcus pyogenes* type II system involving precursor CRISPR RNA (pre-crRNA), trans-activating crRNA (tracrRNA), and Cas9 nuclease was the first system that was specifically used for DNA cleavage in vitro as well as in vivo. Pre-crRNAs is matured into crRNAs which is brought about by tracrRNA (Deltcheva et al. 2011; Gasiunas et al. 2012; Cong et al. 2013). TracrRNA and crRNAs are repurposed as single-guide RNAs (sgRNA) associated with Cas9 nuclease forming Cas9/sgRNA complex. At its 5′ end, sgRNA contains 20 nucleotide sequence complementary to the target DNA adjacent to 5′ -NGG-3′ protospacer-adjacent motifs (PAMs) and a 3′ sequence that binds to Cas9 protein. sgRNA directs the cleavage of cognate DNA by Cas9 (Chen et al. 2019). *Streptococcus pyogenes* Cas9 (SpCas9) is commonly used due to its high efficiency of creating DSB. RuvC and HNH are characteristic nuclease domains of multifunctional protein Cas9 which cleave complementary and noncomplementary DNA strands, respectively (Gasiunas et al. 2012; Vats et al. 2019). Cas9 can be customized into nickase by mutating either RuvC or HNH domains or into dead Cas9 (dCas9) by mutating both domains. The spCas9 PAM region is 5'-NGG-3 (N-any nucleotide, G-guanine); thus, its application is limited in AT-rich regions and susceptible to off-target effects (Xie et al. 2014; Ansari et al. 2020). To deal with this, many Cas9 variants with dissimilar PAM preferences are now used. For example, class II type V nuclease variants from *Prevotella* and *Francisella* Cpf1 (recently called Cas12a) possess the RuvC-like domain naturally and lack the HNH domain. A T-rich PAM sequence is used by Cpf1 for target DNA recognition, conferring a wider range of destined genes (Vats et al. 2019; Ansari et al. 2020).

CRISPR gene editing is not only limited to DSBs in DNA; it can be optimized for precise genome regulation. dCas9 fused with proteins for gene activation (VP64, p65AD, VPR), gene repression (KRAB and SRDX domain), epigenome editing, and base editing (Dominguez et al. 2016). Epigenetic effectors like histone acetyltransferase p300, histone demethylase LSD1, and ten-eleven translocation (TET) proteins results in chromatin alterations resulting in modified gene expression and related biological processes (Chen et al. 2019) (Fig. 4.4).

Base editing is a precise way of editing by CRISPR which creates mutations at specific points without causing DSBs. The base-editing technologies engage fusion

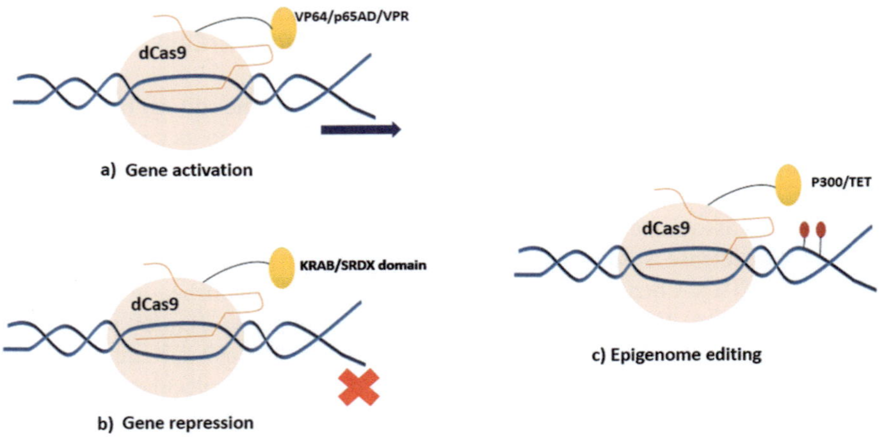

Fig. 4.4 Overview of dCas9-based genome editing. Protein fused dCas9 related to gene activation, repression, and, epigenetic effectors leading to gene expression modification. Image source: Chen et al. (2019)

of enzyme carrying base conversion activity with Cas9 nickase (nCas9) or dead Cas9 (dCas9). The Cas9 base editors (BEs) targeting range is confined by G/C-rich PAM sequences. To deal with this issue, the first base editor (BE3) was developed involving the fusion of rat deaminase "APOBEC1" fused to Cas9 changing cytosine (C) to uracil (U). The U•G mismatches results into U•A and, subsequently, the T•A nucleotide pair during DNA repair and replication (Li et al. 2018). Cytosine base editor (CBE) comprises fusion between a cytidine deaminase and nCas9 (D10A) and a uracil glycosylase inhibitor (UGI) and converts cytosine to uracil thereafter (Fig. 4.5a). BE2 is another more efficient base editor, that uses uracil DNA glycosylase, causing hindrance in the excision repair pathway (Li et al. 2018). Like CBE system that converts C to U, the conversion of A to G in genomic DNA is mediated by adenine base editors (ABEs). A is deaminated to I (Inosine) which is recognized as guanine by polymerases leading to A•T to G•C substitution (Zhang et al. 2018a, b) (Fig. 4.5b).

There are significant studies of crop improvement by CRISPR-mediated gene editing. The increased iron content in wheat through CRISPR-mediated genome editing of TaVIT2 gene was obtained (Connorton et al. 2017). In soybean, carotenoid content was elevated by engineering GmPDS11 and GmPDS18 genes (Du et al. 2016). The phytic acid content in maize was reduced by targeting ZmIPK1A, ZmIPK, and ZmMRP4 genes (Liang et al. 2014). In cassava, *Phytoene desaturase* (MePDS) gene is altered to obtain high carotenoid content (Odipio et al. 2017).

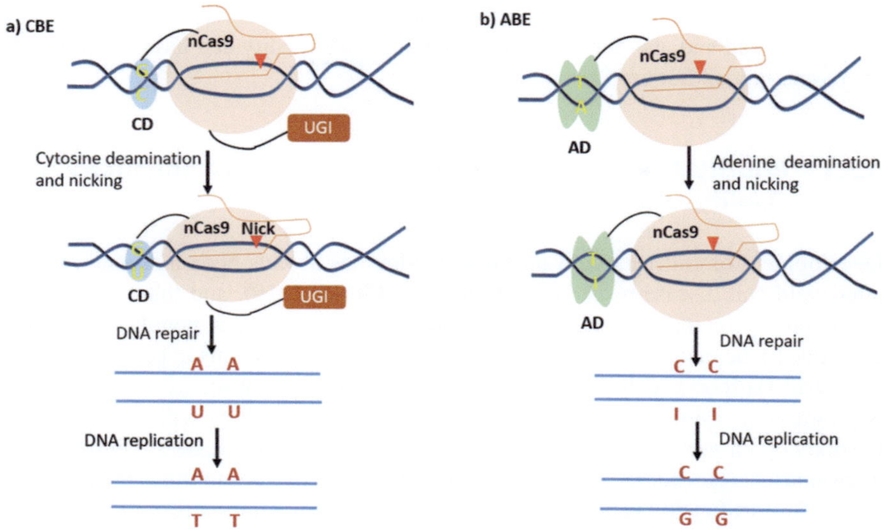

Fig. 4.5 Demonstration of CRISPR-mediated base editing (**a**) CBE system: Cas9 fused with cytosine deaminase (CD) and uracil glycosylase inhibitor (UGI) creating C•G to T•A substitution. (**b**) ABE system: nCas9 fused with adenine deaminases (AD) creating A•T to G•C substitution. Modified from Chen et al. (2019); Zhang et al. (2018a, b); and Li et al. (2018)

4.3 Genome Editing in Plants

Mineral deficiency is causing global worry, especially in developing nations. Though a balanced diet with protein, vegetables, and fruits is beneficial to provide all nutrients needed for physical and mental growth, developing countries still deal with deficiencies in minerals such as Fe, Mg, Zn, I, Se, and Cu due to preferential cereal-based diet (White and Broadley 2005). A total of 22 mineral nutrients are essential for overall growth and development, which can be supplemented with a balanced diet. Many people from Asia, Africa, and Latin America are dependent solely on rice and wheat which are insufficient to provide all essential micronutrients, which results in malnutrition. To overcome the issue of malnutrition, many techniques and tools have emerged including fortifying food crops with mineral nutrients, a transgenic approach, and supplementation of these mineral nutrients to the soil. However, reaching out to fortified foods or supplements is constrained by high prices in economically poor countries (Agarwal et al. 2014).

Biofortification of crops can be achieved through traditional breeding as well as genetic engineering approaches. It is a cost-efficient and sustainable method for providing micronutrients toward an inhabitant where diversification in the diet is limited (Garg et al. 2018). Though traditional approaches of biofortification are also efficient, it is time-consuming, labor-intensive, and costly. Genome engineering through CRISPR/Cas9 emerged as a system for an effective method for gene editing.

Many genes and genomes have been targeted by CRISPR/Cas9 for disease resistance and yield and also for enhancing the nutritional quality and have the power to overcome hidden hunger.

4.3.1 Priority Traits for Genome Editing in Cereals: Enhanced Quality and Yield

Development of cultivars with nutritional value has always been a second priority after yield and disease resistance (Hillary and Ceasar 2019), and this is due to the continually growing demand for food production of emergent populations. However, due to the increased rate of malnutrition and health concerns in people, demand for nutritious food is gaining attention, and biofortification of the essential elements is progressing. Although any essential micronutrient deficiency will cause harm to the body by impairing metabolomics, iron, provitamin A, and iodine are the top three micronutrients considered for biofortification followed by Zn and folate (Mayer et al. 2008). Biofortification is not only limited to increasing the nutrients of crops but can also reduce antinutrients such as phytic acid and gluten.

4.3.2 Biofortification for Micronutrients: Fe, Provitamins, and Low Phytic Acid in Cereal Crops

4.3.2.1 Biofortification for Fe

Iron (Fe) is a desirable component for many different plant cellular processes. In plants, iron is stored in plant vacuoles in the form of ferritin; different crops have adapted various storage mechanisms for storing iron due to the toxicity of excess iron (Briat et al. 2010). For developing seedlings, iron is supplied by the seed until it forms its first roots capable of absorbing iron from the soil; iron (Fe) is of particular importance for biofortification because its deficiency causes anemia in humans which is a critical global health issue. Genetic complexity and homeostasis make it hard to change the Fe content of staple crops, for example, wheat. However, overexpression of iron vacuolar transport was increased by twofold using CRIPSR/Cas9 in wheat and also worked in barley for iron and manganese transport by targeting TaVIT2 gene (Connorton et al. 2017).

4.3.2.2 Biofortification for Provitamin A

Provitamin A like beta-carotene, alpha-carotene, and beta-cryptoxanthin have antioxidant properties and are precursors of vitamin A. These carotenoids help fight oxidative stress caused by free radicals. These are highly volatile molecules and create oxidative stress, which is linked to various diseases like diabetes, cancer, and cardiovascular problems. Vitamin A is essential to maintain our vision and eye health. Due to antioxidant properties, it might help fight a certain type of cancer (nih.gov). Besides these vitamin A is also needed for fetal development and boosting immunity. Vitamin A is involved in the creation of B and T cells which protect us from illnesses. Developed countries have a variety of food that meets the daily need of vitamin A, while developing countries still struggle with deficiency of this element. The biofortification of staple crops will play an important role to overcome malnutrition in developing countries (Maqbool et al. 2018).

Previously, targeted insertion of genes or nucleotides was depending upon markers and conventional breeding with lower efficacy for small fragments of DNA. However, due to CRISPR/Cas9 high efficiency, targeted insertion of 5.2 kb carotenoid cassette was performed successfully in rice without any noticeable change in morphology or yield (Table 4.4) (Dong et al. 2020). Instead of using any traditional transformation method, genomic safe harbor (GSHs) method was used in this study which is indicated as a safe method of transformation without intimidating the host cell. Another successful effort by disrupting OsOr gene in rice calli for beta-carotene biofortification was carried out (Lu et al. 2006). The study reported that the CRISPR/Cas9 can be used as a great alternative to enhance the beta-carotene accumulation in crops, and this will also help overcome the issue of public renounce for transgenics (Endo et al. 2019).

4.3.2.3 Low Phytic Acid Biofortification

Phytic acids (PA) also known as inositol hexakisphosphate (IP6) perform two major functions in plants, one includes storing of phosphorus while the other is regulating some cellular activities. However, the tight-binding affinity of Fe, Zn, and calcium to phytic acid makes it an antinutritional component of the plant. Phytic acid acts as a cation chelator and restricts the absorption of essential nutrients by the small intestine. Developing low phytic acid crops is of interest, and mutants of these traits are in demand (Cominelli et al. 2020). In 2019, Jiang and the group utilized CRISPR/Cas9 technology to generate mutation, which is one of the OsITPK genes based on previous studies (Table 4.4). Though this study was able to reduce the phytic acid content in rice grain, it significantly affected the growth and reproduction of the crop; therefore, further exploration of this gene and its tight linkage to growth and reproduction-related genes will be useful for generating low phytate mutants.

4.3.2.4 Other Traits

Other than the traits mentioned above, amylose, gluten, etc., are also targeted for biofortification through genome-editing approaches. High amylose content is an important resistant starch content that gets directed to the large intestine instead of getting absorbed by the stomach. Therefore, it is helpful to fight some noninfectious diseases linked with diet such as heart disease and colon cancers (Sun et al. 2017). By taking this into account, high amylose content was produced by using CRISPR/Cas9 by targeting SBEIIb gene in rice (Sun et al. 2017). Gluten acts as an antinutrient component for genetically prone individuals, who cannot digest gluten and develop an autoimmune disorder and celiac disease. Gluten can be found in cereals like wheat, barley, and rye, and biofortification for low gluten cereal crops is an efficient way to fight with this issue. A CRISPR/Cas9 technology is used to produce reduced gluten and transgene-free wheat (Sánchez-León et al. 2018). In this effort they have used two sgRNAs to target preserved region close to transcribed

Table 4.4 List of genomic approaches in biofortification in cereals (rice, wheat, and maize)

Crop species	Modification	Genome-editing approach	Nutrients	Gene	Method of transformation	Vectors used	Promoter used	Reference
Rice	Knock in	Crispr/cas9	Carotenoid	–	Particle bombardment	–	Maize Ubi1	Dong et al. 2020
	Knockout	Crispr/cas9	High amylose	SBEIIb	Agrobacterium mediated	pCXUN-Cas9	OsU3	Sun et al. 2017
	Knockout	Crispr/cas9	Low phytic acid	OsITPK6	Agrobacterium mediated	pH_itpk6	–	Jiang et al. 2019
	Knockout	Crispr/cas9	Beta-carotene	Osor	Agrobacterium mediated	–	–	Endo et al. 2019
	Knockout	Crispr/cas9	Amylose	Waxy	Agrobacterium transformation	CRISPR/Cas9 vector	OsU6	Zhang et al. 2018a, b
	Knockout	Crispr/cas9	Sucrose efflux transporter	OsSWEET11, OsSWEET14	Agrobacterium transformation	pTOPO/D	OsU6	Jiang et al. 2019
	Knockout	Crispr/cas9	Amylase synthase	OsU3, OsU6a, OsU6b, OsU6c	Agrobacterium-transformation	pCAMBIA1300	OsU3, OsU6b, and OsU6c	Ma et al. 2015
Wheat	–	Fertilizer application	Zn	–	–	–	–	Cakmak and Kutman 2018
	Knockout	Crispr/cas9	Low gluten	Alpha gliadin	Biolistic transformation	pANIC-6E destination vector	U6 RNA polymerase III promoter (TaU6)	Leon et al. 2018
	Overexpression	Crispr/cas9	Fe, mg	TaVIT2	Agrobacterium mediated	pBract202	HMW-GLU	Connorton et al. 2017

Maize	Knockout	Cripsr/cas9	Carotenoid	Phytoene synthase	Agrobacterium mediated	pMD18-T	ZmU6	Zhu et al. 2016
	Knockout	Cripsr/cas9	Low phytic acid content	Phytic acid synthesis	Agrobacterium transformation	pEasy blunt vector	ZmU3	Liang et al. 2014

region of α-gliadin genes, resulting in a significant decrease in alpha-gliadins in all 21 mutant lines used in the experiment.

4.4 Different Promoters and Transformation Methods for Cereals' Genome Editing

Genes contain three main parts including promotor, coding region, and terminator; each performs a specific function. The promoter handles the regulation of cells, which means it decides the time, location, and amount of gene/protein to be generated. A coding region mostly comprises the mRNA information, and the terminator indicates the gene ends. Promoters in other words regulate gene expression. The key role in the monitoring process of gene regulation is played by regulatory elements (Venter and Botha 2010). There are two ways to obtain promoters; one is from spontaneously occurring genes, and the other is by commercially manufacturing it, depending upon the aim of the study. Based on the gene regulation and expression, these are divided into different types as constitutive, tissue, or developmental stage-specific and inducible. We have discussed some of them briefly here.

4.4.1 Constitutive Promoters

These types of promoters as the name suggest facilitate gene expression regardless of the environment and growth phase of the organism. Therefore, it switches genes in each living cell for the lifetime of that organism. The cauliflower mosaic virus: Examples of this type of promoters comprise cauliflower mosaic virus (CaMV) 35S, opine, ubiquitin (Ubi), actin 1 (Act-1), and maize alcohol dehydrogenase 1 (Adh-1). For dicots, CaMV 35S is a widely utilized constitutive promoter, while in monocots, maize ubiquitin and rice actin are more common where a high gene expression study is conducted.

4.4.2 Tissue-Specific or Developmental Phase-Specific Promoters

These types of promoters, instead of accelerating gene expression all over the lifetime like constitutive promoters, only express at certain stages of development or at certain tissues like flowers, roots, seeds, etc. Pleiotropy of gene makes it harder to know the exact gene responsible for a specific trait. Therefore, such type of promoter is needed for analyzing the functionality of genes. Through the genome-editing lens of CRISPR/Cas, these promoters will have a high impact on functional genomics.

4.4.3 Inducible Promoters

External factors are needed to activate these kinds of promoters. These external factors include heat, water, salinity, chemicals like abiotic stresses, or biotic stress like insect or pathogen attack. As abiotic stress can be biased easily, these promoters are favorable promoters for genetic engineering. These types of promoters can be exploited as tissue and developmental phase-specific. Examples of inducible promoters include glucocorticoid receptor (GR) based, GVG, AlcR/AlcA (ethanol inducible), pOp/LhGR (dexamethasone inducible), and XVE/OlexA (beta-estradiol inducible). However, till now to our best knowledge, these types of promoters are not explored through CRISPR/Cas in plants.

4.4.4 Artificial Promoters

Artificial promoters are promoters created through the unification of many elements from different backgrounds to meet the required type of expression. For creating synthetic promoters, ligation of different regulatory components in various combinations from varied promoters to native cis sequence can be done (Ali and Kim 2019). Regulatory elements containing activators, enhancers, or repressor elements added to upstream or downstream core sequences lead to different types of gene expression and can be shuffled to get different combinations suited for the aimed study. The rationale behind the generation of synthetic promoters is to expand the gene expression characteristics and to avoid unwanted contextual expression. The expression of genes is independent of the physical location of the target gene. When it comes to its use in CRISPR/Cas9, U6 or U3 promoters are of choice because it lacks the downstream transcriptional initiation sets and different transcriptional launch sets (Ansari et al. 2020).

4.5 Transformation Methods for Genome Editing

The effective transformation method determines the success of genome editing for the desired trait. There are three types of transformation methods currently employed in plants, and these are agrobacterium-intermediated, particle bombardment, as well as protoplast transformation techniques. Of all the above-stated methods, the first one is a more convenient and easier method used widely as a successful approach for cereal crops. The other two methods, particle bombardment and protoplast transformation, are having their own advantages but needed great caution and specific facilities. The agrobacterium approach is specific to plants susceptible to it because it is not a method of transformation used for all the plant species especially horticultural crops that are resistant to agrobacterium.

4.5.1 Agrobacterium-Mediated Transport

Agrobacterium tumefaciens is a soil pathogen that can affect plant lesion spots naturally which results in crown gall disease in plants. Agrobacterium then transfers T-DNA to the host organism with the help of microbial type IV secretion structure. Now that we know how this host and bacterial interaction happens, and with the help of molecular advancement, many genes involved in this process are known. The T-DNA contains genes that encode for proteins involved in this host and bacterial interaction and only get activated when inserted in the plant cell. Due to the need of some eukaryotic machinery for expression in the plant cell, these genes contain TATA and CAAT boxes and typical poly-A tail signals (Hwang et al. 2017). Due to all these positive sides of *Agrobacterium tumefaciens* and molecular advancement, it is becoming a popular method for transformation over all other methods. The two key factors here are infecting the plant cell and transporting the T-DNA to the host organism. Although this method is of preference, due to host specificity, its use is restricted to species susceptible to Agrobacterium. The species resistant to these bacteria have other suitable methods for transformation and are discussed below.

4.5.2 Particle Bombardment

The particle bombardment method was developed to penetrate the cell wall to insert the foreign DNA into the cell. The device used to penetrate the cell is also termed a gene gun which is a widely used method for transformation in plants. The steps are similar to Agrobacterium transformation which include the selection of gene of interest, making construct, plasmid insertion along with its incorporation into the cell, and finally regeneration of the whole plant through this single cell or tissue. The final step includes field or greenhouse trials to test the inserted transgenes. This method uses gold or tungsten particles to coat the plasmid DNS, and air pressure is used to conduct the insertion of the plasmid into the cell. This method is also useful for organelle gene transformation like the chloroplast and mitochondria. The disadvantages include the inclusion of many transgene copies at a single site or transgene that gets inserted into multiple sites. This may cause the silencing of genes in subsequent progenies.

4.5.3 Protoplast Transformation

Protoplast transformation is another alternative for transforming genes and is arguably better than the other methods in efficiency transfer. Protoplast is the cell remaining after removal of the cell wall, and this is a good way to transfer genes because cell wall rather prevents foreign material to enter into the cell. Protoplast in case of plant can regenerate into the whole plant and that is why can be an efficient method for transformation.

4.6 Multiplex Gene Editing

As soon as CRISPR/Cas originated as a great tool for editing genomes, many advancements have taken place according to the need for specific studies. One of the great features of CRISPR is multiplex genome/gene editing. Multiplex gene editing allows researchers to edit many genes at once and is very useful for the functional analysis of complex traits. Some of the multiplex gene-editing methods are described briefly here.

4.6.1 Multiplex Editing through Csy4 Nuclease

The Csy4 nucleases are part of RNA cleaving enzymes and from *Psudomonas aeruginosa*. This bacterial origin makes it preferable because it does not obstruct the host machinery of RNA processing. Other uses of these endonucleases are it has high specificity toward specific 28 base nucleotides and cut the fragment at guanine around position 20 to produce many RNA transcripts (Vats et al. 2019). It exists as a type I system, where the Cas operons contain different crRNA intermediated with endoribonuclease spacer motifs like Cys4. The Cys4 associates with Cascade complex and Cas3 to form a fully functional complex and is required to cut the pre-crRNA in guided fashion (Kurata et al. 2018). The Cys4 uses polymerase II enzyme instead of polymerase III which is good for multiplex genome editing (Fig. 4.6). The polymerase III necessitates a precise sequence at 5′ terminal of transcript, and a separate promoter is needed for each gene which is difficult to manage, and therefore Cys4 with polymerase II is beneficial for multiplex editing (Ansari et al. 2020). Another drawback that comes with polymerase III promoter like U6 is that this approach is unable to efficiently deliver multiple plasmids to a single cell at once (Kurata et al. 2018). Cys4's aptitude of RNA processing can be explored more toward its ability for gene deletion or insertion in different crops.

4.6.2 Multiplex Editing Based on Polycistronic T-RNA Transcripts

Multitarget genome editing is a great advantage of CRISPR to edit various genes simultaneously. The cell's tRNA processing system makes it possible in eukaryotes. The tRNA system is important in transcribing RNA and translating RNA into protein. The tRNA is capable of cutting both ends by identifying specific sequence sites with the help of RNase Z and RNase P. With this notion, an RNA processing system was established for tissue-specific expression (Xie et al. 2014). Briefly, in CRISPR/Cas9 system, Cas9 is synthesized by RNA polymerase II promoter, but its expression is dependent on polymerase III promoter, such as U6, and Xie et al. (2014) have studied this and developed a robust endogenous RNA managing structure which can create many gRNA using the specific transcript.

The study conducted revealed that use of synthetic DNA with tRNA-gRNA combination arranged in tandem is an effective way to produce gRNA with required

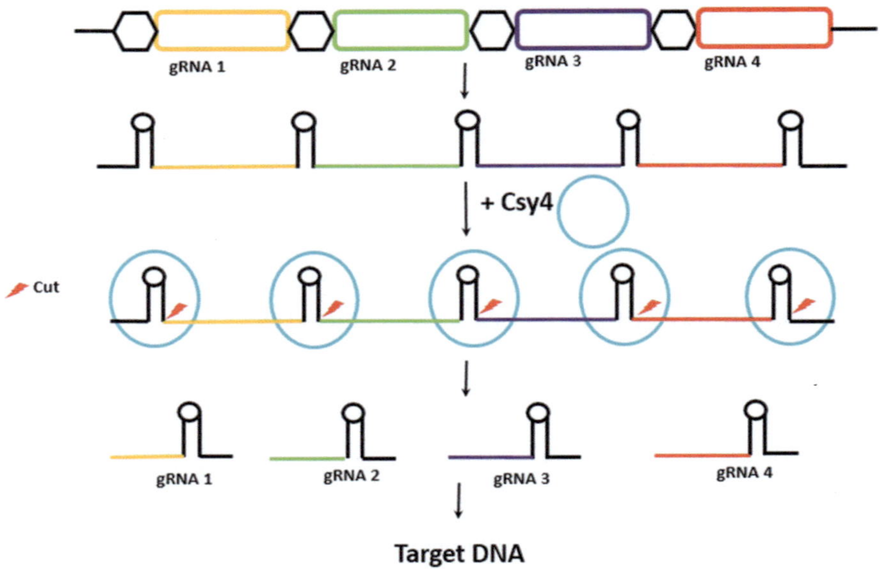

Fig. 4.6 Multiplex genome editing using CRISPR/Csy4 system, Cys4 (Yersinia) endonuclease is shown as blue circles. Csy4 restriction sites are cloned between each gRNA, and Csy4 endonuclease gene is cloned in the same vector. Cys4 expression causes separation of gRNA which then targets their complementary loci in the genome (taken and modified from Vats et al. 2019)

5′ targeting repeats that instructs the Cas9 to edit many targeted sites (Xie et al. 2014; Vats et al. 2019). The tRNA processing system contains RNase Z and RNase P naturally, whose job is to cleave the tRNA at 5′ and 3′ to release specific gRNAs. Another advantage of RNase Z and RNase P is that they are very specific to tRNA, indicating that there is very little chance of off-target mutations (Xie et al. 2016). They proposed this method for efficient multigene editing by developing a polycistronic tRNAs-gRNA (PTG) cassette (Fig. 4.7) (Xie et al. 2014). In other crops like maize, this strategy worked significantly and improved the mutation efficiency (Qi et al. 2016), in which corn glycine-tRNA has been targeted for designing many tRNA-gRNA units, and proposed that it increased the efficiency of mutation with this tRNA-gRNA processing system for maize. There are opportunities to use this system efficiently in other crops for different tissue-specific purposes because the tRNA system is conserved mostly in all organisms (Vats et al. 2019). This system may increase the possibility to explore more regarding biofortification especially interrelated complex traits.

4.6.3 Multiplex Editing Based on Drosha and Dicer

Another method for multiplex genome editing came into site as Drosha-based genome-editing method (Fig. 4.8). This system engages Drosha and Dicer which

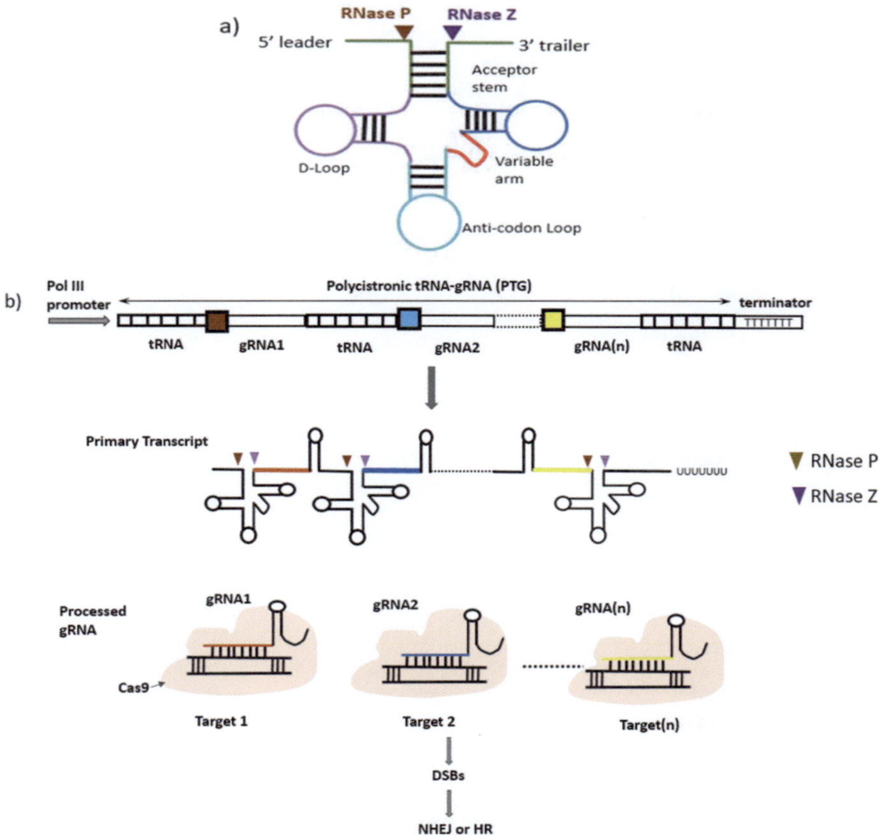

Fig. 4.7 CRISPR/Cas9 multigene targeting using PTG/Cas9 method. a) Eukaryotic pre-tRNA representing post-translational processing by RNase P (brown arrow) and RNase Z (purple arrow) splicing 5′ leader and 3′ trailer, respectively. B) Each gRNA having target sequence (colored squares) and conserved gRNA sequence (rectangles) fused with tRNA coding sequence (rectangles with boxes) which is cleaved post transcription by Rnase P and Rnase Z to release mature tRNAs and gRNAs (lines of same color as squares). Processed gRNA directs Cas9 to target site, which then causes double-stranded break (DSB) and is repaired by NHEJ or HDR (taken and modified from Vats et al. 2019)

are two endonucleases of RNase III type and contains dsRNA binding site and catalytic site. These two endonucleases are required for micro RNA maturation (Ansari et al. 2020). This approach involves tandemly arranged miRNA and sgRNA genes, and the expression of these genes depend upon polymerase II promoter. As mentioned in the previous section, polymerase II has the advantage of tissue-specific expression and flexible but has redundancy in nuclease activity due to 5′ cap (Vats et al. 2019). Drosha-based method with miRNA can be an option to overcome this issue. But still, this technology needs more attention and exploration to be used by many researchers as a robust system for multigene editing.

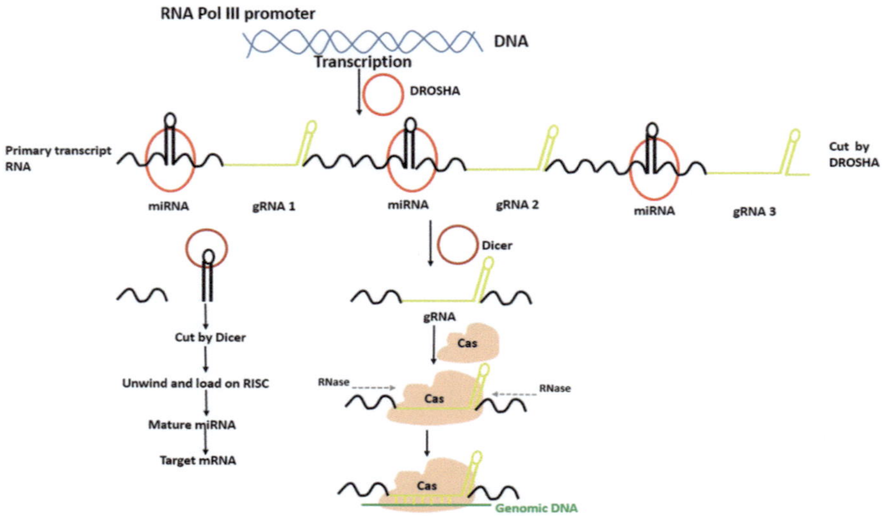

Fig. 4.8 Drosha-based multiplex genome editing in which gRNA and miRNA are cloned in a tandem array. Dicer cleaves miRNA ends, separating gRNA also (taken and modified from Vats et al. 2019)

4.7 Challenges

Advanced genome-editing techniques such as TALEN, ZFN, and CRISPR/Cas have been successfully employed to biofortify cereal crops. However, several technical issues remain to improve the practical application in cereal crop biofortification. First, most cereal grain crops (wheat, barley, sugarcane, oats, and rye) are polyploidy. Polyploidy genomes, including autopolyploids, allopolyploids, and segmental allopolyploids, are difficult to sequence because of their complexity, particularly repetitive sequences and larger genome size (Ansari et al. 2020). Also, paralogous polyploid genes make genome editing inefficient by leading to no phenotypic variations (Ansari et al. 2020). Secondly, genome-editing techniques have methodological concerns, including selecting a target gene location, designing gRNA, off-targeting, and delivery. The off-targeting mutation is a major concern of genome-editing techniques, generating unintended gene editing at the sequence with less than five mismatches (Liang et al. 2018; Pineda et al. 2019; Vats et al. 2019). Although delivering in vitro CRISPR/Cas9 ribonucleoproteins can reduce the risk of off-target mutations, there is still a need to solve several practical issues such as high expense, low stability, and demanding technical requirements (Malnoy et al. 2016; Subburaj et al. 2016; Murovec et al. 2018). Another difficulty in the execution of genome-editing technology is the efficiency of transformation vectors inserted into host cells correctly, followed by the successful regeneration of plants (Vats et al. 2019). There are two plant transformation techniques: transient transformation and stable transformation, distinguished by the inheritance of DNA introduced or integrated into the plant to the next generation. Although there is an impressive advancement in Agrobacterium-mediated and biolistic transformation over the last three decades, they still require further improvements to be more effective due to (1) the low rate in successful stable plant transformation, (2) time-consuming tissue culture, (3) damaging plant tissue caused by the biolistic transformation method, (4) limitation of Agrobacterium-mediated transformation, (5) unwanted symptoms in leaves induced from Agrobacterium, (6) mutations in somatic cells, (7) the inefficiency of the transformation of monocot species with Agrobacterium, (8) the insufficient DNA quantities for efficient HDR, and (9) low precision of gene transfer mediated through biolistic transformation (Vats et al. 2019). In addition, there are methodological concerns related to target gene site selection. Therefore, improvements in proficiency, simplification, labor-saving, and easy usage are required to break through current bottlenecks.

4.8 Future Perspectives

Vitamin and mineral deficiency, known as hidden hunger, has adversely affected the nutritional status of children and women in developing countries due to the limited diversification in diet. In the last decades, genome-editing technologies have been effectively deployed as a game-changer improving a micronutrition (iron, zinc, and vitamin A) deficiency in the edible portions of cereal crops (Vats et al. 2019). Also, genome-editing technologies have continuously improved cost-efficiency, expeditiousness, and precision (Vats et al. 2019). Despite great advancements of genome-editing technologies, there has been a lack of public understanding and acceptance regarding such methodologies mediated by modifying crops. It results in lengthy regulatory processes for the clearance of cultivation and human consumption. To convince the public, it is essential to set guidelines differentiating between genetically modified (GM) organisms and gene-edited cultivars developed by genome-editing technologies including CRISPR-Cas9 (Ansari et al. 2020). The fundamental difference is whether the presence of foreign DNA in the plant. Both methods are leading to genetic modification. However, GM organisms usually obtain genetic materials from different organisms, while CRISPR organisms are only involved in altering original base pair arrangements within their genome. Therefore, CRISPR organisms are indistinguishable from the natural allelic variants, which are mainly utilized to develop new cultivars in conventional breeding programs (Schaeffer and Nakata 2015). However, recently, the Court of Justice of the European Union (ECJ) in Luxembourg has decided that the crops developed with CRISPR technology are subjects of the lengthy approval process in the EU Directive 2001 (Es et al. 2019). Nevertheless, a total of 2052 patents related to CRISPR were already registered (European Patent Office 2018), of which 344 and 1239 patents were in Canada (Government of Canada 2015) and in the United States (US patent collection 2018), respectively. The great success in genome-editing technologies still needs to overcome political issues to be more powerful, efficient, easy, and fast.

While improving yield potential and biotic/abiotic resistance of crops have been major focuses of crop breeding programs, the introduction of nutritional quality traits in crops has been considered as another essential breeding objective in recent years (Khush et al. 2012). To accomplish those breeding objectives effectively and successfully, collaborations among plant breeders, nutrition scientists as well as molecular biologists are important in the needs of the application of genetic engineering technologies in practical implementations. In molecular biology and biochemistry sectors, there have been tremendous efforts to identify candidate genes to understand the synthetic pathway of micronutrients in cereal crops, and these candidate genes can be useful for future application of genome-editing technology (Table 4.5). For sustainable solutions using those candidate genes, the plant breeding program plays a key role to elevate micronutrition concentrations in plants utilizing natural genomic resources without affecting yield or quality. Also, the newly coined terminology "nutrition-sensitive agriculture" showed that the importance of interaction among various factors is tied closely to achieve nutrition security (Fig. 4.9) (Yu and Tan 2018). Plant breeding programs can exploit the multiple-omics and advanced technologies for phenotypic analysis as well as publicly available genotypic data to improve the breeding efficiency in nutrition-sensitive agricultural practice. As compared to breeding programs, micronutrient biofortification with genome-editing approaches can benefit from the advent of synthetic biology tools to speed up the process and the exploitation of previously identified genes associated with micronutrients for future application. Therefore, the collaboration of various research fields and usage of previous studies are necessary not only to improve genome-editing technology but also to achieve the goal of sustainable nutrition security.

Table 4.5 List of possible candidate genes associated with Fe, Zn, and vitamins for future application of biofortification using genome-editing technology in cereals (barley, maize, rice, and wheat)

Crops	Candidate gene	Micronutrient	Gene function	Reference
Barley	*HvHGGT*	Vitamin E	Homogentisate geranylgeranyl transferase	Chen et al. (2017)
	HvMTP1	Zn	Vacuolar heavy metal transporter	Podar and Sanders (2010)
	HVPT1, HVPT2	Zn	Phosphate transporter	Huang et al. (2000)
	HvYS1	Fe	Iron-phytosiderophore transporter	Murata et al. (2006)
	NAS, NAAT, DMAS, IDS2, and IDS3	Fe	Mugineic acid family phytosiderophores synthesis	Masuda et al. (2008)
Maize	*crtRB1*	Vitamin A	β-Carotene hydroxylase 1	Pixley et al. (2012)
	LcyE	Vitamin A	Lycopene epsilon cyclase	Pixley et al. (2012)

(continued)

Table 4.5 (continued)

Crops	Candidate gene	Micronutrient	Gene function	Reference
	Y1/Psy1	Vitamin A	Phytoene	Pixley et al. (2012)
	ZmFer1	Fe	Maize ferritin gene	Petit et al. (2001)
	ZmYS1	Fe	Fe(III)-phytosiderophore transporter	Curie et al. (2001)
	ZmZIP	Fe and Zn	Zinc/iron-regulated transporter-like proteins	Li et al. (2013)
Rice	*Crt1*	Vitamin A	Phytoene desaturase	Ye et al. (2000)
	HvNAS1	Fe and Zn	Nicotianamine synthase	Masuda et al. (2009)
	MxIRT1	Fe and Zn	Fe(II) transporter	Tan et al. (2014)
	OsDMAS1	Fe	Biosynthesis of deoxymugineic acid	Bashir et al. (2017)
	OsFer2	Fe	Ferritin gene	Paul et al. (2012)
	OsIRO2	Fe	Fe-deficiency-inducible transcription factor	Ogo et al. (2011)
	OsIRT1	Fe	Fe(II) transporter	Lee and An (2009)
	OsNAAT1	Fe	Nicotianamine aminotransferase	Inoue et al. (2008)
	OsNAS1, OsNAS2, OsNAS3	Fe	Nicotianamine synthase	Johnson et al. (2011)
	OsVIT1, OsVIT2	Fe	Vacuolar Fe transporter	Zhang et al. (2012)
	OsYSL13	Fe	Metal complex transporter	Itai et al. (2013)
	OsYSL15	Fe	Fe(III)-deoxymugineic acid transporter	Lee et al. (2009)
	OsYSL16	Fe	Fe(III)-deoxymugineic acid transporter	Kakei et al. (2012)
	OsYSL2	Fe	Fe(II)-nicotianamine transporter	Ishimaru et al. (2010)
	OsYSL9	Fe	Fe(II)-nicotianamine/Fe(III)-deoxymugineic acid transporter	Senoura et al. (2017)
	PSY	Vitamin A	Phytoene synthase	Ye et al. (2000)
	TOM1	Fe	Deoxymugineic acid efflux transporter	Nozoye et al. (2011)

(continued)

Table 4.5 (continued)

Crops	Candidate gene	Micronutrient	Gene function	Reference
Wheat	*Crt1*	Vitamin A	Carotene desaturase	Wang et al. (2014)
	CrtB	Vitamin A	Phytoene synthase	Wang et al. (2014)
	Gpc-B1	Zn	Grain protein content	Distelfeld et al. (2007)
	OsNAS2	Fe and Zn	Nicotianamine synthase	Singh et al. (2017)
	PvFERRITIN	Fe and Zn	Nicotianamine synthase 2	Singh et al. (2017)
	TaFer1, TaFer2	Fe	Ferritin gene	Borg et al. (2012)

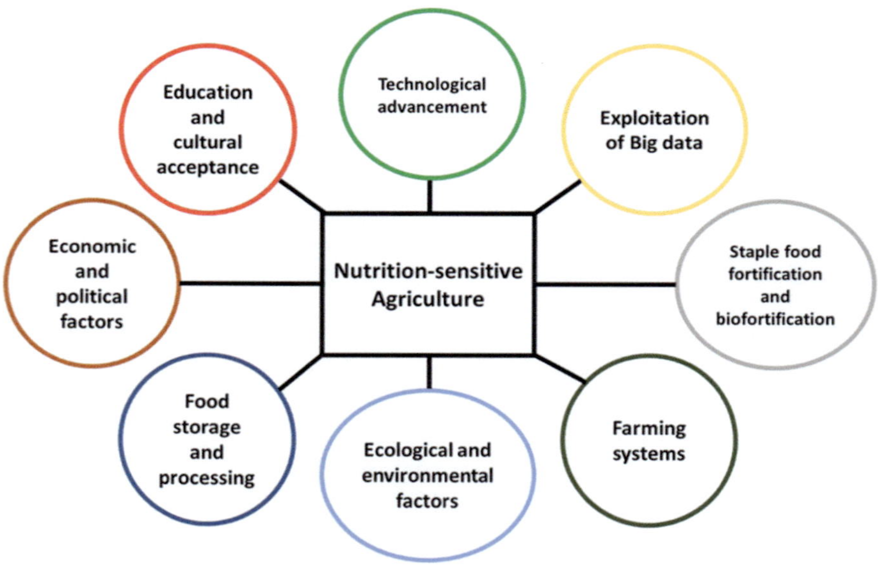

Fig. 4.9 The complex interplay of various factors in the development of nutrition-sensitive agriculture. Image modified from Yu and Tan (2018)

References

Abdallah NA et al (2015) Genome editing for crop improvement: challenges and opportunities. GM Crops & Food 6(4):183–205

Agarwal S et al (2014) Expression patterns of QTL based and other candidate genes in Madhukar× Swarna RILs with contrasting levels of iron and zinc in unpolished rice grains. Gene 546(2):430–436

Ahmed T et al (2012) Global burden of maternal and child undernutrition and micronutrient deficiencies. Ann Nutr Metab 61(Suppl. 1):8–17
Ali S, Kim WC (2019) A fruitful decade using synthetic promoters in the improvement of transgenic plants. Front Plant Sci 10:1433
Amoroso L (2016) The second international conference on nutrition: implications for hidden hunger. Hidden Hunger, Karger Publishers 115:142–152
Ansari WA et al (2020) Genome editing in cereals: approaches, applications and challenges. Int J Mol Sci 21(11):4040
Bashir K et al (2017) Paralogs and mutants show that one DMA synthase functions in iron homeostasis in rice. J Exp Bot 68(7):1785–1795
Bhullar NK, Gruissem W (2013) Nutritional enhancement of rice for human health: the contribution of biotechnology. Biotechnol Adv 31(1):50–57
Biesalski HK et al (2016) Hidden hunger: malnutrition and the first 1,000 days of life: causes, consequences and solutions. Karger Medical and Scientific Publishers
Black MM (1998) Zinc deficiency and child development. Am J Clin Nutr 68(2):464S–469S
Black RE (2003) Zinc deficiency, infectious disease and mortality in the developing world. J Nutr 133(5):1485S–1489S
Borg S et al (2012) Wheat ferritins: improving the iron content of the wheat grain. J Cereal Sci 56(2):204–213
Bortesi L, Fischer R (2015) The CRISPR/Cas9 system for plant genome editing and beyond. Biotechnol Adv 33(1):41–52
Bouis HE (2003) Micronutrient fortification of plants through plant breeding: can it improve nutrition in man at low cost? Proc Nutr Soc 62(2):403–411
Bouis HE (2018) Biofortification: An agricultural tool to address mineral and vitamin deficiencies. Food fortification in a globalized world. Elsevier, pp 69–81
Briat JF et al (2010) Ferritins and iron storage in plants. Biochim Biophy Acta (BBA)-General Subjects 1800(8):806–814
Cai CQ et al (2009) Targeted transgene integration in plant cells using designed zinc finger nucleases. Plant Mol Biol 69(6):699–709
Cakmak I (2008) Enrichment of cereal grains with zinc: agronomic or genetic biofortification? Plant Soil 302(1–2):1–17
Cakmak I (2009) Enrichment of fertilizers with zinc: An excellent investment for humanity and crop production in India. J Trace Elem Med Biol 23(4):281–289
Cakmak I, Kutman UB (2018) Agronomic biofortification of cereals with zinc: a review. Eur J Soil Sci 69(1):172–180
Camaschella C (2019) Iron deficiency. Blood 133(1):30–39
Cantos C et al (2014) Identification of safe harbor loci in indica rice genome by harnessing the property of zinc-finger nucleases to induce DNA damage and repair. Front Plant Sci 5:302
Carroll D (2011) Genome engineering with zinc-finger nucleases. Genetics 188(4):773–782
Čermák T et al (2015) High-frequency, precise modification of the tomato genome. Genome Biol 16(1):232
Chen J et al (2017) Overexpression of HvHGGT enhances Tocotrienol levels and antioxidant activity in barley. J Agric Food Chem 65(25):5181–5187
Chen K et al (2019) CRISPR/Cas genome editing and precision plant breeding in agriculture. Annu Rev Plant Biol 70:667–697
Cominelli E, Pilu R, Sparvoli F (2020) Phytic acid and transporters: what can we learn from low phytic acid mutants. Plan Theory 9(1):69
Cong L et al (2013) Multiplex genome engineering using CRISPR/Cas systems. Science 339(6121):819–823
Connorton JM et al (2017) Wheat vacuolar iron transporter TaVIT2 transports Fe and Mn and is effective for biofortification. Plant Physiol 174(4):2434–2444
Curie C et al (2001) Maize yellow stripe1 encodes a membrane protein directly involved in Fe(III) uptake. Nature 409(6818):346–349

Curtin SJ et al (2011) Targeted mutagenesis of duplicated genes in soybean with zinc-finger nucleases. Plant Physiol 156(2):466–473

Dary O, Hurrell R (2006) Guidelines on food fortification with micronutrients. World Health Organization, Food and Agricultural Organization of the United Nations, Geneva

Deltcheva E et al (2011) CRISPR RNA maturation by trans-encoded small RNA and host factor RNase III. Nature 471(7340):602–607

Distelfeld A et al (2007) Multiple QTL-effects of wheat Gpc-B1 locus on grain protein and micronutrient concentrations. Physiol Plant 129(3):635–643

Dominguez AA et al (2016) Beyond editing: repurposing CRISPR–Cas9 for precision genome regulation and interrogation. Nat Rev Mol Cell Biol 17(1):5

Dong OX et al (2020) Marker-free carotenoid-enriched rice generated through targeted gene insertion using CRISPR-Cas9. Nat Commun 11(1):1–10

DRI (2005) Institute of Medicine, food and nutrition board, dietary reference intakes: energy, carbohydrate, fiber, fat, fatty acids, cholesterol, protein and amino acids. Washington, The National Academy

Du H et al (2016) Efficient targeted mutagenesis in soybean by TALENs and CRISPR/Cas9. J Biotechnol 217:90–97

Dwyer JT (1994) Vegetarian eating patterns: science, values, and food choices—where do we go from here? Am J Clin Nutr 59(5):1255S–1262S

Endo A et al (2019) A novel approach to carotenoid accumulation in rice callus by mimicking the cauliflower Orange mutation via genome editing. Rice 12:81

Es I et al (2019) The application of the CRISPR-Cas9 genome editing machinery in food and agricultural science: current status, future perspectives, and associated challenges. Biotechnol Adv 37:410–421

European Patent Office (EPO). 2018. Available at https://www.epo.org/

Gaj T et al (2013) ZFN, TALEN, and CRISPR/Cas-based methods for genome engineering. Trends Biotechnol 31(7):397–405

Garg M et al (2018) Biofortified crops generated by breeding, agronomy, and transgenic approaches are improving lives of millions of people around the world. Front Nutr 5:12

Gasiunas G et al (2012) Cas9–crRNA ribonucleoprotein complex mediates specific DNA cleavage for adaptive immunity in bacteria. Proc Natl Acad Sci 109(39):E2579–E2586

Gibson RS (2006) Zinc: the missing link in combating micronutrient malnutrition in developing countries. Proc Nutr Soc 65(1):51–60

Gilani GS, Nasim A (2007) Impact of foods nutritionally enhanced through biotechnology in alleviating malnutrition in developing countries. J AOAC Int 90(5):1440–1444

Gödecke T et al (2018) The global burden of chronic and hidden hunger: trends and determinants. Glob Food Sec 17:21–29

Government of Canada. 2015. Available at https://www.canada.ca/en.html

Grusak MA, DellaPenna D (1999) Improving the nutrient composition of plants to enhance human nutrition and health. Annu Rev Plant Biol 50(1):133–161

Gupta RK et al (2015) Reduction of phytic acid and enhancement of bioavailable micronutrients in food grains. J Food Sci Technol 52(2):676–684

Haun W et al (2014) Improved soybean oil quality by targeted mutagenesis of the fatty acid desaturase 2 gene family. Plant Biotechnol J 12(7):934–940

Hillary VE, Ceasar SA (2019) Application of CRISPR/Cas9 genome editing system in cereal crops. The open. Biotechnol J 13, no. 1:173

Huang C et al (2000) Zinc deficiency up-regulates expression of high-affinity phosphate transporter genes in both phosphate-sufficient and -deficient barley roots. Plant Physiol 124(1):415–422

Hunger AH (2012) In: Fan S, Pandya-Lorch R (eds) Biofortification: leveraging agriculture to reduce hidden hunger, p 83

Hwang HH et al (2017) Agrobacterium-mediated plant transformation: biology and applications, vol 15. The Arabidopsis Book, p 15

Ilbert M, Bonnefoy V (2013) Insight into the evolution of the iron oxidation pathways. Biochim Biophys Acta (BBA)-Bioenergetics 1827(2):161–175

Inoue H et al (2008) Identification and localisation of the rice nicotianamine aminotransferase gene OsNAAT1 expression suggests the site of phytosiderophore synthesis in rice. Plant Mol Biol 66(1–2):193–203

Ishimaru Y et al (2010) Rice metal-nicotianamine transporter, OsYSL2, is required for the long-distance transport of iron and manganese. Plant J 62(3):379–390

Itai RN et al (2013) Rice genes involved in phytosiderophore biosynthesis are synchronously regulated during the early stages of iron deficiency in roots. Rice 6(1):16

Jiang M et al (2019) Mutation of inositol 1, 3, 4-trisphosphate 5/6-kinase6 impairs plant growth and phytic acid synthesis in rice. Plan Theory 8(5):114

Johnson AA et al (2011) Constitutive overexpression of the OsNAS gene family reveals single-gene strategies for effective iron- and zinc-biofortification of Rice endosperm. PLoS One 6(9):e24476

Jung Y-J et al (2018) Application of ZFN for site directed mutagenesis of rice SSIVa gene. Biotechnol Bioprocess Eng 23(1):108–115

Kakei Y et al (2012) OsYSL16 plays a role in the allocation of iron. Plant Mol Biol 79(6):583–594

Kennedy G et al (2003) The scourge of hidden hunger : global dimensions of micronutrient deficiencies. Food Nutr Agric 32:8–16

Khandagale K, Nadaf A (2016) Genome editing for targeted improvement of plants. Plant Biotechnol Rep 10(6):327–343

Khush GS et al (2012) Biofortification of crops for reducing malnutrition. Plant Biotechnol Rep 6: 195–202

Killip S et al (2007) Iron deficiency anemia. Am Fam Physician 75(5):671–678

Kurata M et al (2018) Highly multiplexed genome engineering using CRISPR/Cas9 gRNA arrays. PLoS One 13(9):e0198714

LaFountaine JS et al (2015) Delivery and therapeutic applications of gene editing technologies ZFNs, TALENs, and CRISPR/Cas9. Int J Pharm 494(1):180–194

Lásztity R, Lásztity L (1990) Phytic acid in cereal technology. In: Pomeranz Y (ed) Advances in cereal science, vol X. AACC, St. Paul, MI, pp 309–371

Lee S, An G (2009) Over-expression of OsIRT1 leads to increased iron and zinc accumulations in rice. Plant Cell Environ 32(4):408–416

Lee S et al (2009) Disruption of OsYSL15 leads to iron inefficiency in rice plants. Plant Physiol 150(2):786–800

Li Q et al (2016) Development of japonica photo-sensitive genic male sterile rice lines by editing carbon starved anther using CRISPR/Cas9. J Genet Genomics 43(6):415–419

Li S et al (2013) Identification and characterization of the zinc-regulated transporters, iron-regulated transporter-like protein (ZIP) gene family in maize. BMC Plant Biol 13(1):1–14

Li T et al (2012) High-efficiency TALEN-based gene editing produces disease-resistant rice. Nat Biotechnol 30(5):390

Li X et al (2018) Base editing with a Cpf1–cytidine deaminase fusion. Nat Biotechnol 36(4):324–327

Liang Z et al (2014) Targeted mutagenesis in Zea mays using TALENs and the CRISPR/Cas system. J Genet Genomics 41(2):63–68

Liang Z et al (2018) Genome editing of bread wheat using biolistic delivery of CRISPR/Cas9 in vitro transcripts or ribonucleoproteins. Nat Protoc 13(3):413–430

Linder MC (1991) Nutritional biochemistry and metabolism: with clinical applications. Elsevier

Lloyd A et al (2005) Targeted mutagenesis using zinc-finger nucleases in Arabidopsis. Proc Natl Acad Sci 102(6):2232–2237

Lu S et al (2006) The cauliflower or gene encodes a DnaJ cysteine-rich domain-containing protein that mediates high levels of beta-carotene accumulation. Plant Cell 18:3594–3605

Ma D, Liu F (2015) Genome editing and its applications in model organisms. Genomics Proteomics Bioinformatics 13(6):336–344

Ma X et al (2015) A robust CRISPR/Cas9 system for convenient high-efficiency multiplex genome editing in monocot and dicot plants. Mol Plant 8(8):1274–1284
Maberly GF et al (1994) Programs against micronutrient malnutrition: ending hidden hunger. Annu Rev Public Health 15(1):277–301
Malnoy M et al (2016) DNA-free genetically edited grapevine and apple protoplast using CRISPR/Cas9 ribonucleoproteins. Front Plant Sci 7:1904
Maqbool MA et al (2018) Breeding for provitamin a biofortification of maize (Zea mays L.). Plant Breed 137(4):451–469
Marton I et al (2010) Nontransgenic genome modification in plant cells. Plant Physiol 154(3):1079–1087
Masuda H et al (2008) Increase in iron and zinc concentrations in Rice grains via the introduction of barley genes involved in Phytosiderophore synthesis. Rice 1(1):100–108
Masuda H et al (2009) Overexpression of the barley Nicotianamine synthase gene HvNAS1 increases iron and zinc concentrations in Rice grains. Rice 2(4):155–166
Mayer JE et al (2008) Biofortified crops to alleviate micronutrient malnutrition. Curr Opin Plant Biol 11(2):166–170
Mottiar Y (2013) Iodine biofortification through plant biotechnology. Nutrition 29(11/12):1431
Murata Y et al (2006) A specific transporter for iron(III)–phytosiderophore in barley roots. Plant J 46(4):563–572
Murgia I et al (2012) Biofortification for combating 'hidden hunger'for iron. Trends Plant Sci 17(1):47–55
Murovec J et al (2018) DNA-free genome editing of Brassica oleracea and B. rapa protoplasts using CRISPR-Cas9 ribonucleoprotein complexes. Front Plant Sci 9:1594
NairK M, Choudhury D (2013) Zinc nutrition in health and diseases. J SAT Agric 11:1–6
Nekrasov V et al (2013) Targeted mutagenesis in the model plant Nicotiana benthamiana using Cas9 RNA-guided endonuclease. Nat Biotechnol 31(8):691–693
Nozoye T et al (2011) Phytosiderophore efflux transporters are crucial for iron acquisition in graminaceous plants. J Biol Chem 286(7):5446–5454
Nubé M, Voortman R (2011) Human micronutrient deficiencies: linkages with micronutrient deficiencies in soils, crops and animal nutrition. In: Combating micronutrient deficiencies: food-based approaches, vol 7, p 289
O'Dell BL, Sunde RA (1997) Handbook of nutritionally essential mineral elements. CRC Press
Odipio J et al (2017) Efficient CRISPR/Cas9 genome editing of phytoene desaturase in cassava. Front Plant Sci 8:1780
Ogo Y et al (2011) The rice bHLH protein OsIRO2 is an essential regulator of the genes involved in Fe uptake under Fe-deficient conditions. Plant J 51(3):366–377
Ortiz-Monasterio JI et al (2007) Enhancing the mineral and vitamin content of wheat and maize through plant breeding. J Cereal Sci 46(3):293–307
Oski FA (1993) Iron deficiency in infancy and childhood. N Engl J Med 329(3):190–193
Paul S et al (2012) Molecular breeding of Osfer2 gene to increase iron nutrition in rice grain. GM Crops Food 3(4):310–316
Peer R et al (2015) Targeted mutagenesis using zinc-finger nucleases in perennial fruit trees. Planta 241(4):941–951
Petit JM et al (2001) Characterization of an iron-dependent regulatory sequence involved in the transcriptional control of AtFer1and ZmFer1 plant ferritin genes by iron. J Biol Chem 276(8):5584–5590
Petolino JF (2015) Genome editing in plants via designed zinc finger nucleases. In Vitro Cell Develop Biol-Plant 51(1):1–8
Petolino JF et al (2010) Zinc finger nuclease-mediated transgene deletion. Plant Mol Biol 73(6):617–628
Pineda MA et al (2019) Safe CRISPR: challenges and possible solutions. Trends Biotechnol 37: 389–401

Pixley K et al (2012) Biofortification of maize with Provitamin a carotenoids. Carotenoids and human health, vol 2013. Humana Press, Totowa, NJ, pp 271–292

Podar D, Sanders D (2010) Biofortification of barley grains by cell-type-specific expression of a vacuolar metal transporter. Roman Biotechnol Lett 15(2):117–119

Qi W et al (2016) High-efficiency CRISPR/Cas9 multiplex gene editing using the glycine tRNA-processing system-based strategy in maize. BMC Biotechnol 16:85

Rawat N et al (2013) Biofortification of cereals to overcome hidden hunger. Plant Breed 132(5):437–445

Rice AL et al (2004) Vitamin a deficiency. Comparative quantification of health risks: global and regional burden of disease attributes to selected major risk factors. World Health Organization, Geneva, pp 211–256

Sanchez PA, Swaminathan MS (2005) Hunger in Africa: the link between unhealthy people and unhealthy soils. Lancet 365(9457):442–444

Sánchez-León S et al (2018) Low-gluten, nontransgenic wheat engineered with CRISPR/Cas9. Plant Biotechnol J 16(4):902–910

Schaeffer SM, Nakata PA (2015) CRISPR/Cas9-mediated genome editing and gene replacement in plants: transitioning from lab to field. Plant Sci 240:130–142

Senoura T et al (2017) The iron-chelate transporter OsYSL9 plays a role in iron distribution in developing rice grains. Plant Mol Biol 95(4–5):375–387

Shan Q et al (2013) Targeted genome modification of crop plants using a CRISPR-Cas system. Nat Biotechnol 31(8):686–688

Shan Q et al (2015) Creation of fragrant rice by targeted knockout of the Os BADH 2 gene using TALEN technology. Plant Biotechnol J 13(6):791–800

Shukla VK et al (2009) Precise genome modification in the crop species Zea mays using zinc-finger nucleases. Nature 459(7245):437–441

Singer, S. D., et al. (2019). Molecular improvement of grain: target traits for a changing world

Singh SP et al (2017) Rice NICOTIANAMINE SYNTHASE 2 expression improves dietary iron and zinc levels in wheat. Theor Appl Genet 130(2):283–292

Stoltzfus RJ, Dreyfuss ML (1998) Guidelines for the use of iron supplements to prevent and treat iron deficiency anemia. Ilsi Press, Washington, DC

Stoltzfus RJ et al (2004) Iron deficiency anaemia. In: Comparative quantification of health risks: global and regional burden of disease attributable to selected major risk factors, vol 1, pp 163–209

Subburaj S et al (2016) Site-directed mutagenesis in petunia x hybrida protoplast system using direct delivery of purified recombinant Cas9 ribonucleoproteins. Plant Cell Rep 35:1535–1544

Sun Y et al (2017) Generation of high-amylose rice through CRISPR/Cas9-mediated targeted mutagenesis of starch branching enzymes. Front Plant Sci 8:298

Tan S et al (2014) Over-expression of the MxIRT1 gene increases iron and zinc content in rice seeds. Transgenic Res 24(1):109–122

Urnov FD et al (2010) Genome editing with engineered zinc finger nucleases. Nat Rev Genet 11(9):636–646

US Patent Collection (2018). Available at https://www.uspto.gov/patents-application-process/search-patents

Vats S et al (2019) Genome editing in plants: exploration of technological advancements and challenges. Cell 8(11):1386

Venter M, Botha FC (2010) Synthetic promoter engineering, in plant developmental biology-biotechnological perspectives. Springer, Berlin, Heidelberg, pp 393–414

Von Grebmer K et al (2014) Synopsis: 2014 global hunger index: the challenge of hidden hunger. Intl Food Policy Res Inst

Voytas DF (2013) Plant genome engineering with sequence-specific nucleases. Annu Rev Plant Biol 64:327

Wakeel A et al (2018) Micronutrient malnutrition and biofortification: recent advances and future perspectives. In: Plant micronutrient use efficiency. Elsevier, pp 225–243

Wang C et al (2014) Enrichment of provitamin a content in wheat (Triticum aestivum L.) by introduction of the bacterial carotenoid biosynthetic genes CrtB and CrtI. J Exp Bot 65(9):2545–2556

Weinthal D et al (2010) Genome editing in plant cells by zinc finger nucleases. Trends Plant Sci 15(6):308–321

Welch RM, Graham RD (2002) Breeding crops for enhanced micronutrient content. Food security in nutrient-stressed environments. In: Exploiting plants' genetic capabilities, vol Springer, pp 267–276

West KP Jr (2002) Extent of vitamin a deficiency among preschool children and women of reproductive age. J Nutr 132(9):2857S–2866S

White PJ, Broadley MR (2005) Biofortifying crops with essential mineral elements. Trends Plant Sci 10(12):586–593

WHO (2006) Guidelines on food fortification with micronutrients. World Health Organization. Food and Agricultural Organization, Geneva

Xie K et al (2014) Genome-wide prediction of highly specific guide RNA spacers for CRISPR–Cas9-mediated genome editing in model plants and major crops. Mol Plant 7(5):923–926

Ye X et al (2000) Engineering the Provitamin a (β-carotene) biosynthetic pathway into (carotenoid-free) Rice endosperm. Science 287(5451):303–305

Yu S, Tan L (2018) Breeding major cereal grains through the lens of nutrition sensitivity. Mol Plant 11:23–30

Zhang F et al (2010) High frequency targeted mutagenesis in Arabidopsis thaliana using zinc finger nucleases. Proc Natl Acad Sci 107(26):12028–12033

Zhang Y et al (2012) Vacuolar membrane transporters OsVIT1 and OsVIT2 modulate iron translocation between flag leaves and seeds in rice. Plant J 72(3):400–410

Zhang J et al (2018a) Generation of new glutinous rice by CRISPR/Cas9-targeted mutagenesis of the waxy gene in elite rice varieties. J Integr Plant Biol 60:369–375

Zhang Y et al (2018b) Applications and potential of genome editing in crop improvement. Genome Biol 19(1):210

Zhu J et al (2016) Efficiency and inheritance of targeted mutagenesis in maize using CRISPR-Cas9. J Genet Genomics 43(1):25–36

Zimmermann MB (2016) The importance of adequate iodine during pregnancy and infancy. Hidden Hunger 115:118–124

Metabolomic Approaches to Study Nutritional Aspects in Cereal Crops

5

Anshika Tyagi, Nisha Singh, Sajad Ali, Harsha Srivastava, Muntazir Mushtaq, and Zahoor Ahmad Mir

Abstract

Owing to the increasing global population and food demand, cereals are the important source of food supply in the world. Due to global warming and other stresses, the overall nutritional quality and quantity have been severely affected in cereal crops. Various research studies have been conducted for cereal crop improvement but nutrient deficiency remains a major problem to the growing population, especially in developing countries. Hence, more comprehensive scientific tools like multi-omics will provide novel platforms to identify the high-yielding cultivars with improved nutritional quality which is important for food security. In the modern agricultural system, the development of nutrition-rich germplasm is necessary to cope with nutrient deficiencies. Among omics tools "Metabolomics", is a high-throughput and cost-effective approach in improving the nutritional quality in cereals through large-scale metabolic profiling and identifying the complex metabolic pathways. In addition, the integration of metabolomics with other modern techniques like genomics, proteomics, transcriptomics, and phenomics has witnessed an incredible technology not only the nutrition improvement but also in identifying contrasting nutritional genotypes. In this chapter, we have highlighted the application of metabolomics,

A. Tyagi (✉) · H. Srivastava
ICAR-National Institute for Plant Biotechnology, New Delhi, India
e-mail: tyagi.anshika9@gnail.com

N. Singh
Gujarat Biotechnology University, Gandhinagar, Gujarat, India

S. Ali
Department of Biotechnology, Yeungnam University, Gyeongsan, Republic of Korea

M. Mushtaq · Z. A. Mir
ICAR-National Bureau of Plant Genetic Resources, New Delhi, India

R. Deshmukh et al. (eds.), *Biofortification in Cereals*,
https://doi.org/10.1007/978-981-19-4308-9_5

tools, and database used in metabolomics, and analytical techniques for metabolic profiling focusing on nutritional aspects in cereal crops.

5.1 Introduction

Metabolomics is the complete evaluation of metabolites present endogenously from an experimental sample that uses a range of analytical platforms like NMR (nuclear magnetic resonance), MS (mass spectroscopy), GC (gas chromatography), HPLC (high-performance liquid chromatography), UPLC (ultraperformance liquid chromatography), and CE-MS (capillary electronics-mass spectrometry) providing different categories of small compounds counting carbohydrate, lipids, vitamins, and amino and organic acids. The primary metabolites play an important role in plant growth and development as a plant biomolecule or intermediate byproduct, while secondary metabolite takes part in influencing the former (primary metabolite) indirectly. Taking into account the complexity of metabolites in the plant system and the absence of a complete database of all plant metabolites, plant biologists are focusing on a single experimental tool that can isolate every single compound in the plant system. Metabolomics is one of the most important disciplines in high-throughput studies, which has been used for gene identification in several studies. The *Arabidopsis thaliana* has been the most comprehensively investigated because of the availability of huge genomic resources. Moreover, metabolomics has also been used broadly to identify gene functions in other non-model crops. Among different food crops, cereals serve as the main source of nutrition globally due to their seeds being enriched with essential fatty acids, carbohydrates, and macro and microelements for producing energy. Total food grain production of cereals like rice, wheat, barley, millets, and ragi was predicted to be 324 million tonnes (MT) in India (FAOSTAT 2021). The South Asian countries were the second-largest rice and wheat producer across the world during 2002–2020 (production volume of cereals India FY 2002–2020, Statista Research Department, 2020). Among cereals, metabolomics in rice has been used by many research groups to investigate the diversity of metabolites. Additionally, metabolomic research in maize has helped biologists to identify the superior genotypes with improved nutritional value. The metabolomic study has been conducted in several maize and rice varieties and their natural hybrids. The details of a total number of genes, their metabolic pathway, enzymatic reactions, transporters, and the protein features in different crops have been described in Table 5.1.

5.2 Application of Metabolomics in Crops

A metabolomic study identifies the low-molecular-weight molecules or metabolites within the biological systems.It is a novel approach that focuses on the dynamic nature and composition of biochemical within the living system. Intracellular metabolite contents, which are the primary product of protein-mediated cellular processes,

Table 5.1 Details of metabolomic profiling performed in different cereal crops showing different metabolic pathways using Plant Metabolic Network (PMN) database (https://plantcyc.org/)

	T. urartu	T. aestivum	O. brachyantha	O. punctata	O. longistaminata	O. meridionalis	O. barthii	O. glaberrima	O. sativa subspecies spontanea	O. sativa (Japonica)	O. rufipogon	Zea mays	Hordeum vulgare	Sorghum bicolor
Total genes	7185	24,489	6226	6232	7030	5604	6397	6511	6602	13,043	6555	8924	8933	7038
Pathways	431	446	435	431	444	426	440	442	443	545	442	499	464	439
Enzymatic reactions	2824	2980	2784	2765	2816	2722	2832	2785	2812	3547	2842	3089	3167	2805
Transporters	307	1388	247	322	239	376	332	238	348	547	380	416	3730	406
Compounds	2190	2336	2144	2135	2191	2139	2178	2151	2157	2859	2194	2468	2504	2160
GO terms	3	3	3	3	3	3	3	3	3	26	3	72	3	3
Transport reactions	84	84	67	70	70	53	71	65	74	90	72	95	95	79
Polypeptides	7186	30,682	6227	8371	7031	8608	7994	6512	9055	14,682	8833	27,682	57,690	9924
Protein complexes	0	0	0	0	0	0	0	0	0	5	0	13	0	1
Enzymes	7186	30,681	6226	8370	7029	8607	7993	6511	9054	13,043	8831	8695	57,688	9923
tRNA	0	0	0	0	0	0	0	0	0	0	0	6	0	0
Protein features	0	0	0	0	0	0	0	0	0	0	0	1	0	0

could accurately represent cellular physiological changes, especially when coupled with other molecular profiling datasets. This information could be used to construct complex molecular networking by integration of different RNA expression, DNA variation, DNA–protein binding, protein–metabolite interaction, and protein–protein interaction data with metabolite concentration to elucidate cellular regulation within the plant system (Zhu et al. 2012). Investigators can implement gene information onto supplementary sets of data to gain a far more widespread understanding of the disease or any other traits of interest using a multi-omics/systems-level approach, which combines data from the genome (genomics), RNA transcription (transcriptomics), proteins/peptides (proteomics), and metabolites (metabolomics) (Romero et al. 2006). Metabolomics stands out significantly in the sense of environment-gene interaction, mutant characterization, marker recognition, and drug discovery (Razzaq et al. 2019). Metabolomic strategies have the potency to use system biology platform to understand natural product pathways and novel pathways regulated by transcription factors beneficial for elevating trait enhancement in agricultural food and industrial product (Dixon et al. 2006).

Plants can produce over 20,000 metabolites that are involved in diverse resistance and stress tolerance responses or have a specific function in permitting distinct ecological niches to be adapted, as well as contributing to the color, taste, perfume, and scent of fruits and flowers (Oksman-Caldentey et al. 2004; Fiehn 2002; Bino et al. 2004). The customs of agricultural varieties vary from obsolete foods to those

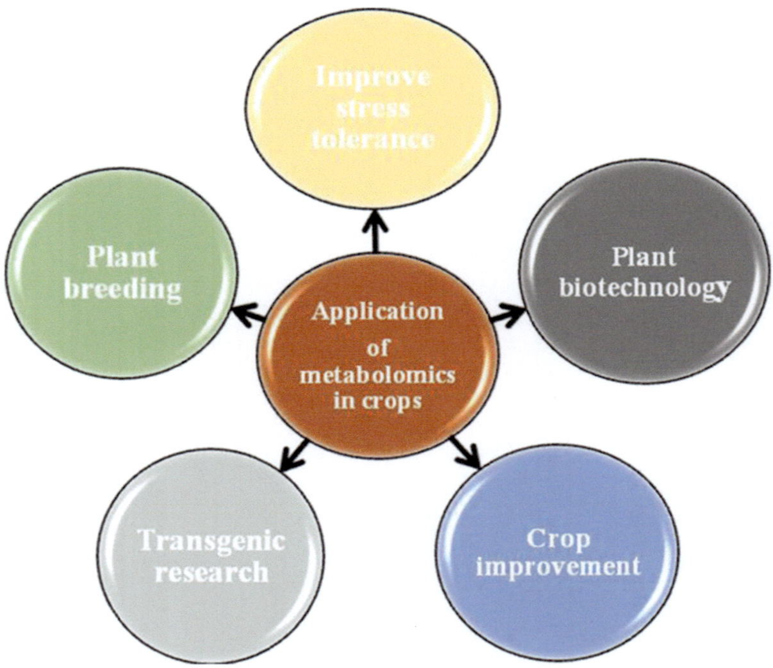

Fig. 5.1 Applications of metabolomics in agriculture crops for improving quality traits

with certain valuable characteristics, such as nutritional values, and industrial goods derived from fibers, latex, packaging material, polymers, and certain essential chemical fuels (Abbas and Cheryan 2002). The goal of the metabolomic approach in agriculture is to evaluate the biochemistry of metabolites and applying this information for food and environmental safety (Dixon et al. 2006) (Fig. 5.1).

5.3 Different Applications of Metabolomics in Crop Production

Metabolomics is an important tool having a role in various aspects of crop improvement from classical breeding to transgenic research, fruit maturation processes, resistance to adverse environmental factors, stress-related issues, and pathogen attacks, for sustainable agriculture. Some applications of plant metabolomics are the following:

5.3.1 Safety Assessment of Genetically Modified (GM) Crops

Food security is one of the most pressing challenges for the world's rising population. Modern agricultural biotechnologies, such as genetic modification, may provide a solution by allowing for increased productivity, more efficient use of natural resources, and lower environmental impacts. New crop varieties with altered genetic elements, on the other hand, may be submitted to safety studies before marketing in order to meet the regulatory criteria (Simó et al. 2014). Due to GM crop production, a huge understanding in economic potential effecting qualitative trait like tolerance to herbicide, resistance to insects, faster or delayed ripening, high level of antioxidant and nutrients, etc., has been performed. Metabolomics therefore provides more scope for the study of GM crops by enabling the id-entification of both intended and unintended effects (such as silencing off-target genes through RNA inference in the case of RNAi-based GM) that may occur in GM crops due to metabolic genetic modification. Some observable changes have been identified in the metabolite of some important crops such as rice, soybean, peas, wheat, potato, barley, and so on due to transgenic modification (Chang et al. 2012). In order to combine the authentic impact of genetic modification, metabolomic studies compare GM crops with their non-GM counterpart line using various cultural conditions (Zhou et al. 2009). The significant metabolic difference between GM and non-GM foods will be considered as healthy metabolic alteration and provide a deeper understanding on GM food composition (Harrigan et al. 2010).

5.3.2 Plant Improvement by Metabolomic Engineering

Plant metabolic pathway engineering combined with current technologies will be advantageous to humanity (food and medicines) as plants are capable to produce varied chemical compounds (Oksman-Caldentey and Saito 2005). Metabolic engineering of large datasets and logical metabolic pathway models through a huge-scale processing and mining of multiple omics data can help to improve the performance of engineered plants (Farre et al. 2015). For instance, plant metabolic engineering has been used to improve vintages of endogenous sugars such as higher-level sugars and simple sugar compounds by discovering sugar biosynthesis and accumulation pathways (Patrick et al. 2013).

5.3.3 Metabolomic Crop Improvement

Crop breeding rely on phenotypic and genomic assortment by genetic markers. However, this causes a great hurdle due to marker effects for selecting complex traits that are commonly different among population. This can be resolve using metabolomic approach combined with other omics technologies, which provide detailed information of crop plants that are performed in larger-scale environment. This mQTL and mGWAS data enables us to examine the existence of quantitative characteristics of interest (Langridge and Fleury 2011). Thus, plant metabolic technologies may contribute to the creation of a more logical models linked to precise metabolite or pathways associated with yield or quality characteristics by providing information on the number of metabolites defined that are also correlated with agronomically significant characteristics (Carreno-Quintero et al. 2013). Remarkably, continuous efforts illuminating the metabolic response to different stresses infer that breeding supported by metabolomics may also be helpful in obtaining more stress-resistant crops (Fernie and Schauer 2009).

5.3.4 Ecological Metabolomics

It deals with the study of plant biochemical relation of plant through discrete temporal (habitat lifespan to the generation time) and spatial (distance between habitat patches) framework. This method helps us to determine the interaction of abiotic factors with intra–interspecific linkages between two trophic levels and multiple impacts. In response to environmental factors, it encounters the effect of abiotic and biotic stressors on any biological activity by metabolite recognition. Biochemical network results from variance in the concentration of multiple metabolites that clarify the phenotype and physiological responses due to changes in the environment (Garcia-Cela et al. 2018).

5.3.5 Biological Control

In agriculture, biopesticides have many advantages, but still their usage is minimal due to unreliable manner, quality, and shelf-life and environmental restrictions (Babalola 2010). We need novel approach like metabolomics, which defines the need for stimuli or gene expression to synthesize metabolites that have already been discovered. Therefore, combination of molecular approach (gene sequencing and detection) and metabolomics can assist in the discovery of novel metabolites and reliable biopesticides for agricultural use (Mishra and Arora 2018).

5.3.6 Metabolomic-Assisted Breeding

For improvement of crop quality, we need the most probable, faster, cheaper, and modern breeding technique other than the standard breeding methods (Gao 2018). Metabolomics is the rightful platform to develop new crop varieties that can overcome the environmental changes, pests, and diseases and maintain their health status without change in its metabolome caused by various environmental factors like season, time, and temperature. The quality of end products involves the use of metabolomics in crop breeding, genetic modification, and biomarker discovery to meet the demand of overgrowing populations (Khakimov et al. 2014). Linkages between quantitative trait loci (QTLs) and phenotype function are a part of metabolic and transduction pathways. Metabolomics was used to classify the methylation qualitative trait loci (mQTLs) and to assess the difference in metabolic adaptation to heat and drought stress (Beckles and Roessner 2012; Templer et al. 2017). These findings revealed an important feature of mQTLs located on genes encoding the pathway of enzymes that generate antioxidant metabolites. For breeders to breed cultivators with abiotic stress tolerance, this can be a useful source. The metabolomic method is known to be a useful tool for plants to deal with environmental stresses (Rouphael et al. 2016). In conjunction with other omics techniques, metabolomic-assisted breeding would thus dramatically increase the accuracy and efficiency required for future breeding (Christ et al. 2018). Metabolomics is an emerging omics tool strategy, which has now been widely used for crop improvement. It is essential for the tolerance of abiotic stress, pathogen resistance, robust ecotype, metabolic-assisted crop reproduction (Shulaev et al. 2008). To understand traditional biological pathways and explorer secret networks that regulate crop growth and development, current metabolomic approaches are demoralized (Deborde et al. 2017). Several metabolome extraction methods and their analysis techniques have been used to assess the complex nature of metabolite and diverse chemical composition (Wishart 2011). Integration of modern plant genomic tools (GBS, genome-wide genetic variants and whole-genome sequencing) with metabolomics reveals exciting horizon for crop improvement (Zivy et al. 2015). In addition, metabolomic tool performs metabolic profiling of biofluid and various cell tissues to reflect the entire physiological composition of the cell (Yang et al. 2018).

The metabolome comprises a huge number of different chemical and physical composition such as pka, stability, molecular weight, size, polarity, and solubility (Villas-Bôas et al. 2007). A variety of analytical technologies were applied for separation, detection, and quantification of these chemicals. Metabolite content in agriculture is related to several different processes, such as fruit maturation, resistance to adverse environmental condition, stress tolerance, and pathogen infection. To analyze these compounds, various analytical techniques are used. For instance, liquid chromatography combine with mass spectrometry can be used to investigate a

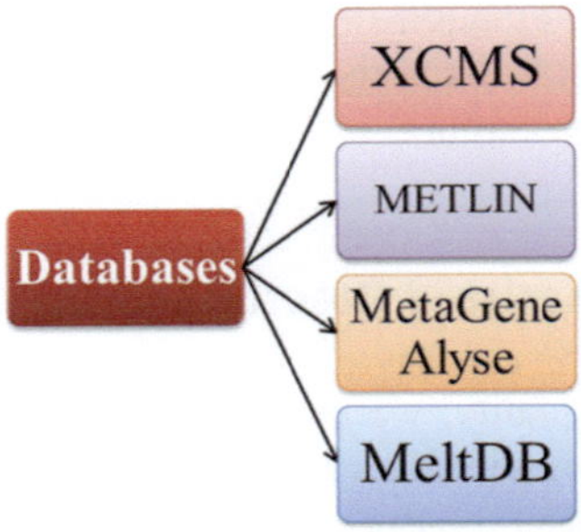

Fig. 5.2 Different databases or platformsavailable online that are useful inmetabolomic profiling for improving important food crop traits in agriculture. (**a**) XCMS (https://xcmsonline.scripps.edu), (**b**) METLIN (https://metlin.scripps.edu), (**c**) MetaGeneAlyse (http://metagenealyse.mpimp-golm.mpg.de/), and (**d**) MeltDB (https://meltdb.cebitec.uni-bielefeld.de)

wide range of compound like vitamins, coenzyme, carbohydrates, amino acids, and many more (Carreno-Quintero et al. 2013).

5.4 Tools and Databases Used in Metabolomics

Computational informatics is needed for metabolomic processes as a result of advancement in modern analytical and technological tools (Wishart 2007). To support metabolomic data mining, data evaluation, and data interpretation, we need to design online-based program. Hence, we have discussed some important tools and databases for metabolic processes (Gardinassi et al. 2017) (Fig. 5.2).

XCMS It is a bioinformatics platform available online (https://xcmsonline.scripps.edu) that enable direct access to raw data and facilitate data processing and statistical analysis. Often due to limited space, it is unfit (Tautenhahn et al. 2012). This software has been recognized for data transfer by LC-MS, which decreases data processing time and increases online system performance (Montenegro-Burke et al. 2017).

METLIN It is an online database used in plants for metabolic profiling of stress response. For metabolite profiling and for data mining, annotation, and processing, this database is the most accurate. It retrieves the results of experiments with LC-MS, FT-IR, and MS by allowing its operators to position queries in the database via the program system (Smith et al. 2005).

MetaGeneAlyse It is an online method (http://metagenealyse.mpimp-golm.mpg.de/) for routine clustering technique implementation, i.e., ICA (independent component analysis) and k-mean. It is also useful for PLS-DA, pathway enrichment analysis, and t-test, in addition to this (Daub et al. 2003).

Table 5.2 Various metabolomic-profiling tools used in analysis workflow, data preprocessing, metabolite annotation, data-post processing, and statistical analysis

	Tool	Operating system (OS)	Software type/ interface	Language	Website	**Referece**
Analysis workflow	Workflow4 metabolomics	Unix/Linux, Mac OS, Windows	Graphical user interface/galaxy-based	Galaxy-based	http://workflow4metabolomics.org	Giacomoni et al. (2014)
	Galaxy-M	-do-	Galaxy-based	R Package, Python, MATLAB	https://github.com/Viant-Metabolomics/Galaxy-M	Davidson et al. (2016)
	XCMS online	-do-	Web user interface	R package	https://xcmsonline.scripps.edu/landing_page.php?pgcontent=mainPage	Tautenhahn et al. (2012)
	Metabo Analyst 3.0	-do-	-do-	Java, R Package	http://www.metaboanalyst.ca	Xia et al. (2009)
	MAVEN	-do-	Graphical user interface	C++	https://maven.apache.org/	Clasquin et al. (2012)
	MAIT	-do-	Command line interface/R Package	R package	https://www.bioconductor.org/packages/release/bioc/html/MAIT.html	Fernández-Albert et al. (2014)
	MZmine 2	-do-	Graphical user interface	Java, R package	http://mzmine.github.io/	Pluskal et al. (2010)
Data pre processing	XCMS	Command line interface	Unix/Linux, Mac OS, Windows	R Package, C ++	http://bioconductor.org/packages/release/bioc/html/xcms.html	Smith et al. (2005)
	MetAlign	Command line interface, graphical user interface	Windows 7 (32 And 64 bit)/XP/NT/2000	Visual C++	http://www.wageningenur.nl/en/show/MetAlign-1.html	Lommen and Kools (2012)
	MS-DIAL	-do-	Windows XP/vista/7/8	C	http://prime.psc.riken.jp/Metabolomics_Software/MS-DIAL/index.html	Tsugawa et al. 2015

(continued)

Table 5.2 (continued)

	Tool	Operating system (OS)	Software type/ interface	Language	Website	**Referece**
	mzMatch	Command line Interface	Unix/Linux, Mac OS, Windows	R Package, Java	http://mzmatch.sourceforge.net/index.php	Daly et al. 2014
	AMDIS	-do-	Windows GUI	–	http://chemdata.nist.gov/dokuwiki/doku.php?id=chemdata:amdis	Meyer et al. (2010)
	Metabolite Detector	-do-	Unix/Linux, Mac OS, Windows	C++	http://md.tu-bs.de	Hiller et al. 2009
	MET-IDEA	-do-	Windows	.NET	http://bioinfo.noble.org/download	
	MeltDB	Web user Interface	Unix/Linux, Mac OS, Windows	Perl, Java, R Package	https://meltdb.cebitec.uni-bielefeld.de/cgi-bin/login.cgi	Kessler et al. (2013)
	metaMS	-do-	-do-	R Package	http://bioconductor.org/packages/release/bioc/html/metaMS.html	Wehrens et al. (2014)
	MSeasy	-do-	-do-	-do-	https://cran.r-project.org/web/packages/MSeasy/index.html	Nicolè et al. (2012)
	SpectConnect	Web user interface	-do-	Python, C	http://spectconnect.mit.edu	Styczynski et al. (2007)
	rNMR	-do-	-do-	R Package	http://rnmr.nmrfam.wisc.edu	Lewis et al. (2009)
Metabolite annotation	CAMERA	Command line interface	Unix/Linux, Mac OS, Windows	-do-	http://bioconductor.org/packages/release/bioc/html/CAMERA.html	Mahieu et al. (2016)
	BATMAN	-do-	-do-	R Package, C ++, MATLAB	http://batman.r-forge.r-project.org	Draper et al. (2009)
	Bayesil	-do-	-do-	R Package	http://bayesil.ca	Ravanbakhsh et al. (2015)
	CFM-ID	Web user nterface	-do-	Ruby, Java, MySQL	https://sourceforge.net/projects/cfm-id/	Allen et al. (2014)
	COLMAR	-do-	-do-	–	http://spin.ccic.ohio-state.edu/index.php/colmar	Zhang et al. (2009)

	FingerID	-do-	-do-	MATLAB, Python	https://github.com/icdishb/fingerid	Heinonen et al. (2012)
	MAGMa	-do-	-do-	-do-		Ridder et al. (2013)
	MetaboMiner	Command line interface	-do-	Java	http://wishart.biology.ualberta.ca/metabominer	Xia et al. (2008)
	MetAssign-mz Match	Graphical user Interface	-do-	Java, C, MATLAB	http://mzmatch.sourceforge.net/index.php	Daly et al. (2014)
	MetFrag	Web user interface, Command line Interface	-do-	Java, R package	http://c-ruttkies.github.io/MetFrag	Ruttkies et al. (2016)
	MI-PACK	Command line Interface	-do-	Python, R package	http://www.biosciences-labs.bham.ac.uk/viant/mipack	
	MyCompound ID	Web user Interface	-do-	Java	http://www.mycompoundid.org	Li et al. (2013)
	MZedDB	-do-	-do-	MySQL, Perl, PHP, R package	http://maltese.dbs.aber.ac.uk:8888/hrmet/index.html	Draper et al. (2009)
	ProbMetab	Command line Interface	-do-	R package	http://labpib.fmrp.usp.br/methods/probmetab	Silva et al. (2014)
	PUTMEDID-LCMS	-do-	-do-	Beanshell (Java)	http://www.mcisb.org/resources/putmedid.html	Brown et al. (2011)
	Rdisop	-do-	-do-	R package	http://bioconductor.org/packages/release/bioc/html/Rdisop.html.	Bioconductor - Rdisop (2016)
	SIRIUS	Command line interface, graphical user interface	-do-	Java	https://bio.informatik.uni-jena.de/software/sirius	Bocker et al. (2009)
	SpinAssign	Web user interface	-do-	Javascript, HTML, PHP, MySQL	http://prime.psc.riken.jp/?action=nmr_search	Chikayama et al. (2010)

(continued)

Table 5.2 (continued)

	Tool	Operating system (OS)	Software type/interface	Language	Website	**Referece**
Data post-processing	batchCorr	Unix/Linux, Mac OS, Windows	Command line interface	R Package	https://gitlab.com/CarlBrunius/batchCorr	Brunius et al. (2016)
	Crmn	-do-	-do-	-do-	https://cran.r-project.org/web/packages/crmn/	Redestig et al. (2009)
	EigenMS	-do-	-do-	R/MATLAB	https://sourceforge.net/projects/eigenms	Karpievitch et al. (2014)
	KMDA	-do-	-do-	R Package	https://cran.r-project.org/web/packages/KMDA/	Zhan et al. (2015)
	Metabolomics	-do-	-do-	-do-	https://cran.r-project.org/web/packages/metabolomics/	De Livera et al. (2012)
	Metabomxtr	-do-	-do-	-do-	https://www.bioconductor.org/packages/release/bioc/html/metabomxtr.html	Nodzenski et al. (2014)
	Metabnorm	-do-	-do-	-do-	https://sourceforge.net/projects/metabnorm	Jauhiainen et al. (2014)
	MetabR	-do-	Graphical user interface	-do-	http://metabr.r-forge.r-project.org/	Ernest et al. (2012)
	MetNorm	-do-	Command line interface/graphical user interface	-do-	https://cran.r-project.org/web/packages/MetNorm/	De Livera et al. (2015)
	MSPrep	-do-	Command line interface	-do-	https://sourceforge.net/projects/msprep/	Hughes et al. (2014)
Statistical analysis	Ionwinze	Windows (32 bit)	-do-	R Package/C ++	https://sourceforge.net/projects/ionwinze	Kokubun and D'Costa (2013)

	Metabol Analyze	Unix/Linux, Mac OS, Windows	-do-	R Package	https://cran.r-project.org/web/packages/MetabolAnalyze	Nyamundanda et al. (2010)
	Metabolomics	-do-	-do-	-do-	https://cran.r-project.org/web/packages/metabolomics/	De Livera et al. (2012)
	MetaboLyzer	-do-	-do-	R Package/ Python	https://sites.google.com/a/georgetown.edu/fornace-lab-informatics/home/metabolyzer	Mak et al. (2014)
	Muma	-do-	-do-	R Package	https://cran.r-project.org/web/packages/muma/	Gaude et al. (2013)
	Ropls	-do-	-do-	-do-	https://www.bioconductor.org/packages/release/bioc/html/ropls.html	Thévenot et al. (2015)
	mOTL.NMR	-do-	-do-	-do-	https://www.bioconductor.org/packages/release/bioc/html/mQTL.NMR	Hedjazi et al. (2015)

MeltDB It is a web-based platform (https://meltdb.cebitec.uni-bielefeld.de) for data assessment, processing, and statistical analysis used in plant metabolomics (Kessler et al. 2013). Other than these, many more databases have been used to analyze and compare different metabolites such as MetPA (http://metpa.metabolomics.ca), MSEA (http://www.msea.ca), iMet-Q (http://ms.iis.sinica.edu.tw/comics/Software_iMet-Q.html), MS-Dial (http://prime.psc.riken.jp/Metabolomics_Software/MS-DIAL/), and MetAlign (www.metalign.nl). (Xia and Wishart 2010; Kessler et al. 2013; Chang et al. 2016; Lommen and Kools 2012; Tsugawa et al. 2015). The detail of various metabolomic tools at different stages has been given in Table 5.2.

5.5 Cutting-Edge/High-Throughput Analytical Techniques in Metabolomic Analysis

Not only is one method used in metabolomics to examine all the metabolites present in a metabolome, but a lot of different technologies are required to manage data blocks (Fig. 5.3).

1. Nuclear magnetic resonance spectroscopy (NMR).
    - It provides rapid, highly reproducible, and nondestructive high-throughput method (Wishart 2019).
    - It provides structural and functional information of biomolecules and also is capable to analyze solid, liquid, and gas samples (Gouilleux et al. 2018).
    - In order to achieve biochemical measurement, it increased the sensitivity and spectral resolution of analytical assays on metabolomic samples.
    - Help in examining food quality and standardization of phytomedical preparations (Ward et al. 2007).
    - Detection in one single study of various groups of metabolites with different chemical and physical properties (Ward et al. 2007).
2. Liquid chromatography mass spectrometry (LC-MS).
    - It detects mass spectrometric of metabolites.
    - Separation of various metabolites based on different mobile phase and stationary phase portioning coefficients (Khakimov et al. 2014).
    - It is best suitable for detection of polar compound and secondary metabolite analysis like vitamins and flavonoids.
    - It utilizes the source of electroscopy ionization (ESI) to analyze metabolites of high molecular weight.
    - It allows direct probing of metabolites in any sample without derivatization (Wang et al. 2017).
3. Gas chromatography-mass spectroscopy (GC-MS).
    - This method is suitable for hydrophobic and polar compound with high resolving power and sensitivity.
    - Electron impact method is used.
    - It is used to classify thermally volatile and unstable compounds and has great power of separation and reproducibility (Jorge et al. 2016).

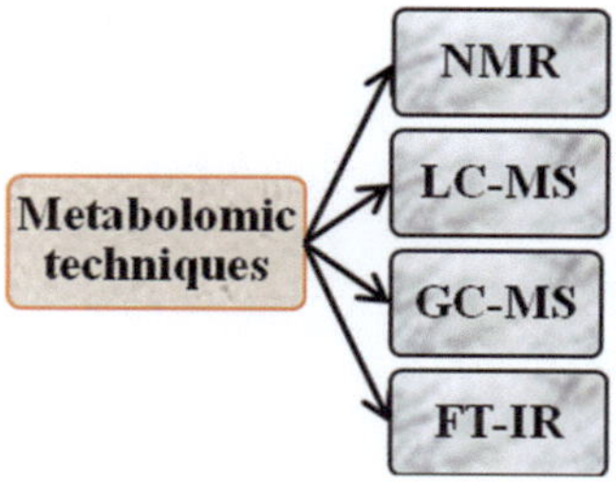

Fig. 5.3 High-throughput metabolomic technique used for metabolic profiling or quantification of metabolomes present in a host system. (**a**) Nuclear magnetic resonance spectroscopy (NMR), (**b**) liquid chromatography mass spectrometry (LC-MS), (**c**) gas chromatography mass spectroscopy (GC-MS), and (**d**) Fourier transform infrared spectroscopy (FT-IR)

- GC-MS metabolomic approach is used to display the effects of drought and heat on metabolite distribution of cultivators at different developmental stages (Lawas et al. 2019).

4. Fourier transform–infrared spectroscopy (FT-IR).
    - It allows the study of unknown metabolites to be identified based on the ratio of mass to charge.
    - In plant science, this approach has also been introduced as a metabolic fingerprinting technique (Kaderbhai et al. 2003).
    - It provides the most reliable information about data.
    - It allows characterization and separation of mixed sample (Vasmatkar et al. 2019).

5.6 Metabolomic Approaches to Improve Nutritional Quality in Major Crops

Cereals or foodomics such as wheat, rice, barley, corn, rye, and oat are agriculturally important food crops that are grown all over the world. These are the largest part of crops having important role in human utilization basically for health beneficial factors. The important characteristics of cereal crops include the carbohydrates, fats, essential and nonessential amino acids, dietary fibers, and micronutrients. Some primary and secondary metabolites often have a major impact on the health and nutritional implications. In an experimental research comparing hypercholesterolemic patients, whole grain barley consumption was shown to reduce the low-density lipoprotein (LDL) cholesterol. In addition, phenolic acids have been identified as essential texturizing agents in food preparation and key antioxidant component of cereals. The variation in phenolic content in different cereal crops has been shown in Fig. 5.4.

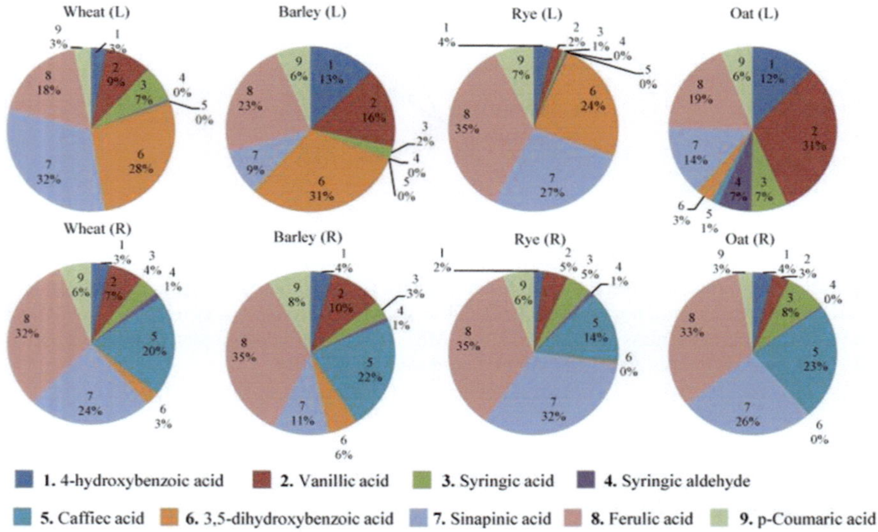

Fig. 5.4 Detailed pie-chart illustration showing percent variation of phenolic acid in different cereal crops including wheat, barley, rye, and oat

To date, several studies have been conducted in cereal crops including corn (count 99), rice and barley (count 103), oat (count 23), wheat (count 4), and rye (count 48). Previous research has established that utility of metabolomics in understanding the molecular mechanisms underlying diverse quality-related attributes. Metabolomics has emerged as a valuable talent for comprehensive characterization and synthesis of diverse metabolites in biological systems, and a number of plant science programs have been documented.

The primary and secondary metabolite compositions in the kernel as well as other aerial components were presented in cultivated rice with the aid of advanced techniques such as GC-MS, LC-MS, and (CE)-MS focusing on nutritionally essential bioactive compounds. Rice bran metabolite profiling was carried out to uncover 209 amino acids, including cofactors and vitamins, as well as other secondary metabolites, in order to improve the present understanding on chemicals delivered during dietary supplementation. Furthermore, certain wild rice species from North America (*Zizania palustris*) and China (*Zizania latifolia*) were compared for secondary metabolite levels, and it was found that the two differed particularly in anthocyanins and catechins among 357 metabolites studied. Similarly, comparative metabolomic profiling of giant vs regular embryo rice suggested the better quality of giant embryo-derived rice grains.

Wheat is the third-largest cereal crop in global production. Metabolic profiling has proven contrasting facts in numerous wheat organs. For example, a total of 118 and 56 metabolites have been identified in durum wheat roots; 111 in spring wheat roots, 53 metabolites in wheat phloem; followed by 51, 93, and

103 metabolites in leaf while 127 and 117 metabolites in spikelet and rachis of wheat, respectively. In the immature grain of bread wheat and durum wheat grain, a total of 74 and 42–50 metabolites were found, respectively (Zhen et al. 2016). Finally, a UPLC-TOF-MS was used to identify 935 ions on the mature grain of diverse genotypes of durum and bread wheat (Matthews et al. 2012).

Numerous metabolomic research have focused on deciphering the complex metabolic pathways involved in abiotic and biotic stress in corns along with the comparative metabolic profiling between genetically manipulated and nongenetically manipulated corn lines. Apart from these various studies targeting nutritional improvement (such as proteins, carbohydrates, fat) in corn along with an emphasis on anthocyanins, primary metabolites and mineral composition have also been presented. Out of 398 genetically distinct colored corn varieties from various regions, a total of 167 were chosen for producing anthocyanins. In addition to this, a total of 210 metabolites including primary metabolites (199), secondary metabolites (9), and phytohormones (2) have been identified in mature Chinese kernel lines. An integrated metabolic map from the identification of essential macronutrients and other important molecules was built having seven key pathways and 23 subpathways of corn kernel metabolism.

The metabolite patterns from three barley lines (*lys3.a*, *lys5.f*, and Bomi) from developing seeds were analyzed for alpha-glucan using GC-MS. Mutation in *lys3.a* and *lys5.f* led to an increase in lysine and ADP-glucose transporter gene production, respectively, regulating the carbohydrate, lipid, tricarboxylic acid cycle (TCA), shikimate–phenylpropanoid (SPP), and mevalonate pathways resulting in the improvement of new highly nutritive foods.

5.7 . Conclusions and Future Perspective

In conclusion, targeting nutritional metabolomics is an increasingly budding area to combine nourishment with multifarious metabolomic data to determine novel genetic markers. In order to incorporate, nutritional metabolomics with dietary supplements and observable traits, a broad range of softwares, repositories, and analytical tools are available. In addition, metabolomics can lead to the market problem analysis, consumer expectation, and food security. The holistic amalgamation of metabolomics with supplementary omic tools such as genomics, transcriptomics, proteomics, and phenomics may open a new opportunity for nutrition improvement as well as identification of nutrition-rich germplasm not only in cereal crops but also in other agriculturally important crops. The combination of metabolomics with high-throughput genotyping tools in the future would provide novel avenues to the plant breeders to develop highly nutritional cereals that adequately meet the expanding population's food requirement and provide food security.

References

Abbas CA, Cheryan M (2002) Biorefinery opportunities. Appl Biochem Biotechnol 98:1147

Allen F, Pon A, Wilson M, Greiner R, Wishart D (2014) CFM-ID: a web server for annotation, spectrum prediction and metabolite identification from tandem mass spectra. Nucleic Acids Res 42:W94–W99

Babalola O (2010) Beneficial bacteria of agricultural importance. Biotechnol Lett 32:1559–1570

Beckles D, Roessner U (2012) Plant metabolomics: applications and opportunities for agricultural biotechnology. In: Plant biotechnology and agriculture. Elsevier, pp 67–81

Bino RJ, Hall RD, Fiehn O et al (2004) Potential of metabolomics as a functional genomics tool. Trends Plant Sci 9(9):418–425

Bioconductor - Rdisop. (2016). Accessed August 18, 2016 from http://bioconductor.org/packages/release/bioc/html/Rdisop.html

Bocker S, Letzel MC, Liptak Z, Pervukhin A (2009) SIRIUS: decomposing isotope patterns for metabolite identification. Bioinformatics 25(2):218–224

Brown M, Wedge DC, Goodacre R, Kell DB, Baker PN, Kenny LC et al (2011) Automated workflows for accurate mass-based putative metabolite identification in LC/MS-derived metabolomic datasets. Bioinformatics 27(8):1108–1112

Brunius C, Shi L, Landberg R (2016) Large-scale untargeted LC-MS metabolomics data correction using between-batch feature alignment and cluster-based within-batch signal intensity drift correction. Metabolomics 12:173. https://doi.org/10.1007/s11306-016-1124-4

Carreno-Quintero N, Bouwmeester HJ, Keurentjes JJ (2013) Genetic analysis of metabolome–phenotype interactions: from model to crop species. Trends Genet 29:41–50

Chang HY, Chen CT, Lih TM, Lynn KS, Juo CG, Hsu WL, Sung TY (2016) iMet-Q: a user-friendly tool for label-free metabolomics quantitation using dynamic peak-width determination. PLoS One 11:e0146112

Chang Y, Zhao C, Zhu Z, Wu Z, Zhou J, Zhao Y, Lu X, Xu G (2012) Metabolic profiling based on LC/MS to evaluate unintended effects of transgenic rice with cry1Ac and sck genes. Plant Mol Biol 78:477–487

Chikayama E, Sekiyama Y, Okamoto M, Nakanishi Y, Tsuboi Y, Akiyama K et al (2010) Statistical indices for simultaneous large-scale metabolite detections for a single NMR spectrum. Anal Chem 82(5):1653–1658

Christ B, Pluskal T, Aubry S, Weng J (2018) Contribution of untargeted metabolomics for future assessment of biotech crops. Trends Plant Sci 23:1047–1056

Clasquin MF, Melamud E, Rabinowitz JD (2012) LC-MS data processing with MAVEN: a metabolomic analysisand visualization engine. Curr Protoc Bioinform 37:14.11.1–14.11.23

Daly R, Rogers S, Wandy J, Jankevics A, Burgess KE, Breitling R (2014) MetAssign: probabilistic annotation of metabolites from LC–MS data using a Bayesian clustering approach. Bioinformatics 30:2764–2771

Daub CO, Kloska S, Selbig J (2003) MetaGeneAlyse: analysis of integrated transcriptional and metabolite data. Bioinformatics 19:2332–2333

Davidson RL, Weber RJ, Liu H, Sharma-Oates A, Viant MR (2016) Galaxy-M: a galaxy workflow forprocessing and analyzing direct infusion and liquid chromatography mass spectrometry-based metabolomicsdata. Gigascience 5:10

De Livera AM, Aho-Sysi M, Jacob L, Gagnon-Bartch J, Castillo S, Simpson J, Speed TP (2015) Statistical methods for handling unwanted variation in metabolomics data. Anal Chem 87(7):3606–3615

De Livera AM, Dias DA, De Souza D, Rupasinghe T, Pyke J, Tull D, Roessner U, McConville M, Speed TP (2012) Normalizing and integrating metabolomics data. Anal Chem 84(24):10768–10776. https://doi.org/10.1021/ac302748b. Epub 2012 Nov 29. PMID: 23150939

Deborde C, Moing A, Roch L, Jacob D, Rolin D, Giraudeau P (2017) Plant metabolism as studied by NMR spectroscopy. Prog Nucl Magn Reson Spectrosc 102:61–97

Dixon RA, Gang DR, Charlton AJ, Fiehn O, Kuiper HA, Reynolds TL et al (2006) Applications of metabolomics in agriculture. J Agric Food Chem 54(24):8984–8994

Draper J, Enot DP, Parker D, Beckmann M, Snowdon S, Lin W, Zubair H (2009) Metabolite signal identification in accurate mass metabolomics data with MZedDB, an interactive m/z annotation tool utilizing predicted ionizationbehavior 'rules'. BMC Bioinform 10:227

Ernest B, Gooding JR, Campagna SR, Saxton AM, Voy BH (2012) MetabR: an R script for linear model analysis of quantitative metabolomic data. BMC Res Notes 5:596

FAOSTAT, Food and Agricultural Organization of the United Nations. 2021. http://www.fao.org/faostat/en/#home.

Farre G, Twyman RM, Christou P, Capell T, Zhu C (2015) Knowledge-driven approaches for engineering complex metabolic pathways in plants. Curr Opin Biotechnol 32:54–60

Fernández-Albert F, Llorach R, Andrés-Lacueva C, Perera A (2014) An R package to analyse LC/MS metabolomic data: MAIT (metabolite automatic identification toolkit). Bioinformatics 30(13):1937–1939. https://doi.org/10.1093/bioinformatics/btu136

Fernie AR, Schauer N (2009) Metabolomics-assisted breeding: a viable option for crop improvement? Trends Genet 25:39–48

Fiehn O (2002) Metabolomics—the link between genotypes and phenotypes. Plant Mol Biol 48(1–2):155–171

Gao C (2018) The future of CRISPR technologies in agriculture. Nat Rev Mol Cell Biol 19:275–276

Garcia-Cela E, Kiaitsi E, Medina A, Sulyok M, Krska R, Magan N (2018) Interacting environmental stress factors affects targeted metabolomic profiles in stored natural wheat and that inoculated with F. graminearum. Toxins 10:56

Gardinassi LG, Xia J, Safo SE, Li S (2017) Bioinformatics tools for the interpretation of metabolomics data. Curr Pharmacol Rep 3:374–383

Gaude E, Chignola F, Spiliotopoulos D, Spitaleri A, Ghitti M, Garcia-Manteiga JM et al (2013) Muma, an R package for metabolomics univariate and multivariate statistical analysis. Curr Metabol 1(2):180–189

Giacomoni F, Le Corguillé G, Monsoor M, Landi M, Pericard P, Pétéra M, Duperier C, Tremblay-Franco M, Martin J-F, Jacob D, Goulitquer S, Thévenot EA, Caron C (2014) Workflow4Metabolomics: a collaborative research infrastructure for computational metabolomics. Bioinformatics 31:1493. https://doi.org/10.1093/bioinformatics/btu813

Gouilleux B, Rouger L, Giraudeau P (2018) Ultrafast 2D NMR: methods and applications. Annu Rep NMR Spectrosc *93*:75–144

Harrigan GG, Lundry D, Drury S, Berman K, Riordan SG, Nemeth MA, Ridley WP, Glenn KC (2010) Natural variation in crop composition and the impact of transgenesis. Nat Biotechnol 28: 402–404

Hedjazi L, Gauguier D, Zalloua PA, Nicholson JK, Dumas M-E, Cazier J-B (2015) mQTL.NMR: an integrated suite for genetic mapping of quantitative variations of (1)HNMR-based metabolic profiles. Anal Chem 87(8):4377–4384

Heinonen M, Shen H, Zamboni N, Rousu J (2012) Metabolite identification and molecular fingerprint prediction through machine learning. Bioinformatics 28(18):2333–2341

Hiller K, Hangebrauk J, Jäger C, Spura J, Schreiber K, Schomburg D (2009) Metabolite detector: comprehensive analysis tool for targeted and non-targeted GC/MS based metabolome analysis. Anal Chem 81(9):3429–3439. https://doi.org/10.1021/ac802689c

Hughes G, Cruickshank-Quinn C, Reisdorph R, Lutz S, Petrache I, Reisdorph N, Bowler R, Kechris K (2014) MSPrep--summarization, normalization and diagnostics for processing of mass spectrometry-based metabolomic data. Bioinformatics 30(1):133–134. https://doi.org/10.1093/bioinformatics/btt589

Jauhiainen A, Madhu B, Narita M, Narita M, Griffiths J, Tavaré S (2014) Normalization of metabolomics data with applications to correlation maps. Bioinformatics 30(15):2155–2161. https://doi.org/10.1093/bioinformatics/btu175

Jorge TF, Rodrigues JA, Caldana C, Schmidt R, van Dongen JT, Thomas-Oates J, António C (2016) Mass spectrometry-based plant metabolomics: metabolite responses to abiotic stress. Mass Spectrom Rev 35:620–649

Kaderbhai NN, Broadhurst DI, Ellis DI, Goodacre R, Kell DB (2003) Functional genomics via metabolic footprinting:Monitoring metabolite secretion by *Escherichia coli* tryptophanmetabolism mutants using FT-IR and direct injectionelectrospray mass spectrometry. Comp Funct Genom 4(4):376–391

Karpievitch YV, Nikolic SB, Wilson R, Sharman JE, Edwards LM (2014) Metabolomics data normalization with EigenMS. PLoS One 9(12):e116221. https://doi.org/10.1371/journal.pone.0116221

Kessler N, Neuweger H, Bonte A, Langenkämper G, Niehaus K, Nattkemper TW, Goesmann A (2013) MeltDB 2.0–advances of the metabolomics software system. Bioinformatics 29:2452–2459

Khakimov B, Bak S, Engelsen S (2014) High-throughput cereal metabolomics: current analytical technologies, challenges and perspectives. J Cereal Sci 59:393–418

Kokubun T, D'Costa L (2013) Direct and unbiased information recovery from liquid chromatography-mass spectrometry raw data for phenotype-differentiating metabolites based on screening window coefficient of ion currents. Anal Chem 85(18):8684–8691

Langridge P, Fleury D (2011) Making the most of 'omics' for crop breeding. Trends Biotechnol 29: 33–40

Lawas LMF, Li X, Erban A, Kopka J, Jagadish SK, Zuther E, Hincha DK (2019) Metabolic responses of ricecultivars with deferent tolerance to combined drought and heat stress under field conditions. Gigascience. 8:50

Lewis IA, Schommer SC, Markley JL (2009) rNMR: open source software for identifying and quantifying metabolites in NMR spectra. MagnReson Chem 47(1):S123–S126. https://doi.org/10.1002/mrc.2526. PMID: 19821464; PMCID: PMC2798074

Li L, Li R, Zhou J, Zuniga A, Stanislaus AE, Wu Y et al (2013) MyCompoundID: using an evidence-based metabolome library for metabolite identification. Anal Chem 85(6):3401–3408

Lommen A, Kools HJ (2012) MetAlign 3.0: performance enhancement by ancient use of advances in computer hardware. Metabolomics 8:719–726

Mahieu NG, Genenbacher JL, Patti GJ (2016) A roadmap for the XCMS family of software solutions in metabolomics. Curr Opin Chem Biol 30:87–93

Mak TD, Laiakis EC, Goudarzi M, Fornace AJ (2014) MetaboLyzer: a novel statistical workflow for analyzing postprocessed LC-MS metabolomics data. Anal Chem 86(1):506–513

Matthews SB, Santra M, Mensack MM, Wolfe P, Byrne PF, Thompson HJ (2012) Metabolite profiling of a diverse collection of wheat lines using ultraperformance liquid chromatography coupled with time-of-flight mass spectrometry. PLoS One 7(8):e44179. https://doi.org/10.1371/journal.pone.0044179

Meyer MR, Peters FT, Maurer HH (2010) Automated mass spectral deconvolution and identification system for GC-MS screening for drugs, poisons, and metabolites in urine. Clin Chem 56(4): 575–584

Mishra J, Arora N (2018) Secondary metabolites of fluorescent pseudomonads in biocontrol of phytopathogens for sustainable agriculture. Appl Soil Ecol 125:35–45

Montenegro-Burke JR, Aisporna AE, Benton HP, Rinehart D, Fang M, Huan T, Warth B, Forsberg E, Abe BT, Ivanisevic J (2017) Data streaming for metabolomics: accelerating data processing and analysis from days to minutes. Anal Chem 89:1254–1259

Nicolè F, Guitton Y, Courtois EA, Moja S, Legendre L, Hossaert-McKey M (2012) MSeasy: unsupervised and untargeted GC-MS data processing. Bioinformatics 28(17):2278–2280. https://doi.org/10.1093/bioinformatics/bts427

Nodzenski M, Muehlbauer MJ, Bain JR, Reisetter AC, Lowe WL Jr, Scholtens DM (2014) Metabomxtr: an R package for mixture-model analysis of non-targeted metabolomics data. Bioinformatics (Oxford, England) 30(22):3287–3288. https://doi.org/10.1093/bioinformatics/btu509

Nyamundanda G, Brennan L, Gormley IC (2010) Probabilistic principal component analysis for metabolomic data. BMC Bioinform 11(1):571

Oksman-Caldentey K-M, Inze D, Ore M (2004) Connecting genes to metabolites by a systems biology approach. Proc Natl Acad Sci U S A 101(27):9949–9950

Oksman-Caldentey KM, Saito K (2005) Integrating genomics and metabolomics for engineering plant metabolic pathways. Curr Opin Biotechnol 16:174–179

Patrick JW, Botha FC, Birch RG (2013) Metabolic engineering of sugars and simple sugar derivatives in plants. Plant Biotechnol J 11:142–156

Pluskal T, Castillo S, Villar-Briones A, Orešiˇc M (2010) MZmine 2: Modular framework for processing,visualizing, and analyzing mass spectrometry-based molecular profile data. BMC Bioinform. 11:395

Ravanbakhsh S, Liu P, Bjorndahl TC, Mandal R, Grant JR, Wilson M, Eisner R, Sinelnikov I, Hu X, Luchinat C, Greiner R, Wishart DS (2015) Accurate, fully-automated NMR spectral profiling for metabolomics. PLoS One 10(5):e0124219

Razzaq A, Sadia B, Raza A, Khalid Hameed M, Saleem F (2019) Metabolomics: a way forward for crop improvement. Meta 9(12):303

Redestig H, Fukushima A, Stenlund H, Moritz T, Arita M, Saito K, Kusano M (2009) Compensation for systematic cross-contribution improves normalization of mass spectrometry based metabolomics data. Anal Chem 81(19):7974–7980

Ridder L, van der Hooft JJJ, Verhoeven S, de Vos RCH, Bino RJ, Vervoort J (2013) Automatic chemical structure annotation of an LC–MSn based metabolic profile from green tea. Anal Chem 85(12):6033–6040

Romero R, Espinoza J, Gotsch F, Kusanovic JP, Friel LA, Erez O et al (2006) The use of high-dimensional biology (genomics, transcriptomics, proteomics, and metabolomics) to understand the preterm parturition syndrome. BJOG 113:118–135

Rouphael Y, Colla G, Bernardo L, Kane D, Trevisan M, Lucini L (2016) Zinc excess triggered polyamines accumulation in lettuce root metabolome, as compared to osmotic stress under high salinity. Front Plant Sci 7:842

Ruttkies C, Schymanski EL, Wolf S, Hollender J, Neumann S (2016) MetFrag relaunched: incorporating strategies beyond in silico fragmentation. J Cheminform 8:3

Shulaev V, Cortes D, Miller G, Mittler R (2008) Metabolomics for plant stress response. Physiol Plant 132:199–208

Silva RR, Jourdan F, Salvanha DM, Letisse F, Jamin EL, Guidetti-Gonzalez S et al (2014) ProbMetab: an R package for Bayesian probabilistic annotation of LC–MS-based metabolomics. Bioinformatics 30(9):1336–1337

Simó C, Ibáez C, Valdés A, Cifuentes A, García-Cañas V (2014) Metabolomics of genetically modified crops. Int J Mol Sci 15:18941–18966

Smith CA, O'Maille G, Want EJ, Qin C, Trauger SA, Brandon TR, Custodio DE, Abagyan R, Siuzdak G (2005) METLIN: a metabolite mass spectral database. Ther Drug Monit 27:747–751

Styczynski MP, Moxley JF, Tong LV, Walther JL, Jensen KL, Stephanopoulos GN (2007) Systematic identification of conserved metabolites in GC/MS data for metabolomics and biomarker discovery. Anal Chem 79(3):966–973. https://doi.org/10.1021/ac0614846. PMID: 17263323

Tautenhahn R, Patti GJ, Rinehart D, Siuzdak G (2012) XCMS online: a web-based platform to process untargeted metabolomic data. Anal Chem 84:5035–5039

Templer S, Ammon A, Pscheidt D, Ciobotea O, Schuy C, McCollum C, Sonnewald U, Hanemann A, Förster J, Ordon F (2017) Metabolite profiling of barley flag leaves under drought and combined heat and drought stress reveals metabolic QTLs for metabolites associated with antioxidant defense. J Exp Bot 68:1697–1713

Thévenot EA, Roux A, Xu Y, Ezan E, Junot C (2015) Analysis of the human adult urinary metabolome variations with age, body mass index, and gender by implementing a comprehensive workflow for univariate and OPLS statistical analyses. J Proteome Res 14(8):3322–3335

Tsugawa H, Cajka T, Kind T, Ma Y, Higgins B, Ikeda K, Kanazawa M, VanderGheynst J, Fiehn O, Arita M (2015) MS-DIAL: data-independent MS/MS deconvolution for comprehensive metabolome analysis. Nat Methods 12:523

Vasmatkar P, Kaur K, Pannu PPS, Kaur G, Kaur H (2019) Unraveling the metabolite signatures of maize genotypes showing differential response towards southern corn leaf blight by 1H-NMRand FTIR spectroscopy. Physiolol Mol Plant Pathol 108:101441

Villas-Bôas SG, Roessner U, Hansen M et al (2007) Metabolome analysis: an introduction. John Wiley & Sons, Inc., New Jersey, NJ

Wang L, Sun X, Weiszmann J, Weckwerth W (2017) System-level and granger network analysis of integrated proteomic and metabolomic dynamics identifies key points of grape berry development at the interface of primary and secondary metabolism. Front Plant Sci 8:1066

Ward JL, Baker JM, Beale MH (2007) Recent applications of NMR spectroscopy in plant metabolomics. FEBS J 274(5):1126–1131

Wehrens R, Weingart G, Mattivi F (2014) metaMS: an open-source pipeline for GC-MS-based untargeted metabolomics. J Chromatogr B Analyt Technol Biomed Life Sci 966:109–116. https://doi.org/10.1016/j.jchromb.2014.02.051

Wishart DS (2007) Current progress in computational metabolomics. Brief Bioinform 8:279–293

Wishart DS (2011) Advances in metabolite identification. Bioanalysis 3:1769–1782

Wishart DS (2019) NMR metabolomics: a look ahead, vol 306. J. Magn, Reson, p 155

Xia J, Bjorndahl TC, Tang P, Wishart DS (2008) MetaboMiner—semi-automated identification of metabolites from 2D NMR spectra of complex biofluids. BMC Bioinformat 9:507

Xia J, Psychogios N, Young N, Wishart DS (2009) MetaboAnalyst: a web server for metabolomic data analysis and interpretation. Nucleic Acids Res 37:W652–W660

Xia J, Wishart DS (2010) MetPA: a web-based metabolomics tool for pathway analysis and visualization. Bioinformatics 26(2342):2344

Yang L, Fountain JC, Ji P, Ni X, Chen S, Lee RD, Kemerait RC, Guo B (2018) Deciphering drought-induced metabolic responses and regulation in developing maize kernels. Plant Biotechnol J 16:1616–1628

Zhan X, Patterson AD, Ghosh D (2015) Kernel approaches for differential expression analysis of mass spectrometry-based metabolomics data. BMC Bioinformat 16(1):77

Zhang F, Robinette SL, Bruschweiler-Li L, Brüschweiler R (2009) Web server suite for complex mixture analysis by covariance NMR. Magn Reson Chem 47(1):1–10

Zhen S, Dong K, Deng X, Zhou J, Xu X, Han C, Zhang W, Xu Y, Wang Z, Yan Y (2016) Dynamic metabolome profiling reveals significant metabolic changes during grain development of bread wheat (*Triticum aestivum* L.). J Sci Food Agric 96(11):3731–3740

Zhou J, Ma C, Xu H, Yuan K, Lu X, Zhu Z, Wu Y, Xu G (2009) Metabolic profiling of transgenic rice with cryIAc and sck genes: an evaluation of unintended effects at metabolic level by using GC-FID and GC–MS. J Chromatogr B 877(8–9):725–732

Zhu J, Sova P, Xu Q, Dombek KM, Xu EY, Vu H, Tu Z, Brem RB, Bumgarner RE, Schadt EE (2012) Stitching together multiple data dimensions reveals interacting metabolomic and transcriptomic networks that modulate cell regulation. PLoS Biol 10(4):e1001301

Zivy M, Wienkoop S, Renaut J, Pinheiro C, Goulas E, Carpentier S (2015) The quest for tolerant varieties: the importance of integrating "omics" techniques to phenotyping. Front Plant Sci 6: 448

Biofortification of Rice (*Oryza sativa* L.) 6

Sushil Kumar Singh, Jyoti Prakash Sahoo, Sandhani Saikia, Pranaya Pradhan, Ambika Prasad Mishra, Laxmipreeya Behera, Devraj Lenka, and Narayan Panda

Abstract

As a result of increased population, rising per capita incomes, and urbanisation, global agricultural production is increasing, and food demand is expected to continue growing over several decades. Approximately 60% of total calories consumed in developing countries come straight from cereals, with values reaching 80% in developing countries. Rice is the essential source of calories for humans amongst grains. Over half of the world's population is fed on rice. More than 2 billion of them suffer from "hidden hunger," as they do not consume enough nutrients or micronutrients in their regular diet. As part of a complete food

S. K. Singh
DBT North East Centre for Agricultural Biotechnology, Assam Agricultural University, Jorhat, India

J. P. Sahoo (✉) · L. Behera
Department of Agricultural Biotechnology, Odisha University of Agriculture and Technology, Bhubaneswar, India

S. Saikia
Department of Agricultural Biotechnology, Assam Agricultural University, Jorhat, India

P. Pradhan
Department of Nematology, Odisha University of Agriculture and Technology, Bhubaneswar, India

A. P. Mishra
Faculty of Agriculture, Sri Sri University, Cuttack, India

D. Lenka
Department of Plant Breeding and Genetics, Odisha University of Agriculture and Technology, Bhubaneswar, India

N. Panda
Department of Soil Science and Agricultural Chemistry, Odisha University of Agriculture and Technology, Bhubaneswar, India

R. Deshmukh et al. (eds.), *Biofortification in Cereals*,
https://doi.org/10.1007/978-981-19-4308-9_6

systems approach, biofortification is an effective technique for nutrition enrichment, which refers to developing a micronutrient-rich diet by utilising traditional breeding practices and sophisticated biotechnological tools. To enhance the profile of rice grain for biofortification-related properties, researchers must first understand the genetics of critical biofortification characteristics. The polishing procedure removes essential nutrients from white milled rice grains. As a result, seed-specific critical nutrient absorption is necessary. Significant increases in iron and zinc and many other essential minerals and provitamins are acquired in rice grain using the biofortification strategy. Most indica and japonica rice types have been biofortified over the world, giving them the titles of "high-iron rice," "low-phytate rice," "high-zinc rice," and "high-carotenoid rice" or "golden rice". Different techniques of rice biofortification, as well as their effects, have been explored in this chapter.

Keywords

Rice · Biofortification · Hidden hunger · Micronutrients · Bioavailability

6.1 Introduction

Oryza sativa L., commonly known as rice, is the most demanding crop worldwide as a staple food. Enriching it with essential nutrients, which are otherwise absent, would solve nutrient deficiencies to a great extent. To achieve successful biofortification, the mechanism of the particular nutrient uptake and the genes involved have to be elucidated and studied. Essentially those varieties are targeted for biofortification with highly dense micronutrient-rich traits which already have highly preferable agronomic characteristics in the genomic background. Supplements or industrially fortified food can deliver a high level of essential micronutrients to human body. Even if the biofortified rice cannot suddenly increase the concentration in the human body, it can increase the daily sufficiency of micronutrient absorption throughout human life cycle (Bouis et al., 2011).

Minerals can be defined as elements present in the food that human body requires for its healthy growth and development. Out of the 16 essential minerals, 11 are either present abundantly in the traditional food sources, or their deficiency is seen in sporadic cases. Five of these crucial minerals, such as iodine (I), iron (Fe), zinc (Zn), calcium (Ca), and selenium (Se), are present in a limited concentration in traditional food sources. Their deficiencies can lead to severe health disorders. Diseases result from a lack of these minerals when a staple food such as ground grains (with low biomineral content) dominates the diet (Christou and Twyman 2004). Developing countries face a significant crisis in mineral deficiency, lacking in availability of fresh and hygienic foods (Gómez-Galera et al. 2010). However, lacking nutrients like calcium is a common health concern even in the developed countries. Providing access to a more nutrient-rich and diverse diet is challenging in developing and less developed countries. Therefore, biofortification can be a sustainable way to eliminate deficiency diseases in these countries (Bouis and Saltzman 2017).

Table 6.1 Details of economic status and hunger index of major rice-growing countries worldwide

Country	Population (crores) (2019–2020)	Rice production[a] (million metric tonnes) (2019–2020)	GDP per capita (2019–2020) (USD)	GDP growth rate (2019–2020) (% annual change)	Hunger index[b] (2019–2020)
China	139.77	148.5	10,261.68	6.1	<5
India	136.64	116.42	2099.60	4.2	27.27
Indonesia	27.06	36.7	4135.57	5.0	19.1
Bangladesh	16.3	34.91	1855.74	8.2	20.4
Vietnam	9.65	27.77	2715.28	7.0	13.6
Thailand	6.96	20.34	7806.74	2.4	10.2
Philippines	10.81	11.73	3485.08	6.0	19.0
Brazil	21.1	7.14	8717.19	1.1	<5

*Data was obtained from
[a]https://www.statista.com/statistics/255945/top-countries-of-destination-for-us-rice-exports-2011/
[b]https://www.globalhungerindex.org/ranking.html

China assembled parboiled rice about 148.5 million metric tonnes in the 2018/2019 crop year, more than almost any other country. In that crop year, by producing 116.42 million metric tonnes of parboiled rice, India came second. In the 2018/2019, the total production of parboiled rice in volume was 495.9 million metric tonnes worldwide. The largest rice-consuming countries are China, India, and Indonesia, respectively. The consumption of rice in 2018/2019 China was 143.79 million metric tons, and per capita intake in the world has remained remarkably stable since 2000, averaging about 53.9 kg per year (https://www.statista.com/statistics/255945/top-countries-of-destination-for-us-rice-export-2011/) (Table 6.1). Hidden starvation, caused by insufficient essential nutrients, is a major problem affecting approximately two billion people worldwide. Iron deficiency leads to anaemia accounting for 30%–40% (Global Burden of Disease 2015, Collaborators in Disease and Injury and its prevalence, 2016). Children and women are most likely to be infected with IDA. IDA has been shown to impair physical development, decrease immunity, and increase the likelihood of maternal and perinatal death.

The breeding target for Fe to meet the approximate average Harvest Plus requirement of 30% for women and children in polished rice is 13 mg/g or a five to six fold increase in grain iron in ordinary rice. Although wheat contains the dry weight of 59 mg/g of iron, which is twice the amount found in other cereals (Bouis et al. 2011), it is challenging to develop a more nutritious diet that can alleviate micronutrient deficiencies in developing and least developed countries. With rice consumed by half the world's population, genetically engineered rice grown explicitly to counteract "hidden hunger" is perhaps the best promising staple crop. Brown rice, the healthiest type of processed rice, is made by hulling raw rice (field harvested paddy). Unpolished brown rice is rich in iron, zinc, copper, calcium, and phosphorous, as well as vitamins such as thiamine-B1, riboflavin-B2, niacin-B3, pantothenic acid-B5, pyridoxine-B6, biotin-B7, folic acid-B9, and tocopherol-E, but is low in vitamins A, C, and D (Ghosh et al. 2019).

On the other hand, the average consumer prefers white rice grains because of their softness, lightness, ease of digestion, better consuming properties, and less time for cooking. The bran layer and the substrate, embryo, and a small part of the endosperm are removed from polished (milled) white rice (Champagne et al. 2004). Milled rice has poor nutritional quality in comparison with brown rice, with the reduced iron content lowered by 2.14 times to 4.75, and the zinc content lowered by 1.83 times, and essential minerals, fats, fibres, proteins, and vitamins lowered by 1.83 times (Masuda et al. 2009). However, the amounts of mineral reductions can vary between rice varieties and grain polishing processes. Whilst greater awareness and education have improved in the consumption of brown rice, the major rice consumers still prefer polished white rice by considering that the polished white rice was developed as nutritionally improved through biofortification (specific endosperm), leaving scientists to reconsider. Improving essential nutrients to be bioavailable in the edible parts of staple foods through conventional breeding, biotechnology techniques, or agricultural strategies can help alleviate deprivation in places where staple foods are the primary source of micronutrients and calories (Bouis and Saltzman 2017).

6.2 Understanding of Essential Mineral Uptake by Plants

6.2.1 Iron

Even though there is abundant iron available in the soil, most remains unused due to its low solubility. Plants have evolved mechanisms for accessing the insoluble iron from the ground in a highly regulated manner. This mechanism has been divided into two strategies (Marschner et al. 1986). By intensifying the phenolic and proton release (mediated by the enzyme H+ ATPase) into the rhizosphere, nongraminaceous plants bring the soil's pH down. Phenolics efflux zero 1 (PEZ1) transports the phenolics to facilitate and use the precipitated Fe (Takahashi et al. 2011). Iron is converted into its more soluble ferrous form by activating the ferric-chelate reductase expression. The soluble ferrous ion is shifted to the plasma membrane via its iron transporter, IRT1. On the opposite hand, graminaceous plants facilitate iron uptake by exuding MA, which are iron (III) chelators from plant roots, which states strategy II (Takagi et al. 1984). Phytosiderophores (PS) are synthesized and secreted due to the activity of nicotianamine synthase (NAS), nicotianamine aminotransferase (NAAT), and deoxymugineic acid synthase (DMAS) (Shojima et al. 1990). A soluble complex called the Fe (III)-the binding of Fe3+ forms PS with PSs (Fig. 6.1). The YSL proteins (yellow stripe-like proteins) facilitate the inhibition of these complexes from the rhizosphere to the root cells. (Nozoye et al. 2011). Rice uses strategy II, but it can also take up iron via IRT1 from the rhizosphere.

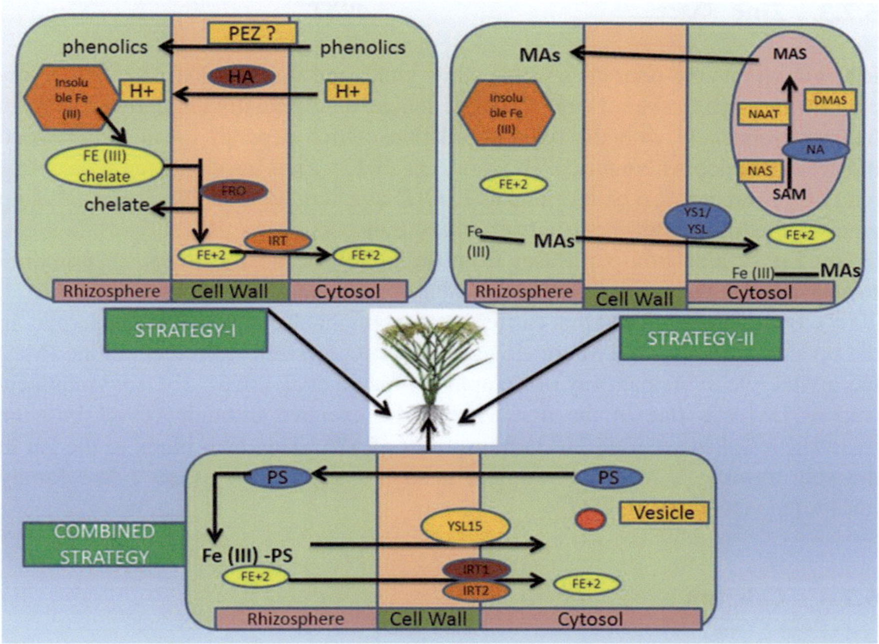

Fig. 6.1 Iron absorption and translocation in rice

6.2.2 Iodine

Iodine makes its way into the plants via two pathways: atmosphere to the plant pathway and soil to the plant pathway. The stomata of the plants and the roots take in iodine from the atmosphere and soil, respectively (Shaw et al. 2007). After iodine has been applied to the plants as IO_3^-, iodate reductase reduces it to I, which responds to iodine absence and presence in the environment (Kato et al. 2013). However, Lawson et al. (2015) reported that plants could readily absorb IO_3^- instead of I^-. Due to the more availability of IO_3^- it in the soil, it has been hypothesized that plants are taken up more efficiently by plants rather than I^- which is relatively more thermodynamically stable. Plants absorb I^- form iodine via chloride transporters and proton pump-actuated ion channels (White and Broadley 2009). In this process, perchlorate, nitrate, thiocyanate, and other anions may cause absorption interference (Voogt and Jackson 2010). The identity of iodine transporters has not been established with certainty, but it has been assumed that they share their activities with other transporters such as Na: K/Co-transporters (Landini et al. 2012).

6.2.3 Zinc

Rice grown on zinc-poor soils yields a low yield, and the nutritional quality is very poor (Welch and Graham 1999). The zinc deficiency eventually leads to weakness in humans, especially with the large population, which depends on the staple food mostly in developing countries (Hussain et al. 2012). Zn is usually taken up as a free ion (Zn^{2+}) and can also be taken up as $ZnOH^{+}$ at a very high pH. The plant takes up zinc via the transporter-mediated secondary transport across the root plasma membrane. ZIP or zinc iron permeases from the family of metal carriers are primarily responsible for the rapid absorption of zinc from the environment (Palmgren et al. 2008). But they speculated that carrier proteins might also have some role to play in the uptake of zinc; likely Mugineic acid (MA) synthesized from methionine (Met) has a vital role in the pathway of Zn uptake (Suzuki et al. 2006). 2′-Deoxymugineic acid or DMA is one of the first MAs to be excreted through TOM1 into the rhizosphere (Nozoye et al. 2011) transporter of MA. This then binds to the Zn in the soil forming a MA-Zn complex, absorbed via yellow stripe 1-like family transporters (Inoue et al. 2009).

6.2.4 Calcium

Calcium is an essential mineral needed for growth, protection from pathogens, and development of the plant. It promotes proper plant cell elongation and is a vital participant for enzymatic and hormonal processes. Plant calcium uptake is mediated in roots by the expressed plasma membrane channels. Scientists classified permeable calcium channels into DACCs (depolarisation-activated channels) and HACCs (hyperpolarized-activated channels) based on their electrophysiological properties (Miedema et al. 2001). Calcium can be taken up through the ischemic pathway as well as the ischemic pathway. However, a higher concentration of calcium causes cell toxicity. Therefore, to maintain a sub-micromolar concentration of calcium in the cytosol, the cell retains the ability to generate a calcium signalling stroke pathway. White (2001) suggested that through the cortex from the epidermis, Ca moves apoplectically till it reaches the endodermis Casparian strip, which is mainly made of suberin and lignin. This strip composition makes it impermeable for the movement of water and solutes (Schreiber et al. 1999). Upon arriving at this layer, Ca move with the help of channels to enter the endodermal cell cytosol. With the help of Ca^{2+}-ATPases or Ca^{2+}/H^{+} antiporters, it is exported to the stele apoplast to be finally loaded into the xylem. After that, the calcium is ultimately distributed to the leaf cells through the shoot (White and Broadley 2003).

6.2.5 Selenium

Incorporation of selenium into two amino acids selenomethionine and selenocysteine is found in cells. The most common form of Se available in soil for uptake by plants is selenate (SeO_4^{2-}), which is more water soluble than that of selenite. Selenium is found in various forms in both acidic and alkaline soils. Selenate is the standard form found in alkaline soils, and acidic soils contain selenite. Both differ in their motility rates and absorption capacity and are subsequently broken down into seleno compounds (Li et al. 2008). The root cell membrane contains selenium uptake transporters. For selenite uptake, the phosphate transport mechanism is responsible (Li et al. 2008), and selenite is transported on the other hand by sulphate transporters (Dong et al. 2003). The nutritional state of the plant, both in and out, is mainly responsible for using transporters for selenite or sulphate uptake (White et al. 2004). Based on the cumulative capacity of plants on Se, they have been classified into over-accumulators, secondary accumulators, and non-compounds (Bodnar et al. 2012). Plants that can accumulate more than 1000 mg of Se/Kg DW and thrive well in selenium-rich soils are called hyperaccumulators. Plants get Se to the extent that the plant shows no sign of toxicity (100–1000 mg of Se/kg DW), and they are called secondary accumulators. Plants that accumulate less than 100 mg citrate/kg of their dry weight are termed non-accumulators. They show retarded growth or fail to survive.

6.3 Transgenic Efforts for the Development of Golden rice

Children who have blindness have been a result of severe vitamin D deficiency. Developing countries face deficiency diseases due to the lack of proper nutrients in their diet. Physiological traits such as cellular differentiation, growth, reproduction, and vision, to name a few, depend on the role played by carotenoids (Wurtzel et al. 2012) as they are known to help tackle several ROS-generated diseases, namely, cancers and neurological and cardiovascular diseases along with eye disorders (Bai et al. 2011). β-Carotene biosynthetic pathways in rice have been the target for the creation of "golden rice." Golden rice enriched with β-carotene was formulated in supplementing provitamin A, which naturally enhances immortality caused by Vitamin a deficiency (VAD). Phytoene synthase (PSY) of daffodil (*Narcissus pseudonarcissus*) and phytoene desaturase (crtI) of pathogenic bacteria (*Erwinia uredovora*) were chosen to be introduced and expressed in the endosperm of IR64 and BR29, which are Asian rice varieties under the endosperm-specific promoter (Datta et al. 2007) to produce golden rice. The β-carotene synthesis pathway is already present in rice.

However, it is only expressed in the leaves and not in the grains. This pathway can be reactivated for β-carotene synthesis to take place in the grains by incorporating the genes mentioned above. β-Carotene (carotenoids with at least one unsubstituted β-ionone ring) possesses provitamin A activity. Rabbani et al. (1998) stated that tissues that had a high level of lipid accumulation could act as a lipophilic sink, which can drive the formation of carotenoids. But rice had a non-carotenogenic endosperm which has low lipid concentration. This leads to the nonavailability of carotenoid deposition in the rice endosperm. There were also doubts on the presence of precursors of the carotenoid biosynthesis pathway in the grains of golden rice. Due to many questions arising on the success of the golden rice project, a lengthy research phase was conducted to answer the questions. Ye et al. (2000) reported that the breakthrough achieved by Prof. Peter Beyer and Prof. Ingo Potrykus led to the discovery that incorporation of only two transgenes was necessary for golden rice to see the light of the day.

Plant phytoene synthase (PSY) was the first transgene to form phytoene from geranylgeranyl-diphosphate (GGPP), which is found endogenously. Bacterial carotenoids (CRTI) are encoded by a second gene that adds four double bonds that lead to conjugation. PSY and CRTI, in combination, form a red compound called lycopene, which has been observed and established in tomatoes. However, in any rice, the transformant accumulation of lycopene has never been observed. Lutein and zeaxanthin (oxygenated carotenoids) have instead been found in association with α- and β-carotene. It was revealed by observing the carotenoid pattern in the grain endosperm that the pathway went beyond the point that was supposed to be the peak of the transgene's action. The wild-type rice endosperm produces lycopene cyclases (LCDs) and α- and β-carotene hydroxylases (HYDs) downstream. However, PSY and the plant carotene desaturases are not present in the same. This results in the use of lycopene produced by PSY and CRTI as substrates for the enzymes mentioned above, leading to the formation of the noticeable products (Fig. 6.2).

6.3.1 Golden Rice: First Attempt

Ye et al. (2000) proved in the early development years that rice grains could produce β-carotene. It has been quite evident that there was a need for only transgenes (PSY and CRTI) for β-carotene production. There was, however, no need for lycopene synthase. Although initial studies were carried out in japonica cultivar, indica varieties were also included later (Hoa et al. 2003). Many ways were developed in rice seeds to improvise with the production of carotenoids to the permissible levels. The first generation of golden rice or GR1 was transformed with two transgenes (from daffodils and bacteria) and placed under an endosperm-specific gt1 promoter. Field carotenoid levels amounted to four times to that of the proposed model.

6.3.2 Golden Rice 2

Golden rice 1 gave us the possibility to produce β-carotene in rice endosperm. It also helped us see that to tackle vitamin A deficiency, β-carotene should be produced in higher amounts. Since only two transgenes have been involved in the production of β-carotene, it was simple to understand that by manipulating the enzymatic activities of the products of the two genes, one can achieve higher β-carotene. Normal pathways usually have specific rate-limiting steps which control the entire path. Overcoming the rate-limiting step simply enhances the concentration of the rate-limiting enzyme or by shifting to a more active enzyme. Different PSY sources were examined to see that the maize and rice genes were more efficient (Paine et al. 2005).

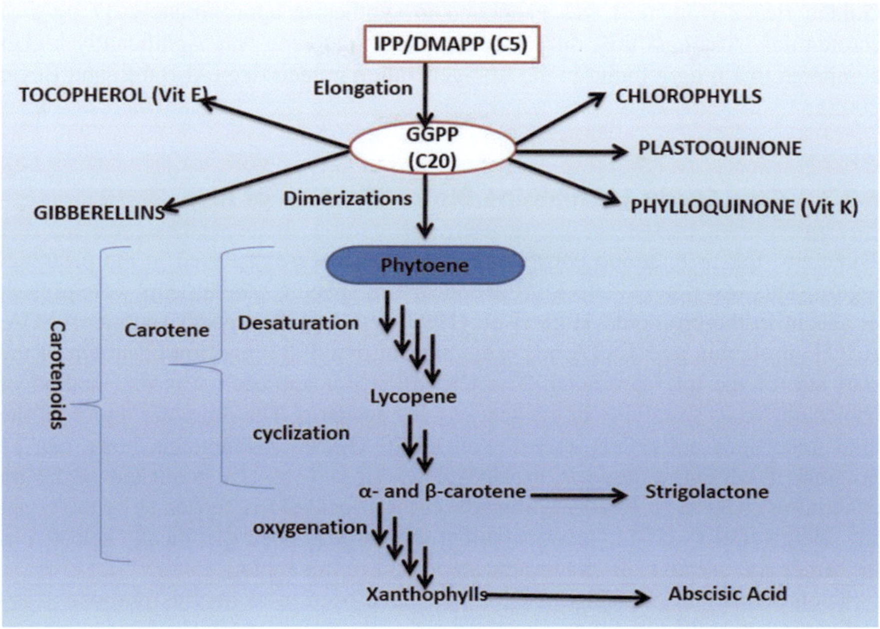

Fig. 6.2 Overview of carotenoid synthesis pathway: isopentenyl-diphosphate (IPP) and dimethylallyl-diphosphate (isomer of IPP-DMAPP), C5 compound is the starting molecule for this pathway. On elongation by C5 units, geranylgeranyl-diphosphate (GGPP) (C20 compound) is formed, a precursor for many biosynthetic compounds. Phytoene, the first carotenoid in the pathway, is formed due to the dimerisation of GGPP. After that, after a series of desaturation events, lycopene, a coloured compound, is included. Further cyclisation events lead to the production of β- and ε-ionone rings. Additional oxygenation of the rings eventually produces xanthophylls (Figure courtesy: www.goldenrice.org)

Fig. 6.3 Comparative visualisation of wild type, golden rice 1, and golden rice 2. (image courtesy: goldenrice.org)

Golden rice 2 (Fig. 6.3) was generated to synthesize approximately 37 μg/g of carotenoids. Amongst this, only 31 μg/g had β -carotene was significantly higher compared to 1.6 μg/g found in the first-generation golden rice (Al-Babili and Beyer 2005).

6.3.3 Case Study 1: Thiamine Biofortification of Rice

Three rice tpk or thiamine phosphokinase variants were analysed in the promoter region. This was done to check if the endosperm-specific *cis*-elements were present or absent in the promoter. Higo et al. (1999) reported that motifs such as AACA, ACGT, prolamin, and TATA box were present in a 300 bp region upstream of the start site of the tpk3 promoter. The CRISPR-Cas9 approach was the method of choice due to its simplicity in the gene-editing scenario. The promoter region of the tpk3 gene does not reflect an essential motif, GCN4. Adding this motif can be possible if editing is done at a selected position. A gRNA sequence of 20 bp containing NGG as an adjacent protospacer motif or PAM is chosen to be the target site. Heigwer et al. (2014) mentioned that this site had been specifically selected as the target site because off-site targets are absent in this region.

Tools like E-CRISP and Cas-OFFinder have been used for this purpose. Cas9 cleaves a part in the DNA 3 to 4 nucleotides upstream of PAM. The introduction of GCN4 box at the site of editing a very stable transformation with Cas9 and gRNA is essential. Homologous recombination promoting oligonucleotide has been used for co-transformation for the introduction of the GCN4 box (Zhang and Huang 2012). The box absence has been synonymous with the expression of tpk3 in rice, which is quite negligible. With the introduction of the edited gene, enhanced propagation of tpk3 is predicted. Higher knowledge of this pathway is needed for opportune enhancing to be performed for higher thiamine manufacturing in rice. A summary of the thiamine biosynthesis pathway in rice is listed in Fig. 6.4. The box absence has been synonymous with negligible expression of tpk3 in rice. With the introduction of the edited gene, enhanced production of tpk3 is expected. A better understanding of this pathway is needed for proper editing for higher thiamine production in rice. A summary of the thiamine biosynthesis pathway in rice is indexed in Fig. 6.4.

6.3.4 Case Study 2: Biofortification of High-Zinc Rice

Zinc is an essential nutrient for the proper absorption of iron in the body. Plants typically take up zinc from the soil. Several genetic approaches have been made to enrich the plants to increase the uptake of zinc from the rhizosphere. Zinc translocation and mobilisation involve a lot of genes. Overexpressing these genes for increased bioavailability may lead to an essential way by enhancing Zn content in rice grains. In rice, overexpressing NA synthase genes via the 35S enhancer element has contributed to its manifold increase. Transgenic rice harbours barley nicotianamine synthase gene HvNAS1, which showed a threefold higher zinc accumulation under the influence of rice actin1 promoter. OsIRT (ZIP family protein

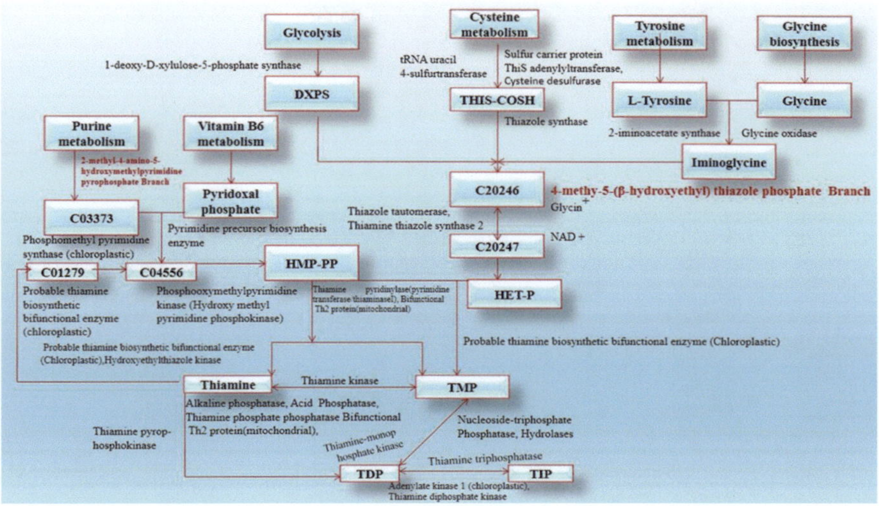

Fig. 6.4 Summary of thiamine biosynthesis pathway (Courtesy Minhas et al. 2018)

in rice) can be overexpressed for a higher concentration of zinc to accumulate in rice. OsZIP1, OsZIP2, OsZIP3, and OsZIP4 have been connected with Zn homeostasis (Ishimaru et al. 2007). Lee et al. 2009 concluded that in GE rice, high amount of Fe and Zn in rice grains resulted from the overexpression of OsIRT and MxIRT genes. Boonyaves et al. (2016) reported that the polished grains of GM rice accumulated at the highest concentration of Fe and Zn as a group of four genes (AtIRT1, Pvferritin, AtNAS1, and Afphytase) were channelled to rice. Many reports have also been published where there was overexpression of rice OsNAS genes, leading to a high accumulation of zinc in grains (Johnson et al. 2011). RNAi silencing was also used to increase zinc concentration in GM rice. Phytic acid metabolism pathway's MIPS gene was silent via this mechanism, and this increased the concentration of zinc, calcium, iron, and magnesium in rice grains. Ali et al. (2013) suppressed IPK1, another gene from the same pathway, to report an increased zinc concentration. Paul et al. (2012) stated that in PSII rice, the ferritin gene (Osfer2), was overexpressed to show a 1.37-fold increase of zinc (Fig. 6.5).

6.3.5 Case Study 3: RNAi Technology Low-Phytate Rice

Overall in most cereals, approximately 80% of the total phytic acid is accumulated at the aleurone layer of the grains except for maize. Phytate, a cumulation of salts, accumulates as phytic acid accumulates. Since phytate contains six negatively charged ions, including Fe^{2+}, Zn^{2+}, Ca^{2+}, and Mg^{2+}, and lowering their bioavailability. Rice mutant varieties (low phytic acid (LPA) phenotype) have been developed in a series of endeavours to truncate phytic acid levels (Hambidge 2000). Despite their

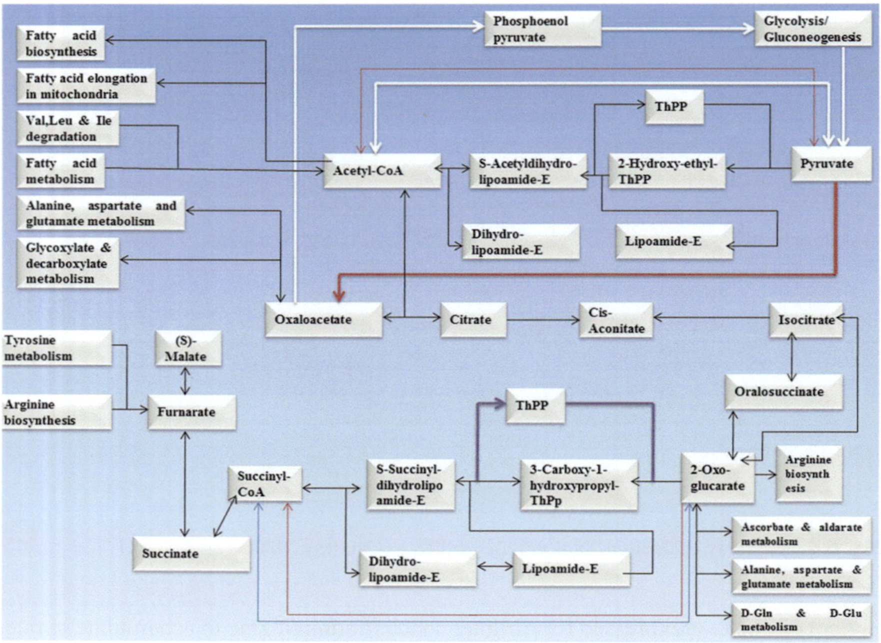

Fig. 6.5 TCA cycle in zinc-fortified rice variety (source: Boonyaves et al. 2016)

efficacy, these mutant lines obstructed crop yield and overall performance (Roda et al. 2020). Transgenic crops were developed as an alternative approach by suppressing the phytic acid biosynthetic pathway (Fig. 6.6) utilising RNA interference (RNAi) (Fig. 6.7).

6.4 Different Approaches for Improvement of Nutraceutical Properties in Rice Grain

For successful biofortification to occur, a proper understanding of nutrient uptake from the soil by the plants is necessary. In these consecutive years, progress has elucidated the pathways associated with various nutrient uptakes by plants from the ground. Hao et al. (2005) stated that rice crops rich in iron accumulate high amounts of Fe and Zn in the grain's endosperm tissues than iron-deficient plants. Molecular genetics has been able to divulge the process of zinc uptake by plants from the soil in various crops, including wheat and rice (Yang et al. 1994). Grotz and Guerinot (2006) have explained the critical process of chelation, distribution patterns, and iron, copper, and zinc transport mechanisms.

Biofortification programs have accommodated the idea of making the nutrients bioavailable in the plant system. It has been proposed to be more important than

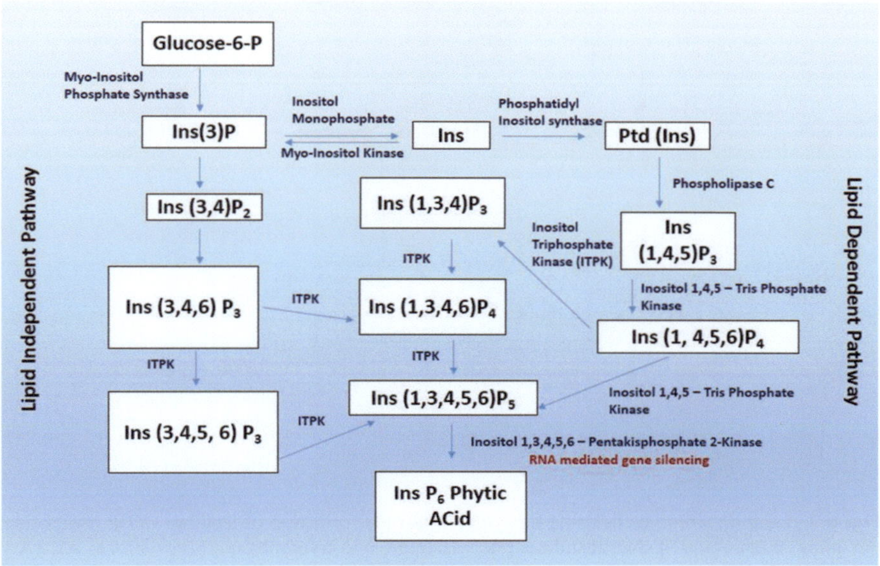

Fig. 6.6 Phytic acid biosynthesis pathway (source: Suzuki et al. 2007)

simply increasing the density of the nutrients. Phytic acid has been found to restrict micronutrient absorption (Nagashima et al. 2014). Thereby, there have been attempts to decrease phytic acid concentration in the plant system through enzymatic pathway modification. However, many antinutrients like phytic acid and polyphenols are required by the plants to build stress resistance and increase vigour (Yang et al. 2007a, b, c). They are also crucial for human health, considering their involvement in reducing heart diseases (Ferguson and Harris 1999). Hence, Welch and Graham (2004) rightly pointed out that antinutrient concentration manipulation via breeding approaches should be done carefully, keeping in mind the side effects associated with it. Various methods have been described for increasing the nutritional value of different crops. Some of them are listed below.

6.5 Breeding Approach

It has been very well documented in genetic diversity found on micronutrients in rice and other food crops (Yang et al. 2007a, b, c). Genetic diversity has been the primary target for developing nutritionally superior varieties by various scientists and breeders (Zapata-Caldas et al. 2009). Many programs create excellent varieties of rice, wheat, potato, bean, etc., with a higher amount of Fe, Zn, vitamin A, etc. (Pfeiffer and McClafferty 2007). In a program (in search for new donors) initiated by IRRI (International Rice Research Institute) in collaboration with the University of Adelaide, Australia, 7000 varieties have been evaluated for zinc and iron

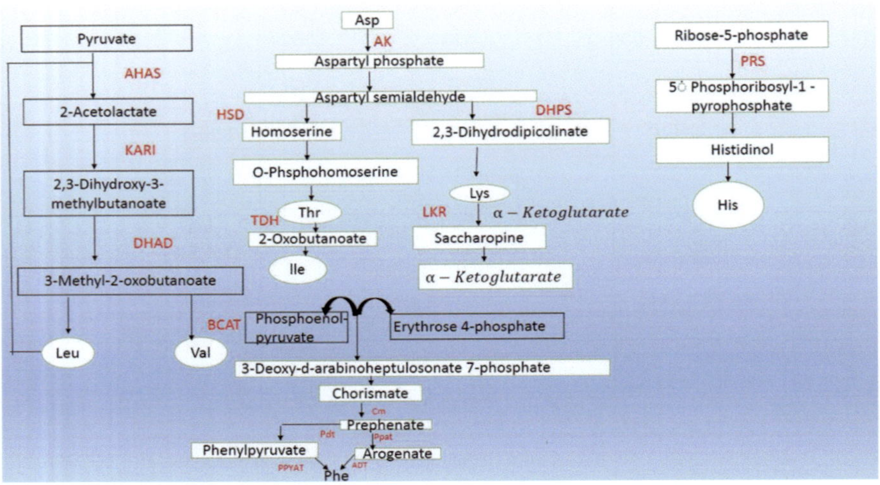

Fig. 6.7 Leading amino acids using biosynthesis for the formation of leucine, valine, isoleucine, threonine, methionine, lysine, histidine, phenylalanine, and tryptophan. Abbreviations: AK (Asp kinase), HSD (homoserine dehydrogenase), DHDPS (dihydrodipicolinate synthase), LKR (lysine ketoglutaric acid reductase), TS (Thr synthase), CGS (cystathionine synthase), TDH (Thr dehydratase), PRS (ribose-phosphate diphosphokinase), MGL (Met synthase), SAMS (S-adenosylmethionine synthase), AHAS (acetohydroxyacid synthase), KARI (ketol acid reductoisomerase), DHAD (dihydroxy-acid dehydratase), BCAT (branched-chain aminotransferase), DAHPS (3-deoxy-D-arabinoheptulosonate 7-phosphate synthase), CM (chorismate mutase), AS (anthranilate synthase), PDT (prephenate dehydratase), PPAT (prephenate aminotransferase), PPYAT (phenylpyruvate aminotransferase), ADT (arogenate dehydratase) (source: Wang et al. 2017)

concentration in the rice grains. Later on, Khush et al. (2012) reported rice grains with a higher zinc and iron concentration. A significant variation in iron concentration has been written for many staple crops like rice, wheat, maize, bean, cassava, etc. (Frossard et al. 2000).

Tiwari et al. (2009) suggested developing nutrient-rich cultivars of different crops, and selective breeding is used as a tool. IR 68144-3B-2-2-3 (IR72 X Zawa Bonday) is identified as an expanded Indian breeding line. Similarly, certain wheat varieties have been found to survive zinc-deficient soil despite maintaining high zinc concentration (Peleg et al. 2008). Through many scientific discoveries and research, it has been found that any food crop can be made nutritionally potent through simple breeding techniques. However, breeding techniques can only be successful if micronutrients are available in the soil for the plant to take in. In the past two decades, above 20 QTL mapping studies have occurred to study the genetic basis of rice protein matter (Mousavi et al. 2013). In addition, to our knowledge, more than 80 stable and reliable QTL GPC have been identified and mapped to all twelve rice chromosomes, with most identified on chromosomes 1, 2, 6, 7, 10, and 11 (Table 6.2).

Table 6.2 List of QTLs identified for biofortification traits in rice (adopted from Sharma et al. 2020)

Purpose	Population type	Cross	Number of QTLs	Chromosome	Phenotypic variability explained
Amino acid	RILs	Zhenshan 97 × Nanyangzhan	2 QTLs	1, 7	4.05–33.3
	RILs	Zhenshan 97 × Minghui 63	10 (His) + 8 (Arg)	1, 2, 3, 6, 7, 10, 11, 12 (His); 2, 3, 5, 6, 7, 10, 11, 12	12–35 (His); 16–33 (Arg)
	RILs	Zhenshan 97 × Minghui 63	12	1, 11	3.4–48.8
	RILs	Zhenshan 97B × Delong 208	3 QTLs	1, 7, 9	4.2–31.7
	RILs	Dasanbyeo × TR22183	6	3	10.2–12.4
Protein	RILs	Zhenshan 97 × Minghui 63)	2	6, 7	6.0–13.0
	DH	Caiapo × IRGC 103544	4	1, 2, 6, 11	4.8–15.0
	DH	Gui630 × Accession 02428	5	1, 4, 5, 6, 7	6.9–35.0
	BC3F1	V20A × Accession 103,544	1	8	9.0–10.0
	RILs	Moritawase × Koshihikari	3	2, 6, 9	2.3–16.3
	BIL	Kasalath × Koshihikari)	2	6, 10	14.3–14.8
	RIL	Chuan × Nanyangzhan	2	6, 7	2.69–4.50
	RILs	Xieqingzao B × Milyang 46		3, 4, 5, 6, 10	3.9–19.3
	RILs	Zhenshan 97 × Minghui 63	9	2, 3, 5, 6, 7, 10, 11, 12	1.60–9.26
	DH	Samyang × Nagdong	3	1, 11	6.92–22.98
	RILs	Asominori) × IR24)	10	1, 3, 4, 6, 7, 8, 9, 10, 12	8.53–23.70
	RILs	Zhenshan 97B × Delong 208	2	1, 7	7.2–25.9
	DH	Cheongcheong × Nagdong	1	2	39–41
	DH	CJ06) × TN1	1	10	12.3–15.8
	DH	Cheongcheong × Nagdong	3	8,9,10	39–40
	RILs	M201 × JY293	5	1, 2, 3, 4	6.74–13.50
	DH	Cheongcheong × Nagdong	1	7	14

(continued)

Table 6.2 (continued)

Purpose	Population type	Cross	Number of QTLs	Chromosome	Phenotypic variability explained
Zinc and iron	DH	IR64 × Azucena	Zn-2; GFe-3	9	Zn (1, 12); Fe (2, 8, 12)
	RILs	Zhengshan 97 × Minghui 63	GZn-3; GFe-2	Zn (5, 7, 11); Fe (1, 9)	Zn (5.3–18.61); Fe (11.11–25.81)
	RILs	Bala × Azucena	GZn-4; GFe-4	Zn (6, 7, 10); Fe (1, 3, 4, 7)	Zn (11.2–14.8); Fe (9.7–21.4)
	DH	ZYQ8 x JX17	GZn-2	4, 6	Zn (10.83–12.38)
	RILs	Madhukar) × Swarna	GZn-6; GFe-7	Zn (3, 7, 12); Fe (1, 5, 7, 12)	Zn (29–35); Fe (69–71)
	F2	PAU201) x Palman 579	GZn-3; GFe- 8	Zn (2, 10); Fe (2, 3, 7, 10, 12)	Zn (4.7–19.1); Fe (2.4–26.8)
	RILs	Swarna X Moroberekan	GFe-1	1	Fe (39)
	F4	PAU201 x Palman	GZn-1; GFe-5	Zn (6); Fe (5, 7, 9)	Zn (25); Fe (34.6–95.2)

6.6 Agronomic Approach

Cakmak (2008) observed that with the varied and worldwide use of high-yielding varieties of food crops, the soil had been stripped of its fertility. The soil deficient in micronutrients cannot help in attaining the biofortification of crops through conventional breeding techniques. In such a scenario, micronutrient fertilisation of such ground is needed for the plants to show any response. There have been reports of nutrient concentration improvement in crops after adopting fertilisation practice. For increasing productivity and at the same time increasing micronutrient density in grains, zinc fertilizer was applied to the rice, pea, and cowpea (Hu and Lutkenhaus 2003; Li and Vasanthan 2003; Fawzi et al. 1993). However, the method of microelement application makes a significant impact on the accumulation of the concerned micronutrient in the parts of the plants that can be consumed.

Missana et al. (2009) conducted an experiment, where it was observed that, if zinc is sprayed on wheat leaves during its early dough or milk stage, the impact is much more than that of fertilizer application to the soil zinc concentration by many folds. One of the most critical consequences of foliar application of zinc in rice is that phosphorus accumulation in the grains is reduced, thereby reducing the amount of phytic acid. It has been observed that iron fertilizers are not effective enough to make a difference (Missana et al. 2009). Even if iron is applied in the foliar form to the plants, it is quickly converted to the insoluble form, which becomes unavailable for the plants to use (Frossard et al. 2000). Suggestions were made to use organic iron as micronutrient fertilizers. It has been reported that, if iron is chelated with other compounds like Ethylene diamine tetra acetic acid (EDTA), Diethylenetriamine pentaacetate (DTPA), or ethylenediamine-N,N'-bis(2-hydroxyphenylacetic acid) (EDDHA).

Waters et al. (2009) noticed that the same genetic process controls nitrogen and iron transportation in the vegetative parts of the crop plant. Hence, to make Fe biofortification a success, special care should be taken for enriching nitrogen in crops. It has been reported by many researchers like Gunes et al. 2007 that intercropping of cereals with dicot plants facilitates the enhancement of iron and nitrogen amount in plants as a result of interspecific root interactions. Selenium fertilizers like Na_2SeO_4 and K_2SeO_4 have proved effective during foliar application and immediately increased selenium in plants (Stroud et al. 2010). However, Yang et al. (2007a, b, c) reported that agronomic approaches were very limiting considering factors like soil-chemical interaction, making the micronutrients less available to the plants, reducing transportation efficiency, reducing the roots' concentration, etc (Miedema et al. 2008).

6.7 Biotechnological Approach

Biotechnology has provided very modern and permanent approaches for biofortifying micronutrients in cereal crops. Researchers like Raboy (2002) and Tucker (2003) have stated that a higher concentration of vitamins and nutrients can be built up by modifying the plant's genetic makeup. The synthesis of antinutrient compounds can be reduced. Transgenics has played an essential part in the biofortification of micronutrients in crops. Agrobacterium-mediated transformation was used to transfer the iron storage protein gene, ferritin, from the French bean to the rice crop (var. Kitaake) (Lucca et al. 2001), increasing the iron concentration.

A synergistic effect was noticed on the storage and uptake of iron concentration in rice endosperm upon evaluation of two transgenes, i.e. NAS (AtNAS1) from *Arabidopsis thaliana* and ferritin (Pvferritin) from *Phaseolus vulgaris* in rice. Phytosiderophore biosynthetic pathway in rice has been modified by introducing nicotinamide aminotransferase gene via transgenic approach. This resulted in transgenic rice with high resistance to growth in iron-deficient soils and maximum yield than those in controls (Takahashi et al. 2001). White and Broadley (2005) reported that these transgenic approaches could be used to increase material accumulation. On the other hand, stimulating agents reduce the concentration of antinutrients.

The Clustered Regularly Interspaced Palindromic Repeats (CRISPR) genome editing tool for a specific change within the genome was recently developed, giving scientists the opportunity to precisely target genes or intergenic regions. Rice has benefited from this technology in yield and stress tolerance (Mishra et al. 2018). Using a CRISPR to knockdown OsVIT2 to increase grain Fe, similar to the reported

T-DNA insertion attempting to silence a gene in different rice cultivars (Bashir et al. 2013), could be a key illustration. Alterations of genes' expression resulted in iron homeostasis by a simple edit to the regulatory element of iron homeostatic genes, which results in iron-enriched rice and wheat could withal benefit from this method (Liu et al. 2004).

6.8 Challenges for the Public Release of Golden Rice

In 1999, the first proof-of-concept β-carotene fortified rice was created. After many modifications made to the prototype golden rice, the inventors gave away the technology and any future version of it for the improvised world to reap the benefits. However, to date, advantages due to this technology have not yet reached the malnourished people in the world. There are varied reasons for the delay. The majority of these reasons are listed below:

1. Golden rice contains genes that have been sourced from other organisms, thereby making it a genetically modified crop. GM crops are those crops that include at least one gene from another species. Different national governments strictly regulate GMO crops. Officials are appointed that keep data of the altered genome structure, with the potential of causing allergies compared with the known allergy-causing agents in the databases, relation, and comparison with the standard genomes of the same crop or any such data that can certify that the GMO is non-invasive and is safe for cultivation and human consumption. Only when the regulatory bodies give such clearances to the applicant after being judged safe, the GMOs can be cultivated and consumed in that country.
2. Money and time invested in creating golden rice could have been better utilized for looking at solutions already available for (VAD). Some of the readily available solutions are vitamin A tablet food fortification to educate people of the weaknesses caused by the lack of proper nutrition. Instead of focusing on the magic of golden rice, many people believe that, concentrating more on other readily available options would be the key to eradicating VAD. There have been apprehensions amongst the farming community about the cultivation of GE golden rice. It may lead to contaminating other non-GE wild-type rice varieties (via cross-pollination). According to a publication by Greenpeace Southeast Asia, GE golden rice can contaminate non-GE rice, thereby negatively affecting traditional and organic farming, which in turn would affect the rural livelihood scenario. If GE golden rice shows adverse effects, then the contamination would be the main bottleneck in countries having rice as a primary or staple food (Cakmak and Kutman 2018).
3. Some section of the society also feels that it is irresponsible for the scientific community to impose the golden rice technology on people who are unwilling to take it for various reasons like going against their religious beliefs or cultural heritage. IRRI or the International Rice Research Institute has also addressed biodiversity loss and maintains that these claims are baseless and non-scientific. The farming community is worried about the introduction of golden rice, pushing

the farmers to enter industrial monoculture production. However, scientists also reported that the practice of monoculture would lead to adverse loss of biodiversity in the local ecosystem (Ishimaru et al. 2011).

6.9 Economic and Social Constraints for the Biofortified Rice

When food crops are enhanced with added nutrition, using modern biotechnological techniques to make the nutrients more readily available to the human population is referred to as biofortification. Even though a diet rich in various types of food is recommended for daily micronutrient intake, the poor section of the society depends on the cheap and reliable source of calories, *i.e.* rice and wheat. Biofortification promises to deliver micronutrients to the impoverished community as a cost-benefit method for its single investment in breeding. The process of biofortification will rely primarily on public funding. Therefore, many factors will come into play, including the R&D resources dedicated to this effort. If private financing is to be considered, then returns would have to be generated either in the form of hybrids or intellectual property rights that prohibits farmers from selling the seeds of their produce (Landberg et al. 2009).

Biofortification does not fall under this model as it has been generated for public use worldwide. Biofortification also faces a challenging limitation. High production cost, i.e. equipment, technology, patenting, etc., is a cause of concern that can hamper this method to go global. Powell (2007) stated that the profit margin for private investors is shallow in the case of biofortified crops. Scarce public funds aggravate the situation. Society's involvement during the designing process also plays a significant role in making any biofortified crop a reality. The existence of a deficiency makes the research process more rational or sensible. If society cooperates, then acceptance becomes a lot easier. There is a cultural or local limitation to golden rice's success, i.e. the value of pure white rice in various cultures and traditions (Thurber and Fahey 2009), the ways of feeding children, and openness towards newer discoveries. Another major constraint is the lack of education which controls the socio-economic status. Consumers should make an informed decision about golden rice, for which education in the necessary field is of utmost importance. The consumers should be in a state to examine critically (Kettenburg et al. 2018). Empowering women and children through education and capacity building would help make golden rice more acceptable. Kettenburg et al. (2018) stated that a complete evaluation of golden rice would need to be conducted on monetary terms to realize its economic efficiency finally. Biofortification can very well help in eradicating malnutrition. However, scientists need to move out and disseminate the knowledge to the masses so that an informed decision can be taken.

6.10 Conclusion

Biofortification is a sustainable agricultural method having minimum cost and has a positive impact on facilitating the well-being of the world's most sizably voluminous undernourished people. Biofortification strategies predicate on crop breeding targeted genetic modification, and the application of mineral fertilizers has an abundance of promise for addressing human mineral malnutrition (Saini et al. 2020). Endeavouring to develop biofortified victuals crops with higher nutritional content, such as Fe, Zn, Se, and provitamin A, ascertains that sufficiency of these and other micronutrients is accessible in the diets of developed and developing countries. In achieving this goal, international initiatives, like CGIAR (Consortium of International Agricultural Research Centers), the Centres Collaboration with HarvestPlus, and national initiatives, serve as pillars. These activities have resulted in crops that can increase both the amounts and bioavailability of essential minerals in human diets, particularly in class grain crops such as rice, wheat, maize, cassava, beans, and sweet potatoes. On the other hand, biofortification of crops is a daunting task. Plant breeders, nutritionists, genetic engineers, and molecular biologists must converge to make this happen. Breeding approaches are generalized and easy to adopt and have been adapted to sustainably improve victorious nutritional qualities. Molecular breeding approaches, which have much higher success rates as genetically fortified crop plants, are facing difficulties due to consumer acceptance and the costly and high regulatory approval processes used between different countries. The use of biofortified crops will create a desirable future as they can kick off malnutrition regarding the micronutrients amongst the poor people worldwide, especially people of developing countries.

Conflict of Interest The authors declare that they have no conflict of interest.

References

Al-Babili S, Beyer P (2005) Golden Rice–five years on the road–five years to go? Trends Plant Sci 10(12):565–573

Ali N, Paul S, Gayen D, Sarkar SN, Datta K, Datta SK (2013) Development of low phytate rice by RNAi mediated seed-specific silencing of inositol 1, 3, 4, 5, 6-pentakisphosphate 2-kinase gene (IPK1). PLoS One 8(7):e68161

Bai C, Twyman RM, Farré G, Sanahuja G, Christou P, Capell T, Zhu C (2011) A golden era—provitamin A enhancement in diverse crops. In Vitro Cell Develop Biol Plant 47(2):205–221

Bashir K, Takahashi R, Akhtar S, Ishimaru Y, Nakanishi H, Nishizawa NK (2013) The knockdown of OsVIT2 and MIT affects iron localization in rice seed. Rice 6:31. https://doi.org/10.1186/1939-8433-6-31

Bodnar M, Konieczka P, Namiesnik J (2012) The properties, functions, and use of selenium compounds in living organisms. J Environ Sci Health C 30(3):225–252

Boonyaves K, Gruissem W, Bhullar NK (2016) NOD promoter-controlled AtIRT1 expression functions synergistically with NAS and FERRITIN genes to increase iron in rice grains. Plant Mol Biol 90(3):207–215

Bouis HE, Hotz C, McClafferty B, Meenakshi JV, Pfeiffer WH (2011) Biofortification: a new tool to reduce micronutrient malnutrition. Food Nutr Bull 32(1_suppl1):S31–S40
Bouis HE, Saltzman A (2017) Improving nutrition through biofortification: a review of evidence from HarvestPlus, 2003 through 2016. Glob Food Secure 12:49–58
Cakmak I (2008) Enrichment of cereal grains with zinc: agronomic or genetic biofortification? Plant Soil 302(1):1–17
Cakmak I, Kutman UÁ (2018) Agronomic biofortification of cereals with zinc: a review. Eur J Soil Sci 69(1):172–180
Champagne ET, Wood DF, Juliano BO, Bechtel DB (2004) The rice grain and its gross composition. In: Champagne ET (ed) Rice Chemistry and Technology, Cereal. Chem, vol 3, pp 93–96
Christou P, Twyman RM (2004) The potential of genetically enhanced plants to address food insecurity. Nutr Res Rev 17(1):23–42
Datta SK, Datta K, Parkhi V, Rai M, Baisakh N, Sahoo G et al (2007) Golden rice: introgression, breeding, and field evaluation. Euphytica 154(3):271–278
Dong Y, Zhang H, Hawthorn L, Ganther HE, Ip C (2003) Delineation of the molecular basis for selenium-induced growth arrest in human prostate cancer cells by oligonucleotide array. Cancer Res 63(1):52–59
Fawzi AFA, El-Fouly MM, Moubarak ZM (1993) The need of grain legumes for iron, manganese, and zinc fertilization under Egyptian soil conditions: effect and uptake of metabolites. J Plant Nutr 16(5):813–823
Ferguson LR, Harris PJ (1999) Protection against cancer by wheat bran: role of dietary fibre and phytochemicals. Eur J Cancer Prev 8:25
Frossard E, Condron LM, Oberson A, Sinaj S, Fardeau JC (2000) Processes governing phosphorus availability in temperate soils. J Environ Qual 29(1):15–23
Ghosh S, Datta K, Datta SK (2019) Rice vitamins. In: Bao J (ed) Rice chemistry and technology, vol 7, 1st edn. Elsevier Inc./AACC International, Amsterdam/AACC International, pp 195–220
Gómez-Galera S, Rojas E, Sudhakar D, Zhu C, Pelacho AM, Capell T, Christou P (2010) Critical evaluation of strategies for mineral fortification of staple food crops. Transgenic Res 19(2): 165–180
Grotz N, Guerinot ML (2006) Molecular aspects of Cu, Fe and Zn homeostasis in plants. BiochimicaetBiophysicaActa (BBA)-Molecular. Cell Res 1763(7):595–608
Gunes A, Inal A, Alpaslan M, Eraslan F, Bagci EG, Cicek N (2007) Salicylic acid-induced changes on some physiological parameters symptomatic for oxidative stress and mineral nutrition in maize (Zea mays L.) grown under salinity. J Plant Physiol 164(6):728–736
Hambidge M (2000) Human zinc deficiency. J Nutr 130(5):1344S–1349S
Hao ZW, Xu XH, Wang DH (2005) Reductive denitrification of nitrate by scrap iron filings. J Zhejiang Univ Sci B 6(3):182
Heigwer F, Kerr G, Boutros M (2014) E-CRISP: fast CRISPR target site identification. Nat Methods 11(2):122–123
Higo K, Ugawa Y, Iwamoto M, Korenaga T (1999) Plant cis-acting regulatory DNA elements (PLACE) database: 1999. Nucleic Acids Res 27(1):297–300
Hoa TTC, Al-Babili S, Schaub P, Potrykus I, Beyer P (2003) Golden Indica and Japonica rice lines amenable to deregulation. Plant Physiol 133(1):161–169
Hu Z, Lutkenhaus J (2003) A conserved sequence at the C-terminus of MinD is required for binding to the membrane and targeting MinC to the septum. Mol Microbiol 47(2):345–355
Hussain S, Maqsood MA, Rengel Z, Aziz T (2012) Biofortification and estimated human bioavailability of zinc in wheat grains as influenced by methods of zinc application. Plant Soil 361(1): 279–290
Inoue H, Kobayashi T, Nozoye T, Takahashi M, Kakei Y, Suzuki K et al (2009) Rice OsYSL15 is an iron-regulated iron (III)-deoxymugineic acid transporter expressed in the roots and is essential for iron uptake in the early growth of the seedlings. J Biol Chem 284(6):3470–3479

Ishimaru Y, Bashir K, Nishizawa NK (2011) Zn uptake and translocation in rice plants. Rice 4(1): 21–27
Ishimaru Y, Masuda H, Suzuki M, Bashir K, Takahashi M, Nakanishi H et al (2007) Overexpression of the OsZIP4 zinc transporter confers disarrangement of zinc distribution in rice plants. J Exp Bot 58(11):2909–2915
Johnson AA, Kyriacou B, Callahan DL, Carruthers L, Stangoulis J, Lombi E, Tester M (2011) Constitutive overexpression of the OsNAS gene family reveals single-gene strategies for effective iron-and zinc-biofortification of rice endosperm. PLoS One 6(9):e24476
Kato S, Wachi T, Yoshihira K, Nakagawa T, Ishikawa A, Takagi D et al (2013) Rice (Oryza Sativa L.) roots have iodate reduction activity in response to iodine. Front Plant Sci 4:227
Kettenburg AJ, Hanspach J, Abson DJ, Fischer J (2018) From disagreements to dialogue: unpacking the Golden Rice debate. Sustain Sci 13(5):1469–1482
Khush GS, Lee S, Cho JI, Jeon JS (2012) Biofortification of crops for reducing malnutrition. Plant Biotechnol Rep 6(3):195–202
Landberg R, Andersson AA, Åman P, Kamal-Eldin A (2009) Comparison of GC and colorimetry for the determination of alkylresorcinol homologues in cereal grains and products. Food Chem 113(4):1363–1369
Landini M, Gonzali S, Kiferle C, Tonacchera M, Agretti P, Dimida A et al (2012) Metabolic engineering of the iodine content in Arabidopsis. Sci Rep 2(1):1–6
Lawson PG, Daum D, Czauderna R, Meuser H, Härtling JW (2015) Soil versus foliar iodine fertilization as a biofortification strategy for field-grown vegetables. Front Plant Sci 6:450
Lee S, Chiecko JC, Kim SA, Walker EL, Lee Y, Guerinot ML, An G (2009) Disruption of OsYSL15 leads to iron inefficiency in rice plants. Plant Physiol 150(2):786–800
Li HF, McGrath SP, Zhao FJ (2008) Selenium uptake, translocation and speciation in wheat supplied with selenate or selenite. New Phytol 178(1):92–102
Li JH, Vasanthan T (2003) Hypochlorite oxidation of field pea starch and its suitability for noodle making using an extrusion cooker. Food Res Int 36(4):381–386
Liu X, Zhao Y, Gao J, Pawlyk B, Starcher B, Spencer JA, Li T (2004) Elastic fiber homeostasis requires lysyl oxidase-like 1 protein. Nat Genet 36(2):178–182
Lucca P, Ye X, Potrykus I (2001) Effective selection and regeneration of transgenic rice plants with mannose as selective agent. Mol Breed 7(1):43–49
Marschner H, Römheld V, Kissel M (1986) Different strategies in higher plants in mobilization and uptake of iron. J Plant Nutr 9(3–7):695–713
Masuda H, Usuda K, Kobayashi T, Ishimaru Y, Kakei Y, Takahashi M, Higuchi K, Nakanishi H, Mori S, Nishizawa NK (2009) Overexpression of the barley nicotianamine synthase gene HvNAS1 increases iron and zinc concentrations in rice grains. Rice 2:155–166
Miedema H, Bothwell JH, Brownlee C, Davies JM (2001) Calcium uptake by plant cells–channels and pumps acting in concert. Trends Plant Sci 6(11):514–519
Miedema H, Demidchik V, Véry AA, Bothwell JH, Brownlee C, Davies JM (2008) Two voltage-dependent calcium channels co-exist in the apical plasma membrane of Arabidopsis thaliana root hairs. New Phytol 179(2):378–385
Minhas AP, Tuli R, Puri S (2018) Pathway editing targets for thiamine biofortification in rice grains. Front Plant Sci 9:975
Mishra R, Joshi RK, Zhao K (2018) Genome editing in rice: recent advances, challenges, and future implications. Front Plant Sci 9:1361
Missana T, Alonso U, García-Gutiérrez M (2009) Experimental study and modelling of selenite sorption onto illite and smectite clays. J Colloid Interface Sci 334(2):132–138
Mousavi SR, Galavi M, Rezaei M (2013) Zinc (Zn) importance for crop production—a review. Int J Agron Plant Prod 4(1):64–68
Nagashima S, Jirintai S, Takahashi M, Kobayashi T, Nishizawa T, Kouki T et al (2014) Hepatitis E virus egress depends on the exosomal pathway, with secretory exosomes derived from multivesicular bodies. J Gen Virol 95(10):2166–2175

Nozoye T, Nagasaka S, Kobayashi T, Takahashi M, Sato Y, Sato Y, Nishizawa NK (2011) Phytosiderophore efflux transporters are crucial for iron acquisition in graminaceous plants. J Biol Chem 286(7):5446–5454

Paine JA, Shipton CA, Chaggar S, Howells RM, Kennedy MJ, Vernon G, Drake R (2005) Improving the nutritional value of Golden Rice through increased provitamin A content. Nat Biotechnol 23(4):482–487

Palmgren MG, Clemens S, Williams LE, Krämer U, Borg S, Schjørring JK, Sanders D (2008) Zinc biofortification of cereals: problems and solutions. Trends Plant Sci 13(9):464–473

Paul S, Ali N, Gayen D, Datta SK, Datta K (2012) Molecular breeding of Osfer2 gene to increase iron nutrition in rice grain. GM Crops Food 3(4):310–316

Peleg Z, Saranga Y, Yazici A, Fahima T, Ozturk L, Cakmak I (2008) Grain zinc, iron and protein concentrations and zinc-efficiency in wild emmer wheat under contrasting irrigation regimes. Plant Soil 306(1):57–67

Pfeiffer WH, McClafferty B (2007) HarvestPlus: breeding crops for better nutrition. Crop Sci 47: S-88

Powell K (2007) Functional foods from biotech—and unappetizing prospect? Nat Biotechnol 25(5): 525–531

Rabbani S, Beyer P, Lintig JV, Hugueney P, Kleinig H (1998) Induced β-carotene synthesis driven by triacylglycerol deposition in the unicellular alga Dunaliella bardawil. Plant Physiol 116(4): 1239–1248

Raboy V (2002) Progress in breeding low phytate crops. J Nutr 132(3):503S–505S

Roda FA, Marques I, Batista-Santos P, Esquível MG, Ndayiragije A, Lidon FC et al (2020) Rice biofortification with Zinc and Selenium: a transcriptomic approach to understand mineral accumulation in flag leaves. Front Genet 11:543

Saini DK, Devi P, Kaushik P (2020) Advances in genomic interventions for wheat biofortification: A review. Agronomy 10(1):62

Schreiber M, Yuan A, Salkoff L (1999) Transplantable sites confer calcium sensitivity to BK channels. Nat Neurosci 2(5):416–421

Sharma V, Saini DK, Kumar A, Kesh H, Kaushik P (2020) Breeding for biofortification traits in rice: means to eradicate hidden hunger. In: Agronomy-climate change & food security, p 35

Shaw J, Lickey EB, Schilling EE, Small RL (2007) Comparison of whole chloroplast genome sequences to choose noncoding regions for phylogenetic studies in angiosperms: the tortoise and the hare III. Am J Bot 94(3):275–288

Shojima S, Nishizawa NK, Fushiya S, Nozoe S, Irifune T, Mori S (1990) Biosynthesis of phytosiderophores: in vitro biosynthesis of 2′-deoxymugineic acid from L-methionine and nicotianamine. Plant Physiol 93(4):1497–1503

Stroud JL, Broadley MR, Foot I, Fairweather-Tait SJ, Hart DJ, Hurst R et al (2010) Soil factors affecting selenium concentration in wheat grain and the fate and speciation of Se fertilisers applied to soil. Plant Soil 332(1):19–30

Suzuki M, Takahashi M, Tsukamoto T, Watanabe S, Matsuhashi S, Yazaki J et al (2006) Biosynthesis and secretion of mugineic acid family phytosiderophores in zinc-deficient barley. Plant J 48(1):85–97

Suzuki M, Tanaka K, Kuwano M, Yoshida KT (2007) Expression pattern of inositol phosphate-related enzymes in rice (Oryza Sativa L.): implications for the phytic acid biosynthetic pathway. Gene 405(1–2):55–64

Takagi SI, Nomoto K, Takemoto T (1984) Physiological aspect of mugineic acid, a possible phytosiderophore of graminaceous plants. J Plant Nutr 7(1–5):469–477

Takahashi R, Ishimaru Y, Senora T, Shimo H, Ishikawa S, Arao T et al (2011) The OsNRAMP1 iron transporter is involved in Cd accumulation in rice. J Exp Bot 62(14):4843–4850

Takahashi Y, Shomura A, Sasaki T, Yano M (2001) Hd6, a rice quantitative trait locus involved in photoperiod sensitivity, encodes the α subunit of protein kinase CK2. Proc Natl Acad Sci 98(14):7922–7927

Thurber MD, Fahey JW (2009) Adoption of Moringa oleifera to combat under-nutrition viewed through the lens of the "Diffusion of Innovations" theory. Ecol Food Nutr 48(3):212–225

Tiwari BK, O'donnell CP, Cullen PJ (2009) Effect of non thermal processing technologies on the anthocyanin content of fruit juices. Trends Food Sci Technol 20(3–4):137–145

Tucker G (2003) Nutritional enhancement of plants. Curr Opin Biotechnol 14(2):221–225

Voogt W, Jackson WA (2010) Perchlorate, nitrate, and iodine uptake and distribution in lettuce (Lactuca Sativa L.) and potential impact on background levels in humans. J Agric Food Chem 58(23):12192–12198

Wang Z, Hong X, Hu K, Wang Y, Wang X, Du S et al (2017) Impaired magnesium protoporphyrin IX methyltransferase (ChlM) impedes chlorophyll synthesis and plant growth in rice. Front Plant Sci 8:1694

Waters BM, Uauy C, Dubcovsky J, Grusak MA (2009) Wheat (Triticum aestivum) NAM proteins regulate the translocation of iron, zinc, and nitrogen compounds from vegetative tissues to grain. J Exp Bot 60(15):4263–4274

Welch RM, Graham RD (1999) A new paradigm for world agriculture: meeting human needs: productive, sustainable, nutritious. Field Crop Res 60(1–2):1–10

Welch RM, Graham RD (2004) Breeding for micronutrients in staple food crops from a human nutrition perspective. J Exp Bot 55(396):353–364

White PJ (2001) The pathways of calcium movement to the xylem. J Exp Bot 52(358):891–899

White PJ, Bowen HC, Parmaguru P, Fritz M, Spracklen WP, Spiby RE et al (2004) Interactions between selenium and sulphur nutrition in Arabidopsis thaliana. J Exp Bot 55(404):1927–1937

White PJ, Broadley MR (2003) Calcium in plants. Ann Bot 92(4):487–511

White PJ, Broadley MR (2005) Biofortifying crops with essential mineral elements. Trends Plant Sci 10(12):586–593

White PJ, Broadley MR (2009) Biofortification of crops with seven mineral elements often lacking in human diets–iron, zinc, copper, calcium, magnesium, selenium and iodine. New Phytol 182(1):49–84

Wurtzel ET, Cuttriss A, Vallabhaneni R (2012) Maize provitamin carotenoids, current resources, and future metabolic engineering challenges. Front Plant Sci 3:29

Yang L, Liu H, Hu N (2007a) Assembly of electroactive layer-by-layer films of myoglobin and small-molecular phytic acid. Electrochem Commun 9(5):1057–1061

Yang P, Li X, Wang X, Chen H, Chen F, Shen S (2007b) Proteomic analysis of rice (Oryza Sativa) seeds during germination. Proteomics 7(18):3358–3368

Yang X, Römheld V, Marschner H (1994) Effect of bicarbonate on root growth and accumulation of organic acids in Zn-inefficient and Zn-efficient rice cultivars (Oryza Sativa L.). Plant Soil 164(1):1–7

Yang XE, Chen WR, Feng Y (2007c) Improving human micronutrient nutrition through biofortification in the soil-plant system: China as a case study. Environ Geochem Health 29(5):413–428

Ye X, Al-Babili S, Klöti A, Zhang J, Lucca P, Beyer P, Potrykus I (2000) Engineering the provitamin (β-carotene) biosynthetic pathway into (carotenoid-free) rice endosperm. Science 287(5451):303–305

Zapata-Caldas E, Hyman G, Pachón H, Monserrate FA, Varela LV (2009) Identifying candidate sites for crop biofortification in Latin America: case studies in Colombia, Nicaragua and Bolivia. Int J Health Geogr 8(1):1–18

Zhang T, Huang Y (2012) Impacts of climate change and inter-annual variability on cereal crops in China from 1980 to 2008. J Sci Food Agric 92(8):1643–1652

Biofortification of Wheat Using Current Resources and Future Challenges

7

Rupesh Tayade, Rolly Kabange, Muhammad Waqas Ali,
Byung-Wook Yun, and Rizwana Begum Syed Nabi

Abstract

Wheat is the second most significant staple food grain crop after rice; however, its grains mostly contain suboptimal levels of provitamins, proteins, and essential micronutrients, including zinc, iron, selenium, and iodine. However, during processing, wheat flour is enriched or fortified with several required nutrients. The most reasonable, long-lasting, and viable solution for this problem is biofortification that can be performed through either agronomic approaches, breeding efforts, or transgenic techniques. Agronomic fertilization techniques for wheat biofortification include basal application, foliar spray, and seed priming with the appropriate nutrient sources. Recently, various potent bacterial strains have been used, and these techniques can be used in combination with agronomic and genetic techniques to significantly enhance the density of the nutrients that require to be supplemented in wheat grains. Compared to agronomic approaches, breading techniques are more sustainable and include conventional and marker-assisted breeding. Transgenic approaches for micronutrient biofortification of

R. Tayade
Laboratory of Plant Breeding, School of Applied Biosciences, Kyungpook National University, Daegu, Republic of Korea

R. Kabange · R. B. S. Nabi (✉)
Department of Southern Area Crop Science, National Institute of Crop Science, Rural Development Administration, Miryang, Republic of Korea
e-mail: rizwananabi@korea.kr

M. W. Ali
School of Biosciences, University of Birmingham, Birmingham, UK

B.-W. Yun
Laboratory of Plant Functional Genomics, School of Applied Biosciences, Kyungpook National University, Daegu, Republic of Korea

R. Deshmukh et al. (eds.), *Biofortification in Cereals*,
https://doi.org/10.1007/978-981-19-4308-9_7

wheat include modulation of the gene expression of transporters to improve the absorption rate and assimilation capability of the wheat plant while lowering antinutrient content. In this chapter, along with the possible techniques of biofortification, we discuss the mode of uptake and deposition of the desired nutrients in the grain at molecular and physiological levels. We discuss the possible wheat genomic obstacles that hinder wheat biofortification as well as the economic and social challenges for the release of biofortified wheat.

Keywords

Wheat biofortification · Micronutrients · Provitamin · Transgenic approaches · Breeding tools · Malnutrition · Hidden hunger

7.1 Introduction

Wheat (*Triticum* spp.) is one of the very important staple cereal grain crop that is produced, consumed, and traded globally. This grain is a major source of food, feed, protein, and nutrition for human beings and animals (Mayer et al. 2014). Therefore, the consumption rate of wheat is on the rise across the globe. China is the largest consumer of wheat, followed by the European Union (EU), and India. The cultivation of wheat has been increasing following its domestication about 8000 years ago; presently, it is widely grown in the world. About 214 million ha area is currently being used for wheat cultivation across the world, and wheat is considered the single largest trade crop in the world, in comparison to all other crops together (FAO 2018). Wheat is a widely consumed crop that can be used for several food products, such as alcohol, bread, and baked goods. Thus, wheat production is an important contributor to the agriculture economy and influences the gross domestic product (GDP) of developing countries (Kiss 2011). However, the world wheat export is dominated by a few countries and is directly connected to the international wheat trade and price determination. Price changes in the wheat market may increase the overall food prices in the future.

The production yield of wheat in the world during the period of 2001–2019 was accessed and showed the declining trend in the cultivation area; however, the production is on the rise, relative to the planted area, probably owing to modern cultivation practices and the use of technology (Fig. 7.1). Figure 7.2 shows information about wheat production in the world's major areas; Asia produces a major proportion (43.7%) of the total wheat produced in the world followed by Europe and America. Wheat cultivation is largely classified in developing and developed countries. When ranked as per the volume of wheat produced during the period from 2001 to 2019, China ranked first (114 M ton), followed by India (83 M ton), and the USA (56 M ton); these three countries account for roughly 54% of the total global wheat production (Fig. 7.3). The next-largest wheat producers are Russia, France, Canada, Germany, Pakistan, Australia, and Ukraine that together account for 45% of the global wheat production (FAO 2020). However, it is noteworthy that

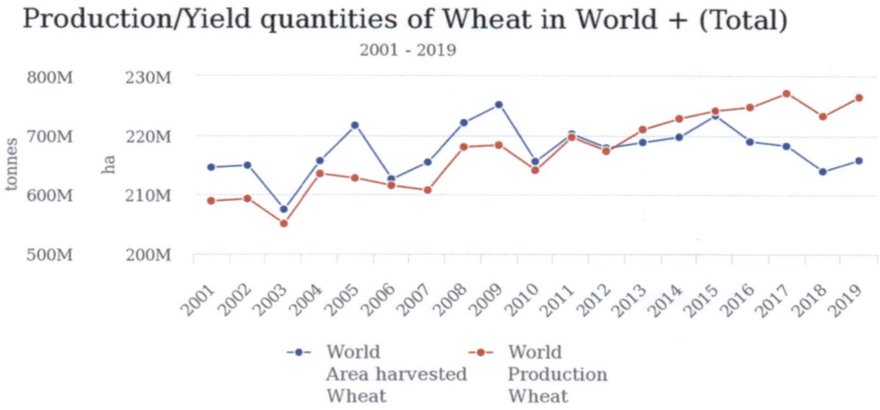

Fig. 7.1 Worldwide wheat production and cultivation area during the period from 2001 to 2019

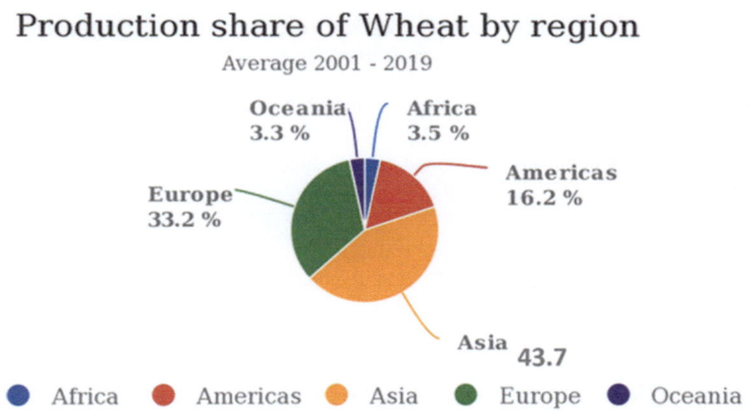

Fig. 7.2 Region-wise average wheat production during the period from 2001 to 2019

wheat production in the USA has reduced significantly during the previous decades. The wheat trade is worth about US $50 billion annually across the world. Among the world regions, in 2019 South East Asian countries earned their most revenue from overall wheat imports, contributing $16 billion (40.6%) of the world total imports. Africa ranked second, followed by Europe with wheat revenue importers at 22.2% and 20.9%, respectively. The per capita GDP of the top 10 wheat-growing countries showed significant change (Table 7.1). The major wheat exporter countries Argentina, Australia, Canada, EU, Russia, Ukraine, and the USA showed higher per capita GDP. However, an increasing trend has also been observed for the per capita GDP of China, India, and Pakistan during the previous two decades (World Bank Data).

Wheat production is divided in developed and underdeveloped countries; irrespective of the wheat-producing group, mostly hunger is a huge issue across

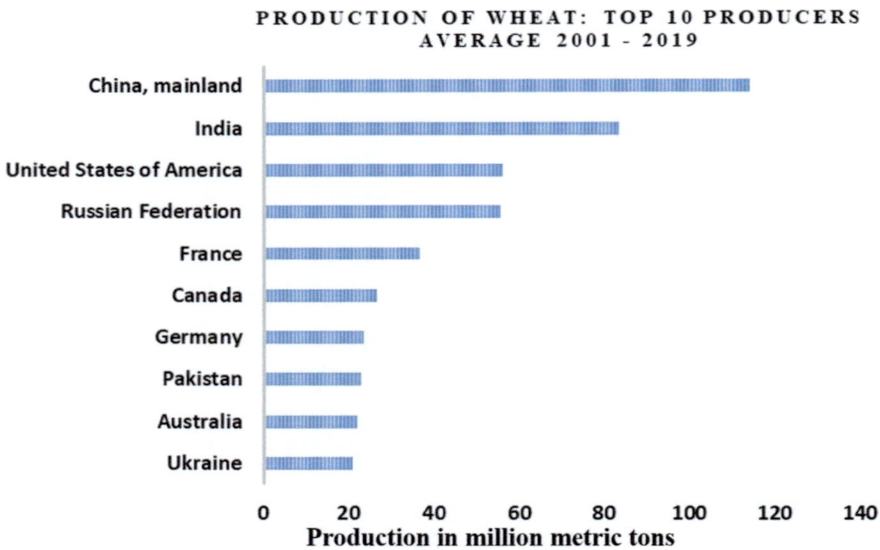

Fig. 7.3 Top ranked average wheat-producing countries during the period from 2001 to 2019

Table 7.1 Details of economic status and hunger index of major wheat-growing countries worldwide

Country name	Country code	Population[a]	Wheat production[b]	GDPA per capita[c]	Hunger index[d]	Child[e]
China, mainland	CHN	1,427,647.79	114,333,057.9	10,261.7	<5	5.2
India	IND	1,352,642.28	83,480,287.68	2104.1	27.2	37.9
United States of America	USA	327,096.27	56,257,303.11	65,118.4	—	—
Russian Federation	RUS	145,734.04	55,764,659.42	11,585.0	5.2	10.7
France	Fra	64,990.51	36,814,143.16	40,493.9	—	—
Canada	CAN	37,074.56	26,772,996.58	46,194.7	—	—
Germany	GER	83,124.42	23,530,102.32	10,006.1	—	—
Pakistan	PAK	212,228.29	22,953,457.68	1284.7	24.6	37.6
Australia	AUS	24,898.15	22,064,214.05	55,060.3	—	—
Ukraine	UKR	44,246.16	20,916,046.53	3659.0	<5	6

Note: —, Data not available or not presented

[a] 1000 persons (unit)

[b] Average wheat production in tons (2001–2019)

[c] GDP per capita (current US$) (2019)

[d] Hunger index (2019)

[e] Child malnutrition index; source of information FAO http://www.fao.org/faostat/en/#data/OA, and https://databank.worldbank.org/

the globe. The severity of this problem is more prominent in developing countries, with a higher impact observed in rural areas (von Grebmer et al. 2019). Poverty is a major cause of hunger in most countries. As per the Global Hunger Index (GHI) 2019, among the top 10 wheat-producing countries, India ranks the first in terms of hunger, with a hunger index of 27.2, followed by Pakistan (Table 7.1) (von Grebmer et al. 2019). Although abundant natural and agricultural resources are available in both of these countries, they face the problem of hunger, owing to several reasons, such as high population, poverty, and less per capita income in the region (Table 7.1). In addition, child mortality and malnutrition are also associated with the hunger index; India and Pakistan have the highest prevalence of child mortality and malnutrition among the top 10 wheat-growing or producing countries (Table 7.1). In contrast, most countries that belong to the wheat export group showed lower hunger index, child mortality, and malnutrition.

About 98% of the population with a higher hunger index belongs to developing regions. Furthermore, <780 million people from the Asia Pacific region and sub-Saharan Africa still face major hunger crises and go to bed without food every night (McGuire 2015). Malnutrition is presently a growing problem affecting children and adults all over the world (Dukhi 2020). Despite the record production of cereal grains, vegetables, and other food products, malnutrition (including overweight, fat, and undernourishment) and hunger remain to be the major problems across the world. In fact, world hunger and malnutrition cannot be resolved only by increasing food production and supply. The eradication of world hunger and malnutrition can only be achieved if we precisely identify the root cause of the problems. The major cause of these issues is poverty. Several efforts, as discussed below, are required to completely eradicate hunger and malnutrition.

It was considered that growth potential is higher for the sector of agriculture than for any other sector for reducing poverty. Therefore, the role of small-scale farmers in agricultural development in concurrence with policies that facilitate the poor increase of access to food will be vitally important for reducing mass poverty and hunger in the world. Furthermore, improving the economic level of people by establishing government policies, developing systems that enable secure access to food, increasing awareness, and changing behaviors toward the use of a nutritional diet can help to reduce hunger in the world. We need to accelerate progress in raising the income of the poor, ending hunger, and ensuring a more sustainable system for food distribution and use. Recently, the World Bank Group presented the action plan to "Ending Poverty and Hunger by 2030: An Agenda for the Global Food System." In addition, food fortification and biofortification are important for handling malnutrition across the globe (Bouis and Saltzman 2017). Moreover, there is a need to enhance the existing health system policies and upgrade and revise the child malnutrition research and interventions, starting from the community leveling up to the regional and national levels. Further, the need of the hour is to close the sex gap, sex-based inequality, and give equal opportunities to women in employment, assets, and decision inputs. The participation of women in agriculture services could increase the farm yield by 20–30% and could eliminate hunger in 12–17% of all hungry people across the world (Doss 2018). In addition, providing health support and micronutrient supplementation to underweight mothers is crucial for healthy childbirth; varied forms of micronutrient supplementation are also crucial for ensuring the health of the population (Caron et al. 2017). Moreover, stable political conditions and leadership with the required political will can influence economic growth and social reforms and help protect targeted vulnerable population groups. Medium- and long-term participation of all the abovementioned stakeholders is an urgent need in tackling the issues of hunger and malnutrition.

Developing countries have a higher population of individuals who are below the poverty line and cannot afford a fortified diet, consisting of fruits and animal food products. Moreover, most of these people consume staple crops, including wheat, rice, and maize. Wheat is consumed by >2.5 billion people worldwide, with most of them being from developing countries (CIMMYT 2017; Lobell et al. 2011). Wheat accounts for almost 20% of the overall calorie intake received by humans (FAO 2017). However, the current high-yielding wheat varieties are deficient in basic and essential elements (minerals or micronutrients), such as zinc (Zn), iron (Fe), selenium (Se), and iodine (I). As per the WHO, one in every three women of reproductive age has Fe deficiency (FAO 2018). During 2010, Fe deficiency alone caused a loss of >45,000 disability-adjusted life years (Murray and Lopez 2013). Furthermore, Zn malnutrition affects around 17% of the worldwide population, subsequently leading to malnutrition and increased child mortality (Wessells and Brown 2012). Therefore, a population where wheat grain is used as the prime source of diet can develop micronutrient deficiencies that subsequently lead to malnutrition. To address this serious issue, biofortification is an attractive, sustainable, and long-term effective strategy that can overcome micronutrient deficiency. This approach is more cost-effective for staple crops than other fortification approaches or dietary supplements (de Valença et al. 2017). Biofortification can be achieved via breeding, genomics, or transgenic strategies (Bouis and Saltzman 2017; Ludwig and Slamet-Loedin 2019). Considering the human health implications, biofortification with micronutrients has become the prime focus of several staple crop breeding research.

7.2 Top Priorities for Wheat Biofortification

All organisms including plants and animals require most micronutrients and proteins because of their pivotal roles as cofactors in enzymatic activities and regulators in a number of significant biochemical activities. The importance of Zn, Fe, Se, I, provitamins, and protein in plants and humans is discussed in the following sections.

7.2.1 Zinc

Basic and essential micronutrients, such as Zn, Fe, Se, and I, are important elements for animals as well as plants. These micronutrients are required by all organisms in small amounts; however, they play a very crucial role via several mechanisms. Zn is the most abundant transition element and is considered a key micronutrient that plays a diverse role in maintaining physical health, growth, development, immunity, reproductive health, and mental health in humans (Dapkekar et al. 2020). In a similar manner, in plants, Zn regulates several metabolic and physiological process, acts as a cofactor in several enzymatic pathways, and participates in lipid, protein, carbohydrate, chloroplast, and nucleic acid synthesis (Palmer and Guerinot 2009; Hänsch and Mendel 2009). Deficiencies of Zn in animals can cause growth stunting, malnutrition in adults and children, and other severe implications, such as weak immunity, learning disabilities, diarrhea, and impaired wound healing. Likewise, it causes a decrease in growth and development, stress tolerance, as well as photosynthesis in plants (Kawachi et al. 2009; Lee et al. 2010).

7.2.2 Iron

Like Zn, Fe is also an abundant mineral element in nature that is required for all organisms for several functions. In both plants and animals, it participates in a variety of biochemical and physiological activities. Being an active redox element, it involves several plant functions, such as plant hormone regulation, nitrogen assimilation, photosynthesis, and mitochondrial respiration. In addition, it participates in electron transport, scavenging, and the production of reactive oxygen species and thereby protects the plants from abiotic stress (Palmer and Guerinot 2009). Fe deficiency in plants can cause leaf chlorosis, inhibit growth, and increase the susceptibility for several plant diseases and yield (Chatterjee et al. 2006; Cakmak 2002). Similarly, in humans, Fe is involved in electron transfer and the transport of oxygen which is important for myoglobin and hemoglobin synthesis. Fe deficiency has several implications in humans and leads to malnutrition, stunted height, reduced learning ability, fatigue, and nutritional anemia in children and adults (Dapkekar et al. 2020; Thomas and Frankenberg 2002). Moreover, reproductive-age women are severely affected by Fe deficiency that may lead to poor pregnancy outcomes and high child mortality (Bailey et al. 2015).

7.2.3 Selenium

Selenium (Se) is another element found in soil (1.0 and 1.5 $\mu g\ g^{-1}$). Commonly, it occurs in the form of organically bound selenium, selenide, and selenite. Although Se is considered nonessential to plants, selenate is a dominant and available form to plants. In several regions of the world, Se deprivation has been recorded (Combs 2001; Fordyce et al. 2000; Lin et al. 2004; Sempértegui et al. 2003). Se insufficiency is not a common or major social problem in humans. However, insufficient Se intake has been identified in several parts of the world, especially Africa, Europe, Finland, New Zealand, Australia, and Russia (Combs 2001; Dorea 2002; Lyons et al. 2003; Rayman 2000). In humans, Se is involved in several metabolic regulatory processes that involve proteins, hormones, and the antioxidant defense system (Fordyce et al. 2000; Rayman 2000; Tan et al. 2002). Inadequate intake of Se may cause several health issues, such as cardiovascular diseases, hypothyroidism, reduced male fertility, cancer, and HIV/AIDS (Rayman 2000, 2002; Kupka et al. 2004). Furthermore, Se consumption is linked to Keshan illness and Kashin-Beck disease (Lyons et al. 2003). In addition to that in humans, other creatures, such as goats, sheep, cows, and pigs, require Se for a variety of functions (Gupta and Gupta 2000; Meschy 2000; Tinggi 2003).

7.2.4 Iodine

Broadly, iodine (I) is a trace mineral that is considered a nonessential element for plants; however, in some aquatic plants, it is involved in antioxidant metabolism. Plants grown in soil that is rich in iodine content can absorb the I and provide it to humans via diet. Sufficient intake of I is very essential for humans because I is required for thyroid metabolism, which is essential for the growth and development. In addition, I shortage causes a wide range of health problems in humans. For example, it has been linked to the development of several types of cancers, nervous system impairments, mental retardation, cretinism diseases, deaf-mutism, stunting, and death in children (Bailey et al. 2015; Cakmak et al. 2017).

7.2.5 Provitamins

In addition to essential micronutrients, provitamins are essential for organisms. A provitamin is a substance that can be converted within the body to a vitamin, such as a plant that synthesizes ergosterol that when eaten as plant food by animals can form ergocalciferol (vitamin D2) under the exposure of ultraviolet light (Brody 1999). Widely known provitamins are "provitamin A" which is a name for β-carotene, "provitamin D2," "provitamin D3," and menadione (vitamin K). Here we discuss vitamin A.

Vitamin A essential nutrient is commonly known as β-carotene. In plants, tocopherols and β-carotene, two precursors of vitamin A, are found in the leaf or photosynthetic organ in most plants. However, other forms (α-carotene and β-cryptoxanthin) and tocotrienols are not consistently found in the plant across the species. It is mainly synthesized in the plastids and plays distinct functions in plants and animals (Asensi-Fabado and Munné-Bosch 2010; Medrano-Macías et al. 2016). All the provitamins, including vitamin A, are important regulators of cellular metabolism in plants, are involved in several enzymatic reactions, and act as antioxidants. Particularly, vitamin A is involved in improving light harvesting and photoprotection, stress signaling, and growth and development of the root and the shoot. Vitamin A function in humans is greatly related to vision, dark adaptation, and enhances immunity (Lima et al. 2010; Wiseman et al. 2017). In humans, deficiency of vitamin A may lead to anemia via infusion of Fe metabolism, reproductive health issues, susceptibility toward several infections, and gastric issues (Wiseman et al.

2017; Brown and Noelle 2015; Hogarth and Griswold 2010; Clagett-Dame and Knutson 2011; West and Mehra 2010; Semba et al. 1992).

7.2.6 Protein

Protein commonly comprises 20 different amino acids (AA) in different concentrations; peptide linkages connect these amino acids (Wu 2013). The word "protein" derived from the Greek word "proteios," which means "principal" or "first." It is a crucial component of both human and animal cells. Protein is considered an essential macronutrient, and unlike minerals and vitamins that can be stored in the human body, proteins cannot be stored based on daily intake. Proteins are sources of AA (indispensable and dispensable) and provide nitrogen, hydrocarbon skeletons, and sulfur. Proteins are needed in larger amounts in the body for the maintenance of homeostasis. The two main sources of protein are animals and plants. Plant protein is relatively less digestible than animal protein. Proteins are important for human cell function; they provide AA (indispensable and dispensable) for humans, for physical and normal metabolic functions. Protein deficiency is associated with several diseases, such as Kwashiorkor, that increase an individual's susceptibility to metabolic and infectious diseases (Dasgupta et al. 2005; Wu 2016). In addition, excess protein intake causes hepatic or renal dysfunction (Hoffer and Bistrian 2012).

Globally, the deficiencies of Zn, Fe, Se, I, provitamin, and protein have serious implications on human health. Malnutrition due to deficiency of these essential micronutrients and protein causes serious health issues in billions of individuals, most predominantly affecting those in developing countries (Ritchie 2017). Individuals in these developing nations mainly consume cereal-based staple food. It has been well documented that wheat, rice, and maize have a low content of micronutrients and inadequate protein concentration (Dapkekar et al. 2020; Zou et al. 2019). Wheat is a prominent staple grain crop; yet, the currently developed varieties of wheat inherently have lower micronutrients and protein concentrations (Cakmak et al. 2017; Zou et al. 2019; Zia et al. 2015). Therefore, enhancing the nutritional composition of wheat as regards Zn, Fe, Se, I, vitamin A, and protein concentrations is widely recognized as the top priority for improving public health. The best suitable approach for improving the concentration of these vital elements is biofortification via conventional breeding or biotechnological and genomic approaches.

7.3 Agronomic Biofortification of Wheat

In the previous years, there has been a surge of enthusiasm in boosting the health advantages of cereal, such as improving its mineral and vitamin content, improving their production, and disease resistance (Vasconcelos et al. 2017). Biofortification is performed to improve the micronutrient content of the grain, rather than the endosperm tissue. The two main biofortification approaches are agronomy and genetics, which comprise traditional breeding, gene editing, and genetic manipulation. Agronomic biofortification can increase Zn content in crops by supplementing with Zn fertilizers; for instance, foliar application of $ZnSO_4$ can increase Zn content in grains by about 60% (Zhang et al. 2012). However, these agronomic techniques are not much beneficial for Fe biofortification, unless supplemented with enhanced chemical fertilizers (Aciksoz et al. 2011) that are not economically and environmentally suitable.

The International Crops Research Institute for the Semi-Arid Tropics (India) is employing traditional breeding to develop cereal (sorghum and pearl millet) cultivars with higher nutritional content, particularly Fe. Moreover, the International Maize and Wheat Improvement Center (Mexico) is working to improve the Zn content of wheat (Velu et al. 2018). The CIMMYT developed Zn biofortified lines that are currently being grown in India and Pakistan. These lines provide around 20–40% higher level of Zn, with overall production comparable to that of the best local cultivars (Velu et al. 2018). Furthermore, in Pakistan, human interference efforts to evaluate the bioavailable Zn in biofortified lines are presently underway (Lowe et al. 2018). However, no improved Fe or other micronutrient wheat lines have been developed thus far with conventional breeding across the globe, including at CIMMYT, irrespective of several research programs having been conducted.

7.4 Breeding Efforts for Wheat Biofortification

Biofortification of essential micronutrients into crops is generally achieved using the following approaches: transgenic, genetic, and agronomic approaches, which use biotechnology, plant breeding, and fertilizer methods, correspondingly (Garg et al. 2018). Latest developments in plant molecular breeding science, as well as the advent of contemporary genomics technologies, have greatly enhanced our knowledge of the genetic components to ultimate crop growth, evolution, and performance, including yield. It is well understood that qualitative traits are controlled by a single gene, whereas quantitative traits, like yield, are typically influenced by multiple genes (Sempértegui et al. 2003; Dorea 2002; Lyons et al. 2003). The Zn, Fe, Se, I, provitamin, and protein content traits in wheat are quantitative. It is easier to breed the crop for qualitative traits than for quantitative traits via conventional breeding. Using conventional breeding methods, enriched Zn content-biofortified wheat varieties have been successfully developed by CIMMYT. These released Zn-biofortified varieties are “Zn Shakti” (Chitra), “Znol 2016,” “WB02,” “HPBW-01,” and “BARI Gom 33” that have been created using different background varieties, such as PBW343 and NARC2011 (Bouis and Saltzman 2017; Das et al. 2019; HarvestPlus 2019).

Several genetic methodologies are well established for wheat biofortification, such as genome-wide association study (GWAS), quantitative trait loci (QTL) mapping, marker-assisted selection (MAS), and omics (Abid et al. 2017; Adams et al. 2002; Bohra et al. 2016, 2019; Mérida-García et al. 2019). In addition, the advent of genome sequencing technologies has offered new opportunities for determining the significance of functional genetic links of plants in phenotype heterogeneity as well as the unpredictability of complex traits across and within species. In wheat, grain is the primary objective component for human utilization and the food business. However, given the micronutrient deficiency or low levels of micronutrients (Fe, Zn, Se, I, etc.) in several modern wheat cultivars, many wheat-breeding programs have employed diverse strategies for wheat biofortification to enhance the nutrient quality and nutraceutical compounds of wheat grain.

To achieve this, plant breeders and other research groups have identified multiple putative QTLs associated with wheat with varying content of micronutrients (Fe, Zn, Se, I, etc.) and proteins (Blanco et al. 1996; Guo et al. 2020; Olmos et al. 2003; Prasad et al. 1999). These QTLs were identified using linkage mapping and the GWAS approach (Mérida-García et al. 2019; Alomari et al. 2018; Bhatta et al. 2018; Crespo-Herrera et al. 2016). These approaches used various types of mapping populations that include recombinant inbred lines (RILs) (Crespo-Herrera et al. 2017; Krishnappa et al. 2017; Wang et al. 2017), with single seed descent (SSD) (Parker et al. 1998), doubled haploid (DH) (Perretant et al. 2000; Shi et al. 2008), and near-isogenic lines (NILs) (Krishnappa et al. 2017). In order to identify the desired traits, researchers used a variety of molecular markers, like restriction fragment length polymorphism (RFLP) (Parker et al. 1998), amplified fragment length polymorphism (AFLP) (Elouafi et al. 2001), single sequence repeat (SSR) (Krishnappa et al. 2017), and single nucleotide polymorphism (SNP) (Alomari et al. 2018; Bhatta et al. 2018). The identified QTLs for micronutrients and grain proteins in wheat are mapped across the genome under specific environments, each with a small effect on the phenotypic variation explained (PVE), as shown in Table 7.2. Furthermore, multiple QTLs in wheat have been found to influence micronutrient absorption by plant roots from the soil and subsequent transfer to the shoot.

Among the numerous QTLs identified, few have been fine mapped, and candidate genes have been proposed (Olmos et al. 2003; Alomari et al. 2018; Uauy et al. 2006) for MAS using specific DNA markers or cloned using the reverse genetic approach (Uauy et al. 2006). Unlike genetic approaches, agronomic biofortification, often known as a fortification, is a fertilizer-based exogenous method for soil or plant foliar utilizing a composite micronutrient mixture (Zou et al. 2019; Cakmak and Kutman 2018; Garcia-Oliveira et al. 2018). This technique is based on inorganic elements being taken up by the roots from soil or the application site and mobilized inside the plants to the source or sink functions (White and Brown 2010). Previously, the potential of agronomic biofortification in alleviating micronutrient inadequacy in wheat was investigated (Ram et al. 2016; Cakmak et al. 2010). However, this approach is regarded as a short-time solution that is used as a complement to genetic biofortification, singularly when the soil in a specific region has poor micronutrient content (Cakmak et al. 2010, 2018). For example, in soil with an alkaline pH, Zn adheres with soil particles, like ferric oxides and calcites, which leads to decreasing Zn availability to plant roots. Similarly, soil with decreased humidity or moisture and organic compound level leads to the decrease of Zn in the root environment significantly (Graham et al. 1992; Marschner 1993; Alloway 2009). Furthermore,

Table 7.2 List of significant studies performed for wheat biofortification

Biofortification trait	Mapping population	QTL ID	Flanking marker[a]	Marker interval/ position (cM)[b]	LOD[c]	PVE (%)[d]	Ref.[e]
Grain iron (Fe) content	RILs	*QGFe.iari-2A*	Xgwm359–Xgwm249	21.9–34.9	4.1	6.8	Kiss (2011)
		QGFe.iari-5A	Xgwm126–Xgwm595	11.3–25.2	3.0	2.3	
		QGFe.iari-7A	Xbarc49–Xwmc525	7.8–29.4	2.9	4.3	
		QGFe.iari-7B	Xgwm364–Xgwm396	44.0–65.4	2.5	3.5	
		QGFe.iari-7B	Xgwm364–Xwmc396	49.5–65.3	2.8	2.5	
		QGFe.iari-7B	Xgwm577–Xbarc264	60.2–74.5	3.4	6.0	
		FeE1-4A	Xwmc468–Xbarc170	50	3.4	10.3	von Grebmer et al. (2019)
		*FeE1-5A**	Xsrap97–Xbarc330	163	3.5	10.4	
		*FeE2-2B**	wPt–7004–wPt–4210	171	4.6	9.5	
		FeE2-4D	Xgwm154–Xbarc108	160	5.0	19.1	
		*FeE2-5B**	Xbarc216–Xbarc74	92	3.4	9.2	
		*FeE2-5B**	wPt–7237–wPt–0708	171	3.8	7.4	
		FeE2-5D	wPt–667104–wPt–731590	8	3.5	8.3	
		FeE2-7D	wPt–663949–wPt–7068	182	2.8	5.4	
		QGFe.co-1A	100008269|F|0–134239|F|0	179.5–184.5	2.6	2.2	McGuire (2015)
		QGFe.co-2A	4993302|F|0–3954215	146.5–147.5	4.1	3.0	
		QGFe.co-3B.1	1089107–1127875|F|0	199.5–200.5	3.7	14.6	
		QGFe.co-3B.2	1233878–4262223|F|0	262.5–263.5	3.1	5.6	
		QGFe.co-3D	1372776–100008980|F|0	124.5–125.5	2.6	2.2	
		QGFe.co-4B	1242543–1125612|F|0	120.5–124.0	5.3	4.4	
		QGFe.co-5A.1	14394095|F|0–4543804	43.5–46.5	2.7	2.1	
		QGFe.co-5A.2	1102433–988523	96.5–97.5	3.1	6.9	
		QGFe.co-6B	5332918–7342703	292.5–295.5	3.2	6.2	
		QGFe-2A	Xwmc382–Xbarc124	23.6	3.7	14.3	Dukhi (2020)
		QGFe-7A	Xgwm473–Xbarc29	153.8	3.3	12.6	

		QFe. Across_7DS	TP43715-TP37547	2.0–5.4	6.6	14.5	Bouis and Saltzman (2017)
		QFe.Y13-14_5BS	TP91631-TP81797	23.1–23.7	5.3	12.0	
		QFe. Across_4BS	TP91631-TP81797	23.1–23.7	5.1	10.7	
		QFe. Across_6AL	TP5797-TP5676	86.9–89.5	4.2	9.6	
		QFe.Y13-14_6BL	TP68423-TP54256	36.6–37.9	3.9	8.9	
		QFe. Across_2DS	TP16058-TP69509	39.4–41.1	3.6	7.8	
		QFe. Across_2BL	TP54992-TP7812	103.5–106.5	3.5	7.5	
		QFe.Y13-14_5BL	TP17691-TP85478	62.7–64.4	3.4	7.7	
		QFe.Y12-13_6DS	TP52164-TP31712	19.5–20.5	3.2	9.1	
		QGFe.cimmyt-4A_P2	3385350–1211533	198.5–199.5	9.7	21.1	Doss (2018)
		QGFe.cimmyt-3A_P1	1234521–3034169F0-11AG	85.5–87.5	7.1	10.4	
		QGFe.cimmyt-3B_2P2	4394657–3947677	511.5–513.5	6.5	7.2	
		QGFe.cimmyt-4D_P2	2363822–3961236	0–2.5	6.5	14.6	
		QGFe.cimmyt-2A_P2	4262668–1226245	420.5–428.5	6.4	14.2	
		QGFe.cimmyt-5B_P2	4407677–1129284	168.5–174.5	5.4	11.6	

(continued)

Table 7.2 (continued)

Biofortification trait	Mapping population	QTL ID	Flanking marker[a]	Marker interval/ position (cM)[b]	LOD[c]	PVE (%)[d]	Ref.[e]
		QGFe.cimmyt-3B_1P2	3533713–1007339	275.5–281.5	5.1	5.8	
		QGFe.cimmyt-2B_P2	wPt–0289–1026059	170.5–173.5	5.0	5.8	
		QGFe.cimmyt-5B_P1	4989996–5410720	96.5–97.5	4.7	5.5	
		QGFe.cimmyt-4B_P1	1008589F0–58TC-2256263	43.5–44.5	4.6	6.7	
Grain zinc (Zn) content	DH	*Zn-2D*	P3470.3–P3176.1	–	5.6	13.4	Caron et al. (2017)
		Zn-3A	Xgwm391–P8422	–	3.0	7.1	
		Zn-4A	P3446–205–CWM145	–	3.2	8.3	
		Zn-4D	WMC331–Xgwm624	–	4.0	8.6	
		Zn-4D	Xgwm192–WMC331	–	4.2	11.9	
		Zn-5A	Xgwm291–Xgwm410	–	3.6	10.9	
		Zn-7A	WMC488–P2071–180	–	6.0	14.6	
	RILs	*QGZn.iari-2A*	Xgwm359–Xwmc407	30.0–38.5	13.5	11.1	Kiss (2011)
		QGZn.iari-2A	Xgwm359–Xgwm249	22.0–34.7	11.8	14.4	
		QGZn.iari-2A	Xgwm359–Xgwm249	23.6–32.9	6.5	8.5	
		QGZn.iari-4A	Xbarc184–Xbarc106	58.0–76.5	2.6	4.7	
		QGZn.iari-5A	Xgwm126–Xbarc144	5.8–17.5	3.7	6.2	
		QGZn.iari-7A	Xbarc49–Xwmc525	7.0–25.2	2.6	3.2	
		*QZnE1-2D**	wPt-730057–wPt-671700	174	5.0	8.6	von Grebmer et al. (2019)
		QZnE1.2-3D	wPt-6191–wPt-8658	218	2.7	7.7	
		*QZnE1-4D**	wPt-671648–wPt-667352	41	3.2	5.5	
		*QZnE2-5B**	wPt-7237–wPt-0708	81	2.8	7.8	

		QZnE2 3D	Xbarc6–Xcfe172	33	4.0	14.5	
		*QZnE2-4D**	Xcfa2149–Xbarc48	194	3.3	15.9	
		*QZnE1-5B**	Xgwm213–Xbarc216	75	3.4	13.8	
		QGZn.co-1B	124470–3028438|F|0	382.5–383.5	2.8	5.4	McGuire (2015)
		QGZn.co-2B	230380–2275590	516.5–517.5	2.8	7.1	
		QGZn.co-3A	3022261–3936326	99.5–100.5	3.6	5.2	
		QGZn.co-3B	1002594|F|0–1103633	315.5–316.5	2.8	3.6	
		QGZn.co-3D	1372776–100008980|F|0	123.5–126.5	2.8	2.7	
		QGZn.co-4B	2277812–1242543	120.5–124	3.3	7.2	
		QGZn.co-5A	124421–1272027|F|0	53.5–54.5	2.7	14.2	
		QGZn.co-6B.1	1252668–100005882|F|0	277.5–278.5	3.6	4.2	
		QGZn.co-6B.2	3941131–990183	177.5–180.5	3.4	4.1	
		QGZn.co-7A	5356706–5325178|F|0	285.5–288.5	5.5	7.8	
		QGZn.cimmyt-1A_P2	4543935–3937719	133.5–136.5	8.9	10.8	Doss (2018)
		QGZn.cimmyt-1B_P1	3934172–3934936	83.5–84.5	8.3	15.1	
		QGZn.cimmyt-1B_P2	4991478–3937490	226.5–228.5	8.6	11.3	
		QGZn.cimmyt-3B_1P2	3533713–1007339	277.5–284.5	5.4	4.4	
		QGZn.cimmyt-3B_2P2	4394657–3947677	512.5–513.5	7.9	10.9	
		QGZn.cimmyt-3D_P2	wPt–741157–1297057	191.5–193.5	5.8	7.5	
		QGZn.cimmyt-4A_P2	4543988–3533871	123.5–124.5	5.2	3.8	

(continued)

Table 7.2 (continued)

Biofortification trait	Mapping population	QTL ID	Flanking marker[a]	Marker interval/ position (cM)[b]	LOD[c]	PVE (%)[d]	Ref.[e]
		QGFe.cimmyt-4D_P2	2363822–3961236	0–2.5	6.5	14.6	
		QGZn.cimmyt-5B_P2	1078595–4538122	102.5–108.5	6.6	5.1	
		QGZn.cimmyt-6A_P1	1238392–4990410	108.5–109.5	7.9	9.7	
		QGZn.cimmyt-6A_P2	1697218–1082136	178.5–179.5	11.9	8.5	
		QGZn.cimmyt-7B_1P1	3945822–1132640F0–5CG	61.5–62.5	7.1	16.8	
		QGZn.cimmyt-7B_1P2	1079651–1262636	43.5–44.5	20.8	32.8	
		QGZn.cimmyt-7B_2P1	989723F0–48CT–1204955F0–26CT	85.5–86.5	5.4	2.9	
		QGZn.cimmyt-7B_2P2	4003947–3532745	95.5–98.5	7.8	3.3	
		QGZn.cimmyt-7B_3P2	4009608–5411574	146.5–148.5	7.0	5.4	
		QGZn.cimmyt-7D_P2	wPt–733859–3033815	41.5–53.5	5.2	5.8	
		QGZn-7A.1	Xcfd31–Xcfa2049	72.6	4.4	21.1	Dukhi (2020)
		QGZn-7A.2	Xgwm473–Xbarc29	153.8	3.5	14.7	
		QZn.Across_4BS	TP91631-TP81797	23.1–23.7	6.6	17.3	Bouis and Saltzman (2017)
		QZn.Across_6AL	TP38159-TP75088	50.2–50.7	3.3	8.4	

		QZn.Y12-13_4BS	TP73864-TP71929	17.1–19.4	4.0	11.7	
		QZn.Y13-14_4BS	TP91631-TP81797	23.1–23.7	7.0	19.6	
		QZn.Y13-14_6BL	TP54256-TP29689	37.9–39.9	3.3	9.0	
Grain selenium (Se) content	RILs	*QGsec-5B*	gwm67Ð–swes100	–	8.0	17.7	Lobell et al. (2011)
		QGsec-5B	BS00082312–51ÐwPt-5120	–	7.0	14.5	
		QGsece-5B.1	D-3953407ÐD–1071681	–	4.6	9.0	
		QGsece-5B.2	D-2289135ÐD–1236560	–	3.8	7.4	
		QGsece-5B.3	RAC875c33387_888ÐD–3936732	–	7.0	13.9	
		QGsece-2B.1	D-3956657ÐS–1105975	–	3.7	12.6	
		QGsece-2B.2	S-1120640ÐD–3024250	–	4.3	10.4	
		QGsec-5D	S-3958480Ð–S-2347952	–	4.4	8.9	
		SeE1-4D	Xbarc241–Xsrap15a	100	10.4	35.1	von Grebmer et al. (2019)
		SeE2-7D	wP-t671684–wPt-667506	215	10.8	28.5	
		SeE1-3D	wPt-6191–wPt-8658	218	5.2	23.4	
		SeE1-5B	wPt-5120–wPt-0708	169	3.8	10.1	
		SeE1-4A	wPt-731120–wP-t1642	91	2.8	6.4	
Grain protein content	RILs	*QGpc.iari-2A*	Xgwm359–Xwmc407	29.6–37.2	14.6	18.8	Kiss (2011)
		QGpc.iari-2A	Xgwm359–Xwmc407	25.3–38.1	7.0	12.1	
		QGpc.iari-2A	*Xgwm359–Xwmc407*	26.1–38.8	5.9	9.9	
		QGpc.iari-2A	Xgwm359–Xwmc407	24.4–39.3	5.8	9.2	
		QGpc.iari-2A	Xgwm359–Xwmc407	22.0–45.1	3.0	5.4	
		QGpc.iari-2A	Xgwm359–Xwmc407	23.5–45.6	2.9	4.3	
		QGpc.iari-3A	Xbarc45–Xbarc57	61.1–81.9	2.9	5.3	

(continued)

Table 7.2 (continued)

Biofortification trait	Mapping population	QTL ID	Flanking marker[a]	Marker interval/ position (cM)[b]	LOD[c]	PVE (%)[d]	Ref.[e]
		QGpc.co-2A	1267600–1138191	302.5–308.5	3.5	4.2	McGuire (2015)
		QGpc.co-2B.1	1083804–1117983	304.5–312.5	3.1	6.3	
		QGpc.co-4A	3942314–5323574lFl0	0–18.5	2.8	3.2	
		QGpc.co-2A	1267600–138191	302.5–308.5	2.8	5.8	
		QGpc.co-2B.2	2303802–2275590	516.5–518.5	2.8	10.8	

[a] DNA markers flanking the detected QTL
[b] The absolute position of the QTL
[c] The logarithm of the odd profile
[d] Phenotypic variation explained by the QTL
[e] Reference

agronomic biofortification is also less effective due to the limited mobility of Fe and Zn in the root system, resulting in reduced concentration in the plant's edible portions (Cakmak et al. 2010; White and Broadley 2005). Several studies have found that the amount of elements like zinc varies across current wheat cultivars and wild ecotypes. For instance, the Zn content in high-yielding wheat cultivars is significantly lower (two- to threefold) than that recorded in the primitive or wild wheat genetic resources. Thus, wild wheat genotypes are widely used as important genetic resources in many breeding programs on wheat biofortification, among other quality traits (Cakmak et al. 2010; Fan et al. 2008; Zhao et al. 2009). Thus, using diverse breeding approaches, a panel of wheat varieties has been developed and released to the public to address the issue of malnutrition or hidden hunger caused by micronutrient deficiency.

7.5 Challenges, Limitations, and Success of Breeding Approaches for Wheat Biofortification

Biofortification of crops has been performed for a long time, and numerous strategies have been considered to grow biofortified crops. Wheat has ample natural wild germplasm and landrace resources with Zn, Fe, and Se content. However, to develop the biofortified wheat grain for micronutrient, provitamin, and protein content is very challenging and has some limitations to accomplish using breeding strategies. First, agronomic or fertilizer approach implementation is costly and non-sustainable and does not offer long-term solutions for the biofortification of wheat. Similarly, the conventional breeding approach of wheat biofortification is a conventional and phenotyping-based long-term, expensive method for increasing micronutrient, provitamin, and protein content. Furthermore, the inverse relationship between grain yield and grain mineral composition is seen as a key barrier to biofortification of wheat using the traditional breeding approach (Garvin et al. 2006; Shi et al. 2013). This limitation could be overcome via the selection of desired traits with MAS or a genetic approach. However, this approach also poses some challenges owing to complex polyploidy; large (16Gb), high riddance sequence (>85%) of a genome; and risk of linkage drag associated with the traits. In order to design genetic markers, map loci that are responsible for micronutrient, vitamin, and protein content prerequisite are well-annotated reference genome sequence, and multiple cultivars need to a sequence.

The availability of multiple diploid, tetraploid progenitors of wheat, the discovery of genes in the Chinese Spring reference landrace, and perhaps other modern genomic resources, will aid efforts to increase wheat grain micronutrient, protein, and provitamin composition. This would also facilitate the mapping of micronutrient, protein, and provitamin composition genomic regions. However, efforts need to be put for the integration of data on standard new reference, and multiple accessions of wild cultivars should be sequenced. In addition, previous studies related to biofortification traits need to combine on a standard reference sequence that would improve the understanding of wheat biofortification traits. Another important challenge is to look beyond higher Zn, Fe, and Se content resources in wheat and to use them in breeding programs to develop new cultivars with accumulated micronutrients. Deficiency of I is also a prevalent issue; however, salt iodization has its limitations, such as volatilization while cooking, inaccessibility of iodized salt for all households, and inadequate intake because of health concerns (Medrano-Macías et al. 2016; White and Broadley 2009). Thus, it's crucial to focus on identifying desirable traits that control uptake, mobilization, and accumulation of I and other minerals in wheat grain. This can be achieved via plant breeding and genetic or biotechnological approaches (Cakmak et al. 2017).

Despite the challenges in the conventional breeding, genomic, and agronomic approaches, some Zn-biofortified wheat varieties have been successfully developed and released ["Zn Shakti" (Chitra), "Znol 2016," "WB02," "HPBW-01," and "BARI Gom 33"] (Bouis and Saltzman 2017; Das et al. 2019; HarvestPlus 2019). However, there is a need to create awareness about commercially available biofortified cultivars, and these cultivars must be accessible to the most vulnerable sections of the world. In this context CIMMYT and HarvestPlus are raising awareness about the importance of biofortified wheat and acting toward the accessibility of these biofortified released varieties in India, Pakistan, and Bangladesh.

Furthermore, it is well recognized that traditional breeding-based fortification increases the nutritive value of rice and wheat (Garg et al. 2018; To 2014). The current trend shows a growing interest in the use of genetic engineering techniques (transgenic, gene editing, or the use of sequenced mutants) that target specific metabolic pathways to improve the genotypes for the desired trait (Vanderschuren et al. 2013) for animal consumption and, progressively, for human diet (Graybosch et al. 2013). In addition, increasing multiple genome sequences and modern genomic or well-annotated gene models offer gene-centric methodologies for wheat biofortification.

7.6 Molecular Understanding of Essential Micronutrient Uptake and Deposition in Wheat Grain

Uptake, transport, mobilization, and deposition of essential micronutrients in cereals are important research topics. Usually, essential minerals, such as Fe and Zn, are available from the soil. In order to increase the essential micronutrient content in wheat grain, it is crucial to precisely understand the absorption and translocation of elements from the soil to the plant and their subsequent accumulation in the grain. Here, we briefly discuss the prospect of Zn and Fe absorption, translocation, and accumulation in wheat grains.

7.6.1 Factors Affecting Micronutrient Availability to Wheat Grains

In cereal cultivation areas, the soil's physical and chemical properties (moisture, temperature, organic matter, and soil) determine the Zn and other metal solubility and the ability of roots to absorb adequately (Cakmak 2008; Nadeem and Farooq 2019). Here, we briefly discuss the prospect of Zn absorption and availability to wheat grains.

Commonly, soil pH ranges between 5.5 and 7.0, every single unit increase in pH that influences available soil Zn concentration decreases by 30–45-folds to plants leading to a deficiency of Zn in plants. Moreover, increasing pH causes Zn adsorption to soil clay minerals, etc. Zn^{2+} concentration in the soil solution is adequately high at a pH of 5. Zn is mainly transported in the soil to the root surface through diffusion. This mineral is very sensitive to soil moisture and pH (Marschner 1993; Cakmak 2008; Lindsay and Collins 1991). Similarly, reduced organic matter content and soil moisture ratio hinder the absorption and availability of Zn in the root environment (Graham et al. 1992; Marschner 1993; Alloway 2009). Plant Zn nutrition is also negatively affected by water deficit, mainly in areas where the topsoil is dry during the reproductive stages of crops. Zn deficiency stress was more severe in rain-fed areas than in irrigated areas in field trial conditions. Thus, the rate of transportation of Zn to the plant roots and Zn solubility is considerably influenced by the percent content of organic matter in the soil (Catlett et al. 2002). A successful genetic crop biofortification program with Zn and other essential micronutrients relies on its concentration of availability for crops in the soil. A continuous supply of sufficient quantities of available essential micronutrients for crops is necessary for significant biological impact in genetic biofortification programs.

7.6.2 Molecular-Level Translocation of Micronutrients from Soil to Grain in the Wheat

The essential elements' uptake in plants from the soil occurs via two methods, direct and indirect. The direct uptake comprises Fe^{2+} and Zn^{2+} of ZRT/IRT-related protein (ZIP) family, while the indirect method occurs via phytosiderophores that chelate Fe cations, with consequent uptake via yellow stripe-like transporters (YSL) (Dempski 2012; Guerinot 2000; Sperotto et al. 2012; Milner et al. 2013). Monocot crops, such as wheat, barley, and maize, usually follow chelation for Fe uptake. Same protein families generally regulate the transport of essential micronutrients (Fe and Zn) to plants. However, plants, with the help of multigene family members, treat different elements differently. In addition, metal chelators, such as metallothioneins, nicotianamine, low-molecular-weight proteins, and non-proteinogenic AA also play a crucial role in chelation, detoxification, and circulation of Fe and Zn from the root to the other parts of plants (Rellán-Álvarez et al. 2010; Deinlein et al. 2012). Vacuolar nicotinamide is observed to play a critical and distinct role in Fe deficiency and for Zn sequestration in *Arabidopsis* (Haydon et al. 2012). Some other transporter proteins are also involved and play important roles in metal and micronutrient cellular homeostasis. These transporter proteins belong to the cation diffusion facilitator (CDF) family, vacuolar iron transporter family, natural resistance-associated macrophage protein (NRAMP) family, P1B-like heavy metal ATPase

(HMA) family, major facilitator superfamily (MFS), ATP-binding cassette protein (ABC) transporter family, and cation exchange (CAX) family (Gustin et al. 2011; Montanini et al. 2007; Connorton et al. 2017; Nevo and Nelson 2006; Rice et al. 2014; Pittman and Hirschi 2016; Verrier et al. 2008). In particular, the ZIP and YSL family proteins facilitate the transport of Fe and Zn from the xylem to the phloem in the root and shoot or during remobilization from the leaves to the grains. Wheat has xylem discontinuity; thus, all minerals and nutrients must pass through the phloem to be obtained in the grains (Zee and O'brien 1970).

The majority of Fe and Zn intakes from the root to the seed grains have been studied in cereal crops (maize, rice, and barley). The information of these cereal crops facilitated the proposed Fe and Zn transport in wheat where information is partial. The uptake and transport of Fe and Zn from the soil to the grains and its homeostasis in plants, including wheat, is reviewed and covered in detail in previous trials (Ludwig and Slamet-Loedin 2019; Borrill et al. 2014; Olsen and Palmgren 2014; Sinclair and Krämer 2012; Nakagawa et al. 2018; Sperotto et al. 2018; Curie and Mari 2017). Based on these data, we illustrated the putative framework of protein families/genes involved in micronutrient (Fe/Zn) translocation from the roots to the seed grains in wheat (Fig. 7.4).

The literature survey determined the probable groups of transport proteins, which are represented in red font. Unknown transporters are represented by question marks. Bioavailable elements in the soil (free Zn^{2+} and phytosiderophore (PS)-bound Fe and Zn) are assimilated in the root epidermal cells. Fe and Zn travel to the pericycle via apoplast and symplast; on the way they may be trapped in vacuoles. Further, Fe and Zn are transported into the xylem and subsequently transferred into the phloem in the root, basal shoot, or leaf tissues (not shown). Remobilization of Fe and Zn takes place from leaf cell plastids (P) → vacuoles (V) → phloem→ ear. Further, Fe and Zn are transferred into the embryonic cavity from parent tissue. After absorption into the aleurone layer, mostly Fe and Zn are sequestered in protein storage vacuoles (PSVs) attached to phytate (Phy). A limited amount of Fe and Zn may penetrate the endosperm and be stored or aggregated to bind with ferritin (Fer) in starch bodies, also known as amyloplasts (A). ZIP, ZRT-, IRT-like protein; YSL, yellow stripe-like transporter; MFS, major facilitator superfamily transporter; MTP, metal tolerance protein; HMA, heavy metal ATPase; FPN, ferroportin; NRAMP, natural resistance-associated macrophage protein; VIT, vacuolar iron transporter; NA, nicotianamine; Cit, citrate; SP, small proteins.

7.7 Transgenic Efforts for the Development of Biofortified Wheat

The nutritional value of food crops can be enhanced with transgenic approaches. Transgenic approaches differ from other strategies. New specific gene is added directly into the organism or plants using transgenic approaches, and plants can be converted for the production of preferred compounds. This strategy is dependent on the compound source and whether it is produced by plants or obtained from the

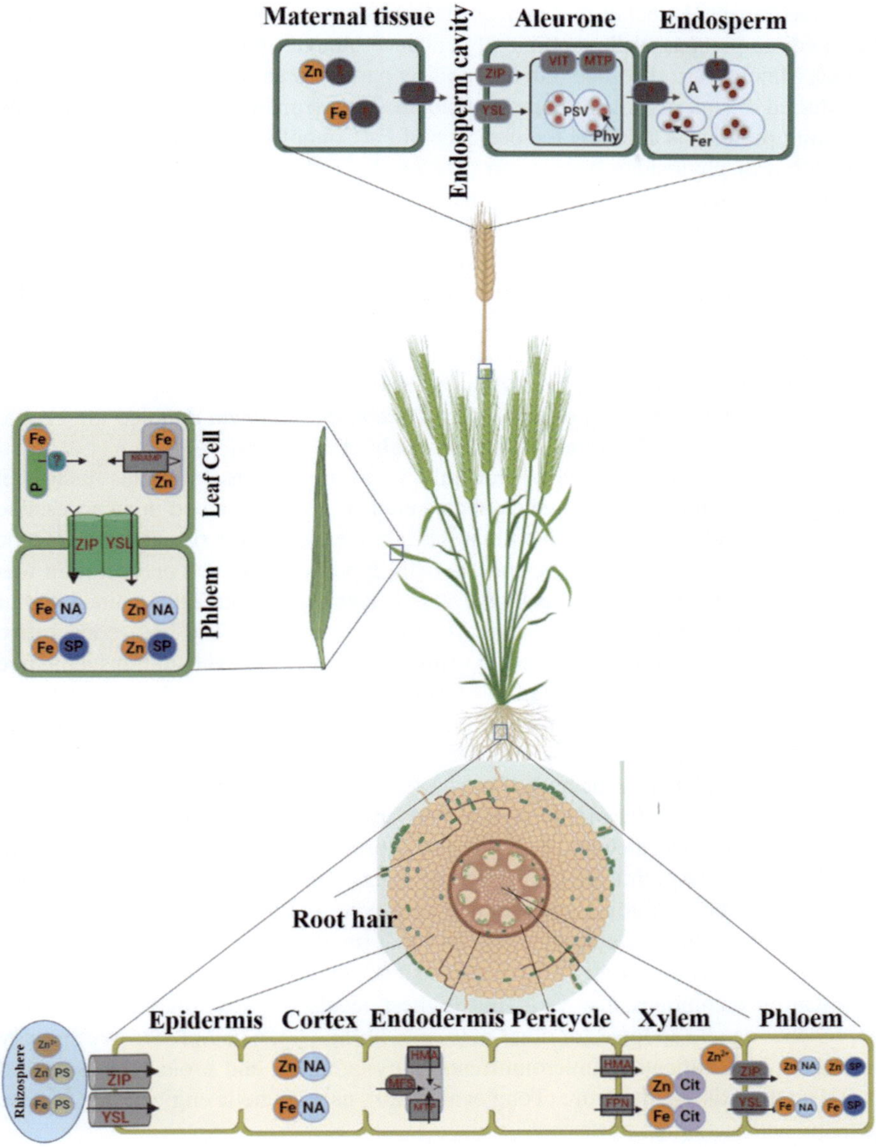

Fig. 7.4 Micronutrient (Fe and Zn) uptake and translocation from the root to the grain in wheat is depicted schematically. (Modified from Borrill et al. 2014)

atmosphere. The amino acid, micronutrients, essential fats or lipids, and vitamins are produced by plants, and these molecules can increase the nutrient content. However, to increase the cumulative amount of these required nutrients, metabolic engineering is necessary.

Some transgenic techniques increase total absorption and transport of the Zn and Fe to edible parts as well as efficiently enhance bioavailable minerals added to the plants. Minerals are absorbed into the roots by particular enzymes or proteins and are transferred to sink organs through the phloem. The formation of these transport and chelating proteins is aided by the presence of Fe(Connolly 2002). Alternatively, the generation of phytosiderophores can promote Fe concentration and accumulation; for example, the expression of the barley genes (naat-A and naat-B) encoding nicotinamide aminotransferases in rice crops resulted in higher Fe uptake. Remarkably, mutants and transgenic plants with overexpressed Fe reductases have shown that the Fe and Zn transport networks share some interaction. Fe transporters also show improved Zn accumulation, leading to increased synthesis of nicotinamide that improves the mobilization of metals in the vascular tissue. Accordingly, the overexpression of nicotianamine synthases also indicates Fe and Zn accumulation; for example, the expression of HvNAS1 in *Nicotiana tabacum* doubled the Fe and Zn content of leaves (Ma et al. 2001; Takahashi et al. 2003).

Another way for improving minerals is to overexpress proteins using an endosperm-specific promoter which allows elements to be retained in an accessible form, such as soybean ferritin (ferritin is a Fe storage protein) in rice crops. This type of rice has threefold the amount than the wild-type rice. The Fe concentration was also measured in polished grains; however, the levels of Fe and Zn remained higher than those of unpolished rice grains. A comparison of the use of a constitutive promoter to initiate ferritin expression resulted in higher Fe levels in transgenic rice leaves but not in the grains owing to higher ferritin expression in the tissues (Goto et al. 1999). An additional bioavailability is another difficulty with nutritional availability; minerals must be accessible in some kind of a category that could be eaten then assimilated by the body. Phytate seems to be an antinutritional component that chelates Fe and Zn and decreases their absorption in the human gut. Consequently, a transgenic method involving the expression of both phytase and ferritin was developed. Experiments on transgenic maize and rice crops showed how rice grains had twofold more Fe content than wild type. Further, predictions of absorption consuming the maize kernels revealed that the quantity of bioavailable Fe had improved (Drakakaki et al. 2000, 2005). Thus, the combined use of different Fe fortification techniques can yield the highest levels of bioavailable Fe. Similar approaches can be adopted for other nutrients. In Fig. 7.5, a comprehensive strategy for wheat biofortification (micronutrients, provitamin A, and protein) is proposed based on the existing literature (Tong et al. 2020), using genetic engineering, reverse genetic, and breeding approaches.

7.7.1 Challenges for the Public Release of Transgenic Wheat

In wheat, very few agronomical genes have been introduced via the transgenic approach. Although the development of transgenic wheat is increasing rapidly, some important challenges remain. Primarily, there is considerable misinformation in the general public about the consumption of transgenic food grains, such as it may

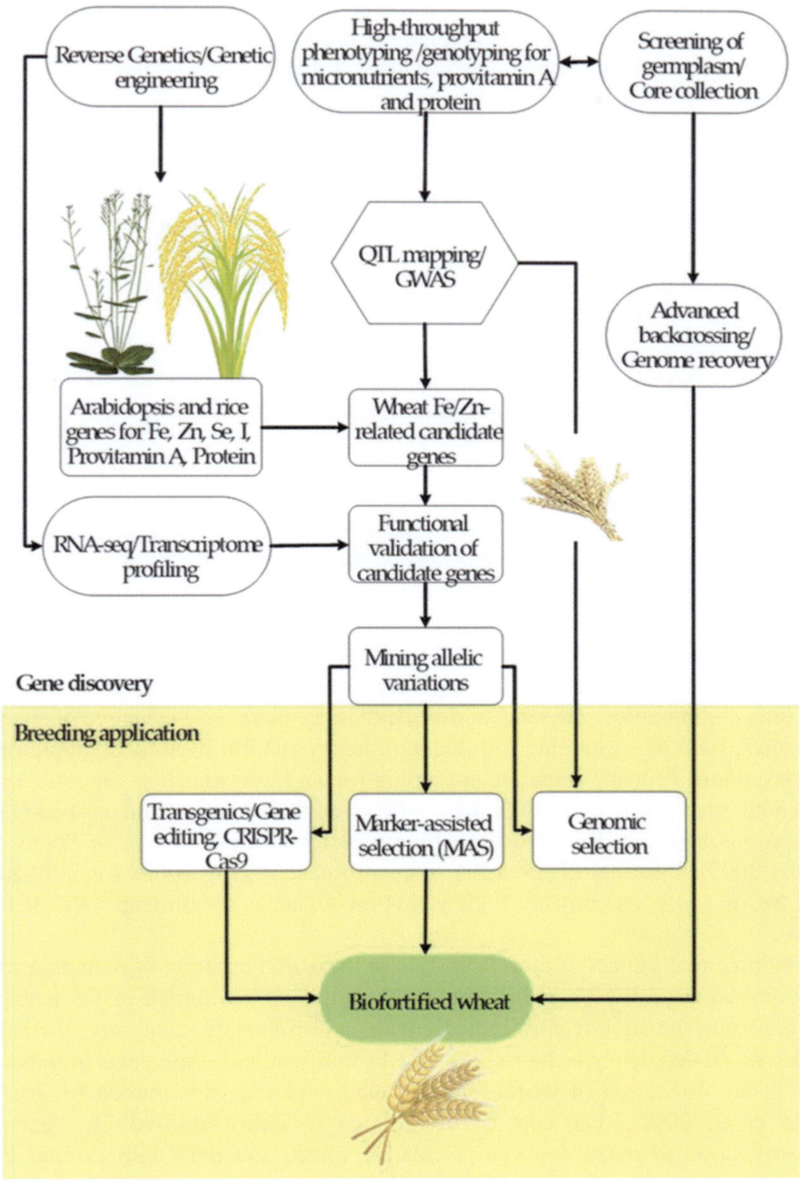

Fig. 7.5 A comprehensive strategy for wheat biofortification

harm human health (Dale 1999). Transgenic wheat developed in Bobwhite or Fielder background which is not a commercially preferable transgenic wheat (Li et al. 2012). Secondly, not providing better consumer benefits, all the outlook is controlled by multinational private organizations of the main crops, leading to rejection of genetically modified (GM) food by the regulatory authorities of most European countries (Bhalla 2006). There are complex or confusing regulations and government policies regarding transgenic crop utilization in several countries. Lack of political wills toward transgenic or GM wheat and other food crops is another crucial factor involved in the release of transgenic wheat.

Despite all the challenges, the majority of the scientific community believes that transgenic technology can offer a very exciting future to plant breeders, farmers, consumers, and the general public. The acceptance and mainstream use of transgenic wheat can lead to substantial benefits and provide numerous opportunities for crop improvement.

7.8 Economic and Social Constraints for the Use of Biofortified Wheat

The micronutrient deficiency confronted by the poor all over the world and particularly in Asia and Africa revealed that scientists have to work out the resolution for the eradication of micronutrient deficiency. Biofortification of food crops resulting from genetic techniques has the potential to be cost-effective and provides numerous advantages to 40% of the population that rely solely on food for nourishment. It is recommended that a one-time financial investment be made in the production of food crop seeds that uptake micronutrients efficiently, ensuring adequate Zn and Fe intake by rural communities. Genetic biofortification is more cost-effective than other strategies, such as agronomic fortification, dietary modification, and supplementation programs. Primary commercial training for Zn biofortification suggests a cost-to-benefit proportion of greater than 20% over two decades and cost-to-benefit proportions between 20% and 30% for Fe biofortification of rice in South Asia (Bouis 2003). Informal assessments of cost-to-benefit proportions for fertilization with Se or I also recommend high yields on financial investments (Nestel et al. 2006).

The potential impact of biofortification is measured in terms of reduction in the disability-adjusted life years in India. Further, the annual burden of Fe deficiency (anemia) may be significantly reduced with biofortification. Similarly, the annual burden of Zn deficiency is 2.8 million lost disability-adjusted life years in India, and the Zn biofortification of wheat and rice may overcome this burden by 20–51% (Stein et al. 2007). The cost of saving one disability-adjusted life year with biofortification of major food crops (maize, wheat, and rice) with Zn and Fe in African countries is estimated to be US $2–20.

7.9 Genome Editing Approaches for Wheat Biofortification

In the lack of stable genetic diversity for the particular trait in a single plant species, genetic engineering technologies offer a feasible edge over traditional breeding strategies (Bhalla 2006; Hu et al. 2003; Vasil et al. 1992). The use of genome resequencing and genome manipulation have contributed to the understanding of functional genetic components underlying the biochemical regulation and physiological and molecular developments in plants, including the nutritional and nutraceutical properties of food crops. Although there has been considerable progress in the use of QTL mapping and the development of gene-transformation technologies, this remains an important challenge for wheat biofortification. Recent advancements in plant biotechnology have offered various possibilities to uphold the level of biofortification programs. The availability of complete newly sequenced genomes in staple food crops and novel gene-editing techniques, such as transcription activator-like effector nucleases and cluster regularly interspaced short palindromic repeats (CRISPR)/Cas9, have opened new doors for the biofortification of cereal crops (Ricroch et al. 2017). In wheat, CRISPR/Cas9 has been employed in various studies, to improve abiotic stress tolerance (Kim et al. 2018) and biotic stress resistance (Shan et al. 2014; Wang et al. 2014; Zhang et al. 2017). However, few studies targeting micronutrients or protein contents have been reported (Ludwig and Slamet-Loedin 2019; Connorton et al. 2017; Liang et al. 2017; Sánchez-León et al. 2018; Zhang and Gao 2017). Although the CRISPR/Cas9 system is a powerful tool for crop genome editing, achieving highly efficient and specific editing in polyploidy species, such as wheat, this technology can be challenging because its efficiency depends on the gRNA used, as reported previously (Arndell et al. 2019). Moreover, the progress in the CRISPR technique (Kim et al. 2006; Lapinskas et al. 1996; Li et al. 2001) is underutilized for the editing of the genes associated with micronutrient levels. This might be due to the uncertain stand on its regulation and political will in many countries and international organizations.

7.10 Improving the Nutraceutical Properties of Wheat

Recent studies that have investigated the health benefits of functional products in wheat have shown the importance of introducing phytochemicals with a high nutraceutical potential using different varieties and ecotypes. Thus, there is renewed interest in the ancient genetic resources of wheat, specifically those with high nutraceutical properties (Dinelli et al. 2007, 2011; Adom et al. 2003; Heimler et al. 2010; Behall et al. 2006; Fardet 2010; He et al. 2010). Phenolic compounds have emerged as health-promoting phytochemicals in wheat grains and have gained attention owing to their high antioxidant capacity and can protect against various degenerative diseases (Heimler et al. 2010; Laddomada et al. 2015). A study that aimed to investigate the phenolic profile in grains of modern and old common wheat varieties suggested that the flavonoid (both bound and free) content and the ratio between flavonoids and polyphenols influence the antioxidant activity of FRAP and

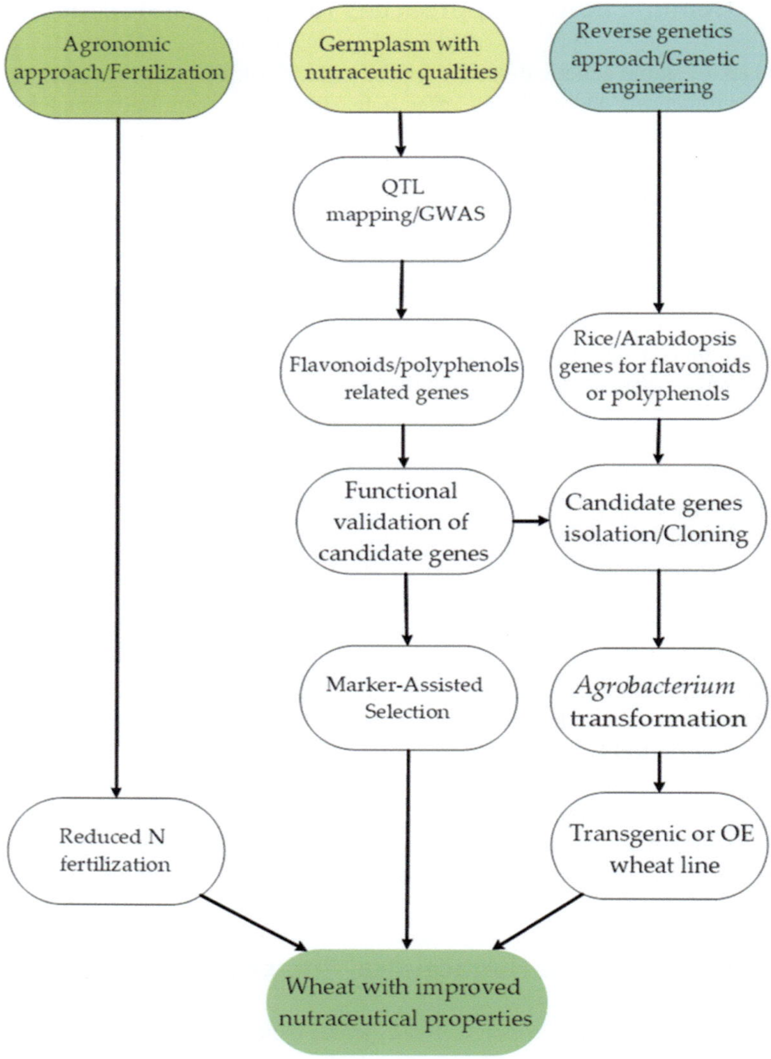

Fig. 7.6 Different approaches for the improvement of nutraceutical properties of wheat grain

DPPH (Leoncini et al. 2012). The authors also supported that phenolic extracts from wheat exerted a cytoprotective and antiproliferative effect on cardiomyocytes and leukemic cell (HL60) line, respectively, suggesting that an increased intake of wheat grain-derived products could help achieve both chemoprevention and protection against oxidative stress-related diseases. Thus, to improve the nutraceutical properties of wheat varieties, while maintaining an optimum level of productivity, different approaches can be used, as illustrated in Fig. 7.6. These approaches include agronomic (reduced nitrogen supply without hampering plant growth and development and yield), conventional breeding (employing the available genetic resources with a high potential for improving the nutraceutical qualities of wheat), and genetic engineering targeting specific genes.

7.11 Conclusion and Future Prospects

Biofortification approaches are based on plant breeding and targeted genetic manipulation to address malnutrition in humans. Genotype and micronutrient interactions in grain production and nutrient density remain unclear. In most studies, enhancements in nutrient use effectiveness are limited by the costly and laborious phenotyping. Moreover, the bioavailability of minerals is another important factor to measure grain quality. Because more starch accumulates in grains, a quest for better yield is typically followed by a diluting effect of minerals. Therefore, more focused efforts are needed to achieve these goals. In this situation, it is recommended that researchers and breeders work very closely not only to attain the maintained Fe and Zn contents but also to improve Fe, Zn, Se, I, provitamin, and protein contents required in the edible parts. Considering the problems outlined above, we believe that wheat researchers have the tools and resources necessary to greatly improve the concentration of important micronutrients, provitamins, and proteins in wheat grains as well as transfer these developed varieties to fields. These improved varieties with higher nutrient contents could make a vital contribution to the health of the global population by preventing micronutrient deficiency, but still nearly 30% of the global population face deficiency for one or more crucial micronutrients.

Ultimately, the human diet depends on the sufficient consumption of several different minerals and other bioactive compounds in combinations and within ranges that are not entirely understood. Therefore, the best solution to eradicate malnutrition as a community health issue in developing countries is a more appropriate consumption of a variety of foodstuff. However, achieving this requires several years for awareness creations, the formation of informed management policies, and a comparatively huge investment in agricultural research and on-farm infrastructure. Human nutrition scientists need to be informed about the levels to which the mineral concentrations of the edible parts of plants and compounds that stimulate and hinder their bioavailability can be amended via the processes of plant breeding and genetics. Plant geneticists need to be knowledgeable about the major effects that plant research may have previously had on micronutrient utilization. Further, they should also be aware of the differences in mineral bioavailability between current varieties and traditional cultivars, as well as the potential of plant genomes to improve human nutrition in the future.

References

Abid N et al (2017) Transgenic expression of phytase in wheat endosperm increases bioavailability of iron and zinc in grains. Transgenic Res 26:109–122

Aciksoz SB, Yazici A, Ozturk L, Cakmak I (2011) Biofortification of wheat with iron through soil and foliar application of nitrogen and iron fertilizers. Plant Soil 349:215–225

Adams ML, Lombi E, Zhao F, McGrath SP (2002) Evidence of low selenium concentrations in UK bread-making wheat grain. J Sci Food Agric 82:1160–1165

Adom KK, Sorrells ME, Liu RH (2003) Phytochemical profiles and antioxidant activity of wheat varieties. J Agric Food Chem 51:7825–7834

Alloway BJ (2009) Soil factors associated with zinc deficiency in crops and humans. Environ Geochem Health 31:537–548

Alomari DZ et al (2018) Identifying candidate genes for enhancing grain Zn concentration in wheat. Front Plant Sci 9:1313

Arndell T et al (2019) gRNA validation for wheat genome editing with the CRISPR-Cas9 system. BMC Biotechnol 19:1–12

Asensi-Fabado MA, Munné-Bosch S (2010) Vitamins in plants: occurrence, biosynthesis and antioxidant function. Trends Plant Sci 15:582–592

Bailey RL, West KP, Black RE (2015) The epidemiology of global micronutrient deficiencies. Ann Nutr Metab 66:22–33

Behall KM, Scholfield DJ, Hallfrisch J (2006) Whole-grain diets reduce blood pressure in mildly hypercholesterolemic men and women. J Am Diet Assoc 106:1445–1449

Bhalla PL (2006) Genetic engineering of wheat–current challenges and opportunities. Trends Biotechnol 24:305–311

Bhatta M et al (2018) Genome-wide association study reveals novel genomic regions associated with 10 grain minerals in synthetic hexaploid wheat. Int J Mol Sci 19:3237

Blanco A et al (1996) Quantitative trait loci influencing grain protein content in tetraploid wheats. Plant Breed 115:310–316

Bohra A, Jha UC, Kumar S (2016) Enriching nutrient density in staple crops using modern "-Omics" tools. In: Biofortification of food crops. Springer, Cham, pp 85–103

Bohra A et al (2019) Genomic interventions for biofortification of food crops. In: Quality breeding in field crops. Springer, Cham, pp 1–21

Borrill P et al (2014) Biofortification of wheat grain with iron and zinc: integrating novel genomic resources and knowledge from model crops. Front Plant Sci 5:1–8

Bouis HE (2003) Micronutrient fortification of plants through plant breeding: can it improve nutrition in man at low cost? Proc Nutr Soc 62:403–411

Bouis HE, Saltzman A (2017) Improving nutrition through biofortification: a review of evidence from HarvestPlus, 2003 through 2016. Glob Food Sec 12:49–58

Brody T (1999) Vitamins. In: Nutritional biochemistry, vol 53. Elsevier, Amsterdam, pp 491–692

Brown CC, Noelle RJ (2015) Seeing through the dark: new insights into the immune regulatory functions of vitamin A. Eur J Immunol 45:1287–1295

Cakmak I (2002) Plant nutrition research: priorities to meet human needs for food in sustainable ways BT. In: Progress in plant nutrition: plenary lectures of the XIV international plant nutrition colloquium: food security and sustainability of agro-ecosystems through basi, vol 247, pp 3–24

Cakmak I (2008) Enrichment of cereal grains with zinc: agronomic or genetic biofortification? Plant Soil 302:1–17

Cakmak I, Kutman UB (2018) Agronomic biofortification of cereals with zinc: a review. Eur J Soil Sci 69:172–180

Cakmak I, Pfeiffer WH, McClafferty B (2010) Biofortification of durum wheat with zinc and iron. Cereal Chem 87:10–20

Cakmak I et al (2017) Iodine biofortification of wheat, rice and maize through fertilizer strategy. Plant Soil 418:319–335

Caron P et al (2017) High level panel of experts on food security and nutrition 2nd note on critical and emerging issues for food security and nutrition prepared for the Committee on World Food Security HLPE Steering Committee members (May 2017) Coordinator of the HLPE Intro, pp 1–23

Catlett KM, Heil DM, Lindsay WL, Ebinger MH (2002) Soil chemical properties controlling zinc 2+ activity in 18 colorado soils. Soil Sci Soc Am J 66:1182–1189

Chatterjee C, Gopal R, Dube BK (2006) Impact of iron stress on biomass, yield, metabolism and quality of potato (Solanum tuberosum L.). Sci Hortic 108:1–6

Clagett-Dame M, Knutson D (2011) Vitamin A in reproduction and development. Nutrients 3:385–428

Combs GF (2001) Selenium in global food systems. Br J Nutr 85:517–547

Connolly EL (2002) Expression of the IRT1 metal transporter is controlled by metals at the levels of transcript and protein accumulation. Plant Cell Online 14:1347–1357

Connorton JM et al (2017) Wheat vacuolar iron transporter TaVIT2 transports Fe and Mn and is effective for biofortification. Plant Physiol 174:2434–2444

Crespo-Herrera LA, Velu G, Singh RP (2016) Quantitative trait loci mapping reveals pleiotropic effect for grain iron and zinc concentrations in wheat. Ann Appl Biol 169:27–35

Crespo-Herrera LA, Govindan V, Stangoulis J, Hao Y, Singh RP (2017) QTL mapping of grain Zn and Fe concentrations in two hexaploid wheat RIL populations with ample transgressive segregation. Front Plant Sci 8:1–12

Curie C, Mari S (2017) New routes for plant iron mining. New Phytol 214:521–525

Dale PJ (1999) Public concerns over transgenic crops. Genome Res 9:1159–1162

Dapkekar A, Deshpande P, Oak MD, Paknikar KM, Rajwade JM (2020) Getting more micronutrients from wheat and barley through agronomic biofortification. In: Wheat and barley grain biofortification. Elsevier, Amsterdam. https://doi.org/10.1016/b978-0-12-818444-8.00003-1

Das S, Chaki AK, Hossain A (2019) Breeding and agronomic approaches for the biofortification of zinc in wheat (Triticum aestivum L.) to combat zinc deficiency in millions of a population: a Bangladesh perspective. Acta Agrobot 72:1–13

Dasgupta M, Sharkey JR, Wu G (2005) Inadequate intakes of indispensable amino acids among homebound older adults. J Nutr Elder 24:85–99

de Valença AW, Bake A, Brouwer ID, Giller KE (2017) Agronomic biofortification of crops to fight hidden hunger in sub-Saharan Africa. Glob Food Sec 12:8–14

Deinlein U et al (2012) Elevated nicotianamine levels in Arabidopsis halleri roots play a key role in zinc hyperaccumulation. Plant Cell 24:708–723

Dempski RE (2012) The cation selectivity of the ZIP transporters. Curr Top Membr 69:221–245

Dinelli G et al (2007) Lignan profile in seeds of modern and old Italian soft wheat (Triticum aestivum L.) cultivars as revealed by CE-MS analyses. Electrophoresis 28:4212–4219

Dinelli G et al (2011) Profiles of phenolic compounds in modern and old common wheat varieties determined by liquid chromatography coupled with time-of-flight mass spectrometry. J Chromatogr A 1218:7670–7681

Dorea JG (2002) Selenium and breast-feeding. Br J Nutr 88:443–461

Doss CR (2018) Women and agricultural productivity: reframing the issues. Dev Policy Rev 36:35–50

Drakakaki G, Christou P, Stöger E (2000) Constitutive expression of soybean ferritin cDNA in transgenic wheat and rice results in increased iron levels in vegetative tissues but not in seeds. Transgenic Res 9:445–452

Drakakaki G et al (2005) Endosperm-specific co-expression of recombinant soybean ferritin and Aspergillus phytase in maize results in significant increases in the levels of bioavailable iron. Plant Mol Biol 59:869–880

Dukhi N (2020) Global prevalence of malnutrition: evidence from literature. In: Malnutrition. IntechOpen, London, p 38

Elouafi I, Nachit MM, Martin LM (2001) Identification of a microsatellite on chromosome 7B showing a strong linkage with yellow pigment in durum wheat (Triticum turgidum L. var. durum). Hereditas 135:255–261

Fan M-S et al (2008) Evidence of decreasing mineral density in wheat grain over the last 160 years. J Trace Elem Med Biol 22:315–324

FAO (2017) The future of food and agriculture – trends and challenges, Rome

FAO (2020) World Food and Agriculture – Statistical Yearbook 2020, Rome. https://doi.org/10.4060/cb1329en

FAO, IFAD, UNICEF, WFP and WHO (2018) The State of Food Security and Nutrition in the World 2018. Building climate resilience for food security and nutrition. FAO, Rome

Fardet A (2010) New hypotheses for the health-protective mechanisms of whole-grain cereals: what is beyond fibre? Nutr Res Rev 23:65–134

Fordyce FM, Zhang G, Green K, Liu X (2000) Soil, grain and water chemistry in relation to human selenium-responsive diseases in Enshi District, China. Appl Geochem 15(1):117–132. https://doi.org/10.1016/S0883-2927(99)00035-9Get

Garcia-Oliveira AL, Chander S, Ortiz R, Menkir A, Gedil M (2018) Genetic basis and breeding perspectives of grain iron and zinc enrichment in cereals. Front Plant Sci 9:937

Garg M et al (2018) Biofortified crops generated by breeding, agronomy, and transgenic approaches are improving lives of millions of people around the world. Front Nutr 5:12

Garvin DF, Welch RM, Finley JW (2006) Historical shifts in the seed mineral micronutrient concentration of US hard red winter wheat germplasm. J Sci Food Agric 86:2213–2220

Goto F, Yoshihara T, Shigemoto N, Toki S, Takaiwa F (1999) Iron fortification of rice seed by the soybean ferritin gene. Nat Biotechnol 17:282–286

Graham RD, Ascher JS, Hynes SC (1992) Selecting zinc-efficient cereal genotypes for soils of low zinc status. Plant Soil 146:241–250

Graybosch RA, Seabourn B, Chen YR, Blechl AE (2013) Transgenic enhancement of high-molecular-weight glutenin subunit 1Dy10 concentration: effects in wheat flour blends and sponge and dough baking. Cereal Chem 90:164–168

Guerinot M (2000) The ZIP family of metal transporters. Biochim Biophys Acta 1465:190–198

Guo Y et al (2020) QTL mapping for quality traits using a high density genetic map of wheat. PLoS One 15:1–18

Gupta UC, Gupta SC (2000) Selenium in soils and crops, its deficiencies in livestock and humans: implications for management. Commun Soil Sci Plant Anal 31:1791–1807

Gustin JL, Zanis MJ, Salt DE (2011) Structure and evolution of the plant cation diffusion facilitator family of ion transporters. BMC Evol Biol 11:76

Hänsch R, Mendel RR (2009) Physiological functions of mineral micronutrients (Cu, Zn, Mn, Fe, Ni, Mo, B, Cl). Curr Opin Plant Biol 12:259–266

HarvestPlus (2019) Getting biofortified food on everyone's plate—2019 annual report, vol 19

Haydon MJ et al (2012) Vacuolar nicotianamine has critical and distinct roles under iron deficiency and for zinc sequestration in Arabidopsis. Plant Cell 24:724–737

He M, Van Dam RM, Rimm E, Hu FB, Qi L (2010) Whole-grain, cereal fiber, bran, and germ intake and the risks of all-cause and cardiovascular disease-specific mortality among women with type 2 diabetes mellitus. Circulation 121:2162–2168

Heimler D et al (2010) Polyphenol content of modern and old varieties of Triticum aestivum L. and T. durum Desf. grains in two years of production. J Agric Food Chem 58:7329–7334

Hoffer LJ, Bistrian BR (2012) Appropriate protein provision in critical illness: a systematic and narrative review. Am J Clin Nutr 96:591–600

Hogarth CA, Griswold MD (2010) The key role of vitamin A in spermatogenesis. J Clin Invest 120: 956–962

Hu T et al (2003) Agrobacterium-mediated large-scale transformation of wheat (Triticum aestivum L.) using glyphosate selection. Plant Cell Rep 21:1010–1019

Kawachi M et al (2009) A mutant strain Arabidopsis thaliana that lacks vacuolar membrane zinc transporter MTP1 revealed the latent tolerance to excessive zinc. Plant Cell Physiol 50:1156–1170

Kim SA et al (2006) Localization of iron in Arabidopsis seed requires the vacuolar membrane transporter VIT1. Science 314:1295–1298

Kim D, Alptekin B, Budak H (2018) CRISPR/Cas9 genome editing in wheat. Funct Integr Genomics 18:31–41

Kiss I (2011) Significance of wheat production in world economy and position of Hungary in it. Appl Stud Agribus Commer 5:115–119

Krishnappa G et al (2017) Molecular mapping of the grain iron and zinc concentration, protein content and thousand kernel weight in wheat (Triticum aestivum L.). PLoS One 12:e0174972

Kupka R et al (2004) Selenium status is associated with accelerated HIV disease progression among HIV-1-infected pregnant women in Tanzania. J Nutr 134:2556–2560

Laddomada B, Caretto S, Mita G (2015) Wheat bran phenolic acids: bioavailability and stability in whole wheat-based foods. Molecules 20:15666–15685

Lapinskas PJ, Lin S, Culotta VC (1996) The role of the Saccharomyces cerevisiae CCC1 gene in the homeostasis of manganese ions. Mol Microbiol 21:519–528

Lee S, Kim SA, Lee J, Guerinot ML, An G (2010) Zinc deficiency-inducible OsZIP8 encodes a plasma membrane-localized zinc transporter in rice. Mol Cells 29:551–558

Leoncini E et al (2012) Phytochemical profile and nutraceutical value of old and modern common wheat cultivars. PLoS One 7(9):e45997

Li J, Saxena S, Pain D, Dancis A (2001) Adrenodoxin reductase homolog (Arh1p) of yeast mitochondria required for iron homeostasis. J Biol Chem 276:1503–1509

Li J, Ye X, An B, Du L, Xu H (2012) Genetic transformation of wheat: current status and future prospects. Plant Biotechnol Rep 6:183–193

Liang Z et al (2017) Efficient DNA-free genome editing of bread wheat using CRISPR/Cas9 ribonucleoprotein complexes. Nat Commun 8:1–5

Lima AAM et al (2010) Effects of vitamin A supplementation on intestinal barrier function, growth, total parasitic, and specific giardia spp. infections in Brazilian children: a prospective randomized, double-blind, placebo-controlled trial. J Pediatr Gastroenterol Nutr 50:309–315

Lin NF, Tang J, Bian JM (2004) Geochemical environment and health problems in China. Environ Geochem Health 26:81–88

Lindsay WL, Collins F (1991) Chapter 4: Inorganic equilibria affecting micronutrients in soils. In: Micronutrients in agriculture. Soil Science Society of America, Madison

Lobell DB, Schlenker W, Costa-Roberts J (2011) Climate trends and global crop production since 1980. Science 333(6042):616–620. https://doi.org/10.1126/science.1204531

Lowe NM et al (2018) Examining the effectiveness of consuming flour made from agronomically biofortified wheat (Zincol-2016/NR-421) for improving Zn status in women in a low-resource setting in Pakistan: Study protocol for a randomised, double-blind, controlled cross-over tri. BMJ Open 8:e021364

Ludwig Y, Slamet-Loedin IH (2019) Genetic biofortification to enrich rice and wheat grain iron: from genes to product. Front Plant Sci 10:1–10

Lyons G, Stangoulis J, Graham R (2003) High-selenium wheat: biofortification for better health. Nutr Res Rev 16:45

Ma JF, Miyak Y, Takahashi E (2001) Silicon as a beneficial element for crop plants. In: Studies in plant science. Elsevier, Amsterdam. https://doi.org/10.1016/S0928-3420(01)80006-9

Marschner H (1993) Zinc uptake from soils. In: Zinc in soils and plants. In: Robson AD (ed) Proceedings of the international symposium on 'zinc in soils and plants' held at the University of Western Australia, 27-28 September, Springer Netherlands, Dordrecht, pp 59–77. https://doi.org/10.1007/978-94-011-0878-2_5

Mayer KFX et al (2014) A chromosome-based draft sequence of the hexaploid bread wheat (Triticum aestivum) genome ancient hybridizations among the ancestral genomes of bread

wheat Genome interplay in the grain transcriptome of hexaploid bread wheat structural and functional pa. Science e345:1250092
McGuire S (2015) FAO, IFAD, and WFP. The state of food insecurity in the world 2015: meeting the 2015 international hunger targets: taking stock of uneven progress. Rome: FAO, 2015. Adv Nutr 6:623–624
Medrano-Macías J, Leija-Martínez P, González-Morales S, Juárez-Maldonado A, Benavides-Mendoza A (2016) Use of iodine to biofortify and promote growth and stress tolerance in crops. Front Plant Sci 7:1–20
Mérida-García R et al (2019) Genetic dissection of agronomic and quality traits based on association mapping and genomic selection approaches in durum wheat grown in Southern Spain. PLoS One 14:1–24
Meschy F (2000) Recent progress in the assessment of mineral requirements of goats. Livest Prod Sci 64:9–14
Milner MJ, Seamon J, Craft E, Kochian LV (2013) Transport properties of members of the ZIP family in plants and their role in Zn and Mn homeostasis. J Exp Bot 64:369–381
Montanini B, Blaudez D, Jeandroz S, Sanders D, Chalot M (2007) Phylogenetic and functional analysis of the Cation Diffusion Facilitator (CDF) family: improved signature and prediction of substrate specificity. BMC Genomics 8:107
Murray CJL, Lopez AD (2013) Measuring the global burden of disease. N Engl J Med 369:448–457
Nadeem F, Farooq M (2019) Application of micronutrients in rice-wheat cropping system of South Asia. Rice Sci 26:356–371
Nakagawa ACS et al (2018) Drought stress during soybean seed filling affects storage compounds through regulation of lipid and protein metabolism. Acta Physiol Plant 40:111
Nestel P, Bouis HE, Meenakshi JV, Pfeiffer W (2006) Symposium: food fortification in developing countries biofortification of staple food crops. J Nutr 136:1064–1067
Nevo Y, Nelson N (2006) The NRAMP family of metal-ion transporters. Biochim Biophys Acta 1763:609–620
Olmos S et al (2003) Precise mapping of a locus affecting grain protein content in durum wheat. Theor Appl Genet 107:1243–1251
Olsen LI, Palmgren MG (2014) Many rivers to cross: the journey of zinc from soil to seed. Front Plant Sci 5:1–6
Palmer CM, Guerinot ML (2009) Facing the challenges of Cu, Fe and Zn homeostasis in plants. Nat Chem Biol 5:333–340
Parker GD, Chalmers KJ, Rathjen AJ, Langridge P (1998) Mapping loci associated with flour colour in wheat (Triticum aestivum L.). Theor Appl Genet 97:238–245
Perretant MR et al (2000) QTL analysis of bread-making quality in wheat using a doubled haploid population. Theor Appl Genet 100:1167–1175
Pittman JK, Hirschi KD (2016) CAX-ing a wide net: cation/H(+) transporters in metal remediation and abiotic stress signalling. Plant Biol (Stuttg) 18:741–749
Prasad M et al (1999) A microsatellite marker associated with a QTL for grain protein content on chromosome arm 2DL of bread wheat. Theor Appl Genet 99:341–345
Ram H et al (2016) Biofortification of wheat, rice and common bean by applying foliar zinc fertilizer along with pesticides in seven countries. Plant Soil 403:389–401
Rayman MP (2000) The importance of selenium to human health. Lancet 356:233–241
Rayman MP (2002) The argument for increasing selenium intake. Proc Nutr Soc 61:203–215
Rellán-Álvarez R et al (2010) Identification of a tri-iron(III), tri-citrate complex in the xylem sap of iron-deficient tomato resupplied with iron: new insights into plant iron long-distance transport. Plant Cell Physiol 51:91–102
Rice AJ, Park A, Pinkett HW (2014) Diversity in ABC transporters: type I, II and III importers. Crit Rev Biochem Mol Biol 49:426–437
Ricroch A, Clairand P, Harwood W (2017) Use of CRISPR systems in plant genome editing: toward new opportunities in agriculture. Emerg Top Life Sci 1:169–182
Ritchie H (2017) Micronutrient deficiency. Our world data

Sánchez-León S et al (2018) Low-gluten, nontransgenic wheat engineered with CRISPR/Cas9. Plant Biotechnol J 16:902–910

Semba RD, Muhilal, West KP Jr, Winger M, Natadisastra G, Scott A, Sommer A (1992) Impact of vitamin a supplementation on hematological indicators of iron metabolism and protein status in children. Nutr Res 53:1689–1699

Sempértegui F et al (2003) Selenium serum concentrations in malnourished Ecuadorian children: a case-control study. Int J Vitam Nutr Res 73:181–186

Shan Q, Wang Y, Li J, Gao C (2014) Genome editing in rice and wheat using the CRISPR/Cas system. Nat Protoc 9:2395–2410

Shi R et al (2008) Identification of quantitative trait locus of zinc and phosphorus density in wheat (Triticum aestivum L.) grain. Plant Soil 306:95–104

Shi R, Tong Y, Jing R, Zhang F, Zou C (2013) Characterization of quantitative trait loci for grain minerals in hexaploid wheat (Triticum aestivum L.). J Integr Agric 12:1512–1521

Sinclair SA, Krämer U (2012) The zinc homeostasis network of land plants. Biochim Biophys Acta 1823:1553–1567

Sperotto RA, Ricachenevsky FK, Waldow VA, Fett JP (2012) Iron biofortification in rice: it's a long way to the top. Plant Sci 190:24–39

Sperotto RA, Menguer PK, Ricachenevsky FK (2018) Molecular bases of iron accumulation towards the development of iron-enriched crops. In: Plant micronutrient use efficiency: molecular and genomic perspectives in crop plants. Elsevier, Amsterdam. https://doi.org/10.1016/B978-0-12-812104-7.00003-4

Stein AJ et al (2007) Plant breeding to control zinc deficiency in India: how cost-effective is biofortification? Public Health Nutr 10:492–501

Takahashi M et al (2003) Role of nicotianamine in the intracellular delivery of metals and plant reproductive development. Plant Cell 15:1263–1280

Tan J et al (2002) Selenium in soil and endemic diseases in China. Sci Total Environ 284:227–235

Thomas D, Frankenberg E (2002) Health, nutrition and prosperity: a microeconomic perspective. Bull World Health Organ 80:106–113

Tinggi U (2003) Essentiality and toxicity of selenium and its status in Australia: a review. Toxicol Lett 137:103–110

To B (2014) Biofortification progress briefs

Tong J et al (2020) Dissection of molecular processes and genetic architecture underlying iron and zinc homeostasis for biofortification: from model plants to common wheat. Int J Mol Sci 21: 9280

Uauy C, Distelfeld A, Fahima T, Blechl A, Dubcovsky J (2006) A NAC gene regulating senescence improves grain protein, zinc, and iron content in wheat. Science 314:1298–1301

Vanderschuren H et al (2013) Strategies for vitamin B6 biofortification of plants: a dual role as a micronutrient and a stress protectant. Front Plant Sci 4:143

Vasconcelos MW, Gruissem W, Bhullar NK (2017) Iron biofortification in the 21st century: setting realistic targets, overcoming obstacles, and new strategies for healthy nutrition. Curr Opin Biotechnol 44:8–15

Vasil V, Castillo AM, Fromm ME, Vasil IK (1992) Herbicide resistant fertile transgenic wheat plants obtained by microprojectile bombardment of regenerable embryogenic callus. Bio/Technology 10:667–674

Velu G et al (2018) Genetic dissection of grain zinc concentration in spring wheat for mainstreaming biofortification in CIMMYT wheat breeding. Sci Rep 8:1–10

Verrier PJ et al (2008) Plant ABC proteins—a unified nomenclature and updated inventory. Trends Plant Sci 13:151–159

von Grebmer K et al (2019) 2019 global hunger index: the challenge of hunger and climate change. 2019 global hunger index l

Wang Y et al (2014) Simultaneous editing of three homoeoalleles in hexaploid bread wheat confers heritable resistance to powdery mildew. Nat Biotechnol 32:947–951

Wang P et al (2017) QTL mapping of selenium content using a RIL population in wheat. PLoS One 12:1–10

Wessells KR, Brown KH (2012) Estimating the global prevalence of zinc deficiency: results based on zinc availability in national food supplies and the prevalence of stunting. PLoS One 7:e50568

West KP, Mehra S (2010) Vitamin A intake and status in populations facing economic stress. J Nutr 140:201S–207S

White PJ, Broadley MR (2005) Biofortifying crops with essential mineral elements. Trends Plant Sci 10:586–593

White PJ, Broadley MR (2009) Biofortification of crops with seven mineral elements often lacking in human diets—iron, zinc, copper, calcium, magnesium, selenium and iodine. New Phytol 182: 49–84

White PJ, Brown PH (2010) Plant nutrition for sustainable development and global health. Ann Bot 105:1073–1080

Wiseman EM, Bar-El Dadon S, Reifen R (2017) The vicious cycle of vitamin a deficiency: a review. Crit Rev Food Sci Nutr 57:3703–3714

Wu G (2013) Amino acids. Fundamentals of protein structure and function 52003. CRC Press, Boca Raton

Wu G (2016) Dietary protein intake and human health. Food Funct 7:1251–1265

Zee S, O'brien T (1970) A special type of tracheary element associated with 'xylem discontinuity' in the floral axis of wheat. Aust J Biol Sci 23:783

Zhang Y, Gao C (2017) Recent advances in DNA-free editing and precise base editing in plants. Emerg Top Life Sci 1:161–168

Zhang Y et al (2012) Field crops research zinc biofortification of wheat through fertilizer applications in different locations of China. Fields Crop Res 125:1–7

Zhang Y et al (2017) Simultaneous modification of three homoeologs of Ta EDR 1 by genome editing enhances powdery mildew resistance in wheat. Plant J 91:714–724

Zhao F-J et al (2009) Variation in mineral micronutrient concentrations in grain of wheat lines of diverse origin. J Cereal Sci 49:290–295

Zia MH, Watts MJ, Gardner A, Chenery SR (2015) Iodine status of soils, grain crops, and irrigation waters in Pakistan. Environ Earth Sci 73:7995–8008

Zou C et al (2019) Simultaneous biofortification of wheat with zinc, iodine, selenium, and iron through foliar treatment of a micronutrient cocktail in six countries. J Agric Food Chem 67:8096–8106

Biofortification of Maize (*Zea mays*) 8

Bharti Aggarwal, Sanskriti Vats, Laveena Kaushal, Aman Singh, Gunashri Padalkar, Himanshu Yadav, Virender Kumar, Sweta Sinha, and Sunil M. Umate

Abstract

The production and consumption of maize (*Zea mays*) are the highest for human food as well as animal feed. Maize lacks essential amino acids (lysine, tryptophan), micronutrients (Fe, Zn, Mn, Se, I), and various vitamins (vitamin A). Therefore, the issue of malnutrition is still prevalent in nations where maize is the staple crop and has led to the impairment in the growth and development of the ever-increasing population, ultimately causing significant economic losses. A combinatorial approach is being used to overcome this issue which includes dietary diversification and food supplements. Regardless of these approaches, maize biofortification holds immense potential, proving to be a sustainable and cost-effective solution to malnutrition. Here, we describe conceivable methods for developing biofortified maize cultivars to curb global malnutrition. Agronomic biofortification includes applying fertilizers having concoctions of sulfates, hydrates, and oxides to fortify maize in its edible parts. GWAS and QTL mapping have so far helped a lot in identifying genetic elements controlling a particular phenotype. Hence, modern breeding techniques have improved the maize germplasm but still face the huge challenge of low yield potential and

B. Aggarwal · S. Vats (✉) · L. Kaushal · G. Padalkar
Department of Biotechnology, Panjab University, Chandigarh, India

National Agri-Food Biotechnology Institute (NABI), Mohali, Punjab, India

A. Singh · H. Yadav · V. Kumar
Department of Biotechnology, Panjab University, Chandigarh, India

S. Sinha
Bihar Agricultural University, Bhagalpur, Bihar, India

S. M. Umate
Wheat and Maize Research Unit, Agriculture Botany, Vasantrao Naik Marathwada Krishi Vidyapeeth, Parbhani, India

R. Deshmukh et al. (eds.), *Biofortification in Cereals*,
https://doi.org/10.1007/978-981-19-4308-9_8

environmental impacts. The underlying molecular mechanisms which help in the uptake and deposition of these essential micronutrients in the maize grain have been studied in detail. Recently, the transgenic approach has been most widely used as it gives a comprehensive analysis of the regulatory pathways. Bacterial as well as plant genes have been expressed in maize, but despite the massive potential, commercial production through this approach has not been put to reality yet. The most advanced and flexible approach which uses CRISPR/Cas9 has opened up new avenues of genome editing for multiple purposes. The economic and social constraints of every approach have also been discussed.

Keywords

Biofortification · Malnutrition · Micronutrients · Marker-assisted breeding · QTL mapping · CRISPR/Cas9

8.1 Introduction

Maize (*Zea mays*) is one of the most produced as well as consumed food grains in the world, with the production reaching an estimated 1.1 million thousand tons, and the consumption of maize is not limited to just human food (https://knoema.com/atlas/World/topics/Agriculture/Crops-Production-Quantity-tonnes/Maize-production). Maize is a vital component of animal feed, where hybrid varieties have been specifically developed for this particular purpose. Another major use of maize is in wet milling where it is processed to recover starch and subjected to fermentation to produce fructose and ethanol.

As far as demography is concerned, the USA is the leading producer as well as exporter of maize throughout the world (Table 8.1). In the years 2019–2020 alone, the USA accounted for 346 million metric tons of maize production, out of which nearly 95% was used as food grain and nearly 14% of its production was being exported to more than 73 countries (https://grains.org/buying-selling/corn). Corn production has therefore dominated the agro-economy of the USA, giving rise to the famous "Corn Belt" region in the Midwestern United States (https://www.investopedia.com/articles/markets-economy/090316/6-countries-produce-most-corn.asp). In that particular area, the corn industry employs over 10,000 people directly, with salaries crossing $900 million (https://www.proquest.com/docview/2449286317?pq-origsite=gscholar&fromopenview=true). With value addition in the job market, it further supports an additional income of nearly $10 billion. As such, in terms of monetary value, the total output reaches almost $47 billion dollars every year (https://corn.org/economic-impact/). Other major producers of corn include China, Brazil, Argentina, Ukraine, and India, but the production here is aimed primarily toward domestic consumption.

Cereal grains have low nutritional value to be considered as a part of a sustainable diet (Rana et al. 2020). With the world population increasing at an alarming rate and expected to reach nine billion by the year 2050, biofortification is indispensable to

Table 8.1 Details of economic status and hunger index in major maize-growing countries worldwide

Country	Population	Maize production[a] (ton)	GDPA per capita USD	Hunger index[b]	Child mortality rate
USA	32.82 crores	34,70,47,570	65,297.52	Not applicable	6
China	139.77 crores	26,09,57,662	10,261.68	Ranked between 1 and 17	7
Brazil	21.1 crores	10,11,38,617	8717.19	Ranked between 1 and 17	12
Argentina	4.49 crores	5,68,60,704	9912.28	20	8
Ukraine	4.44 crores	3,58,80,050	3659.03	Ranked between 1 and 17	7
India	136.64 crores	2,77,15,100	2099.60	94	28

[a] Data was obtained from FAO
[b] Data was obtained from Global Hunger-Index

meet the growing nutrition requirements. Overreliance on cereal crops in several underdeveloped and developing countries is one of the main causes of malnutrition. Estimates show that nearly two billion people in this world have a mineral deficiency of one form or another (http://www.fao.org/3/i6583e/i6583e.pdf). Though required in trace quantities, the deficiency can lead to the disruption of several metabolic functions leading to severe illnesses, disabilities, and obstructed mental and physical growth. Although the requirement of other essential nutrients can be compensated by a well-balanced diet containing meat, poultry, and other fruits and vegetables, these commodities are far from the reach of poor and low-income households, making low-cost grains crucial (Wakeel et al. 2018). The ever-increasing population is difficult to support with the available means of food crops, and efforts to increase the nutrient content without having to increase the food intake are currently necessary. Several organizations and countries are working in unison at both domestic and international levels to tackle this issue. "Zero hunger" is 1 of the 17 sustainable development goals accepted by the United Nations which the member countries aim to achieve by the year 2030 (https://sustainabledevelopment.un.org/content/documents/23576ISRAEL_13191_SDGISRAEL). Other programs such as Food for Peace, the World Food Programme, and "She Feeds the World" are working to provide nutrient-rich food to the poor population across the globe. Major strategies to help improve nutritional quality include dietary diversification, food supplements, food fortification, and biofortification. Dietary diversification is a combinational approach. When it comes to food intake, it can improve our nutritional intake significantly. Prolonged dietary diversification ensures that micronutrients, fibers, vitamins, and other essential components are continuously supplied to the body.

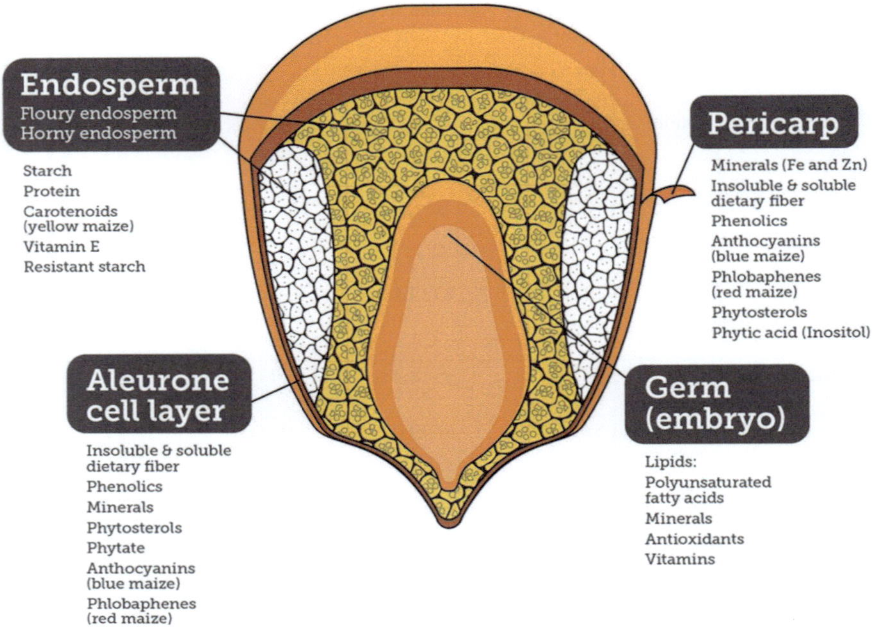

Fig. 8.1 Nutrient content in different parts of maize kernel. (The figure is reproduced from Prasanna et al. 2020 which is available under a Creative Commons Attribution 4.0 (CC BY 4.0) International License, which permits reproduction)

Diversification may also depend on the age and condition of the person where they might require specific nutrients at a specific growth stage. Further, using food supplements that are rich in vitamins and micronutrients is a good way to make up for the lack of nutrients in our diet. But this method is not suitable for developing and underdeveloped nations on a large scale due to commercial infeasibility. Food fortification, the addition of nutrients directly to the food in order to improve their nutritional quality, is a good method to tackle malnutrition. Trace amounts of vital nutrients are added to the main food during processing which helps to achieve the desired level of nutrition in our diet. The addition of iodine in table salt is a good example of food fortification in developing countries such as India. Finally, biofortification, the process of deliberately improving the nutritional content of plants through agronomic practices, breeding, or biotechnological means, aims to improve the nutritional content during plant growth and may be able to improve the quality of food where conventional fortification methods may prove to be cumbersome to introduce (Wakeel et al. 2018).

Regardless of the approach, the major goal is to increase the nutritional value of the food which is being consumed. Maize kernel consists of approximately 73% starch, 10% protein, and 5% oil. The remaining portion consists of fibers, vitamins, and minerals (Pan et al. 1996) (Fig. 8.1). Humans require 0.66 g protein/kg body weight/day as well as an addition of essential amino acids in order to meet their proper growth and development. Essential amino acids like lysine and tryptophan are not synthesized in the human body. These amino acids not only aid in protein synthesis in the body but also act as a major precursor for neurotransmitters and metabolic regulators (Lopez and Mohiuddin 2022). Maize is not a suitable choice here, since it lacks these essential amino acids. Only 1.5–2% of the maize protein is lysine, which is less than half of the recommended dose for humans. Moreover, 60% of maize protein is zein which is deficient in lysine and tryptophan (Gupta et al. 2015).

Being the staple diet in many countries and predominating the diet of poor and low-income households, corn has the potential to increase the nutrient uptake of individuals if biofortified. Targeting maize for biofortification is an implicit strategy to reach a large number of people in one step. Moreover, the benefits of fortified maize comply with economic and environment-friendly agricultural practices.

8.2 Priorities for Maize Biofortification

From the perspective of biofortification, certain nutrients are prioritized by considering their necessity and roles in the human body and can be broadly classified into the following categories:

8.2.1 Essential Micronutrients/Metals

Essential micronutrients are not synthesized in the human body and have to be supplemented through the diet. The metals iron (Fe), zinc (Zn), and manganese (Mn) are of extreme importance, as they play a vital role in several metabolic pathways of our body. Fe is an important cofactor for the proper functioning of some vital enzymes of our body. The Fe-containing enzymes serve as oxygen transporters and electron carriers. Further, they are involved in the production of steroid hormones, removal of foreign substances, and controlling signals in some neurotransmitters (Gupta 2014). The deficiency of Fe causes anemia in the body and disturbs the vital enzymatic pathways mentioned above, which directly affects the immune system and well-being of the individual.

Zn is needed for the functioning of more than 300 enzymes during various growth stages of our body. Zn deficiency has been correlated with the emergence of several renal, liver, gastrointestinal, neoplastic, skin, and diabetic disorders as well as an increase in parasitic infections and pregnancy-related disorders (Prasad 2014). The deficiency of both Fe and Zn is the leading cause of micronutrient deficiency disorders, which affects more than two billion people and causes millions of deaths worldwide (Bashir et al. 2013). Mn deficiency can lead to asthma and severe birth defects and is less prevalent than the deficiency of Fe and Zn. But the combined deficiency of these three essential micronutrients can lead to catastrophic situations in the human body and cause serious harm to our health.

8.2.2 Basic Micronutrients

Basic nutrients such as selenium (Se) and iodine are required in trace amounts in our body but are equally important in several metabolic pathways. Se is present in the active component of the glutathione peroxidase pathway which helps us protect from the H_2O_2. Deficiency of Se may increase our risk of several neoplastic diseases of the colon, prostate, and gastrointestinal tract.

8.3 Protein

The human body is an array of structural and functional proteins, coupled with other regulatory and metabolic functions. WHO guidelines have set 0.83 g/kg of daily protein intake for an average human being (Paddon-Jones et al. 2008). Although meat and poultry products are good sources of proteins, availability, cost, and dietary preferences heavily restrict the protein intake of a normal human being. Biofortification strategies aim at increasing the amino acid pool in the crops which directly increases the protein content in the plants. Further approaches include increasing the expression of storage proteins in plants. This combination can significantly improve the protein concentration inside the plant and compensate for its deficiency in the normal diet. This strategy has been implied in cassava and rice and therefore can also improve the nutritional quality of maize (Leyva-Guerrero et al. 2012; Wong et al. 2015).

The few examples mentioned above further augment the desire to improve the nutritional quality of the maize. We also need to understand that we can only increase the micronutrient content up to a certain degree in a plant without compromising its structural and functional integrity. A better biofortification strategy aimed toward reducing worldwide hunger could be developed based on prioritizing the nutrients that require immediate attention and leaving out the others which can be compensated through other sources.

8.4 Agronomic Biofortification of Maize

Agronomic biofortification aims at improving the mineral content through direct application of mineral fertilizers into the soil which are then absorbed by the plant and stored in its edible parts. The fertilizers are applied in the form of different concoctions primarily as sulfates, hydrates, and oxides. In maize, agronomic biofortification has increased the concentrations of Zn and Se in its edible parts (Table 8.2).

Table 8.2 List of significant studies performed for the maize biofortification

Biofortification trait	Agronomical practices	Success/level of improvement	References
Provitamin A	Hybridization	Production of maize with more than 15 μg/g	Pixley et al. (2013)
Se	Selenate-Se fertilizer	Increase in Se content by an average of 19 μg/kg in each gram	Chilimba et al. (2012)
Zn	Seed priming, soil and foliar applications, genetic engineering	Increase in yield of grain by 27%, 40% increase in Zn concentration in maize kernels	Maqbool and Beshir (2019)
Iron	Fertilization through foliage	51.8.5 increase in Fe content in maize kernel	Saleem et al. (2016)

Zn in the form of $ZnSO_4{\cdot}7H_2O$ has been applied with the results showing a significant increase in the Zn concentration in the plant parts (Imran and Rehim 2017). Moreover, the maize plants were superior in terms of size and other growth characteristics when compared with the control. The increase in Zn concentrations also increased the protein concentrations in the plant, further adding value to the product. The application of these fertilizers also impacts the concentration of Zn. Different techniques such as foliar spray, surface broadcasting, subsurface banding, as well as a combination of these techniques are utilized while applying these fertilizers (Bruulsema et al. 2012). The stage and time of plant growth while applying these fertilizers also determine the result.

Similarly, for Se, Na_2SeO_4(aq) was applied in an experiment that spanned over 2 years. The fertilizer was applied at early stem extension and was sprayed as a high-volume drench. The experiment reported an increase in Se concentration which was directly proportional to the Se fertilization rate. But unlike Zn, there were no other changes in grain yield in any of the experiments (Chilimba et al. 2012).

Another mineral that has been introduced through agronomic biofortification is Fe. $FeS0_4$-enriched soil was used to grow maize which resulted in an increased grain yield when compared with control. There was an increase in cob length, cob size, grains per cob, and weight of the grain. The increased availability of nutrients helps in the formation of stable organometallic complexes which are coupled with the organic matter present in the plant (Kumar and Salakinkop 2018).

While agronomic biofortification seems an alluring option, its limitation is quite apparent in such a way that there are only a few minerals that can be added through this method. Another limitation is that the fertilizers are quite expensive and cannot be used indiscriminately. External application of the fertilizers causes extensive wastage and leaching. Leaching of these minerals can also cause severe damage to the environment. With an external application, it becomes difficult to control the metal concentration in the plant, which, if exceeded, can be dangerous for human consumption.

8.5 Breeding Efforts for Maize Biofortification

Macro- and micronutrient deficiencies are seen in a larger section of population, thereby posing serious worldwide health challenges. In the past few decades, the process of enhancement of nutrient and mineral content of crops at the industrial level, also known as crop fortification or industrial fortification, has become a prominent method to produce nutrient-enriched crops that would help to overcome malnutrition problem (Liberato and Pinheiro-Sant'Ana 2006). But, because of the expensive industrial processing, access to such food products remains restricted, thereby failing in the complete eradication of the hidden hunger among the global population. Thus, biofortification for the development of crop varieties with naturally high bioavailable mineral and nutrient content in them using molecular breeding techniques is the need of the hour. Staple crops such as maize fulfill the carbohydrate requirement of the body, but essential amino acids such as lysine and tryptophan are the major limiting factors in maize-based diets. In addition, maize-based diets also lack micronutrients and minerals such as vitamin A, Fe, and Zn. Major maize growing countries and hunger index along with child mortality rate are provided in Table 8.1. Multiple strategies are available for maize biofortification, breeding being one of them (Fig. 8.2).

Plant breeders and geneticists have adopted various breeding strategies using nutrient reference values known as Dietary Reference Index (DRI) to enhance the nutritional index of crops. DRI determines the minimum intake value of nutrients that is essential to maintain an appropriate level of nutrients in an individual (Murphy et al. 2016). Breeding strategies for the development of biofortified

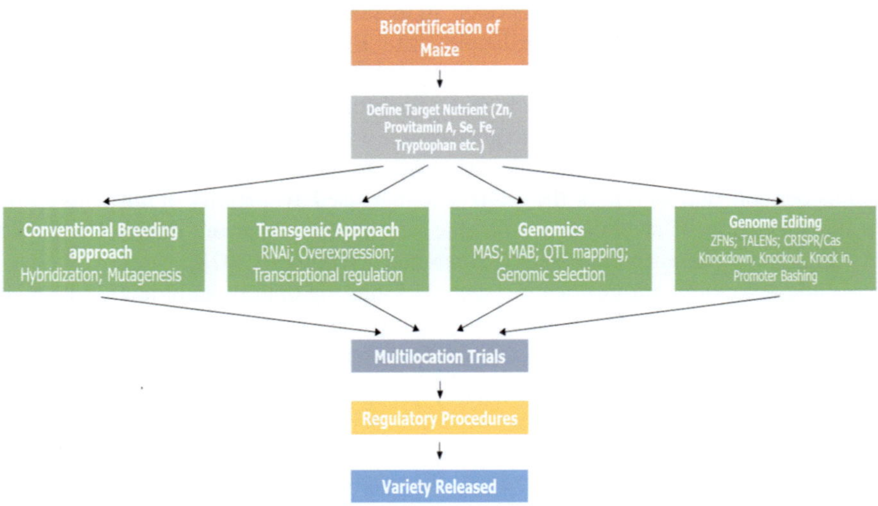

Fig. 8.2 Strategies for the development of biofortified maize. *RNAi* RNA interference; *MAS* marker-assisted selection, *MAB* marker-assisted breeding, *ZFNs* Zn finger nucleases, *TALENs* transcription activator-like effector nucleases

maize include conventional selection and hybridization (heterosis breeding) and molecular breeding methods which imply genetic engineering. Heterosis breeding involves the selection of desirable traits in two different parent crops (usually pure-bred lines with superior characters) and their hybridization to develop a superior plant variety with a high nutritional index, yield, and resistance to pests and unfavorable abiotic conditions (Labroo et al. 2021). Molecular breeding methods involve genetic manipulation (GMOs) of the plants for increasing the nutritional content of crops. In marker-assisted breeding (MAB), the molecular markers are linked to a specific phenotype called quantitative trait loci (QTL) that are subject to genetic rearrangement with recombination and transposition. Genome-wide association studies (GWAS) and the identification of maize allow the estimation of the location and effect of genetic elements that control a particular phenotype.

QTL mapping for biofortification traits has been widely utilized to integrate agronomical important traits and marker-assisted selection procedure and improve quantitative traits using wild germplasm (Hu et al. 2016). Molecular-level understanding of mineral elements and identification of significant QTL help to speed up the biofortified varieties' development. In recent times, provitamin A, Zn, Fe, Se, and iodine-rich maize varieties have been developed with the aid of molecular breeding and conventional breeding techniques. The International Maize and Wheat Improvement Center (CIMMYT) is a leading organization working on the development of biofortified maize. In India, about 14 QPM varieties have been released by CIMMYT. In this row, they have launched more than 50 QPM varieties in areas such as Latin America, the Caribbean, and sub-Saharan Africa. CIMMYT is also conducting field trials of Zn-biofortified maize to analyze the agronomic performance under different environmental conditions (Maqbool et al. 2021). HarvestPlus supports various maize biofortification programs by CIMMYT and has contributed in the development of various nutrient- and mineral-enriched maize varieties. Provitamin A deficiency results in blindness in children; therefore increasing the provitamin A content in maize-based diets can be an effective method to combat this problem globally (Table 8.3). More than 40 provitamin A-biofortified maize genotypes have been released in different African countries (Ekpa et al. 2018). The magnitude of genetic variation for provitamin A content present in maize lines released under HarvestPlus program has been shown in Fig. 8.3. Zn biofortification of white maize is also a project under CIMMYT and IITA which aims to increase the Zn content of maize (Maqbool and Beshir 2019).

8.6 QTL Introgression Exploring Wild Resources

GWAS and QTL mapping are used to identify the loci and genes responsible for desired phenotypes. With the help of QTL mapping, the genetic loci governing the accumulation of micronutrients in maize can be identified. The introgression of highly effective loci can be performed with the help of marker-assisted selection (MAS). The wild germplasm can be explored to improve quantitative traits with breeding strategies. Biofortified maize varieties developed through the breeding

Table 8.3 List of significant studies performed for the identification of QTL that can be used for maize biofortification

Biofortified trait	Mapping population	QTL ID	Marker interval	LOD	Phenotypic variation (R^2) %	References
High provitamin A	By 804 × B73 RIL population	qac6-1	PZE-106025043–PZE-106037935	6.67	11.83	Jittham et al. (2017)
		qbc6-1	PZE-106027535–PZE-106037122	12.25	17.07	
		qbc10-1	PZE-110078694–SYN18728	9.42	12.53	
		qbc/bcry10-1	PZE-110080305–PZE-110082173	11.52	19.84	
		qbc/tc10-1	PZE-110080467–SYN18728	17.93	31.85	
		qbcry10-1	PZE-110015036–PZE-103066235	4.95	8.87	
		qlut6-1	SYN2895–PZE-106039415	17.99	28.88	
		qtc6-1	PZE-106025043–PZE-106037935	18.65	33.50	
		qtva6-1	PZE-106027535–PZE-106037122	14.21	23.56	
		qzea6-1	PZE-106027535–PZE-106039942	8.39	12.45	
		qzea10-1	PZE-110082173–SYN18728	12.26	19.38	
		qzea/tc10-1	PZE-110082173–PZE-110083604	10.72	18.64	
Zn	218 F2:3 mapping population	qZn2-1	bnlg1633–bnlg1138	3.01	6.29	Jin et al. (2013)
		qZn2-2	umc1536–bnlg1633	3.17	5.85	
		qZn5	umc1429–umc1060	5.58	17.57	
		qZn10	umc1506–umc2350	4.23	7.10	
	294 F4 lines of biparental population	Zn4-08	ZM1362	5.19	7.8	Šimić et al. (2012)
		Zn/P 3-05	bnlg1456	5.21	7.8	
Iron	218 F2:3 mapping population	qFe5	umc1429–umc1060	3.49	16.89	Jin et al. (2013)
	B73 × Mo17 (IBM) recombinant inbred (RI) population	FeGB-3.1	PSR754B	3.54	0.078	Lung'aho et al. (2011)
		FeGB-6.1	PHP20528	5.82	0.135	
		FeGB-9.1	UMC2134	3.70	0.103	
		FeGC-2.1	MMP144	6.21	0.101	
		FeGC-5.1	RZ87	7.694	0.12	
		FeGC-9.1	SH1	4.61	0.093	

Magnesium	294 F4 lines (biparental population)	Mg 5-03	bnlg1046	4.44	6.7	Šimić et al. (2012)
		Mg 6-01	bnlg426	4.65	7.0	
		Mg 8-05	bnlg1782	5.03	7.6	
		Mg 9-07	bnlg0128	7.96	12.1	
		Mg/P 1-01	bnlg1014	6.00	9.3	
Phosphorus	294 F4 lines of biparental population	P3-05	umc59e	5.98	8.9	Šimić et al. (2012)
		P3-09	bnlg1257	6.48	9.7	
		P4-08	ZM0819	6.69	10.2	
		P6-03	umc1887	4.89	7.4	
		P6-05	ZM1367	5.42	8.4	
		P8-05	ZM0353	4.50	6.8	
		P9-02	bnlg0244	8.39	12.3	
		P10-07	bnlg1839	8.67	12.7	

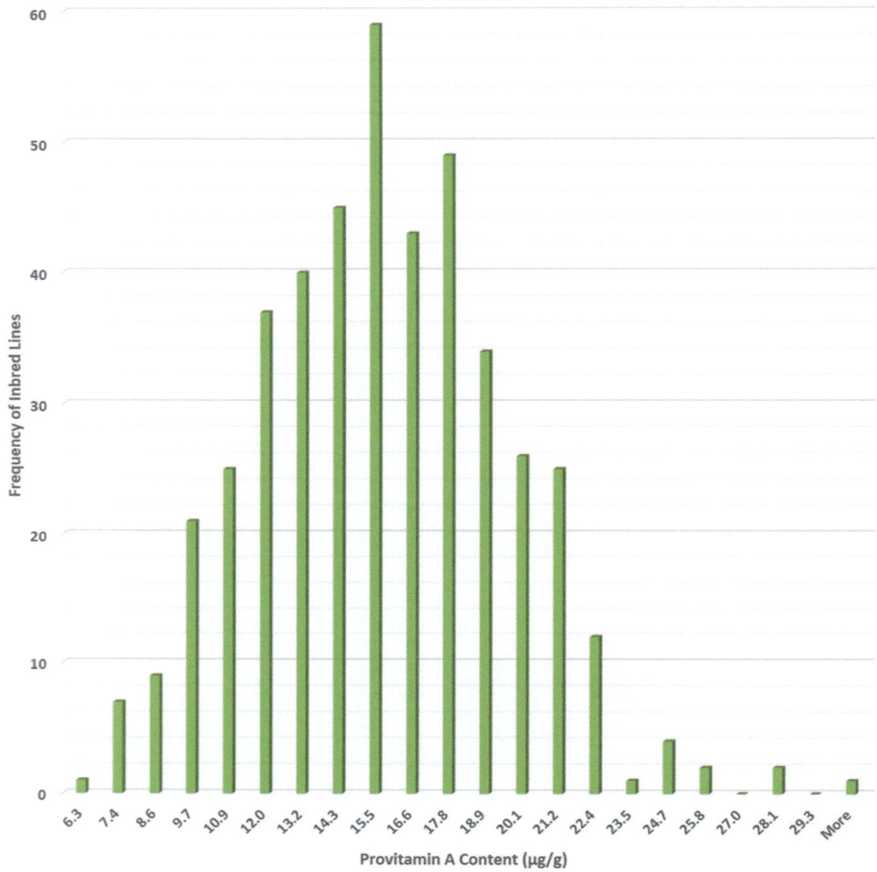

Fig. 8.3 The variation of provitamin A content (in μg/g) present in maize varieties developed under HarvestPlus program by the International Maize and Wheat Improvement Center (CIMMYT). (The figure is reproduced from Prasanna et al. 2020 which is available under a Creative Commons Attribution 4.0 (CC BY 4.0) International License, which permits reproduction)

approach show a higher nutritional index which can be an effective method to globally eradicate malnutrition. Various efforts have been made by different institutions worldwide to develop biofortified varieties with increased micronutrients. These hybrid varieties have been generated using conventional breeding strategies as well as MAS strategies. Genome-wide association studies have a great impact on the selection of genes contributing to a particular phenotype and have been used in order to develop superior crop varieties. Over the past few decades, improvised plant breeding strategies have been successful in increasing crop productivity, enhancing nutritional quality, and developing resistant varieties under various biotic and abiotic stress conditions.

Conventional breeding efforts focusing on the selection of two superior varieties with high nutritional index and hybrid generation have been widely integrated with breeding strategies to improve the crop's nutritional index (Rana et al. 2020; Patil et al. 2018). The efficiency of biofortified varieties has been increased using metabolic engineering methods wherein the biosynthesis pathways of various micronutrients such as carotenoids have been engineered (Van Der Straeten et al. 2020). A program under the Consultative Group on International Agricultural Research (CGIAR) Research Program on Agriculture for Nutrition and Health known as HarvestPlus is successfully working on the development of biofortified crop varieties in order to support the population's nutrient demands. The most popular crops that have been targeted for nutritional enhancement by institutions worldwide are Zn rice, provitamin A cassava, provitamin A maize, Fe beans, Fe pearl millet, etc. These collaborating institutions have been promoting the uptake of such varieties by farmers and consumers, thereby making them aware of the beneficial effects of such biofortified crops.

8.7 Challenges, Limitations, and Success of Breeding Approaches for Maize Biofortification

Nutrient enrichment of maize through biofortification is a potential method to eradicate malnutrition. However, challenges such as the low yield potential of nutritionally enriched maize limit their mass production. Multiple studies targeting maize biofortification have been reported (Table 8.4). Studies that have explored the potential of Zn maize, provitamin A maize, and QPM have reported to exhibit lower yield. On the other hand, some studies have claimed that biofortified maize is high yielding as compared to non-biofortified maize (Prasanna et al. 2020). Therefore, a dilemma exists if the yield of a biofortified crop would be promising to the farmers or not. Figure 8.4 displays the release status of provitamin A- and Zn-biofortified maize varieties in sub-Saharan Africa, Latin America, and Asia.

Apart from the yield as a challenge, the quality assurance of the biofortified variety is a major concern that hinders the public acceptance of biofortified varieties. Since there is a little difference in the genetic base of a biofortified and non-biofortified maize, thus any type of physical or genetic contamination in the germplasm can result in undesired produce. Management of line maintenance and seed production requires extra care for the development of pure lines with desired traits. Nutritional quality check is also required at various stages during the breeding program to ensure the genetic gains in the produce. Thus, the establishment of nutrient quality and content check laboratories is essential to prevent the loss of resources.

HarvestPlus has been working in the field of biofortified crop production in collaboration with research institutes such as CIMMYT and has been successful in bridging the gap between the genetic dissection of biofortified crops and their production on the field. HarvestPlus has successfully launched about 25 state-of-the-art nutrient analytical laboratories that aim to keep a check on the nutrient quality

Table 8.4 List of biofortified maize varieties developed by different institutions worldwide

Biofortified maize varieties	Method of variety development	Improved traits	Other note	References
ICTA Hb 18	Conventional hybridization (heterosis breeding)	15% more Zn	Developed by International Maize and Wheat Improvement Center (CIMMYT) in collaboration with CGIAR Research Program on Maize (MAIZE), CGIAR Research Program on Agriculture for Nutrition and Health (A4NH), and Guatemala's Institute for Agricultural Science and Technology (ICTA) with support of HarvestPlus	https://lac.harvestplus.org/en/icta-hb-18-first-Zn-maize-hybrid-released-and-icta-b-15/
ICTA B15	Conventional hybridization	Tortillas made with ICTA B15 has 60% more Zn	Developed by CIMMYT in collaboration with HarvestPlus	
BIO-MZN01	Conventional hybridization	Increased Zn content by 36% than other varieties	Developed by International Maize and Wheat Improvement Center (CIMMYT) with the support of HarvestPlus	
		Yield increased by 3–8 tons per hectare	Tolerant to maize diseases including rust, leaf blight, and gray spots on leaves	
Vivek QPM 9	Hybrid	Rich in lysine (4.19%) and tryptophan (0.83%)	Developed by ICAR-Vivekananda Parvatiya Krishi Anusandhan Sansthan, Almora	https://iimr.icar.gov.in/wp-content/uploads/2020/12/Biofortified-Varieties-Book_V3_ICAR.pdf
Pusa HM4 improved	Hybrid	Lysine (3.62%) and tryptophan (0.91%)	Developed by ICAR-Indian Agricultural Research Institute, New Delhi	
Pusa Vivek QPM9 improved	Hybrid	Rich in provitamin A (8.15 ppm), lysine (2.67% in protein), and tryptophan (0.74% in protein)	India's first provitamin A-rich maize Developed by ICAR, New Delhi	
Pusa HQPM7 improved	Hybrid	Rich in provitamin A (7.10 ppm), lysine (4.19% in protein), and tryptophan (0.93% in protein)	ICAR, New Delhi	
IQMH 201 (LQMH1)	Hybrid	Rich in lysine (3.04% in protein) and tryptophan (0.66% in protein)	ICAR, New Delhi	

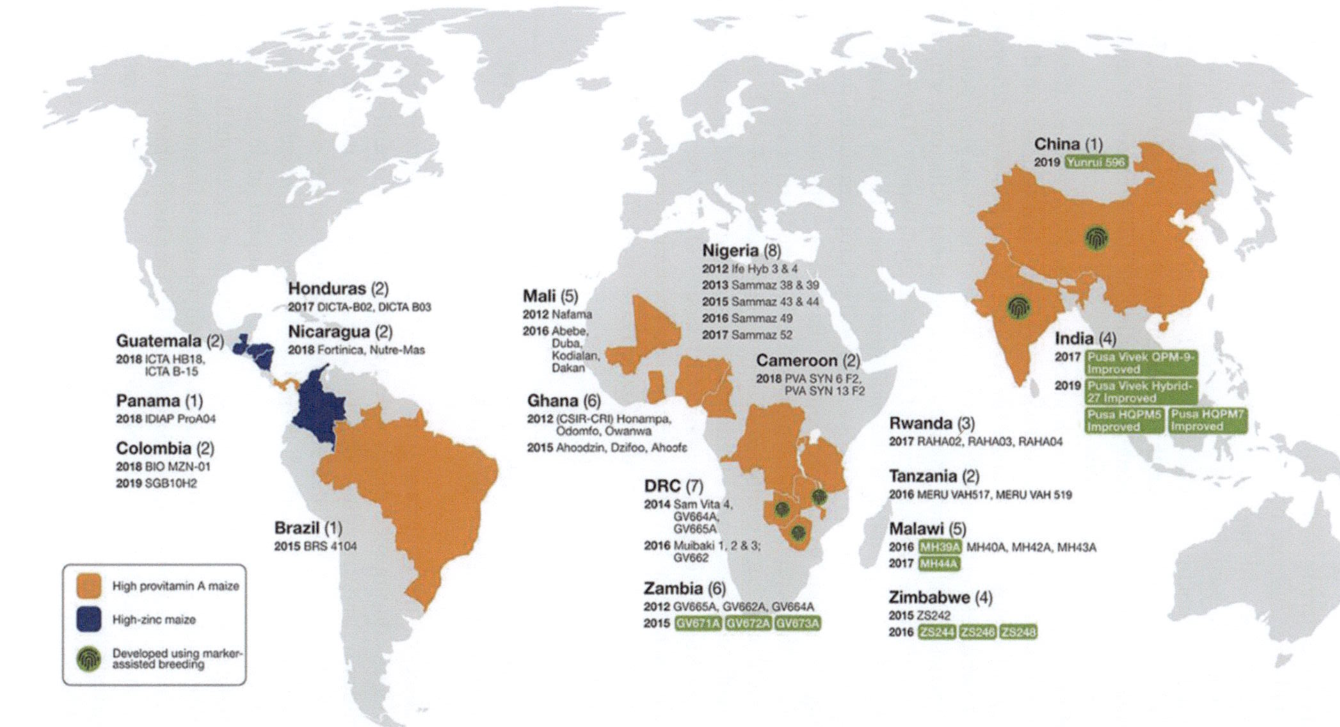

Fig. 8.4 The extent of provitamin A- and Zn-biofortified maize developed via conventional and marker-assisted breeding released in Latin America, Asia, and sub-Saharan Africa. The figure is reproduced from Prasanna et al. (2020) which is available under a Creative Commons Attribution 4.0 (CC BY 4.0) International License, which permits reproduction

of the crops. As there are no visible changes in some biofortified maize varieties such as Zn-biofortified maize, thus it becomes difficult to convince the farmers and traders to buy an expensive variety with a high nutrient index. To overcome this hurdle, HarvestPlus with the support of other organizations has been working in the installment of techniques for precise mineral quantification such as X-ray fluorescence spectrometry, inductively coupled plasma (ICP) mass spectrometry, etc. (https://www.harvestplus.org/sites/default/files/Biofortification_Progress_Briefs_August2014_WEB_0.pdf). However, the development of portable and cost-effective devices for mineral and nutrient content estimation is still required to make the process much easier and more feasible.

Despite the development of nutrient-enriched and enhanced varieties, the success of meeting essential nutrient requirements in every individual depends on the level of acceptability of the biofortified maize. The acceptability depends on various other factors such as awareness in the masses, visible effects in the crop, yield potential of the nutrient-enriched crop, and any change in taste or aroma of the crop which may further reduce the chances of acceptance of the biofortified variety. The method of breeding used in the development of the biofortified variety is also a major determining factor for the acceptability of crops. Biofortified crops produced by conventional breeding are more readily acceptable than those produced by genetic engineering. Thus, the acceptability of genetically modified maize even with high nutrients and yield is a huge challenge. Though identification of QTL and their introgression with marker-assisted breeding is an effective method to improve the nutrient-targeting traits in crops, environmental factors also play a major role in determining the accumulated nutrient content. Zn accumulation in maize is highly influenced by environmental factors; therefore QTL are not effective and reliable in such cases (Prasanna et al. 2020).

8.8 Molecular Understanding of Essential Micronutrient Uptake and Deposition in Maize Grain

The major source of micronutrient uptake in plants is the rhizosphere. Seeds accumulate minerals during seed development stages and subsequently pass to the plant. An in-depth understanding of genes responsible for nutrient uptake, translocation, and loading into the endosperm can aid in the development of nutrient-enriched varieties through nutrient-specific tailored breeding strategies. Chelation-based and reduction-based strategies are adopted in higher plants to increase the mineral uptake from the rhizosphere. Chelation-based method is based on the principle of mobilizing the minerals in the soil, therefore increasing their flow into the plant. For this purpose, phytosiderophores (PS) are used which mobilize Fe and Zn in the soil, making them suitable for uptake by the roots and resultantly aiding in increasing the mineral content in the plant. Complexes of PS with Fe and Zn are formed which are taken up by the plant with YS1/YSL (Yellow Stripe-Like) gene family. A maize variety with mutant YS1 gene was found to be Fe deficient (von Wirén et al. 1994). Zhang et al. (1991) used PS labeled with ^{65}Zn to understand the process of uptake of Zn in maize. The study reported the increase in uptake of Zn (II) phytosiderophores under Zn deficiency conditions in the plant (Zhang et al. 1991). The movement of nutrients and their storage in maize seeds largely depends on the chemical form in which the micronutrients are present.

8.9 Nutrient Accumulation, Remobilization, and Autophagy Recycling

According to a study by Cakmak et al., the probability of micronutrient deficiency in plants increases when the macronutrients are excessively supplied in plants (Cakmak 2002). During early vegetative stages, the developing roots and shoots act as a sink for various inorganic compounds and micronutrients essential for the growth and development of maize. These molecules are accumulated in various parts of the plant and are further translocated through vascular bundles to reach higher plant parts. The dry weight of micronutrient content in maize is directly proportional to the soil microenvironment. In the reproductive stage, the nutrients are further remobilized to grains and other essential reproductive plant parts.

Mineral assimilation in maize is highly nutrient-specific and dependent on the rate of uptake as well as the mechanism of the partitioning of nutrients to different tissues. Sayre (1948) have observed the difference in the rate of nitrogen uptake before and after grain filling. They have reported the rapid uptake of nitrogen during vegetative stages and slow nitrogen uptake during grain fill (Sayre 1948). However, Karlen et al. (1988) have reported a different pattern of nitrogen uptake in high-yielding maize with two distinct accumulation periods. This uptake mechanism includes a lag phase where only a small amount of nutrient uptake takes place. In case of some minerals such as K, Ca, and Mn, the uptake is rapid and usually coincides with the vegetative growth of the plant (Karlen et al. 1988). The accumulation of Zn in maize shows a variegated pattern following uptake at both vegetative and grain-filling stages with a relatively short lag phase (Karlen et al. 1988).

Autophagy is an important factor in nutrient recycling in maize under nitrogen stress conditions. Li et al. (2015) demonstrated that ATG12 which is a key component essential for the functioning of ATG8 (autophagy-related 8) when compromised in maize resulted in low productivity in maize (Li et al. 2015). They observed that autophagy, though is the nonessential process for nutrient recycling and remobilization, plays a crucial role under nitrogen stress conditions in maize. The mobility and translocation characteristics are different for various nutrients. N, P, and Zn are highly plant mobile while micronutrients such as Fe and Mn possess limited remobilization characteristics (Karlen et al. 1988). An in-depth understanding of the remobilization characteristics of nutrient uptake can aid in the optimization of the timing of nutrient applications (Bender et al. 2013; Shao et al. 2021).

8.10 Transgenic Efforts for the Development of Biofortified Maize

Conventional breeding techniques have so far helped in the production of high beta-carotene maize lines, but transgenic approaches can help understand the regulation of the carotenoid metabolism pathway in a more comprehensive manner. The expression of bacterial *crtB* (phytoene synthase) and/or plant *PSY* genes under the control of endosperm-specific zein promoter results in the accumulation of phytoene which is further responsible for the increased level of provitamin A in maize (Aluru et al. 2008). Despite all these efforts, commercial production of such maize is yet to become a reality. The reports of various researchers have generated high hopes for the development of biofortified maize. Carotenoids (C40 tetraterpenoids) are the secondary metabolites that provide abundant health benefits in humans like prevention of cancer, maintenance of the immune system, and buildup of an effective antioxidant system to prevent cell injury. The cereal crops are generally deficient in these carotenoids; therefore, the carotenoid biosynthesis pathway in plants must be engineered to increase their concentration in major cereal crops, especially in the grains. But the problem is that these pathways are quite complex having multiple branches, multifunctional enzymes, and complex feedback mechanisms. So, a combinatorial nuclear transformation strategy has been developed in maize which facilitates the engineering of an entire pathway producing multiplex transgenic plants with a specific combination of carotenoids (Zhu et al. 2008). White maize embryos were transformed with five carotenogenic transgenes producing diverse populations with distinct phenotypes out of which specific carotenoid-producing transgenic maize was screened.

Another success story of transgenic maize fortified with bioavailable Fe has also been reported using *Aspergillus* phytase (*phyA*) either alone or in combination with Fe-binding protein ferritin from soybean with their restricted expression in the maize endosperm. The main advantage of producing phytase in maize seeds was that it facilitated the cellular Fe uptake, and the rate of Fe uptake was enhanced by phytase expression and by the addition of ascorbic acid (Drakakaki et al. 2005).

Tocotrienol is the primary form of vitamin E which has an important role as an antioxidant in maintaining human health. To increase the level of tocotrienol and tocopherol in maize, an enzyme homogentisic acid geranylgeranyl transferase (HGGT) was over-expressed using a transgenic approach (Cahoon et al. 2003). The transgenic approach has also been used to manipulate amino acid composition in maize since zeins are the major seed storage proteins in maize, but they lack essential amino acids like lysine and tryptophan leading to poor nutritional quality of maize, as mentioned before. Mutational studies have proved that the expression level of zein has an antagonistic effect on the lysine and tryptophan content. Efforts were made to elevate the level of lysine content in maize by the expression of lysine-rich *sb401* gene from potato (*Solanum berthaultii*) using the microprojectile bombardment method (Yu et al. 2005; Tang et al. 2013). This transgenic event leads to the production of maize with improved amino acid score and additionally improved nutritive content of maize. Various lysine-rich maize varieties have been released by Monsanto in Japan and Mexico under the trade name Mavera™YieldGard and Mavera™ Maize (LY038) released by Renessen LLC (Netherlands) in Australia, Columbia, Canada, Japan, Mexico, New Zealand, Taiwan, and the USA (Garg et al. 2018). Another approach to elevate the lysine and tryptophan content was used where maize was transformed with constructs expressing chimeric double-stranded RNA. Surprisingly, this approach led to the accumulation of free amino acids as well which included asparagine, aspartate, and glutamate predominantly in the zein-reduced kernels (Huang et al. 2006). Since maize seeds are a part of animal feed; therefore, a different approach has been developed for using maize as a source of methionine. In this, the *Dzs10* gene (encoding a seed-specific high methionine storage protein) is regulated at the posttranscriptional level (Lai and Messing 2002).

8.11 Challenges for the Public Release of Transgenic Maize

Biofortification of maize with essential amino acids, iron, Zn, provitamin A, and vitamin E possesses the massive potential to eradicate malnutrition and diseases caused due to poor dietary intake. However, various factors influence the commercial production and market release of such varieties. The major challenges for commercial production of biofortified maize include increasing the level of lysine, tryptophan, Zn, iron, provitamin A carotenoids (pVAC), and precursors of vitamin A in the cereals without sacrificing essential agronomic traits of the plant (including high yield and resistance to environmental stress). Secondly, as carotenoids are heat-labile compounds, it is essential to reduce their loss during postharvest handling and processing stages and storage (De Moura et al. 2015). However, the loss of carotenoids is greatly influenced by their genetic makeup; therefore, a variant named CCD1 (carotenoid cleavage dioxygenase 1) has been identified to lower the loss of pVAC during storage (Suwarno et al. 2015). Thirdly, the micronutrients fortified in maize must be digestible and present in their bioavailable form, and most importantly, the amino acid balance must be maintained as per the requirements of the consumers. Milk protein lactalbumin can be expressed in transgenic maize to maintain this amino acid balance (Yang et al. 2002).

Till now, efforts have been made to produce maize cultivars fortified with a single micronutrient, hence, named as "first generation." Now the approaches should be directed toward "second-generation" biofortified maize which provides more combinations of enhanced micronutrients. The multi-nutrient-rich maize cultivars would resolve the malnutrition issues in a holistic manner (Gupta et al. 2015). Moreover, it is impractical to use transgenic plants as a forward genetic tool as it is a very tedious job to screen the mutagenized populations for specific mutations; therefore, it is only possible to use the transgenic approach for reverse genetic experiments (Lung'aho et al. 2011).

8.12 Economic and Social Constraints for the Biofortified Maize

Crops fortified using transgenic technology have difficulty getting societal acceptance. In the process of enhancing the nutritional quality of maize, there are changes in the appearance of grains which may affect consumers' purchase decision as they will not accept modified grains and, ultimately, farmers will not be willing to grow these biofortified crops. For instance, in Africa, white grains (traditional ones) are thought to be the only ones suitable for consumption rather than yellow/orange maize despite the latter one having more content of provitamin A as compared to white maize (Gupta et al. 2015). Awareness programs must be conducted for the dissemination of information of biofortified maize including nutritional benefits of nutrients in maize and developing markets where specifically biofortified products can be sold to the common public.

In the areas like Southeast Asia, the climatic conditions are hot and humid which facilitate the development of several diseases with different pathogenic nature, decreasing the grain yield of traditional maize genotypes. This further imposes various economic constraints to deploy some effective protective measures as no single management strategy is enough to control such complex diseases (Chaudhary et al. 2014). The majority of the maize produced in India is utilized as poultry feed, but the use of biofortified maize in the poultry sector is limited due to the availability of low-cost protein from other alternative sources which is economical for the firms. The biofortified maize rich in essential amino acids holds immense potential as poultry feed, but due to lack of awareness about the nutritional requirements for poultry, additional constraints have been imposed for the development of maize-poultry value chains (Krishna et al. 2014). Methionine is additionally added to the feed prepared from traditional maize which adds to the cost; therefore, biofortified maize holds great promises. Phillips et al. (2008) have reported the methionine-rich versions of maize, i.e., A362, B73, and Mo17, which can be used for the preparation of poultry feed. Moreover, the poultry birds are not able to digest phytate which further leads to the deficiency of phosphorous as it is excreted out of the body. This further raises some serious environmental issues, such as eutrophication and scale formation. Biofortified maize would be a popular choice when the small-scale firms will be disseminated with all the requisite information.

8.13 Genome Editing Approaches for Biofortification of Maize

Genome editing based on CRISPR/Cas, utilizing sequence-specific nucleases for targeted double-strand breaks (DSBs) and the inherent repair mechanisms for these DSBs to generate desired mutations, has become a regular technique for plant researchers (Vats et al. 2019; Ansari et al. 2020). The ease, rapidity, and straightforwardness associated with CRISPR have led to its widespread use across plant species. Moreover, maize is an important crop for genetic studies, and hence, all the genome editing techniques, including meganucleases, Zn finger nuclease (ZFN), transcription activator-like effector nuclease (TALEN), and CRISPR/Cas, have been applied on maize (Shukla et al. 2009; Gao et al. 2010; Djukanovic et al. 2013; Char et al. 2015). ZFNs have been used for trait stacking for herbicide resistance (Ainley et al. 2013) and TALENs for the production of haploid maize lines (Kelliher et al. 2019).

However, the field of genome engineering was truly revolutionized with the advent of the CRISPR/Cas technique owing to its ease of application and remarkable results. The basic CRISPR/Cas technique has been successfully utilized in maize for multiple purposes. The DSBs generated can be repaired by the error-prone nonhomologous end joining (NHEJ) or homology-directed repair (HDR). The HDR can produce precise changes in the genome and requires a donor template to complete its purpose, even though with low efficiencies. This HDR-based precise genome editing was used in maize for the generation of drought-resistant plants, with constitutive expression of the ARGOS8 gene. The overexpression of ARGOS8 results in increased yield during drought resistance. Therefore, HDR was used to replace the native promoter of ARGOS8 with a high activity promoter, resulting in lines performing better under drought conditions (Shi et al. 2017). Other examples of CRISPR application are also there for maize, including targeting male fertility genes (Ms26 and Ms45), liguleless1 (LIG1) gene, and acetolactate synthase (ALS) (Svitashev et al. 2015), and MS8 (Chen et al. 2018).

Chen et al. (2018) also reported the production of transgene-free maize plants via segregation in the F2 generation. However, recovering transgene-free maize plants via general CRISPR/Cas and then waiting for subsequent generations is a long process. For faster recovery of transgene-free edited plants, RNPs-based protoplast transfection system has been employed. In fact, maize is one of the few plant species, where whole edited plants have been regenerated from protoplasts (Svitashev et al. 2016). This is a desirable attribute in crop plants for faster production of transgene-free lines to ease regulatory guidelines and has not been achieved in major crop species to date.

Further, Haploid-Inducer Mediated Genome Editing (IMGE) approach, which integrates double haploid with CRISPR/Cas technologies, can generate homozygous edited lines in the shortest time possible (Wang et al. 2019). HI-Edit is another platform for direct modification in the plant genome. They even recovered edited wheat embryos via pollination with Cas9 harboring maize pollens (Kelliher et al. 2019). Maize is also one of the first plants where prime editing has been applied (Jiang et al. 2020). Prime editing is a kind of advanced precise editing method comprising prime editing guide RNA (pegRNA) and reverse transcriptase fused Cas nickase protein (Anzalone et al. 2019). Multiplexing is another advantageous application of CRISPR/Cas, making it a tool of choice for biofortification in various crops including maize (Li et al. 2018). CRISPR has been used by Liang et al. (2014) to reduce anti-nutrition elements by targeting genes such as IPK. Zeins are the major proteins present in corn that are deficient in tryptophan and lysine, resulting in the poor nutritional value of maize plants (Geraghty et al. 1981). These have also been targeted to improve the properties of zein proteins via tRNA-based multiplexed CRISPR/Cas system (Qi et al. 2016).

References

Ainley WM, Sastry-Dent L, Welter ME et al (2013) Trait stacking via targeted genome editing. Plant Biotechnol J 11(9):1126–1134

Aluru M, Xu Y, Guo R et al (2008) Generation of transgenic maize with enhanced provitamin A content. J Exp Bot 59(13):3551–3562

Ansari WA, Chandanshive SU, Bhatt et al (2020) Genome editing in cereals: approaches, applications and challenges. Int J Mol Sci 21(11):4040

Anzalone AV, Randolph PB, Davis JR et al (2019) Search-and-replace genome editing without double-strand breaks or donor DNA. Nature 576(7785):149–157

Bashir K, Takahashi R, Nakanishi H et al (2013) The road to micronutrient biofortification of rice: progress and prospects. Front Plant Sci 4:15

Bender RR, Haegele JW, Ruffo ML, Below FE (2013) Nutrient uptake, partitioning, and remobilization in modern, transgenic insect-protected maize hybrids. Agron J 105(1):161–170

Bruulsema TW, Heffer P, Welch RM, Cakmak I, Moran KK (2012) Fertilizing crops to improve human health: a scientific review volume 1: food and nutrition security

Cahoon EB, Hall SE, Ripp KG et al (2003) Metabolic redesign of vitamin E biosynthesis in plants for tocotrienol production and increased antioxidant content. Nat Biotechnol 21(9):1082–1087

Cakmak I (2002) Plant nutrition research: priorities to meet human needs for food in sustainable ways. Plant Soil 247(1):3–24

Char SN, Unger-Wallace E, Frame B et al (2015) Heritable site-specific mutagenesis using TALENs in maize. Plant Biotechnol J 13(7):1002–1010

Chaudhary D, Kumar S, Yadav O (2014) Nutritive value of maize: improvements, applications and constraints. In: Maize: nutrition dynamics and novel uses. Springer, Cham, pp 3–17

Chen R, Xu Q, Liu Y et al (2018) Generation of transgene-free maize male sterile lines using the CRISPR/Cas9 system. Front Plant Sci 9:1180

Chilimba ADC, Young SD, Black CR et al (2012) Agronomic biofortification of maize with Se (Se) in Malawi. Field Crops Res 125:118–128

De Moura FF, Miloff A, Boy E (2015) Retention of provitamin A carotenoids in staple crops targeted for biofortification in Africa: cassava, maize and sweet potato. Crit Rev Food Sci Nutr 55(9):1246–1269

Djukanovic V, Smith J, Lowe K et al (2013) Male-sterile maize plants produced by targeted mutagenesis of the cytochrome P450-like gene (MS26) using a re-designed I–CreI homing endonuclease. Plant J 76(5):888–899

Drakakaki G, Marcel S, Glahn RP et al (2005) Endosperm-specific co-expression of recombinant soybean ferritin and Aspergillus phytase in maize results in significant increases in the levels of bioavailable iron. Plant Mol Biol 59(6):869–880

Ekpa O, Palacios-Rojas N, Kruseman G, Fogliano V, Linnemann AR (2018) Sub-Saharan African maize-based foods: technological perspectives to increase the food and nutrition security impacts of maize breeding programmes. Glob Food Sec 17:48–56

Gao H, Smith J, Yang M et al (2010) Heritable targeted mutagenesis in maize using a designed endonuclease. Plant J 61(1):176–187

Garg M, Sharma N, Sharma S et al (2018) Biofortified crops generated by breeding, agronomy, and transgenic approaches are improving lives of millions of people around the world. Front Nutr 5: 12

Geraghty D, Peifer MA, Rubenstein I et al (1981) The primary structure of a plant storage protein: zein. Nucleic Acids Res 9(19):5163–5174

Gupta C (2014) Role of Fe(Fe) in body. IOSR J Appl Chem 7(11):38–46

Gupta HS, Hossain F, Muthusamy V (2015) Biofortification of maize: an Indian perspective. Indian J Genet 75(1):1–22

Hu S, Lübberstedt T, Zhao G, Lee M (2016) QTL mapping of low-temperature germination ability in the maize IBM Syn4 RIL population. PLoS One 11(3):e0152795

Huang S, Frizzi A, Florida CA et al (2006) High lysine and high tryptophan transgenic maize resulting from the reduction of both 19- and 22-kD α-zeins. Plant Mol Biol 61(3):525–535
Imran M, Rehim A (2017) Zinc fertilization approaches for agronomic biofortification and estimated human bioavailability of zinc in maize grain. Arch Agron Soil Sci 63(1):106–116. https://doi.org/10.1080/03650340.2016.1185660
Jiang Y-Y, Chai Y-P, Lu M-H et al (2020) Prime editing efficiently generates W542L and S621I double mutations in two ALS genes in maize. Genome Biol 21(1):1–10
Jin T, Zhou J, Chen J, Zhu L, Zhao Y, Huang Y (2013) The genetic architecture of zinc and iron content in maize grains as revealed by QTL mapping and meta-analysis. Breed Sci 63(3):317–324
Jittham O, Fu X, Xu J, Chander S, Li J, Yang X (2017) Genetic dissection of carotenoids in maize kernels using high-density single nucleotide polymorphism markers in a recombinant inbred line population. Crop J 5(1):63–72
Karlen DL, Flannery RA, Sadler EJ (1988) Aerial accumulation and partitioning of nutrients by corn. Agron J 80(2):232–242
Kelliher T, Starr D, Su X et al (2019) One-step genome editing of elite crop germplasm during haploid induction. Nat Biotechnol 37(3):287–292
Krishna VV, Erenstein O, Sadashivappa P et al (2014) Potential economic impact of biofortified maize in the Indian poultry sector. Int Food Agribus Manag Rev 17(4):111–140
Kumar N, Salakinkop SR (2018) Agronomic biofortification of maize with zinc and iron micronutrients. Mod Concepts Dev Agron 1:1–4
Labroo MR, Studer AJ, Rutkoski JE (2021) Heterosis and hybrid crop breeding: a multidisciplinary review. Front Genet 12:643761
Lai J, Messing J (2002) Increasing maize seed methionine by mRNA stability. Plant J 30(4): 395–402
Leyva-Guerrero E, Narayanan NN, Ihemere U, Sayre RT (2012) Iron and protein biofortification of cassava: lessons learned. Curr Opin Biotechnol 23(2):257–264. https://doi.org/10.1016/j.copbio.2011.12.009. Epub 2012 Jan 4. PMID: 22226461
Li F, Chung T, Pennington JG, Federico ML, Kaeppler HF, Kaeppler SM et al (2015) Autophagic recycling plays a central role in maize nitrogen remobilization. Plant Cell 27(5):1389–1408
Li X, Wang Y, Chen S et al (2018) Lycopene is enriched in tomato fruit by CRISPR/Cas9-mediated multiplex genome editing. Front Plant Sci 9:559
Liang Z, Zhang K, Chen K et al (2014) Targeted mutagenesis in Zea mays using TALENs and the CRISPR/Cas System. J Genet Genomics 41(2):63–68
Liberato SC, Pinheiro-Sant'Ana HM (2006) Fortification of industrialized foods with vitamins. Rev Nutr 19:215–231
Lopez MJ, Mohiuddin SS (2022) Biochemistry, essential amino acids. In: StatPearls. Stat Pearls Publishing, Treasure Island (FL). PMID: 32496725
Lung'aho MG, Mwaniki AM, Szalma SJ et al (2011) Genetic and physiological analysis of Fe biofortification in maize kernels. PLoS One 6(6):e20429
Maqbool MA, Beshir A (2019) Zn biofortification of maize (Zea mays L.): status and challenges. Plant Breed 138(1):1–28
Maqbool MA, Beshir Issa A, Khokhar ES (2021) Quality protein maize (QPM): importance, genetics, timeline of different events, breeding strategies and varietal adoption. Plant Breed 140(3):375–399
Murphy SP, Yates AA, Atkinson SA, Barr SI, Dwyer J (2016) History of nutrition: the long road leading to the dietary reference intakes for the United States and Canada. Adv Nutr 7(1):157–168
Paddon-Jones D, Westman E, Mattes RD, Wolfe RR, Astrup A, Westerterp-Plantenga M (2008) Protein, weight management, and satiety. Am J Clin Nutr 87(5):1558S–1561S
Pan Z, Eckhoff S, Paulsen M et al (1996) Physical properties and dry-milling characteristics of six selected high-oil maize hybrids. Cereal Chem 73(5):517–520

Patil G, Vuong TD, Kale S et al (2018) Dissecting genomic hotspots underlying seed protein, oil, and sucrose content in an interspecific mapping population of soybean using high-density linkage mapping. Plant Biotechnol J 16(11):1939–1953

Phillips R, Suresh J, Olsen M et al (2008) Registration of high-methionine versions of maize inbreds A632, B73, and Mo17. J Plant Regist 2(3):243–245

Pixley K, Rojas NP, Babu R et al (2013) Biofortification of maize with provitamin A carotenoids. In: Carotenoids and human health. Springer, Cham, pp 271–292

Prasad AS (2014) Zn: an antioxidant and anti-inflammatory agent: role of Zn in degenerative disorders of aging. J Trace Elem Med Biol 28(4):364–371

Prasanna BM, Palacios-Rojas N, Hossain F et al (2020) Molecular breeding for nutritionally enriched maize: status and prospects. Front Genet 10:1392

Qi W, Zhu T, Tian Z et al (2016) High-efficiency CRISPR/Cas9 multiplex gene editing using the glycine tRNA-processing system-based strategy in maize. BMC Biotechnol 16(1):1–8

Rana N, Rahim MS, Kaur G et al (2020) Applications and challenges for efficient exploration of omics interventions for the enhancement of nutritional quality in rice (Oryza sativa L.). Crit Rev Food Sci Nutr 60(19):3304–3320

Saleem I, Javid S, Bibi F et al (2016) Biofortification of maize grain with Zn and Fe by using fertilizing approach. J Agric Ecol Res Int 7:1–6

Sayre JD (1948) Mineral accumulation in corn. Plant Physiol 23(3):267

Shao H, Shi D, Shi W, Ban X, Chen Y, Ren W et al (2021) Nutrient accumulation and remobilization in relation to yield formation at high planting density in maize hybrids with different senescent characters. Arch Agron Soil Sci 67(4):487–503

Shi J, Gao H, Wang H et al (2017) ARGOS 8 variants generated by CRISPR-Cas9 improve maize grain yield under field drought stress conditions. Plant Biotechnol J 15(2):207–216

Shukla VK, Doyon Y, Miller JC et al (2009) Precise genome modification in the crop species Zea mays using Zn-finger nucleases. Nature 459(7245):437–441

Šimić D, Mladenović Drinić S, Zdunić Z, Jambrović A, Ledenčan T, Brkić J et al (2012) Quantitative trait loci for biofortification traits in maize grain. J Hered 103(1):47–54

Suwarno WB, Pixley KV, Palacios-Rojas N et al (2015) Genome-wide association analysis reveals new targets for carotenoid biofortification in maize. Theor Appl Genet 128(5):851–864

Svitashev S, Young JK, Schwartz C et al (2015) Targeted mutagenesis, precise gene editing, and site-specific gene insertion in maize using Cas9 and guide RNA. Plant Physiol 169(2):931–945

Svitashev S, Schwartz C, Lenderts B et al (2016) Genome editing in maize directed by CRISPR–Cas9 ribonucleoprotein complexes. Nat Commun 7(1):1–7

Tang M, He X, Luo Y et al (2013) Nutritional assessment of transgenic lysine-rich maize compared with conventional quality protein maize. J Sci Food Agric 93(5):1049–1054

Van Der Straeten D, Bhullar NK, De Steur H, Gruissem W, MacKenzie D, Pfeiffer W et al (2020) Multiplying the efficiency and impact of biofortification through metabolic engineering. Nat Commun 11(1):1–10

Vats S, Kumawat S, Kumar V, Patil GB, Joshi T, Sonah H, Nadaf A, Sharma TR, Deshmukh R (2019) Genome editing in plants: exploration of technological advancements and challenges. Cell 8(11):1386

von Wirén N, Mori S, Marschner H, Romheld V (1994) Iron inefficiency in maize mutant ys1 (Zea mays L. cv Yellow-Stripe) is caused by a defect in uptake of iron phytosiderophores. Plant Physiol 106(1):71–77

Wakeel A, Farooq M, Bashir K et al (2018) Micronutrient malnutrition and biofortification: recent advances and future perspectives. In: Plant micronutrient use efficiency. Elsevier, Amsterdam, pp 225–243

Wang B, Zhu L, Zhao B et al (2019) Development of a haploid-inducer mediated genome editing system for accelerating maize breeding. Mol Plant 12(4):597–602

Wong HW, Liu Q, Sun SSM (2015) Biofortification of rice with lysine using endogenous histones. Plant Mol Biol 87:235–248. https://doi.org/10.1007/s11103-014-0272-z

Yang S-H, Moran DL, Jia H-W et al (2002) Expression of a synthetic porcine α-lactalbumin gene in the kernels of transgenic maize. Transgenic Res 11(1):11–20

Yu J, Peng P, Zhang X et al (2005) Seed-specific expression of the lysine-rich protein gene sb401 significantly increases both lysine and total protein content in maize seeds. Food Nutr Bull 26(4 Suppl 3):S312–S316

Zhang FS, Römheld V, Marschner H (1991) Diurnal rhythm of release of phytosiderophores and uptake rate of zinc in iron-deficient wheat. Soil Sci Plant Nutr 37(4):671–678

Zhu C, Naqvi S, Breitenbach J et al (2008) Combinatorial genetic transformation generates a library of metabolic phenotypes for the carotenoid pathway in maize. Proc Natl Acad Sci 105(47): 18232–18237

Biofortification of Barley for Nutritional Security

9

Kiran Khandagale, Dhananjay Shirsat, and Avinash Ade

9.1 Introduction

The worldwide continuous growth of the human population resulted in increased demands for food. The green revolution has increased food production significantly (Evenson and Gollin 2003). But this increase in yield was often accompanied by reduced nutritional quality (Simmonds 1995; Oury et al. 2003). More than one billion people suffer from the low intake of proteins, minerals, and vitamins especially in developing and underdeveloped countries (WHO 2016), and thus biofortification of crops is a very important approach to overcome it (Wiegmann et al. 2019). Biofortification is the practice of enhancing the amount or bioavailability of vital nutrients in food using agronomic, genetic, and biotechnological methods (Bouis et al. 2011).

Generally, staple food crops are targeted for biofortification as they are the major portion of the diet in poor people. Vitamins and minerals are required by humans in very minute amounts (less than 1 mg/day) and thus are the main focus of the biofortification program. These compounds govern several vital biological processes in the body, and therefore biofortification approaches can improve the content and availability of nutrients in the human diet to improve the nutritional security of vulnerable communities around the globe. Three main strategies were followed for biofortification: conventional breeding, agronomic, and biotechnological/transgenic approach. Plant breeding strategy involves crossing of elite variety with genotype having higher micronutrient content, and after several generations, we get the ideal genotype with higher mineral level and other desired characters. Further, agronomic

K. Khandagale · A. Ade (✉)
Department of Botany, Savitribai Phule Pune University, Pune, India

D. Shirsat
ICAR-Directorate of Onion and Garlic Research, Rajgurunagar, India

R. Deshmukh et al. (eds.), *Biofortification in Cereals*,
https://doi.org/10.1007/978-981-19-4308-9_9

methods are comprised of applying fertilizer to soil or foliar application to improve the level of a particular nutrient in the edible part of the crop. The transgenic approach is used where a specific nutrient doesn't exist naturally in that crop or breeding for that trait is not effective. Gene for that trait can be sourced from any organism from bacteria to animals and inserted in the desired crop to get the particular nutrient at a higher level. Acceptability of GM crops, the stability of gene insert, and biosafety regulations are the main hurdles in the transgenic approach.

Barley (*Hordeum vulgare* ssp. *vulgare*) is the fourth most important cereal crop globally. The average annual production of barley is more than 140 million tons from the 50 million hectares of area. About 70% of barley produced is used as animal feed, 21% is used for beer making in distilleries, 6% is used as food for human beings, and the remaining is used in biofuel production (www.fao.org/faostat/en/#data/QC). Nutritionally, barley grain is comprised of 70% starch, 10–20% protein, 2–3% lipids, 5–10% β-glucan, 2–5% minerals, and 11–34% dietary fibers (Sullivan et al. 2013). Barley was the earliest cereal that was domesticated and used for the preparation of bread (Tiwari 2010). Barley is also used as model species for the members of *Triticeae* such as soft wheat, durum wheat, and rye; as these species are closely related, genetic information from barley can be used for the research in these *Triticeae* species (Sreenivasulu et al. 2008). Barley grain harbors several bioactive compounds like β-glucans, lignans, tocotrienols, folate, fructans, phytosterols, polyphenols, policosanol, phytates, etc.; therefore, consumers show interest in barley as a food. It is a rich source of dietary fiber and functional food; β-glucans in barley are known to lower the blood cholesterol and has low glycemic index (Baik and Ullrich 2008). Further biofortification of barley with different nutrients will increase its nutritional value and will help in overcoming malnutrition.

9.2 Biofortification Approaches

Generally, biofortification strategies are comprised of these main approaches: genetic/breeding, biotechnological/transgenic, and agronomic approaches (Fig. 9.1).

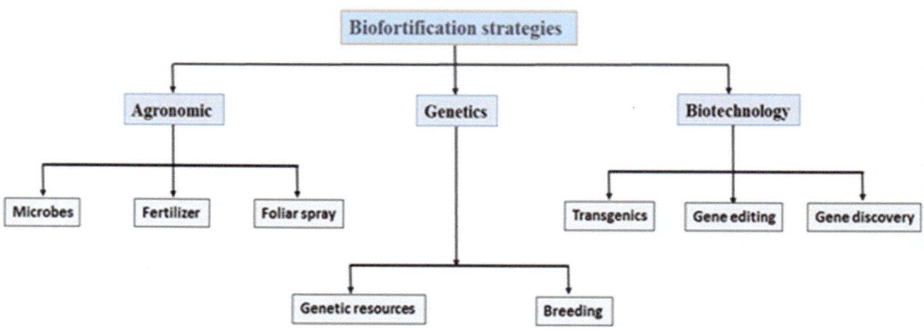

Fig. 9.1 Different strategies for biofortification of barley

9.3 Genetic and Plant Breeding Approach

Genetic and plant breeding is the most believed approach which is more sustainable and economical compared to the other two approaches. Screening of available germplasm for a trait of interest is performed, and the success relies on the availability of enough diversity in that trait (Velu et al. 2014). Further, biofortification by the genetic approach is also influenced by accurate phenotyping and gene-environment interactions. Genetically fortified plants need to be supported with optimal agronomic practices to get the best out of them.

For any breeding experiment, one should have suitable parents for crossing, and getting suitable parents depends upon the availability of diversity in germplasm and the extent of screening. Selection for mineral content is a complex process as this nutrient content is affected by many physiological factors (Narwal et al. 2020). QTL mapping in crops is done to understand the complex biological traits. Transport of Zn from vegetative parts to grain via phloem is a major barrier in the loading of Zn in grain endosperm. Hussain et al. (2016) performed double haploid mapping to genetic characterization of Zn remobilization. They found a large variation in Zn content from 27 to 75 $\mu g\ g^{-1}$, and this variation was correlated with the remobilization of Zn to grain. Three QTLs associated with leaf and two QTLs linked with stem were found to involve in Zn remobilization. Such studies will help in Zn biofortification of barley. Further, QTLs for mineral content were detected using 193 recombinant inbred lines. Seventeen QTLs were detected in barley grain which contributed 6.36–64.08% diversity in Zn, Mg, Ca, K, Na, Mn, Fe, and P. They also detected pleiotropic QTLs having an additive effect on mineral content (Zeng et al. 2016). Such efforts further will help us with marker-assisted selection for mineral biofortification. QTLs for grain and malt β-glucan amounts were mapped with the help of the 123 marker linkage map. They identified three QTLs for grain β-glucan, six for malt β-glucan, and eight for malt β-glucanase using interval mapping (Han et al. 1995). Such dissection of β-glucan content is further useful in the breeding as well as the selection of high- and low-β-glucan-containing genotypes (Li et al. 2008). Genome-wide association studies (GWAS) were performed using 336 spring barley accessions for element amounts in the grain with the help of 6519 SNP markers and TASSEL software. Several QTLs for different minerals were identified that will be used in the future for breeding nutrient-rich barley (Gyawali et al. 2019). Grain protein content (GPC) is the main grain quality character. *Gpc-B1*, a wheat GPC QTL, is an NAC transcription factor (TtNAM-B1) involved in higher levels of protein, zinc, and iron in grain. A similar QTL in barley was identified on chromosome 6H. The colinearity of GPC regions indicated its role in GPC QTL in barley (Distelfeld et al. 2008). Xue et al. (2016) studied the nutrient level in barley grain under different environmental conditions and reported high genetic and environmental interactions. Total nitrogen exhibited high genotype contribution; thus nitrogen remobilization might have increased the Zn and Fe transport to grain indicating the genetic effect of GPC locus on Zn and Fe translocation. Further, Fan et al. (2017) identified environmentally stable QTLs using SNP and SSR marker 190 recombinant inbred lines. These QTLs were identified on chromosomes 2H (1), 4H (1), 6H (1), and 7H (3). These studies could be used in developing high-grain protein barley genotypes.

9.4 Transgenic and Biotechnological Approach

Trangenic and biotechnological methods for biofortification are generally used in such conditions where limited or no variation in the trait of interest was present in germplasm (Zhu et al. 2007). It involves genetic engineering where genes from different sources for a trait of interest are introduced in the target variety of crop. This approach enables us to transfer traits across the species boundaries independent of taxonomic status. Similarly, transgenics is the only method to fortify any crop with a micronutrient that is not present naturally in crops (Perez-Massot et al. 2013); thus transgenic approach of biofortification has the potential to significantly contribute toward the improvement of nutrition and health.

9.5 Transgenics

Transgenic strategies generally target the enhancement of nutrient uptake, biosynthesis of nutrients, and bioavailability of nutrients. It can be achieved by following these key steps: enhancing uptake, increasing translocation to grain, targeting storage toward endosperm, decreasing anti-nutritional factors, and increasing the bioavailability of nutrients of interest (Mulualem 2015). There are only a few attempts taken in barley biofortification using a transgenic approach. Ramesh et al. (2004) reported that overexpression of the known zinc transporters from *Arabidopsis* in barley under ubiquitin promoter increased Zn concentration in transgenic barley. Menguer et al. (2018) improved the zinc content in grains by overexpressing a transition metal transporter (*HvMTP1*) gene under the endosperm-specific promoter. They found higher grain content in the endosperm of transgenic lines of barley. Similarly, transgenic barley expressing cytokinin oxidase/dehydrogenase (CKX) gene in roots led to the development of a larger root system which also accumulated a higher amount of zinc in barley grain than wild type (Ramireddy et al. 2018). Cereal grains are containing an insufficient amount of essential amino acid-like lysine. Ohnoutkova et al. (2012) developed transgenic barley expressing dihydrodipicolinate synthase from *E. coli*. The resulted T1 generation plants were having a more than 50% increase in lysine over the wild type. C-hordein in barley is a member of prolamin protein families and is composed of mainly nonessential amino acids like proline and glutamine and thus has low nutritional value. Therefore, Sikdar et al. (2016) silenced the C-hordein gene in barley using RNAi technology, and quadrupole-time-of-flight mass spectrometry analyses of protein fraction revealed a reduction in C-hordein, and the level of essential amino acids was increased. Earlier, Lange et al. (2007) suppressed the C-hordeins in barley using antisense construct, and amino acid analyses revealed that levels of nonessential amino acids (proline, glutamic acid/glutamine, and phenylalanine) were decreased by 12%, 6%, and 9%, respectively, while the amount of essential amino acids like lysine, threonine, and methionine was elevated by 16%, 13%, and 11%, respectively. Therefore modulation of prolamin levels in barley grains is a promising way to improve protein quality. Further, overexpression of homogentisate geranylgeranyl transferase (*HvHGGT*) gene in barley resulted in an increase of tocotrienol content and antioxidant activity in barley grain (Chen et al. 2017). The cisgenesis concept was used in barley to increase phytase activity in grain (Holme et al. 2012). They expressed a barley phytase gene (*HvPAPhy_a*) during grain filling stages, and homozygous lines showed more than 2.5-fold increase in phytase activity. This enhanced phytase activity was stable for three generations analyzed. The marker elimination method was used in this study to obtain marker-free transgenic plants. Cisgenesis along with marker-free technique might increase the acceptability of genetically engineered crops. The polysaccharides like (1,3;1,4)-b-D-glucans are useful components in the diet of a human being, which decreases the risk of diabetes, obesity, and cancer. Overexpression of barley cellulose synthase-like family (*CslF6*) gene under endosperm-specific promoter led to an 80% increase in (1,3;1,4)-b-D-glucan content in transgenic barley grain (Burton et al. 2011).

9.6 Genome Editing

Recently, genome editing techniques are also being used for the creation of new alleles and gene editing independent of genome sequences (Khandagale and Nadaf 2016) which could be used for biofortification of barley. CRISPR/Cas9 and TALENs were used for the evaluation of the *HvPAPhy_a* gene in barley. It was found that *HvPAPhy_a* is the main contributor to mature grain phytase activity. Thus higher expression of *HvPAPhy_a* led to fast germination as well as higher phosphate utilization (Holme et al. 2017). Pathway of vitamin E biosynthesis in monocots was not studied in detail due to the lack of functional mutants. Zeng et al. (2020) used CRISPR for the generation and characterization of the functional mutants of barley for *HvHPT* and *HvHGGT* genes which revealed that in barley, *HvHGGT* is the only major gene for the biosynthesis of tocotrienols and *HvHPT* plays a minor role. Inositol trisphosphate 5/6 kinases (ITPK) is an enzyme involved in the production of inositol hexakisphosphate which is the main form of storage phosphate in cereal grains. The creation of lines containing less inositol hexakisphosphate would increase the phosphate and mineral bioavailability. CRISPR/Cas9-mediated editing of *HvITPK1* increased phosphate in grains by 65–174% over wild type. In barley, D-hordein is one of the storage proteins in the barley which negatively impacts malting quality. Li et al. (2020) used CRISPR/Cas9 genome editing and created new alleles of D-hordein gene; transcriptome analysis and SDS-PAGE revealed reduced D-hordein content in mutant lines. These new alleles provided the new germplasm resource for breeding barley for malt quality.

9.7 Omics in Better Understanding Nutrient Uptake, Storage, and Bioavailability

For successful biofortification of any crop, a thorough understanding of nutrient homeostasis is needed. Omics approaches such as genomics, transcriptomics, proteomics, metabolomics, and metagenomics will help us to elucidate the complex phenomenon behind mineral homeostasis. Darbani et al. (2015) attempted the elucidation of the mineral homeostasis in barley seed transfer cells using the RNA-seq approach. Seed transfer cells were isolated using laser capture microdissection from the grain cryosections. The number of genes such as auxin and ethylene signaling factors, sulfur homeostasis components, mineral trafficking components, vacuole organization factors, protein sorting, and recycling factors, etc. were differentially expressed in changes in mineral content. Earlier, Tauris et al. (2009) also demonstrated the road map for zinc transport in the developing grain with the help of barley microarray, Affymetrix 22k GeneChip and proposed a model for zinc trafficking from the phloem to the developing grains.

9.8 Agronomic Approach

Agronomic strategies for biofortification involve the application of nutrients to soil or plants to enhance the content of particular nutrient in the edible part of that crop so that after consumption it will improve human nutrition. Micro minerals, like Zn, Fe, Se, copper, manganese, I, Mo, etc. when applied in the soil to improve the nutritional status of soil, are absorbed by plants which results in alleviating the micronutrient deficiency in humans. This approach is simple but provides short-term solutions, and care should be taken for the selection of the source of nutrients, application method, and effects on the environment. Along with chemical fertilizers, plant growth-promoting microbes like N2 bacteria and P- and K-solubilizing microbes and other microbes which enhance the phyto-availability of nutrients are also used in this approach (Garg et al. 2018).

9.9 Fertilization Application

Agronomic biofortification involves the application of micronutrients in the form of fertilizer to elevate the micronutrient level in grains or edible parts.

Micronutrient fertilizers when applied in combination with NPK and organic fertilizers showed a good response in the uptake of micronutrients (De Valenca et al. 2017). The low solubility of Zn in the soil is the main reason for Zn deficiency in plants; thus sufficient amount of available zinc needs to be maintained in the soil during the grain filling stage to achieve a higher level of Zn in grain. Yadav and Sharma (2018) reported that the application of zinc sulfate along with NH_4NO_3 increased yield as well as Zn content in grain of barley; it was due to the acidifying effect of NH_4NO_3. It was found that the foliar-applied Zn gets easily translocated to grain during development and the localization studies also showed the interaction between Zn and grain proteins (Cakmak and Kutman 2018). Several studies showed that the use of Zn fertilizers elevated the Zn level in grains of some cereals including barley (Cakmak 2010). Similarly, zinc fertilizer applications have been reported to increase the Zn content in barley grain (Uddin et al. 2014). Sulfur is an important element that plays a role in plant development and biotic stress. Its application also increases the uptake of Mn, Fe, Zn, and Cu. The application of sulfur in the form of ammonium and potassium sulfate was found effective in the elevation of microelements in barley (Barczak et al. 2019). For selenium biofortification, Rodrigo et al. (2013) sprayed two-rowed barley with four different concentrations of sodium selenate and sodium selenite for two seasons. It was found that sodium selenite was more efficiently absorbed by the plant. For every gram of sodium selenite and selenate sprayed, the Se concentration in grain was increased by 9 and 44 μg/kg dry weights, respectively.

9.10 Microbes in Biofortification

The continuous chemical fertilizer application in high doses leads to soil and environmental pollution as well as toxicity to plants and animals. Further transgenic crops are not easily accepted by the public which is seen in the case of GM brinjal and mustard. An alternate way is the use of microbe for biofortification; despite its huge potential this approach did not receive enough attention. Rhizospheric or endophytic microbes are known to increase the availability and absorption of micronutrients by plants and ultimately led to the enhancement of micronutrient content in the edible part of the crop (Ku et al. 2019). Soil microbes play a key role in maintaining soil health and fertility (Barret et al. 2011). Farmers are using N-fixing and P- and K-solubilizing microbes to increase the availability of major nutrients to increase the yield. These microbes could be used for biofortification as present in soil and increase the availability of nutrients to crops (Prasanna et al. 2016). Several micronutrients are present in fixed form as a precipitate or adsorbed on soil mineral and organic surfaces. These nutrients are solubilized by PGPR by secreting some enzymes. In wheat, it was found that the application of biofertilizers enhanced the acquisition of minerals (Rana et al. 2012). The iron level in paddy was elevated by application of PGPR inoculum comprised of *P. putida*, *P. fluorescens*, and *Azospirillum lipoferum* (Sharma et al. 2013a). Similarly, Fe content in lentils was found to be doubled by the treatment of biofertilizer containing *Pseudomonas* species (Mishra et al. 2011). Inoculation of *Pseudomonas* and *Acinetobacter* strains significantly increased the Zn, Fe, Mg, Ca, K, and P in crops (Tariq et al. 2007; Khan 2005). These PGPRs could be used in barley to achieve sustainable enrichment of micronutrients.

Arbuscular mycorrhizal fungi (AMF) also solubilize different minerals in the soil and have potential use in biofortification (Martino et al. 2003). Ingra et al. (2019) studied the eight different species of AMF in wheat and reported an increase in the uptake of P, Fe, and Zn along with better root lengths and density. Further selenium level was found to be increased in wheat grain after co-inoculation of *Glomus clarideum*, *Pseudomonas* sp., and *Bacillus* sp. (Duran et al. 2013). The use of PGPRs and AMF for biofortification has been attempted in several crops, but microbe-mediated biofortification is little studied in barley. Watts-Williams and Cavagnaro (2018) demonstrated the increased grain and straw zinc concentration in modern barley after inoculation with AMF *Rhizophagus irregularis*. This increase in Zn concentration was due to the increased uptake of Zn from the soil under the upregulation of ZIP transporters; it interestingly did not increase the yield of grains. Similarly, Coccina et al. (2019) showed AMF-mediated Zn uptake in wheat and barley.

9.11 Biofortification for Minerals

Micronutrient deficiencies are an important form of human malnutrition, known as hidden hunger. Globally micronutrient malnutrition is recognized as an enormous and speedily growing public health problem, especially in developing countries (Zou et al. 2019; Sazawal et al. 2018). Deficiencies of mineral micronutrient such as zinc (Zn), iodine (I), selenium (Se), and iron (Fe) denote the global health problems because these affect more than one-third of the world population (Zou et al. 2019; Lyons 2018). The zero hunger is the Sustainable Development Goal 2 which aims to end hunger through enhanced food and nutritional security, and biofortification of food crops is the most sustainable and cost-effective method to provide nutrition to the target population in natural form fulfilling this goal (Yadava et al. 2018).

9.12 Genetic Diversity for Mineral Content in Barley

Besides "calories," various essential micronutrients are important in the health and nutrition of organisms. These nutrients are divided into *macro*nutrients and *micro*nutrients. Micronutrients are needed in minute quantity which makes up only 0.05% of human body, whereas macroelements constitute 99.5% of human body (Kotz et al. 2006). Several micronutrients function as a cofactor of enzymes that regulate crucial life processes in the organism. Genetic variation of micronutrient content is crucial for the breeding of high-nutrient crops. Wild barley is known to harbor the highest amount of micronutrients. Iron content in wild barley ranged from 10.8 to 329.1 mg kg^{-1}, and zinc content was 66.3–493.9 mg kg^{-1} (Yan et al. 2012). Recently, the International Center for Agricultural Research in the Dry Areas (ICARDA) has analyzed 336 accessions for 13 different micronutrients. Some genotypes showed a high amount of these minerals, which are suitable candidates for the biofortification program in barley (Gyawali et al. 2019).

These micronutrients are not evenly distributed in grain; some are concentrated in husk and aleurone layers which get removed during the milling and polishing process in many cereals including barley. This distribution of nutrients is genotype-dependent. Therefore, one should have a thorough knowledge of diversity and the mechanism of micronutrient uptake and accumulation for biofortification. Detterbeck et al. (2016) studied the micronutrient diversity and distribution and found that more than 120 lines showed good variation in Zn content, and the majority of this diversity is due to genetic differences. Micro-proton-induced X-ray emission (l-PIXE) was used for a detailed study of micronutrient distribution within the grains' four tissues: embryo, aleurone, endosperm, and husk. Further, it is also found that the cultivation of high Zn lines in Cd-contaminated soils resulted in higher Cd accumulation which exceeded the *Codex Alimentarius* threshold. Thus, along with genetic variations for the desired micronutrient, one should consider the levels of toxic elements while planning for the biofortification of barley.

Domestication and repetitive selections led to genetic erosion in several modern crops (Zamir 2001; van de Wouw et al. 2010). A wild relative could be used to replenish the gene pools of modern crops. Wild barley (*Hordeum vulgare* ssp. *spontaneum*) can be used for introgression of fertile barley cultivars (Morrell and Clegg 2007). Such a successful example is the introgression of *Gpc-B1* locus from wild emmer into bread wheat through chromosomal substitution technique (Distelfeld et al. 2006). This locus elevated the micronutrient levels in mature grains (Distelfeld et al. 2007). Wild barley also has huge variations for agronomic traits but has been paid limited attention as a source for biofortification. Wiegmann et al. (2019) studied the interplay between plant development, yield, and nutrient concentrations in wild barley nested association mapping population HEB-25. They observed a huge variation in nutrient concentrations; some lines have more than 50% higher levels of protein, iron, and zinc than a recurrent parent. It was found that grain yield and nutritional value are negatively correlated in barley. Analyses of genetic elements in nutrient content revealed that wild alleles were often linked with the higher nutrient level which indicated that the targeted introgression of wild barley alleles may help us in the biofortification of barley (Wiegmann et al. 2019).

Selenium is a vital trace element important for the health of humans, and the main source of selenium is a plant-based diet (Rayman 2000). Most of the soils are deficient in selenium content, and thus ultimately selenium in the food system is also low, and thus a large number of people suffer from Se deficiency (Combs 2001; Jones et al. 2017). Jun et al. (2011) studied the diversity in grain Se concentration of 92 *H. spontaneum* genotypes representing different habitats in Israel. The grain selenium content ranged from 0 to 0.387 mg kg^{-1}. *H. spontaneum* populations exhibited higher Se content due to their abilities for Se uptake and accumulation.

9.13 Transporters for Mineral Uptake and Transport

The plant roots have an important role in the uptake of essential nutrients from the soil and are used in growth and development functions. The mineral uptake is facilitated by the different transporters in plants belonging to the different transporter families (Sasaki et al. 2016). Fe and Zn are known as essential cellular element which plays a critical role in metabolic processes in all living organisms (Darbani et al. 2015). However many of the metabolic pathways are activated by iron; also it is a prosthetic group constituent of many enzymes (Rout and Sahoo 2015). To date, a large number of iron transporters in plants are known. That involves the yellow stripe 1-like (YSL) subfamily of the oligopeptide transporter (OPT) superfamily, the copper transporter (COPT) family, the natural resistance-associated macrophage protein (NRAMP) family, the zinc-/iron-regulated transporter-like protein (ZIP) family, the Ca^{2+}-sensitive cross complementer 1 (CCC1) family, and the iron-regulated protein (IREG) family (Borg et al. 2009). Darbani et al. (2015) reported vacuolar zinc transporter of Cation Diffusion Facilitator *Mtp*1 with higher expression in barley plants treated by zinc.

Manganese (Mn) is another essential mineral element for plants. Mn deficiency is a serious problem of crop productivity worldwide, which leads to reduced photosynthetic activity and lowers the lignin content, and other structural carbohydrates, ultimately hampering plant growth (Long et al. 2018). However, it is also essential for human health where it regulates the enzymes of glucose and lipid metabolism (Li and Yang 2018). Transporters for Mn absorption and Mn homeostasis are much less known for barley. However, Long et al. (2018) reported the role of the iron-regulated transporter 1 for absorption and transport of Mn in barley. Pedas et al. (2008) reported that *HvIRT1* contributes to genetic diversity in Mn kinetics.

In spite of the mineral uptake, some transporters are involved in mineral translocation activities. Fe deficiency activates Fe^{3+}-mugineic acid family phytosiderophores (MAs) transporter in barley (Murata et al. 2006) and rice (Inoue et al. 2009). These transporters are involved both in Fe uptake and translocation under Fe deficiency (Tsukamoto et al. 2009). ZIP family proteins have a specific role in Zn^{2+} uptake and translocations to the specific organelles of plants and are reported to have essential roles in rice and barley *OsZIP4* having an important role in the translocation of Zn^{2+} from roots to developing young leaves and in long-distance transport of Zn^{2+} between old and young leaves of rice. HvZIP proteins are also found to have very specific roles in translocation of Zn^{2+} to specific organelles in barley (Pedas et al. 2009).

9.14 Zinc

In developing countries, cereals comprise a large portion that contains a low amount of and has less bioavailability of zinc. The inadequate dietary intake of Zn signifies major health problems in the population (Cakmak and Kutman 2018). Zn acts as a prosthetic group for more than 3000 proteins (Sharma et al. 2013a, b) for their activity and thus is essential for growth and cell division (Brown et al. 2004). More than 25% of the human population is facing Zn deficiency across the world (Maret and Sandstead 2006). In animals, Zn is required for normal development and proper function of the immune system. Similarly, in plants, it plays a key role in vital developmental processes (Chattha et al. 2017). The foliar application of Zn on wheat at a late stage of growth recorded increased grain zinc concentration by 61% and 65% with foliar application of micronutrient cocktail (Zou et al. 2019). Wheat is inherently low in zinc concentration and high in phytate, which further limits zinc

Table 9.1 Biofortification of barley to increase the micronutrient concentration in grain with the application of mineral fertilizers under field conditions

Sr. no.	Biofortification of element	Agronomic practices	Success/level of improvement	References
1	Zn	Fertilizer application of ZnCHE	Zn content increased in grains up to 30%	Almendros et al. (2019)
		Application of $ZnSO_4$ foliar spray	33.12% increase in grain Zn content	Gonzalez et al. (2019)
2	Mn	Fertilizer application of ammonium sulfate ($(NH_4)_2SO_4$)	Mn content increased in grains up to 19.2%	Barczak et al. (2019)
3	Fe	Fertilizer application of ammonium sulfate ($(NH_4)_2SO_4$)	Fe content increased in grains up to 19.5%	Barczak et al. (2019)
4	Cu	Fertilizer application of ammonium sulfate ($(NH_4)_2SO_4$)	Cu content increased in grains up to 6.5%	Barczak et al. (2019)
5	Se	Sodium selenate foliar spray	Se concentration in grain increased up to 44 μg/kg dry weight	Rodrigo et al. (2013)
		Sodium selenite foliar spray	Se concentration in grain increased up to 9 μg/kg dry weight	Rodrigo et al. (2013)
		Soil application of Se at 10 to 20 g Se ha^{-1} as selenate	Barley grain Se concentration increase from 100 to 200 μg kg^{-1}	Ylaranta (1985)
		Soil application of 30–60 g ha^{-1} selenate and 4.5–10 g ha^{-1} foliar selenate	Increase in barley grain Se from 7 to 100 μg kg^{-1}	Ros et al. (2016)

bioavailability, which is the major reason for zinc malnutrition in humans where wheat is a staple food (Welch and Graham 2004). Chattha et al. (2017) recorded an increase in the wheat grain zinc concentration by 48.33% under the soil and foliar application and a 47.20% increase in only foliar application. Saha et al. (2017) reported an increase in the Zn concentration in rice by soil and foliar application. However, the loss of Zn on the processing of rice grains increased because of the preferential allocation of applied Zn into bran and aleurone of the rice grains. Despite such losses, Zn application increased bioavailability by 52.2% in cooked rice. Similarly, zinc-containing fertilizers have been used in increasing the Zn level in barley also (Yadav and Sharma 2018; Cakmak and Kutman 2018; Cakmak 2010; Gonzalez et al. 2019; Uddin et al. 2014). Further Watts-Williams and Cavagnaro (2018) used mycorrhizal fungi for Zn enrichment of barley which proved the agronomic approach as an efficient method for Zn fortifications in cereal crops (Table 9.1).

9.15 Iodine

Iodine is an essential micronutrient for humans, involved in functioning of the thyroid system. Salt iodization was inadequate to ensure global iodine adequacy as one-third of the world population may face hypothyroidism and iodine deficiency (Lyons 2018). Agronomic biofortification of cereal crops, which are consumed widely as a staple food, is an effective approach to reduce iodine deficiency (De Valenca et al. 2017). Zou et al. (2019) reported an increase in grain iodine concentration by 13.1-fold on foliar iodine spray; however increase of 10.3-fold was recorded in foliar micronutrient cocktail spray in wheat. Iodine in plants is transported mostly in xylem tissues; hence it is relatively easy to biofortify the leafy vegetable crops (Smolen et al. 2014). Comandini et al. (2013) reported vegetables biofortified with foliar I showed a high I stability during cooking. Till today there is no single report on iodine biofortification in barley.

9.16 Selenium

In the view of global health issues, selenium (Se) deficiency in the diet is the major problem as it is an essential element for mammals. Plants represent a major source of selenium as it is a beneficial element for them as an antioxidant and a growth promoter (Schiavon et al. 2020; Garcia-Banuelos et al. 2011). Methyl-selenocysteine (MeSeCys), the organic form of Se, appears to be a predominantly effective source of dietary Se; however, Se is incorporated as selenocysteine (SeCys) at the active site of selenoproteins involved in major metabolic pathways such as antioxidant defense and immune functions (Malagoli et al. 2015). The ability of the plants to accumulate and transform the Se into bioactive compounds tends to be an important implication for human nutrition. Se deficiency is a global problem, and plants are the essential source of dietary Se that can help to solve this problem (Garcia-Banuelos et al. 2011). Agronomic biofortification intends to enrich crops with Se (Schiavon et al. 2020); it can be done by applying the foliar application of sodium selenite or selenite fertilizers (Lidon et al. 2018).

Foliar application of Se was proved very effective to increase wheat grain Se from 90 to 338 μg kg^{-1} (Zou et al. 2019). Lyons et al. (2005) reported Se applied as sodium selenate at rates of 120 g Se/ha sprayed on the soil at seedling stage increased grain Se concentration up to 133-fold, while it increased up to 20-fold when applied as a foliar spray after flowering. The application of sodium selenite 25 g Se/ha was found to increase from 0.02 to 0.38 mg kg^{-1} in rice grains (Reis et al. 2018). Two-rowed barley was sprayed with different concentrations of sodium selenite; selenium level was found to be significantly increased in barley grain (Rodrigo et al. 2013). Sodium selenate application during the anthesis and germination stage of the malting enhanced level of Se in barley grains as well as in final beer products (Gibson et al. 2006).

9.17 Iron

Iron deficiency is the sixth most serious problem for human health in the world and is also known as hidden hunger (Masuda et al. 2020). Fe deficiency causes anemia, poor pregnancy outcomes, and lower immunity (Connorton and Balk 2019). Most of the world's population have monotonous diets consisting mainly of cereals predominantly starch-rich but nutrient-poor such as rice, corn, wheat, and the tubers like potato and cassava which are deficit in iron, which have affected two billion population globally (Connorton and Balk 2019; Sperotto et al. 2012). For humans, plants are the ultimate source of the Fe either directly as a staple food or indirectly from animal fodder. Fe fertilization to crops is not a very effective technique to enrich Fe in crops as Fe is insoluble in soil; therefore Fe biofortification is the most suitable alternative solution to enrich Fe in food grains. It is possible by generating cultivars that can efficiently mobilize, uptake, and translocate Fe to its edible parts (Sperotto et al. 2012). The foliar application of Fe takes 10–20 days to absorb 50% of the micronutrient as it is affected by different factors such as endogenous (leaf anatomy), exogenous (soil, pH), and environmental factors (Ludwig and Slamet-Loedin 2019). Therefore the foliar application to reach a significant enrichment in grain Fe for biofortification remains quite challenging. Dragicevic et al. (2016) demonstrated an increase in bioavailability of iron in barley after foliar spray of nonstandard fertilizers, hormonal growth stimulators.

Genetic engineering has been used in various crops to enrich mineral content like Fe and Zn. The transgenic strategies for the enrichment of Fe content have been focused on the intake and utilization efficiency of the plants by regulating and modulating the expression of the transporter (Kumar et al. 2019). Takahashi et al. (2001) reported enhanced iron uptake in low iron availability in soil by transgenic rice with two *naat* genes, coding for crucial enzymes for phytosiderophores. Drakakaki et al. (2000) reported that recombinant ferritin significantly increases iron in rice and wheat. However, the ferritin hyper-expressing rice lines were reported with a 30% higher iron rice (Qu et al. 2005). Such attempts can be made in barley to improve grain iron content in barley.

9.18 Biofortification for Antioxidants and Vitamins

Antioxidants are health-promoting molecules that nullify the reactive oxygen species and protect the cellular components and nucleic acids from oxidative damage. During the metabolic process and stress, reactive oxygen species and free radicals are produced; antioxidant molecules present in natural foods like fruits, vegetables, and grains scavenge these reactive oxygen species protecting our body (Zhu et al. 2013). Phytochemicals like flavonoids, carotenoids, phenolics, lignans, vitamins, minerals, and phytates present in food act as an antioxidant. Antioxidants are grouped into two categories: lipophilic (carotenoids, tocochromanols, coenzyme Q10, etc.) and hydrophilic (ascorbate, flavonoids, melatonin, etc.).

Vitamin E is one of the lipid-soluble antioxidants essential for human health. The seeds of most monocots contain the majority of vitamin E in the form of tocotrienols. Chen et al. (2017) produced transgenic barley overexpressing HvHGGT under endogenous D-hordein promoter (proHor) which led to an increase in tocotrienol content by 10–15% in seeds of transgenic lines. The radical scavenging activity of transgenic seed extracts was also enhanced by 17–18% over wild type. Similarly, other vitamins and antioxidant levels could be enhanced in barley by regulating individual rate-limiting steps or key branch points or modification of regulatory elements that may help in the biofortification of some antioxidants. Naqvi et al. (2009) developed modified three distinct metabolic pathways and developed multivitamin-rich corn. The levels of the β-carotene, ascorbate, and folate were increased in transgenic kernels by 169-fold, 6-fold, and 2-fold, respectively. Such attempts need to be done in barley to develop vitamin- and antioxidant-rich barley. Ascorbate or vitamin C is a potent water-soluble antioxidant. Hormones can also regulate ascorbate biosynthesis. The increased level of abscisic acid reduces the expression of NADPH oxidases which is the main producer of ROS in seeds (Ishibashi et al. 2017). Biofortification of durum wheat for provitamin A was performed using a tilling approach which resulted in an increase of roughly 75% in β-carotene in the grains (Sestili et al. 2019). The amino acid sequences of lycopene epsilon cyclase of wheat revealed that it has great homology with barley but differs from other cereals. Wicker et al. (2009) showed that the gene structure and order are strongly conserved in wheat and barley despite their divergence about 11 million years ago. Thus, this strategy for provitamin A biofortification could be used in barley.

9.19 Factors Affecting Biofortification

Several pre-harvest and postharvest factors affect the success of the biofortification program (Fig. 9.2).

9.20 Mineral-Deficient Soil

Mineral nutrient-deficient soil is the major factor affecting the biofortification in crop plants. More than 90% of the zinc (Zn) in the soils exists as an insoluble Zn, therefore poorly available to the plants (Singh 2011). The rapid absorption of Zn on clay minerals reduces the mobility of Zn in soil by making it unavailable to plants. However, in India, Arunachalam et al. (2013) reported that 49% of lands under cultivation are having Zn-deficient soils together with 12% deficiency in iron, 3% in copper, 5% in manganese, 33% in boron, and 13% in molybdenum. It limits the crop productivity and nutritional quality of the crops.

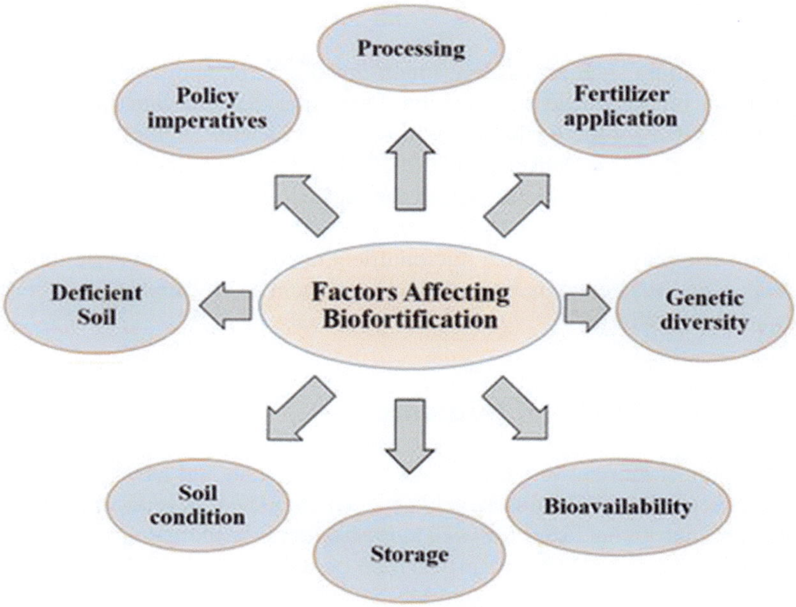

Fig. 9.2 Factors affecting the biofortification

9.21 Soil Condition

Soil pH, moisture status, organic matter content, salinity, and other factors affect micronutrient uptake by roots from the soil. A change in soil pH from 6 to 7 resulted in a 30-fold decrease in the chemical solubility of Zn in soil (Marschner 1993). Similarly, salinity (Cakmak 2008) and moisture stress also negatively affect the mineral availability and uptake by crop plants (Waters and Grusak 2008). Therefore, to achieve efficient and sustainable biofortification of crops, we need to maintain good fertility as well as physical-chemical properties of soil.

9.22 Fertilizer Application

The optimum amount of fertilizer containing the desired nutrient is needed to be applied before or during the growth of plants. These fertilizers can be applied directly to the soil or in the form of foliar applications. Micronutrient uptake was found to be increased when fertilizer containing micronutrients was supplied in combination with NPK and organic fertilizers (De Valenca et al. 2017). Therefore integrated practices are needed to follow to increase soil fertility to achieve successful biofortification using an agronomic approach.

9.23 Soil Microflora

The soil microflora such as bacteria, fungi, mycobacteria, and cyanobacteria helps the plants for precise nutrient acquisition. Increasing the soil microbial diversity is the best approach for the fortification of essential elements such as zinc, iron, and selenium in crop plants (Dapkekar et al. 2020). The long-term and excessive input of chemical fertilizers and pesticides cause chemicals to persist in the soil, which is so bound to affect the soil microflora (Prashar and Shah 2016). As discussed earlier soil microflora also plays a key role in micronutrient uptake by the root system from the soil. Therefore, we need to maintain rich microflora in soil by application of organic inputs and biofertilizers.

9.24 Bioavailability of Nutrients

Accumulation of nutrient in the crop is not enough; it should have good bioavailability character. Bioavailability largely depends upon the biochemical nature of the nutrient, anti-nutritional factors, as well as the health of the individual consuming the biofortified food; these factors can promote or delay the absorption of nutrients (Diaz-Gomez et al. 2017). Pfeiffer and McClafferty (2007) reported that 5% iron and 25% zinc are present in a bioavailable form in several crops. Phytate and phenolic compounds act as anti-nutritional factors that need to be considered during the biofortification of barley. Thus, the bioavailability of nutrients in biofortified crops needs to be assessed before large-scale adoption.

9.25 Storage and Processing

The stability of biofortified nutrients in the storage period is an important factor governing the success of biofortification. It has been reported that in provitamin A-biofortified maize, a large number of carotenoids were lost during storage (Mugode et al. 2014). The nutritional quality of barley was also reduced during the storage period by 1.74% and 2.82% (Polat 2015). Malting quality is known to increase in storage for 1 year.

A biofortified nutrient should be stored in grain endosperm to avoid or minimize losses during postharvest processing like pearling and milling. Generally, cereals are dehulled and polished which results in substantial losses of nutrients like minerals and vitamins (Raes et al. 2014; Dunn et al. 2014). Therefore, the accumulation of biofortified nutrients in endosperm with the help of genetic engineering will be useful in maintaining the quality of grain after milling (De Steur et al. 2015). β-glucan levels in barley were not affected after pearling as it is concentrated in endosperm; also thus even 30% pearling has no significant effect on the β-glucan content. Further, heat treatment during processing also adversely affects nutrients like vitamins and phenolics (Sharma and Gujral 2010). To take good benefit of biofortified crops, the storage and postharvest processing conditions should be optimized to maintain the higher level of nutrient in the processed final product.

9.26 Advantage of Biofortification

There are three main advantages of biofortification of crops: effective outreach, cost-effectiveness, and sustainability.

Biofortification of staple food will help to reach nutritious food to the targeted population as staple foods like cereals are a major part of their diet. Seeds of biofortified high-yielding varieties developed by research institutes and universities can be distributed to the targeted population or poor farmers at a reasonable cost to achieve the nutritional security of targeted people (Bouis et al. 2011). Several processed and fortified foods are available in the market but are unavailable to poor people due to high cost, lack of awareness, and lack of education.

Biofortification is a cost-effective strategy to lift the large population over the threshold from malnourishment to micronutrient sufficiency. A high-yielding variety is fortified with particular micronutrients like Fe and Zn; it will continuously produce nutritionally rich food for several generations in a cost-effective manner compared to processed and physically fortified food. The benefits of this biofortification will be far higher than the cost of development of biofortified variety.

Sustainability is another advantage of biofortification as once the gene or trait of interest is transferred in a variety, it will continue to be grown by farmers and consumed by the needy population year after year. There will be very less investment needed to monitor and maintain biofortified traits as compared to other fortification programs.

9.27 Conclusion

Hidden hunger or micronutrient malnutrition has a severe impact on the health of the population. Therefore, the prevention of micronutrient malnutrition is one of the major goals of scientists and policymakers worldwide. Thus biofortification of staple crops like cereals is considered a sustainable strategy for delivering nutritional food to target at-risk population.

Barley is the fourth most important cereal crop, and thus increasing its nutritional value by biofortification will play important role in the reduction of hidden hunger. Enrichment of barley with minerals like Zn, Fe, and Se and vitamins will help provide sufficient amount of trace elements and vitamins to the target population. Biofortification of barley with essential amino acids like lysine and β-glucan increased the nutritional quality and health-promoting nature of barley. Strategies for biofortification are based on breeding, agronomic practices, and genetic engineering that will help in addressing malnutrition. Transgenics is a promising method of biofortification but the acceptance and biosafety issues are the major hindrance. This hindrance can be overcome by the use of genome editing tools like CRISPR;

these tools are very precise and rapid and produce stable mutants for sustainable use. Further, enrichment with nutrients is not sufficient, the bioavailability of that nutrient is also an important factor, and it depends upon the intrinsic qualities of nutrients, food matrix, and health of the consumer. Adoption of biofortified varieties by the farmers is only possible if the cultivation of these varieties does not require additional inputs and has no yield penalty and if farmers get a premium price for their harvest. Finally, we need to run awareness programs for the target population about the benefits of these biofortified crops.

References

Almendros P, Obrador A, Alvarez JM, Gonzalez D (2019) Zn-DTPA-HEDTA-EDTA application: a strategy to improve the yield and plant quality of a barley crop while reducing the N application rate. J Soil Sci Plant Nutr 19(4):920–934

Arunachalam P, Kannan P, Prabukumar G, Govindaraj M (2013) Zinc deficiency in Indian soils with special focus to enrich zinc in peanut. Afr J Agric Res 8(50):6681–6688

Baik BK, Ullrich SE (2008) Barley for food: characteristics, improvement, and renewed interest. J Cereal Sci 48:233–242. https://doi.org/10.1016/j.jcs.2008.02.002

Barczak B, Jastrzębska M, Kostrzewska MK (2019) Biofortification of spring barley grain with microelements through sulfur fertilization. J Chem 2019:8214298

Barret M, Morrissey JP, O'Gara F (2011) Functional genomics analysis of plant growth-promoting rhizobacterial traits involved in rhizosphere competence. Biol Fertil Soils 47:729–743

Borg S, Brinch-Pedersen H, Tauris B, Holm PB (2009) Iron transport, deposition and bioavailability in the wheat and barley grain. Plant Soil 325(1–2):15–24

Bouis HE, Hotz C, McClafferty B, Meenakshi JV, Pfeiffer WH (2011) Biofortification: a new tool to reduce micronutrient malnutrition. Food Nutr Bull 32:S31–S40

Brown KH, Rivera JA, Bhutta ZA, Gibson RS, King JC, Lonnerdal B et al (2004) International Zinc Nutrition Consultative Group (IZiNCG) technical document no.1. Assessment of zinc deficiency in populations and options for its control. Food Nutr Bull 25:S99–S199

Burton RA, Collins HM, Kibble NA, Smith JA, Shirley NJ, Jobling SA, Henderson M, Singh RR, Pettolino F, Wilson SM, Bird AR (2011) Over-expression of specific HvCslF cellulose synthase-like genes in transgenic barley increases the levels of cell wall (1, 3; 1, 4)-β-d-glucans and alters their fine structure. Plant Biotechnol J 9(2):117–135

Cakmak I (2008) Enrichment of cereal grains with zinc: agronomic or genetic biofortification? Plant Soil 302(1–2):1–17. https://doi.org/10.1007/s11104-008-9584-6

Cakmak I (2010) Biofortification of cereals with zinc and iron through fertilization strategy. In: 19th world congress soil sci soil solutions changing world, vol 5, pp 1–6

Cakmak I, Kutman UÁ (2018) Agronomic biofortification of cereals with zinc: a review. Eur J Soil Sci 69(1):172–180

Chattha MU, Hassan MU, Khan I, Chattha MB, Mahmood A, Nawaz M, Subhani MN, Kharal M, Khan S (2017) Biofortification of wheat cultivars to combat zinc deficiency. Front Plant Sci 8: 281

Chen J, Liu C, Shi B, Chai Y, Han N, Zhu M, Bian H (2017) Overexpression of HvHGGT enhances tocotrienol levels and antioxidant activity in barley. J Agric Food Chem 65(25):5181–5187

Coccina A, Cavagnaro TR, Pellegrino E, Ercoli L, McLaughlin MJ, Watts-Williams SJ (2019) The mycorrhizal pathway of zinc uptake contributes to zinc accumulation in barley and wheat grain. BMC Plant Biol 19:133

Comandini P, Cerretani L, Rinaldi M, Cichelli A, Chiavaro E (2013) Stability of iodine during cooking: investigation on biofortified and not fortified vegetables. Int J Food Sci Nutr 64(7): 857–861

Combs GF (2001) Selenium in global food systems. Br J Nutr 85:517–547

Connorton JM, Balk J (2019) Iron biofortification of staple crops: lessons and challenges in plant genetics. Plant Cell Physiol 60(7):1447–1456

Dapkekar A, Deshpande P, Oak MD, Paknikar KM, Rajwade JM (2020) Getting more micronutrients from wheat and barley through agronomic biofortification. In: Wheat and barley grain biofortification. Elsevier, Amsterdam. https://doi.org/10.1016/B978-0-12-818444-8.00003-1

Darbani B, Noeparvar S, Borg S (2015) Deciphering mineral homeostasis in barley seed transfer cells at transcriptional level. PLoS One 10(11):e0141398. https://doi.org/10.1371/journal.pone.0141398

De Steur H, Blancquaert D, Strobbe S, Lambert W, Gellynck X, Van Der Straeten D (2015) Status and market potential of transgenic biofortified crops. Nat Biotechnol 33:25–29

De Valenca AW, Bake A, Brouwer ID, Giller KE (2017) Agronomic biofortification of crops to fight hidden hunger in sub-Saharan Africa. Glob Food Sec 12:8–14

Detterbeck A, Pongrac P, Rensch S, Reuscher S et al (2016) Spatially resolved analysis of variation in barley (*Hordeum vulgare*) grain micronutrient accumulation. New Phytol 211:1241–1254

Diaz-Gomez J, Twyman RM, Zhu C, Farré G, Serrano JC, Portero-Otin M et al (2017) Biofortification of crops with nutrients: factors affecting utilization and storage. Curr Opin Biotechnol 44:115–123

Distelfeld C, Uauy T, Fahima J (2006) Dubcovsky, physical map of the wheat high grain protein content gene Gpc-B1 and development of a high-throughput molecular marker. New Phytol 169:753–763

Distelfeld I, Cakmak Z, Peleg L, Ozturk AM, Yazici H, Budak Y, Saranga T (2007) Fahima, multiple QTL-effects of wheat Gpc-B1 locus on grain protein and micronutrient concentrations. Physiol Plant 129:635–643

Distelfeld A, Korol A, Dubcovsky J, Uauy C, Blake T, Fahima T (2008) Colinearity between the barley grain protein content (GPC) QTL on chromosome arm 6HS and the wheat Gpc-B1 region. Mol Breed 22(1):25–38

Dragicevic V, Nikolic B, Radosavljevic M, Djuric N, Dodig D, Stoiljkovic M et al (2016) Barley grain enrichment with essential elements by agronomic biofortification. Acta Periodica Technologica 47:1–9

Drakakaki G, Christou P, Stöger E (2000) Constitutive expression of soybean ferritin cDNA intransgenic wheat and rice results in increased iron levels in vegetative tissues but not in seeds. Transgenic Res 9(6):445–452

Dunn ML, Jain V, Klein BP (2014) Stability of key micronutrients added to fortified maize flours and corn meal. Ann N Y Acad Sci 1312:15–25

Duran P, Acuna JJ, Jorquera MA, Azcón R, Borie F, Cornejo P, Mora ML (2013) Enhanced selenium content in wheat grain by co-inoculation of selenobacteria and arbuscular mycorrhizal fungi: a preliminary study as a potential Se biofortification strategy. J Cereal Sci 57:275–280

Evenson RE, Gollin D (2003) Assessing the impact of the green revolution, 1960 to 2000. Science 300:758–762

Fan C, Zhai H, Wang H, Yue Y, Zhang M, Li J et al (2017) Identification of QTLs controlling grain protein concentration using a high-density SNP and SSR linkage map in barley (Hordeum vulgare L.). BMC Plant Biol 17(1):122

Garcia-Banuelos ML, Hermosillo-Cereceres MA, Sanchez E (2011) The importance of selenium biofortification in food crops. Curr Nutr Food Sci 7(3):181–190

Garg M, Sharma N, Sharma S, Kapoor P, Kumar A, Chunduri V, Arora P (2018) Biofortified crops generated by breeding, agronomy, and transgenic approaches are improving lives of millions of people around the world. Front Nutr 5:12. https://doi.org/10.3389/fnut.2018.00012

Gibson C, Park YH, Myoung KH, Suh MK, McArthur T, Lyons G, Stewart D (2006) The biofortification of barley with selenium. In: Proceedings of the Institute of Brewery & Distillating (Asia-Pacific Section) Carventron, pp 19–24

Gonzalez D, Almendros P, Obrador A, Alvarez JM (2019) Zinc application in conjunction with urea as a fertilization strategy for improving both nitrogen use efficiency and the zinc biofortification of barley. J Sci Food Agric 99(9):4445–4451

Gyawali S, Ottec ML, Jacob DL, Abderrazek J, Verma RPS (2019) Multiple element concentration in the grain of spring barley (*Hordeum vulgare* L.) collection. J Plant Nutr 42(9):1036–1046

Han F, Ullrich SE, Chirat S, Menteur S, Jestin L, Sarrafi A, Hayes PM, Jones BL, Blake TK, Wesenberg DM, Kleinhofs A (1995) Mapping of β-glucan content and β-glucanase activity loci in barley grain and malt. Theor Appl Genet 91(6–7):921–927

Holme IB, Dionisio G, Brinch-Pedersen H, Wendt T, Madsen CK, Vincze E, Holm PB (2012) Cisgenic barley with improved phytase activity. Plant Biotechnol J 10(2):237–247

Holme IB, Wendt T, Gil-Humanes J, Deleuran LC, Starker CG, Voytas DF, Brinch-Pedersen H (2017) Evaluation of the mature grain phytase candidate HvPAPhy_a gene in barley (*Hordeum vulgare* L.) using CRISPR/Cas9 and TALENs. Plant Mol Biol 95:111. https://doi.org/10.1007/s11103-017-0640-6

Hussain S, Rengel Z, Mohammadi SA, Ebadi-Segherloo A, Maqsood MA (2016) Mapping QTL associated with remobilization of zinc from vegetative tissues into grains of barley (Hordeum vulgare). Plant Soil 399(1–2):193–208

Ingra R, Amato G, Frenda AS, Giambalvo D (2019) Impacts of arbuscular mycorrhizal fungi on nutrient uptake, N2 fixation, N transfer, and growth in a wheat/faba bean intercropping system. PLoS One 14:e0213672

Inoue H, Kobayashi T, Nozoye T, Takahashi M, Kakei Y, Suzuki K, Nakazono M, Nakanishi H, Mori S, Nishizawa NK (2009) Rice OsYSL15 is an iron-regulated iron (III)-deoxymugineic acid transporter expressed in the roots and is essential for iron uptake in early growth of the seedlings. J Biol Chem 284(6):3470–3479

Ishibashi Y, Aoki N, Kasa S, Sakamoto M, Kai K, Tomokiyo R et al (2017) The interrelationship between abscisic acid and reactive oxygen species plays a key role in barley seed dormancy and germination. Front Plant Sci 8:275. https://doi.org/10.3389/fpls.2017.00275

Jones GD, Droz B, Greve P, Gottschalk P, Poffet D et al (2017) Selenium deficiency risk predicted to increase under future climate change. Proc Natl Acad Sci 114(11):2848–2853

Jun Y, Fang W, Haibo Q, Guoxiong C, Eviatar N, Fahima T, Jianping C (2011) Natural variation in grain selenium concentration of wild barley, *Hordeum spontaneum*, populations from Israel. Biol Trace Elem Res 142(3):773–786

Khan AG (2005) Role of soil microbes in the rhizospheres of plants growing on trace metal contaminated soils in phytoremediation. J Trace Elem Med Biol 18:355–364

Khandagale K, Nadaf A (2016) Genome editing for targeted improvement of plants. Plant Biotechnol Rep 10:327–343. https://doi.org/10.1007/s11816-016-0417-4

Kotz JC, Treichell PM, Weaver G (2006) Chemistry and chemical activity, 6th edn. Thomas Books, Cole

Ku YS, Rehman HM, Lam HM (2019) Possible roles of rhizospheric and endophytic microbes to provide a safe and affordable means of crop biofortification. Agronomy 9(11):764

Kumar S, Palve A, Joshi C, Srivastava RK (2019) Crop biofortification for iron (Fe), zinc (Zn) and vitamin A with transgenic approaches. Heliyon 5(6):e01914

Lange M, Vincze E, Wieser H, Schjoerring JK, Holm PB (2007) Suppression of C-hordein synthesis in barley by antisense constructs results in a more balanced amino acid composition. J Agric Food Chem 55(15):6074–6081

Li L, Yang X (2018) The essential element manganese, oxidative stress, and metabolic diseases: links and interactions. Oxidative Med Cell Longev 2018:1

Li J, Baga M, Rossnagel BG, Legge WG, Chibbar RN (2008) Identification of quantitative trait loci for β-glucan concentration in barley grain. J Cereal Sci 48(3):647–655

Li Y, Liu D, Zong Y, Jiang L, Xi X, Cao D et al (2020) New D hordein alleles were created in barley using CRISPR/Cas9 genome editing. Cereal Res Commun 48(2):131–138

Lidon FC, Oliveira K, Ribeiro MM, Pelica J, Pataco I, Ramalho JC, Leitao AE, Almeida AS, Campos PS, Ribeiro-Barros AI, Pais IP (2018) Selenium biofortification of rice grains and implications on macronutrients quality. J Cereal Sci 81:22–29

Long L, Persson DP, Duan F, Jorgensen K, Yuan L, Schjoerring JK, Pedas PR (2018) The iron-regulated transporter 1 plays an essential role in uptake, translocation and grain-loading of manganese, but not iron, in barley. New Phytol 217(4):1640–1653

Ludwig Y, Slamet-Loedin IH (2019) Genetic biofortification to enrich rice and wheat grain iron: from genes to product. Front Plant Sci 10:833

Lyons G (2018) Biofortification of cereals with foliar selenium and iodine could reduce hypothyroidism. Front Plant Sci 9:730

Lyons GH, Judson GJ, Ortiz-Monasterio I, Genc Y, Stangoulis JC, Graham RD (2005) Selenium in Australia: selenium status and biofortification of wheat for better health. J Trace Elem Med Biol 19(1):75–82

Malagoli M, Schiavon M, Pilon-Smits EA (2015) Effects of selenium biofortification on crop nutritional quality. Front Plant Sci 6:280

Maret W, Sandstead HH (2006) Zinc requirements and the risks and benefits of zinc supplementation. J Trace Elem Med Biol 20:3–18

Marschner H (1993) Zinc uptake from soils. In: Robson AD (ed) Zinc in soils, plants. Kluwer Academic, Dordrecht, pp 59–78

Martino E, Perotto S, Parsons R, Gadd GM (2003) Solubilization of insoluble inorganic zinc compounds by ericoid mycorrhizal fungi derived from heavy metal polluted sites. Soil Biol Biochem 35:133–141

Masuda H, Aung MS, Kobayashi T, Nishizawa NK (2020) Iron biofortification: the gateway to overcoming hidden hunger. In: The future of rice demand: quality beyond productivity. Springer, Cham, pp 149–177

Menguer PK, Vincent T, Miller AJ, Brown JK, Vincze E, Borg S, Holm PB, Sanders D, Podar D (2018) Improving zinc accumulation in cereal endosperm using *Hv MTP 1*, a transition metal transporter. Plant Biotechnol J 16(1):63–71

Mishra PK, Bisht SC, Ruwari P, Joshi GK, Singh G, Bisht JK, Bhatt JC (2011) Bio associative effect of cold tolerant Pseudomonas spp. and rhizobium leguminosarum-PR1 on iron acquisition, nutrient uptake and growth of lentil (Lens culinaris L.). Eur J Soil Biol 47:35–43

Morrell PL, Clegg MT (2007) Genetic evidence for a second domestication of barley (*Hordeum vulgare)* east of the Fertile Crescent. Proc Natl Acad Sci U S A 104:3289–3294

Mugode L, Ha B, Kaunda A, Sikombe T, Phiri S, Mutale R, Davis C, Tanumihardjo S, De Moura FF (2014) Carotenoid retention of biofortified provitamin A maize (*Zea mays* L.) after Zambian traditional methods of milling, cooking and storage. J Agric Food Chem 62:6317–6325

Mulualem T (2015) Application of biofortification through plant breeding to improve the value of staple crops. Biomed Biotechnol 3:11–19

Murata Y, Ma JF, Yamaji N, Ueno D, Nomoto K, Iwashita T (2006) A specific transporter for iron (III)–phytosiderophore in barley roots. Plant J 46(4):563–572

Naqvi S, Zhu C, Farre G, Ramessar K, Bassie L, Breitenbach J, Conesa DP, Ros G, Sandmann G, Capell T, Christou P (2009) Transgenic multivitamin corn through biofortification of endosperm with three vitamins representing three distinct metabolic pathways. Proc Natl Acad Sci 106(19): 7762–7767

Narwal S, Gupta OP, Pandey V, Kumar D, Ram S (2020) Effect of storage and processing conditions on nutrient composition of wheat and barley. In: Wheat and barley grain biofortification. Woodhead Publishing, Sawston, pp 229–256

Ohnoutkova L, Zitka O, Mrizova K, Vaskova J, Galuszka P, Cernei N, Smedley MA, Harwood WA, Adam V, Kizek R (2012) Electrophoretic and chromatographic evaluation of transgenic barley expressing a bacterial dihydrodipicolinate synthase. Electrophoresis 33(15):2365–2373

Oury FX, Berard P, Brancourt-Hulmel M, Depatureaux C, Doussineaux G, Galic N, Giraud A, Heumez E, Lecomte C, Pluchard P, Rousset M, Trottet M (2003) Yield and grain protein

concentration in bread wheat: a review and a study of multi-annual data from a French breeding program. J Genet Breed 57(1):59–68
Pedas P, Ytting CK, Fuglsang AT, Jahn TP, Schjoerring JK, Husted S (2008) Manganese efficiency in barley: identification and characterization of the metal ion transporter HvIRT1. Plant Physiol 148(1):455–466
Pedas P, Schjoerring JK, Husted S (2009) Identification and characterization of zinc-starvation-induced ZIP transporters from barley roots. Plant Physiol Biochem 47(5):377–383
Perez-Massot E, Banakar R, Gomez-Galera S, Zorrilla-Lopez U, Sanahuja G, Arjo G et al (2013) The contribution of transgenic plants to better health through improved nutrition: opportunities and constraints. Genes Nutr 8(1):29–41. https://doi.org/10.1007/s12263-012-0315-5
Pfeiffer WH, McClafferty B (2007) Harvest plus: breeding crops for better nutrition. Crop Sci 47: S88–S105
Polat HE (2015) Effects of different storage conditions on the nutritional qualities of barley and maize. J Agric Sci 21:246–255
Prasanna R, Nain L, Rana A, Shivay YS (2016) Biofortification with microorganisms: present status and future challenges. In: Biofortification of food crops. Springer, New Delhi, pp 249–262
Prashar P, Shah S (2016) Impact of fertilizers and pesticides on soil microflora in agriculture. In: Sustainable agriculture reviews. Springer, Cham, pp 331–361
Qu QL, Yoshihara T, Ooyama A, Goto F, Takaiwa F (2005) Iron accumulation does not parallel the high expression level of ferritin in transgenic rice seeds. Planta 222(2):225–233
Raes K, Knockaert D, Struijs K, Van Camp J (2014) Role of processing on bioaccessibility of minerals: influence of localization of minerals and anti-nutritional factors in the plant. Trends Food Sci Technol 37:32–41
Ramesh SA, Choimes S, Schachtman DP (2004) Over-expression of an Arabidopsis zinc transporter in *Hordeum Vulgare* increases short-term zinc uptake after zinc deprivation and seed zinc content. Plant Mol Biol 54:373–385. https://doi.org/10.1023/B:PLAN.0000036370.70912.34
Ramireddy E, Galuszka P, Schmülling T (2018) Zn-fortified cereal grains in field-grown barley by enhanced root cytokinin breakdown. Plant Signal Behav 13(11):e1530023
Rana A, Joshi M, Prasanna R, Shivay YS, Nain L (2012) Biofortification of wheat through inoculation of plant growth promoting rhizobacteria and cyanobacteria. Eur J Soil Biol 50: 118–126
Rayman MP (2000) The importance of selenium to human health. Lancet 356(9225):233–241
Reis HPG, de Queiroz Barcelos JP, Junior EF, Santos EF, Silva VM, Moraes MF, Putti FF, dos Reis AR (2018) Agronomic biofortification of upland rice with selenium and nitrogen and its relation to grain quality. J Cereal Sci 79:508–515
Rodrigo S, Santamaría O, López-Bellido FJ, Poblaciones MJ (2013) Agronomic selenium biofortification of two-rowed barley under Mediterranean conditions. Plant Soil Environ 59(3):115–120
Ros GH, Van Rotterdam AMD, Bussink DW, Bindraban PS (2016) Selenium fertilization strategies for bio-fortification of food: an agro-ecosystem approach. Plant Soil 404(1–2):99–112
Rout GR, Sahoo S (2015) Role of iron in plant growth and metabolism. Rev Agric Sci 3:1–24
Saha S, Chakraborty M, Padhan D, Saha B, Murmu S, Batabyal K, Seth A, Hazra GC, Mandal B, Bell RW (2017) Agronomic biofortification of zinc in rice: influence of cultivars and zinc application methods on grain yield and zinc bioavailability. Field Crop Res 210:52–60
Sasaki A, Yamaji N, Ma JF (2016) Transporters involved in mineral nutrient uptake in rice. J Exp Bot 67(12):3645–3653
Sazawal S, Dhingra U, Dhingra P, Dutta A, Deb S, Kumar J, Devi P, Prakash A (2018) Efficacy of high zinc biofortified wheat in improvement of micronutrient status, and prevention of morbidity among preschool children and women—a double masked, randomized, controlled trial. Nutr J 17(1):86
Schiavon M, Nardi S, Dalla Vecchia F, Ertani A (2020) Selenium biofortification in the 21st century: status and challenges for healthy human nutrition. Plant Soil 453:1–26

Sestili F, Garcia-Molina MD, Gambacorta G, Beleggia R, Botticella E, De Vita P, Savatin DV, Masci S, Lafiandra D (2019) Provitamin A biofortification of durum wheat through a TILLING approach. Int J Mol Sci 20(22):5703. https://doi.org/10.3390/ijms20225703

Sharma P, Gujral HS (2010) Antioxidant of polyphenol oxidase activity of germinated barley its milling fractions. J Food Chem 120:673–678

Sharma A, Patni B, Shankhdhar D, Shankhdhar SC (2013a) Zinc—an indispensable micronutrient. Physiol Mol Biol Plants 19:11–20

Sharma A, Shankhdhar D, Shankhdhar SC (2013b) Enhancing grain iron content of rice by the application of plant growth promoting rhizobacteria. Plant Soil Environ 59:89–94

Sikdar MSI, Bowra S, Schmidt D, Dionisio G, Holm PB, Vincze E (2016) Targeted modification of storage protein content resulting in improved amino acid composition of barley grain. Transgenic Res 25:19–31

Simmonds NW (1995) The relation between yield and protein in cereal grain. J Sci Food Agric 67: 309–315

Singh MV (2011) Assessing extent of zinc deficiency for soil fertility mapping and nutrition security in humans and animals. Indian J Fertilizer 7:36–43

Smolen S, Sady W, Ledwozyw-Smolen I, Strzetelski P, Liszka-Skoczylas M, Rożek S (2014) Quality of fresh and stored carrots depending on iodine and nitrogen fertilization. Food Chem 159:316–322

Sperotto RA, Ricachenevsky FK, de Abreu Waldow V, Fett JP (2012) Iron biofortification in rice: it's a long way to the top. Plant Sci 190:24–39

Sreenivasulu N, Graner A, Wobus U (2008) Barley genomics: an overview. Int J Plant Genomics 2008:486258

Sullivan P, Arendt E, Gallagher E (2013) The increasing use of barley and barley by-products in the production of healthier baked goods. Trends Food Sci Technol 29:124–134

Takahashi M, Nakanishi H, Kawasaki S, Nishizawa NK, Mori S (2001) Enhanced tolerance of rice to low iron availability in alkaline soils using barley nicotianamine aminotransferase genes. Nat Biotechnol 19(5):466–469

Tariq M, Hameed S, Malik KA, Hafeez FY (2007) Plant root associated bacteria for zinc mobilization in rice. Pak J Bot 39:245–253

Tauris B, Borg S, Gregersen PL, Holm PB (2009) A roadmap for zinc trafficking in the developing barley grain based on laser capture microdissection and gene expression profiling. J Exp Bot 60(4):1333–1347

Tiwari V (2010) Growth and production of barley. In: Verheye W (ed) Soils, plant growth and crop production, Encyclopedia of life support systems (EOLSS), vol 1. UNESCO-EOLSS, Oxford, pp 267–297. http://www.eloss.net

Tsukamoto T, Nakanishi H, Uchida H, Watanabe S, Matsuhashi S, Mori S, Nishizawa NK (2009) 52Fe translocation in barley as monitored by a positron-emitting tracer imaging system (PETIS): evidence for the direct translocation of Fe from roots to young leaves via phloem. Plant Cell Physiol 50(1):48–57

Uddin MN, Kaczmarczyk A, Vincze E (2014) Effects of Zn fertilization on Hordein transcripts at early developmental stage of barley grain and correlation with increased Zn concentration in the mature grain. PLoS One 9(9):e108546. https://doi.org/10.1371/journal.pone.0108546

van de Wouw M, Kik C, van Hintum T, van Treuren R, Visser B (2010) Genetic erosion in crops: concept, research results and challenges. Plant Genet Resour 8:1–15

Velu G, Ortiz-Monasterio I, Cakmak I, Hao Y, Singh RÁ (2014) Biofortification strategies to increase grain zinc and iron concentrations in wheat. J Cereal Sci 59(3):365–372

Waters BM, Grusak MA (2008) Whole-plant mineral partitioning throughout the life cycle in *Arabidopsis thaliana* ecotypes Columbia, *Landsberg erecta,* Cape Verde Islands, and the mutant line ysl1ys. New Phytol 177:389–405

Watts-Williams SJ, Cavagnaro TR (2018) Arbuscular mycorrhizal fungi increase grain zinc concentration and modify the expression of root ZIP transporter genes in a modern barley (*Hordeum vulgare*) cultivar. Plant Sci 274(2018):163–170

Welch RM, Graham RD (2004) Breeding for micronutrients in staple food crops from a human nutrition perspective. J Exp Bot 55(396):353–364

WHO (2016) Vitamin and mineral nutrition information system. World Health Organization. www.who.int

Wicker T, Taudien S, Houben A, Keller B, Graner A, Platzer M, Stein N (2009) A whole-genome snapshot of 454 sequences exposes the composition of the barley genome and provides evidence for parallel evolution of genome size in wheat and barley. Plant J 59:712–722. https://doi.org/10.1111/j.1365-313X.2009.03911.x

Wiegmann M, Thomas WTB, Bull HJ, Flavell AJ, Annette Zeyner A et al (2019) Wild barley serves as a source for biofortification of barley grains. Plant Sci 283:83–84

Xue WT, Gianinetti A, Wang R, Zhan ZJ, Yan J, Jiang Y, Fahima T, Zhao G, Cheng JP (2016) Characterizing barley seed macro- and micro-nutrients under multiple environmental conditions. Cereal Res Comm 44(4):639–649

Yadav N, Sharma YK (2018) Enhancement of Zn density in barley (*Hordeum vulgare*) grain: a physiological approach. Asian J Plant Sci Res 8(4):13–17

Yadava DK, Hossain F, Mohapatra T (2018) Nutritional security through crop biofortification in India: status & future prospects. Indian J Med Res 148(5):621

Yan J, Wang F, Yang R, Xiao T, Fahima T et al (2012) Natural variation in grain iron and zinc concentrations of wild barley, Hordeum spontaneum, populations from Israel. In: Zhang G, Li C, Liu X (eds) Advance in barley sciences. Proceedings of 11th international barley genetics symposium. Springer, Cham, pp 169–183

Ylaranta T (1985) Increasing the selenium content of cereals and grass crops in Finland. Academic dissertation. Agricultural Research Centre, Institute of Soil Science, Jokioinen, p 72

Zamir D (2001) Improving plant breeding with exotic genetic libraries. Nat Rev Genet 2:983–989

Zeng YW, Du J, Yang XM, Pu XY, Wang LX, Yang JZ, Du LJ, Yang T, Yang SM, Sun ZH (2016) Identification of quantitative trait loci for mineral elements in grains and grass powder of barley. Genet Mol Res 15(4)

Zeng Z, Han N, Liu C, Buerte B, Zhou C, Chen J et al (2020) Functional dissection of HGGT and HPT in barley vitamin E biosynthesis via CRISPR/Cas9-enabled genome editing. Ann Bot 126: 929

Zhu C, Naqvi S, Gomez-Galera S, Pelacho AM, Capell T, Christou P (2007) Transgenic strategies for the nutritional enhancement of plants. Trends Plant Sci 12:548–555. https://doi.org/10.1016/j.tplants.2007.09.007

Zhu C, Sanahuja G, Yuan D, Farré G, Arjó G, Berman J, Zorrilla-López U, Banakar R, Bai C, Pérez-Massot E, Bassie L (2013) Biofortification of plants with altered antioxidant content and composition: genetic engineering strategies. Plant Biotechnol J 11(2):129–141

Zou C, Du Y, Rashid A, Ram H, Savasli E, Pieterse PJ, Ortiz-Monasterio I, Yazici A, Kaur C, Mahmood K, Singh S (2019) Simultaneous biofortification of wheat with zinc, iodine, selenium, and iron through foliar treatment of a micronutrient cocktail in six countries. J Agric Food Chem 67(29):8096–8106

Biofortification of Sorghum (*Sorghum bicolor*) 10

Gunjan Guleria, Maneet Rana, Parichita Priyadarshini, Rahul Kaldate, Neeraj Kumar, Rachna Rana, Ramesh Chauhan, Rahul Gajghate, and Shahid Ahmed

Abstract

Malnutrition due to intake of nutritionally poor food is a serious problem among the developing nation. It affects the lives of around two billion people globally, of which most are children and women of reproductive age. Biofortification serves as an excellent, feasible, and cost-effective tool to meet the micronutrient requirement of the populations with limited access to nutrient-rich diets. This strategy not only increases the concentration of essential micronutrients but also enhanced their bioavailability. Sorghum is an important crop of arid and semiarid regions of the world and feeds the poor population of underprivileged countries. Its C_4 nature with intrinsic high photosynthetic rate and biomass potential makes it more tolerant to adverse environmental stresses like heat and drought. Being the cheapest source of micronutrients, it is the most preferred crop for biofortification. Current chapter reviews the nutritional importance of sorghum along with various techniques including agronomic, breeding, transgenic, and genome editing approaches to augment the desired micronutrient in the crop. The

G. Guleria (✉)
Department of Agronomy and Agrometeorology, Rani Lakshmi Bai Central Agricultural University, Jhansi, UP, India

M. Rana · P. Priyadarshini · N. Kumar · R. Gajghate · S. Ahmed
Division of Crop Improvement, ICAR-Indian Grassland and Fodder Research Institute, Jhansi, UP, India

R. Kaldate
Department of Agricultural Biotechnology, Assam Agricultural University, Jorhat, India

R. Rana
Amity Institute of Organic Agriculture, Amity University, Noida, UP, India

R. Chauhan
CSIR-Institute of Himalayan Bioresource Technology, Palampur, HP, India

R. Deshmukh et al. (eds.), *Biofortification in Cereals*,
https://doi.org/10.1007/978-981-19-4308-9_10

limitations and the areas that needed intervention are also discussed along with the challenges that biofortified sorghum holds to address the malnutrition.

Keywords

Agronomic · Breeding · Biofortification · Hidden hunger · Micronutrients · Sorghum

10.1 Introduction

Micronutrient deficiency leading to malnutrition is a worldwide problem but more rampant in developing nations (Ruel-Bergeron et al. 2015). It is estimated that around two billion people across the globe suffer from malnutrition (Hodge 2016; Sumithra et al. 2013). Women along with children below 5 years are most affected (Bailey et al. 2015). This situation will be worsened with the addition of 83–132 million people because of the COVID-19 pandemic, which illustrates the sensitivity of fragile food and agriculture systems. Until now the agriculture system focused only on increasing yield and productivity which currently needs a shift in producing crops having adequate amounts of micronutrients. This will help in countering the effect of micronutrient malnutrition among the population (Khush et al. 2012). Therefore, fortification of crops with essential micronutrients is prerequisite to curb malnutrition among the target population. Biofortification is the nutritional enrichment of food crops with increased bioavailability to humans and can be developed by conventional plant breeding, modern biotechnological techniques, and agronomic practices. It offers a sustainable and long-term solution for human consumption as well as fodder crop for animals (Meenakshi et al. 2010; Hefferon 2016).

Sorghum (*Sorghum bicolor* (L.) Moench) is the fifth most important cereal crop in relation to area and production. It is a crucial and staple food crop for habitants living in semiarid areas of the world, particularly in West Africa, sub-Saharan Africa, and semiarid zones of South Asia (Kumar et al. 2013a, 2013b), while in developed countries, it is used as livestock feed along with several industrial uses. It is cultivated in more than 100 countries with the USA, Mexico, Sudan, India, Nigeria, Niger, Ethiopia, Australia, China, and Brazil together contributing 77% of the world's total production (Aruna and Cheruku 2019). In 2019, about 59 million tonnes of sorghum were produced in the world, with an average yield of ~1.49 MT/ha (FAOSTAT 2020). Table 10.1 presents the details of economic status, hunger index, child mortality, human population, and sorghum production of major sorghum-growing countries worldwide. The child mortality status (deaths per 1000 live births) of top sorghum-growing countries is in the range of 3–13% and the hunger index in the range of moderate (10–19.9) to serious (20–34.9). This necessitates it to be biofortified to curb malnutrition and ultimately lower child mortality and hunger index rate.

Table 10.1 Details of economic status and hunger index in major sorghum-growing countries worldwide

Country	Population (thousands)[a]	Sorghum production (tonne)[b]	GDP per capita (current USD)[c]	Hunger index[d]	Child mortality[e]
Nigeria	206,140	6,665,000	2208.50	29.20	129.13
Ethiopia	114,964	5,265,580	917.88	26.20	55.00
Sudan	43,849	3,714,000	713.79	27.20	66.92
China	1,439,324	3,602,268	11712.85	<5	8.40
India	1,380,004	3,475,410	2030.62	27.20	36.00
Brazil	212,559	2,672,245	6728.17	<5	12.26
Niger	24,207	1,896,638	567.40	–	125.98
Burkina Faso	20,903	1,871,791	850.79	25.80	106.35
Argentina	45,196	1,601,435	9095.10	5.30	9.68
Mali	20,251	1,511,110	992.31	22.90	106.16
Cameroon	26,546	1,216,926	1657.08	19.10	75.70
Chad	16,426	972,516	710.18	44.70	131.68
Bolivia	11,673	949,039	3618.18	14.00	28.58
United Republic of Tanzania	59,734	731,877	1132.13	25.00	66.28
Uganda	45,741	400,000	971.28	–	72.54
Yemen	29,826	230,766	572.56	–	47.48

[a]World Population Prospects 2019
[b]FAOSTAT 2021
[c]World Economic Outlook Database 2021
[d]Global Hunger Index 2020
[e]UN IGME Database 2020

Sorghum is remarkably unique and vital for poor farmers because of its acclimation under drought and heat-prone environments. Being a C_4 plant with the ability to adapt in soils with low phosphorus availability, it is more attractive among farmers of arid and semiarid regions (Leiser et al. 2012; Haussmann et al. 2012). Apart from the food source, sorghum has several other uses such as feed, forage, fuel, and beverage and has phytoremediation potential (Liu et al. 2020) and therefore can be categorized as grain, forage, sweet, and broom type (Batey 2017). Sorghum grain is enriched with starch, protein, micronutrients, and crude fibers (Chavan and Patil 2010) and thus has the potential to provide more than half of the dietary micronutrients to families with low income (Rao et al. 2006, 2010).

Sorghum crop has been utilized for biofortification with various micronutrient concentration strategies, like provitamin A (beta-carotene) by expressing *Homo188-A* gene (Lipkie et al. 2013); enhanced protein content by expressing high lysine protein (Zhao et al. 2003); and digestibility improvement by silencing the *γ-kafirin* using RNAi (Grootboom et al. 2014; Elkonin et al. 2016). There is great interest in sorghum biofortification for Fe and Zn (Pfeiffer and McClafferty 2007; Zhao. 2008), and in line with this in 2018, India released its first biofortified sorghum variety ICSR 14001 also called 'Parbhani Shakti' via conventional breeding technique, having iron (Fe) and zinc (Zn) concentration of 45 ppm and 32 ppm, respectively. Besides this, it has higher protein (11.9%) and low phytate content (4.14 mg/100 g) ("http://www.icrisat.org/india-gets-its-first-biofortified-sorghum"). This chapter reviews the role of various micronutrients in the human diet along with their augmentation in sorghum using various biological activities, such as classical plant breeding, agronomic biofortification, and genetic engineering, or with the latest genome editing tools.

10.2 Top Priorities for Sorghum Biofortification

Nearly 0.5 billion people in 30 nations consume sorghum as a cereal crop (Kumar et al. 2013a, 2013b). It is a highly heat- and drought-tolerant crop with good nitrogen use efficiency. In addition, it is one of the low-cost sources of energy, protein, fat, carbohydrates, Fe, and Zn (Kumar et al. 2015). Further, its gluten-free nature (Ciacci et al. 2007), low glycemic index, and antioxidant properties make it a favorable diabetic cereal (Serna-Saldivar and Espinosa-Ramírez 2019). It supplies more than 50 percent of the dietary micronutrients to rural peoples with low income (Rao et al. 2006; Rao et al. 2010). However, some studies reported limited mineral content and bioavailability in cooked grains of sorghum (Kayodé et al. 2006). Human needs micronutrients for their proper metabolic needs and to stay healthy; however, their deficiency causes malnutrition or hidden hunger. These deficiencies can be overcome by various ways, viz., genetic ways to improve nutrient content, by taking supplements, dietary diversification, and biofortification. Among all these, biofortification is the only cost-efficient and sustainable method to eradicate this malnutrition.

10.2.1 Essential Micronutrients/Metals: Zn and Fe

Soil micronutrient deficiencies affect crop productivity more prominently due to the higher use of chemical fertilizers (Sanchez and Swaminathan 2005). Micronutrients like Fe and Zn are important for the human body but are found to be deficient in human diets these days. More than two billion people are affected due to micronutrient deficiencies, mostly from low-income families in developing nations in which Fe, Zn, vitamin A, and B9 deficiencies are most common (Kennedy et al. 2003; Bailey et al. 2015). It has been reported that during 2008, more than 4.5 lakh children below 5 years died because of diarrhea caused by Zn deficiency (Black et al. 2008). Its deficiency also leads to pneumonia and dwarfism in children (Cakmak et al. 1999; Walker et al. 2009). Zinc is an important micronutrient for growth and development having a role in plant vital functions (photosynthesis and respiration) and is also important for the nutritional value of feed and food-based plant products (Epstein and Bloom 2005).

Fe deficiency is the most prominent and prevalent nutrient deficiency. As per WHO, Fe deficiency affected 38%, 29%, and 43% of pregnant women, non-pregnant women, and preschool children, respectively (WHO 2017). It is assessed that approximately 50% cases of anemia are caused due to inadequate Fe intake and are responsible for poor immunity and lower pregnancy outcomes (Stevens et al. 2013; WHO 2017). It resulted in impaired cognitive development, low productivity, and growth retardation, with complications in health and economic systems (Bailey 2015). Anemia disease is becoming a serious issue these days worldwide. Both zinc and iron deficiency among children create a threat to the physical and mental health of human beings (Bains et al. 2015).

10.2.2 Basic Micronutrients: Selenium and Iodine

Selenium (Se) and iodine (I) are basic micronutrients not essential for plant metabolism. Both of these micronutrients are basically required for humans and animals and that's why these should be present in the diet. Se is important for human health as it has a vital role in the brain, thyroid, gonads, and heart. It has antioxidant, antibacterial, anticancer, and antiviral activities (Lyons et al. 2009), which thus helps in fighting cancer, reducing asthma symptoms, improving immunity, and reducing skin disorders. Its deficiency causes chronic diseases. Selenium also provides stimulating effect on plant growth and development (Abbas 2012). Minimal concentrations of selenium give a favorable effect on growth and development and also increase antioxidative capacity by increasing stress tolerance (Kong et al. 2005). In addition, selenium also helps in protecting DNA against damage and slows down aging of cells.

Iodine is also very important for human life; it helps in preventing chronic diseases. Iodine deficiency among human beings is a big problem nowadays. It is highly common among people from both developing and developed nations (Cakmak et al. 1999). The recommended dietary allowance (RDA) of I for adults is 150–200 μg per day and for lactating or pregnant females is 230 to 260 μg per day (Lawson et al. 2015). However, due to its insufficiency, it causes different health-related problems like goiter, mental disability, growth retardation, and increased miscarriage and infant mortality (Pearce et al. 2013; Lazarus 2015). Earlier reports showed that even mild iodine deficiency affects pregnant women and is associated with cognitive impairment in their children (Pearce et al. 2013). There are so many methods to overcome these deficiency problems. Biofortification of plants is one of the best methods to overcome I and Se deficiency in humans and animals (Smoleń et al. 2016).

10.2.3 Provitamins

Micronutrient malnutrition, mainly the consequence of poor bioavailability of vitamins and minerals in diets, causes blindness, anemia, beriberi, pellagra, scurvy,

Table 10.2 Comparative profiling of sorghum grains with other cereal grains

	Sorghum	Rice	Wheat	Maize	Barley	RDA
Minerals						
Iron (mg)	3.36	0.20	3.71	1.74	2.68	7–18
Zinc (mg)	1.67	0.49	2.96	2.24	2.0	3–11
Calcium (mg)	13.0	10.0	33.0	5.0	32.0	700–1300
Selenium (μg)	12.2	7.5	12.7	2.2	37.7	20–55
Iodine (μg)	0	0	0	0	0	90–250
Provitamins						
Vitamin A (IU)	0	0	0	214	0	300–900
Thiamin; B1 (mg)	0.332	0.02	0.297	0.16	0.37	0.5–1.2
Riboflavin; B2 (mg)	0.096	0.013	0.188	0.23	0.114	0.5–1.3
Niacin; B3 (mg)	3.69	0.4	5.35	2.6	6.27	6.0–16
Pantothenic acid; B5 (mg)	0.367	0.39	1.01	0.55	0.145	6.0–16
Pyridoxine; B6 (mg)	0.443	0.093	0.191	0.47	0.396	0.5–1.7
Folate; B9 (μg)	20.0	3.0	28.0	19.0	8.0	150–400
Vitamin B12	0	0	0	0	0	0.9–2.4
Vitamin C	0	0	0	0	0	15–90
Vitamin D	0	0	0	0	0	15–20
Alpha-tocopherol; E (mg)	0.5	0.04	0.53	0.49	0.57	6.0–15
Vitamin K1 (μg)	0	0	1.9	0.3	2.2	30–120
Beta-carotene (μg)	0	0	5	97	0	–
Lutein + zeaxanthin (μg)	0	0	220	1355	160	–
Protein						
Protein (g)	10.6	2.69	9.61	8.75	10.5	13–56

Data was obtained from USDA database; Dietary Reference Intakes 2011, 2019 [children (1–8 yrs), male and female (8 to >70)]

and rickets in more than 50% of the global population, particularly common among pregnant and lactating women and preschool children (Underwood 2000; Welch and Graham 2004; Asensi-Fabado and Munné-Bosch 2010). It has been reported that vitamin A deficiency leads to the blindness of up to 5 lakh children and the death of 6 lakh women because of pregnancy complications, which can be reduced by the consumption of vitamin A-enriched diets.

Sorghum grain contains several vitamins such as thiamine (vitamin B1), riboflavin (vitamin B2), niacin (vitamin B3), pantothenic acid (vitamin B5), pyridoxine (vitamin B6), and folate (vitamin B9) (Table 10.2). But in the RDA, these vitamins are insufficient to supply the nutritional requirement for children and adults; thus these need to be augmented using various biofortification approaches, viz., conventional or classic plant breeding and agronomic and genetic engineering. Further, some important vitamins such vitamin A, vitamin B12, and vitamins C, D, and E are deficient in sorghum grains but can be enhanced using genetic engineering or transgenic approaches.

10.2.4 Proteins

Sorghum grain lacks gluten content; thus, it serves as a good source of protein for gluten-sensitive individuals. As per data obtained from the USDA database, sorghum grains contain 10.6 g of proteins. However, RDA for children (1–8 yr) is 13–19 g, for males (9 to >70 yr), it is 34–56 g, for females (9 to >70 yr), it is 34–56 g, and during pregnancy 71 g per day is recommended. To combat this malnutrition, protein biofortification is an important and sustainable measure to enhance its bioavailability in staple plant foods (Taylor and Taylor 2011). Lower digestibility of seed storage proteins (SSP) and starch and lower nutritional grain value are some important factors to be addressed for sorghum biofortification along with other micronutrients. These seed storage proteins have low lysine and threonine content which are among the essential amino acids (Mudge et al. 2016). Further, most sorghum food is cooked or heated during preparation; this heat treatment resulted in up to 50% reduced digestibility compared to other cereal grains.

In recent years, biofortification of sorghum grains with proteins has been accomplished with the help of genomic tools, chemically induced mutations, and genetic engineering. These include impaired synthesis of kafirins (Mehlo et al. 2013); identification of natural allelic variants of kafirins (Mudge et al. 2016; Laidlaw et al. 2010; Cremer et al. 2014); increasing the lysine content (Zhao et al. 2003); silencing of γ- and/or α-kafirins genes (Kumar et al. 2012, Grootboom et al. 2014, Elkonin et al. 2016); and knockout strategy (Li et al. 2018).

10.3 Agronomic Biofortification of Sorghum

Biofortification mitigates the hidden hunger by increasing micronutrient concentrations and bioavailability in the food grain crops (Wakeel et al. 2018). Enhancement of Fe and Zn concentration in the grains of particular crops through application of Fe- and Zn-containing fertilizers is known as agronomic biofortification. Zuo and Zhang (2009) reported that any method that could intensify the root growth can result in higher uptake from the soil and can play a vital role in biofortification. Agronomic practices are also known as pre-harvest practices which enhance the nutrient content in crops. Food is categorized as biofortified if practices are pre-harvest and otherwise categorized as fortified if practices are postharvest. Some important agronomic biofortification approaches are the soil or foliar application of organic fertilizer, inorganic fertilizers, and biofertilizer and nutrient priming. For different mineral micronutrients, soil and foliar application of micronutrient fertilizer were found to be effective. Foliar application is found to be highly effective for zinc and selenium as Zn responds best and quick by agronomic biofortification methods mainly for cereal crops (Cakmak 2014). Organic manures, those that contain animal or plant sources such as vermicompost, farmyard and poultry manure, etc., are considered as a cost-efficient, environment-friendly alternative approach for inorganic synthetic fertilizers. Organic manures help in maintaining soil fertility. For instance, vermicompost increases Zn and Fe content by 4% and 7%, respectively, in barley crops (Maleki et al. 2011); poultry manure increases Fe content in rice and wheat by 10 and 15%, respectively (Ramzani et al. 2016, 2017). In addition to these, biofertilizers consisting of microorganisms enhance the productivity and growth of plants by increasing the supply or availability of nutrients (Barbosa et al. 2015; Bhardwaj et al. 2014). These include mycorrhizal fungi, blue-green algae, and cyanobacteria. Some biofertilizers like cyanobacteria were used in the Zn biofortification of wheat (Prasanna et al. 2015). Apart from these, nutri-priming is another approach whereby seeds were treated in micro- and macronutrient solutions before sowing (Farooq et al. 2011, 2019). With nutri-priming only grain zinc content of chickpea was increased by 29% (Farooq et al. 2019), thus considered to be low-cost method for nutrient enrichment (Poblaciones and Rengel 2016).

Only few studies use agronomic approaches to biofortify sorghum (Table 10.3). These include work of Mishra et al. (2015) who recommended the production of micronutrients (Fe and Zn) in post-rainy sorghum cultivar. Further, when soil application of $ZnSO_4$ + $FeSO_4$ (50 kg/ha of each) followed by foliar application (0.50% + 1.0%) was applied to Phule Maulee cultivar of sorghum at 45 DAS, Zn content of plant was increased up to 37.79–37.85 mg/kg along with increased green fodder yield and quality of fodder (Ahmad et al. 2018). More recently, application of a recommended dose of fertilizer with enriched vermicompost [(50 kg vermicompost/ha + 15 kg ZnSO4/ha) + (50 kg vermicompost/ha + 15 kg FeSO4/ha)] was also shown to increase the contents of Zn and Fe (Kumar and Kubsad 2017; Maganur and Kubsad 2020). In addition to these, high Zn content was obtained in CSV-31 genotype of sorghum using soil application of $ZnSO_4$ @ 50 kg/ha + foliar spray @ 0.2% at the knee-high stage and flowering stage (Markole et al. 2020).

10.4 Breeding Efforts for Sorghum Biofortification

Previous trait inheritance studies indicated that Fe and Zn are multigenic traits, hence controlled by many genes (Gregorio et al. (2000) in rice; Distelfeld et al. (2007) in wheat; Lonergan et al. (2009) in barley; Lung'aho et al. (2011) in maize). Thus, to map such genes, quantitative trait loci (QTL) mapping strategy is employed. In this strategy, contrasting parents for the trait of interest say lines with high and low Fe or Zn content will be crossed to generate F_1 hybrid. Then mapping population segregating for these traits will be developed after continuous selfing for over seven to eight generations. After this, phenotyping and genotyping of this population lead to the mapping of traits of interest. Using this strategy, Kotla et al. (2019) have recently mapped Fe and Zn QTLs in F_6 recombinant inbred line (RIL) population of sorghum developed by the crossing of 296B x PVK 801 contrasting parents. The details of QTLs, their marker interval, LOD score, and percent phenotypic variance explained have been summarized in Table 10.4.

The other strategy for mapping QTLs is through genome-wide association studies (GWAS). In this, breeders can directly use a natural structured population or sorghum germplasm for mapping of these genes (Fe, Zn, Se, I, etc.) using an association mapping panel. Association mapping panel or core collection can be

Table 10.3 List of significant studies performed for the sorghum biofortification

Biofortification trait	Agronomic practices	Success/level of improvement	Other note	Reference
High zinc	Fertilization	Soil application of $ZnSO_4$ @ 50 kg ha-1 + foliar spray of $ZnSO_4$ @ 0.2% at the knee-high and flowering stage significantly increase Zn concentration in grain and stover	Genotypes of sorghum, CSV-31, recorded higher stover zinc concentration and uptake (19.00 ppm and 103.67 kg/ha, respectively)	Markole et al. 2020
High zinc and Fe	Fertilization	Application of $ZnSO_4$ + $FeSO_4$ @ 50 kg ha^{-1} *fb.* and foliar application (0.50% + 1.0%) at 45 DAS with RDF (80: 40:40, N/P/K) resulted in Zn- and Fe-rich rainy sorghum	The genotype, Phule Maulee, recorded highest Fe (41.59 mg/kg) and Zn (20.80 and 26.42 mg/kg) over CSH 15R, M 35–1, Phule Chitra, Phule Yashoda	Mishra et al. 2015
High zinc and N	Fertilization	Application of Zn and N at rate of 10 and 120 kg/ha increased the Zn content of fodder sorghum	Zn content of plant also increased up to 37.79 to 37.85 mg kg^{-1} in 2 years	Ahmad et al. 2018
High zinc and iron	Fertilization	Application of recommended dose of fertilizer with enriched vermicompost [(50 kg vermicompost/ ha + 15 kg ZnSO4/ ha) + (50 kg vermicompost/ ha + 15 kg FeSO4/ha)] increased the contents of Zn and Fe	Increase in Fe and zinc contents of grain up to 39.52 and 28.44 mg/kg, respectively	Kumar and Kubsad 2017
High zinc and iron	Fertilization	Application of $ZnSO_4$ + $FeSO_4$ @ 15 kg ha_{-1} each enriched with FYM to kharif sorghum enhanced Zn and Fe content in grain	Increase in Zn and Fe contents of grain up to 23.43 and 33.89 mg/kg, respectively	Maganur and Kubsad 2020

developed after studying genetic diversity, population structure, and removing relative kinship between individuals. This will not only save time but also help in the identification of strongly linked markers to the targeted trait. More recently, with this GWAS strategy, Cruet-Burgos et al. (2020) has mapped provitamin QTLs, the

Table 10.4 List of significant studies performed for the identification of QTL that can be used for sorghum biofortification

Biofortification trait	Mapping population	QTL ID	Marker interval (cM)	LOD	PVE (%)	Reference
High Fe	F_6 RILs (296B x PVK 801)	qfe6.1(E1)	Sn2647940-Sn2657501 (90.3–91.5)	4.6	5.44	Kotla et al. (2019)
		qfe7.1(E1)	Dt3625344-Sn2644846 (36.8–49.6)	3.6	5.82	
		qfe7.2(E1)	Sn2653248-Xtxp525 (123.4–126.7)	4.3	5.09	
		qfe7.1(E3)	Dt3627294-Dt2648007 (65.4–65.6)	3.9	5.19	
		qfe1.1(E4)	Sn1933402-Sn2650907 (107.3–107.6)	5.5	6.80	
		qfe7.1(E6)	Sn2645682-Sn2653275 (60.9–61.4)	4.4	5.66	
		qfe7.2	Xtxp525-Sn2653248 (123.4–126.5)	5.6	6.7	
High Zn	F_6 RILs (296B x PVK 801)	qzn7.1 (E1)	Dt3625344-Sn2644846 (36.8–49.4)	5.6	9.42	Kotla et al. (2019)
		qzn7.2 (E1)	Dt2646020-Sn1925068 (53.0–54.7)	6.8	8.80	
		qzn7.3 (E1)	Sn1895281-Sn2650637 (57.0–57.9)	5.9	6.96	
		qzn7.1 (E3)	Sn1875097-Xiabtp360 (63.5–64.2)	4.9	6.29	
		qzn7.2 (E3)	Dt3628977-Dt2649259 (67.1–67.6)	4.3	5.83	
		qzn7.4	Dt2649259-Dt3628977 (67.1–67.6)	4.5	5.7	
		qzn7.1	Dt2645576-Sn2033434 (55.4–56.2)	4.5	5.7	
Provitamin A	Diverse panel of 403 accessions	*ZEP*	Sobic.006G097500-Sobic.004G281900	–	–	Cruet-Burgos et al. (2020)

details of which are summarized in Table 10.4. Once identified, these QTLs or genes can be used for pyramiding or introgression studies using marker-assisted breeding and for elucidation of their biochemical pathways.

In addition to the breeding approach, chemical mutagenesis (Taylor and Taylor 2011; Teferra et al. 2019) and transgenics (Zhao et al. 2003; Lipkie et al. 2013; Grootboom et al. 2014; Che et al. 2016; Elkonin et al. 2016) approach has been used to generate biofortified sorghum enriched with high protein, vitamin A, Fe, and Zn. The details of biofortified sorghum varieties developed to date are summarized in Table 10.5. In addition to this, Parbhani Shakti (ICSR 14001) biofortified sorghum variety has been developed in India using conventional breeding approaches whereby the sorghum line was enriched with high Fe and Zn content (Kumar et al. 2018).

10.5 Challenges, Limitations, and Success of Breeding Approaches for Sorghum Biofortification

Genetic variation is critical to any plant breeding program, as well as for sorghum biofortification. Conventional plant breeding can only be useful if an ample amount of genetic variability exists in the germplasm. In sorghum, significant genetic variability exists in nature for Fe, Zn, and phytate concentration, and the same has been extensively studied and improved by various workers (Reddy et al. 2005; Kumar et al. 2009, 2012); however for β-carotene low variability exists. Thus, only Fe and Zn can be enhanced using conventional plant breeding approaches. This is the major limitation of conventional breeding approaches.

Further, sorghum also contains some anti-nutritional factors, which makes it inferior to the other major cereals; these include lower digestibility and nutritional value of kafirins (a type of seed storage protein having a low content of lysine and threonine) and lower digestibility of starch. In addition to this, sorghum grain contains phosphorus in the form of phytic acid. The negative charge of this has a strong affinity to micronutrients especially Fe and Zn, thus making them inaccessible to humans and animals even in high concentrations. However, this can be overcome by using transgenics and the latest genome editing tools. So, with the availability of recent genomics tools, any trait can be bred in any crop, and the same has been successfully done in sorghum also.

Table 10.5 List of biofortified sorghum varieties developed by different institutions worldwide

Biofortified sorghum varieties	Method of variety development	Improved traits	Other note	Reference
Protein biofortified sorghum	Chemically induced mutation and genetic engineering	Grain protein quality	Achieved twice protein digestibility-corrected amino acid score than the null controls	Taylor and Taylor (2011)
Protein biofortified sorghum	Agrobacterium co-transformation	Protein quality; enriched lysine	Transformed with lysine-rich *HT12* gene; hemizygous seeds showed 40–60% increase in lysine	Zhao et al. (2003)
Provitamin A biofortified sorghum	Genetic modification	β-carotene	Genetically modified sorghum event Homo188-A shows largest bio-accessibility of β-carotene content, with a four- to eightfold increase from non-transgenic	Lipkie et al. (2013)
Protein biofortified sorghum	Genetic transformation	Protein digestibility	Co-suppression of three genes (γ kafirin-1, γ-kafirin-2, α-kafirin A1) significantly increases digestibility	Grootboom et al. 2014
Biofortified sorghum	Genetic engineering	β-carotene	Co-expression of *homogentisate geranylgeranyl transferase* (*HGGT*) extended half-life of β-carotene from less than 4 week. to 10 week. on average	Che et al. (2016)
Protein biofortified sorghum	*Agrobacterium*-mediated genetic transformation	High protein digestibility	RNAi silencing of the γ-kafirin gene resulted in enhanced digestibility index up to 85–88% compared with 59% in the control line	Elkonin et al. (2016)
Parbhani Shakti (ICSR 14001)	Conventional breeding	Fe, Zn, protein	Higher Fe (45 ppm) with Zn (32 ppm) and increased protein content (11.9%) and decrease in phytates (4.1 mg/100 g)	Kumar et al. (2018)
Protein biofortified sorghum	Mutation breeding	Protein digestibility	Mutant lines showed more protein digestibility (69.4% raw, 57.6% cooked) compared to wild-type lines (61.7% raw, 45.6% cooked)	Teferra et al. (2019)

10.6 Molecular Understanding of Essential Micronutrient Uptake and Deposition in Sorghum Grain

10.6.1 Iron (Fe)

10.6.1.1 Iron Uptake and Transport

Plants opt two types of strategies for Fe uptake from the soil. Strategy 1 (reduction-based) is common in dicotyledons and non-Poaceae monocotyledons. Under this strategy, the plant inaccessible Fe^{3+}, the predominant ionic form of Fe in the soil, is reduced at root surface to plant-accessible Fe^{2+} form (Zhang et al. 2019). Under iron deficiency, the uptake of chelated Fe^{3+} is facilitated by H + -ATPases (AHAs) localized in plasma membrane which aid in the reduction of rhizospheric pH and thereby increasing the solubility of Fe^{3+}. The FRO2 (ferric chelate reductase oxidase) gene catalyzes the reduction of Fe^{3+} to Fe^{2+} which is then imported into the root cells by high-affinity iron transporters, iron-regulated transporter (IRT1). Both FRO2 and IRT1 genes were first isolated and cloned from *Arabidopsis thaliana* (Eide et al. 1996; Robinson et al. 1999).

Strategy 2, also known as chelation-based, is mainly observed in graminaceous species. Plants in this category secrete phytosiderophores (PS), organic compounds belonging to the family of mugineic acids, acting as Fe^{3+} chelators (Rehman et al. 2021). The chelated Fe^{3+} is then transported to roots by yellow stripe-like (YSL) transporters (Curie et al. 2001). Sorghum is a strategy 2 plant, and recent reports suggest that arbuscular mycorrhizal fungi (AMF) can alleviate the iron deficiency symptoms via PS-mediated iron mobilization. Gene expression studies by quantitative real-time PCR revealed upregulation of *SbDMAS2* (deoxymugineic acid synthase 2), *SbNAS2* (nicotianamine synthase 2), and *SbYS1* (Fe-phytosiderophore transporter yellow stripe) in roots due to AMF in Fe-deficient sorghum (Prity et al. 2020). In another study, it was demonstrated that sorghum plants can recognize the volatile compounds released by bacteria and can induce Fe uptake mechanisms like *Arabidopsis* (Zhang et al. 2009; Hernández-Calderón et al. 2018). Of late, it has also been reported that rice plants use a combined strategy iron uptake comprising the components and strategies of both strategies 1 and 2 (Wairich et al. 2019). These iron-related genes are in turn regulated by various transcription factors, for example, basic helix-loop-helix (bHLH), FER-like iron deficiency-induced transcription factor (FIT) in *Arabidopsis*, has been found to regulate *FRO2* and *IRT1* genes for iron acquisition under iron deficiency condition (Bauer et al. 2007). Similarly, another bHLH transcription factor, POPEYE (PYE), regulates growth and development under iron deficiency (Long et al. 2010). After Fe acquisition, the ions get transported and translocated to different organs which are facilitated by two iron efflux transporters (IRON REGULATED1/Ferroportin 1 (IREG1/FPN1) and IREG2/FPN2) identified in *Arabidopsis* (Colangelo and Guerinot 2004). However, the molecular mechanism behind the long-distance iron is still under gray area. The iron ion is highly reactive and less soluble inside the plant environment, so in order to avoid precipitation and toxic effect, the ions are translocated inside the plant as complexes with citrate, mugineic acid, nicotinamine, and phenolic compounds.

Inside xylem, iron complexes with citrate at pH 5.5 and transmembrane protein ferric reductase defective 3 (*AtFRD3*) in *Arabidopsis* and the rice ortholog *OsFRDL1* (FRD-Like) help the transport of these complexes from root to shoot (Rehman et al. 2021). Iron translocation into actively growing plant sites such as shoot apex, root tips, and seeds and remobilization of iron from old parts to new ones occur via phloem. Inside phloem tissue, the iron complexes with the nicotinamine at pH 7.5 and the resulting complex transport in phloem with the help of yellow stripe-like (YSL) transporter family. This YSL transporter gene family is known to aid in unloading iron from xylem to phloem and loading it into developing seeds (Jeong and Guerinot 2009).

10.6.1.2 Fe Deposition in Grains

The distribution of iron in grains has been reported to be heterogeneous in nature. In rice, iron is mainly present in the aleurone layer, scutellum, and integument, whereas in peas the site of accumulation is mainly in the inner and outer epidermal layers of the embryo. Deposition of iron in the developing seed is mainly facilitated by *osYSL2* in rice and *YSL1* and *YSL3* in *Arabidopsis* (Rehman et al. 2021; Tong et al. 2020).

10.6.2 Zinc (Zn)

10.6.2.1 Zn Uptake and Transport

Under high pH conditions, zinc is tightly bound to the soil making it inaccessible for plant uptake. The Zn^{2+} uptake is facilitated by acidification and production of organic chelators like citrate and malate in the plant rhizosphere. The transporter family that contributes to this process belongs to the zinc import protein (ZIP) family (Tong et al. 2020). The zinc ion inside the plant root then makes complexes with nicotinamine and gets radially transported across different root layers which is facilitated by *metal tolerance protein 2 (MTP2)* in *Arabidopsis* (Sinclair et al. 2018). Zinc transport inside the xylem is facilitated by members of the *heavy metal ATPase (HMA)* family of P_{1B}-type ATPases, like *HMA2* and *HMA9* in rice. Once Zn^{2+} gets loaded into the xylem, it then moves to phloem tissues for long-distance Zn transport into the sink. In rice, *OsZIP3/OsHMA2*, *YSL* family transporters, and OsHMA9 are responsible for Zn xylem-to-phloem transport, phloem-to-organs transport, and remobilization, respectively (Tong et al. 2020).

10.6.2.2 Zn Deposition in Grains

Like iron, zinc is concentrated in small vacuoles in aleurone and sub-aleurone layers of the grain. Transcriptional microarray analysis of grain tissues in barley showed expression of heavy metal ATPases (HMAs), metal tolerance proteins (MTPs), and natural resistance-associated macrophage proteins (Nramps), hinting at their role in Zn deposition within the grain (Tauris et al. 2009). Particularly, MTPs, a member of the cation diffusion facilitator (CDF) transporter family, are shown to localize in vacuolar membrane and transport zinc ions to the vacuole (Podar et al. 2012).

Recently, *HvMTP1* has been characterized by overexpression studies in the endosperm of barley grains using endosperm-specific promoters (Menguer et al. 2018). The upregulation of this transporter led to increased zinc concentration in endosperm which opened a new strategy for zinc enrichment in the endosperm of cereal grains.

The molecular mechanism behind Fe and Zn uptake, transport, deposition, and homeostasis has been well characterized in many cereals but not in sorghum. Anuradha et al. (2013) have attempted in silico identification of candidate genes involved in Fe and Zn concentration in grains using reported cereal gene homologs. This study can aid in functional marker development and QTL mapping of grain Fe and Zn concentration in sorghum. Furthermore, the candidate genes can be functionally characterized using overexpression and gene silencing studies to understand their role in case of sorghum.

10.6.3 Provitamin A

The rate-limiting step in isoprenoid biosynthesis is first catalyzed by deoxyxylulose 5-phosphate synthase (DXS). Phytoene synthase (PSY) catalyzes the formation of phytoene from two molecules of geranylgeranyl pyrophosphate. Phytoene is then converted to lycopene by carotene desaturase (CRT-I). Lycopene cyclases ß-LCY and ε-LCY produce ß-carotene (ß,ß-carotene) and α-carotene (ß,ε-carotene). Carotene hydroxylases (CRT-RB) convert a- and ß-carotene to α- and ß-cryptoxanthin and then subsequently to non-provitamin A species like lutein and zeaxanthin. Carotenoids with an unsubstituted ß-ionone ring and all-trans configuration have the potential for conversion to retinol (provitamin A activity) (Lipkie et al. 2013).

10.7 Transgenic Efforts for the Development of Biofortified Sorghum

For the development of transgenics in sorghum, various studies have been carried out to detect the type and mode of transformation. Although several explants like immature zygotic embryos, mature embryos, immature inflorescence, and leaf fragments have been suggested, calli derived from immature zygotic embryos have been the explant of choice for the development of sorghum transgenics. Both biolistic and *Agrobacterium*-mediated gene transformation have been employed for the production of transgenic sorghum, but *Agrobacterium*-mediated transformation is preferred to direct transfer methods because of the added advantages in the former (Kennedy et al. 2003). Sorghum is considered as a staple food for sub-Saharan African (SSA) countries because of its drought and heat tolerance nature; however, sorghum lacks important amino acid like lysine, has poor protein digestibility on cooking, and also lacks micronutrients like provitamin A, zinc, and iron. The cases of blindness and anemia are increasing trends in SSA countries. To address these issues and achieve the goal of enriching multiple nutrients in single staple food, the Grand Challenges in Global Health initiative was launched in 2003 funded by the Bill & Melinda Gates Foundation, in association with the National Institutes of Health (NIH). This led to the initiation of the project entitled "nutritionally enhanced sorghum for arid and semi-arid tropical areas of Africa" (Henley et al. 2010; Zhao et al. 2019). The main objectives were (1) to increase iron and zinc bioavailability by 50%, (2) to increase provitamin A levels to up to 20 mg/kg, (3) to increase lysine content by 80–100%, and (4) to improve protein digestibility by 60–80% (Grand Challenges in Global Health 2021). Various efforts were undertaken to achieve these goals, for example, lysine-enriched genetically modified sorghum was reported by Zhao et al. (2003), where they have overexpressed lysine-rich proteins, such as HT12, an analog of barley *hordothionin*, and suppressed a lysine catabolism enzyme, *lysine ketoreductase*, using super binary vectors which enhanced the lysine content by 40–60%. Similarly, for improving protein digestibility of sorghum grain, the seed storage protein, the protease-resistant kafirin, has been targeted. The protein digestibility has been increased by RNAi silencing of the *ϒ-kafirin* (Elkonin et al. 2016) and simultaneous suppression of three genes: *ϒ-kafirin-1*, *ϒ -Kafirin-2*, and *α-Kafiirin -A1* (Grootboom et al. 2014). Furthermore, efforts were taken to increase provitamin A in sorghum for which scientists overexpressed the genes involved in the β-carotene synthesis pathway in the sorghum line (Tx430). The gene constructs encoding the enzymes like 1-deoxyxylulose 5-phosphate synthase (DXS), *Zea mays* phytoene synthase 1 (PSY1), and the *Pantoea ananatis* carotene desaturase (CRTI) were introduced into the sorghum line which resulted in increased ß-carotene level in transgenic plants (up to 9.1 μg/g vs. 0.5 μg/g in non-transgenic control seeds) (Lipkie et al. 2013; Elkonin et al. 2018). However, it was found that the ß-carotene undergoes degradation due to oxidation under storage. To address this challenge, researchers introduced the barley HGGT gene encoding *homogentisate geranylgeranyl transferase*, associated with the synthesis of the vitamin E (antioxidant in nature), along the same gene construct used for ß-carotene enrichment. The co-expression of HGGT and carotenoid biosynthesis genes increased all-trans ß-carotene accumulation (7.3–12.3 μg/g) and alleviate ß-carotene oxidative degradation, resulting in stable provitamin A in transgenic sorghum seeds (Che et al. 2016). Attempts were also made to increase the bioavailability of important micronutrients like Fe and Zn, for which the phytase enzyme was introduced into the sorghum line (Tx430) in order to degrade the phytic acid that acts as a chelating agent for divalent ions. Multidrug resistance-associated protein ATP-binding cassette transporter encoding gene was silenced resulting in lower phytate content (80–86%) compared to their non-transgenic control plants and increased zinc and iron bioavailability (Kruger et al. 2013). In Africa, Biofortified Sorghum lines with enhanced provitamin A, increased Fe and Zn bioavailability, and improved protein digestibility-corrected amino acid score (PDCAAS) have undergone over seven field trials. Additionally, efforts were done to address the key aspects involved in commercialization (Obukosia 2014) (Fig. 10.1).

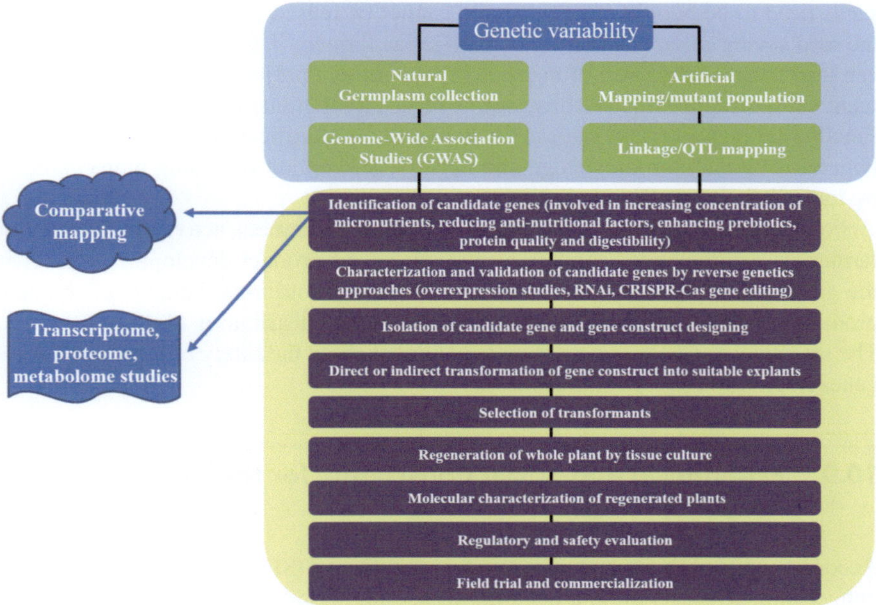

Fig. 10.1 Strategy used for the development of transgenic sorghum

10.8 Challenges for the Public Release of Transgenic Sorghum

The commercialization of transgenics has always been a controversial topic in many countries to date. Some believe that GM has the potential to solve various global challenges, while others pursue GM crops as a risk to the environment and humankind. The ABS sorghum lines are mainly targeted to release in Kenya (Eastern Africa) and Nigeria and Burkina Faso (Western Africa) because of the worst effects of micronutrient deficiencies. A study in Burkina Faso revealed that farmers are quite open to the addition of micronutrients to sorghum since sorghum is a subsistence crop in that region and, at the same time, they want to get rid of the severe micronutrient deficiencies prevailing in that area (Cardona et al. 2018; Chinedu et al. 2018). A market survey showed that farmers are ready to pay more for biofortified sorghum provided it performs better than the local varieties. Moreover, the study pointed out that the farmers who have experienced the benefits of first-generation GM crop (Bt cotton) are more likely to adopt second-generation GM crop, biofortified sorghum (Cardona et al. 2018). A study regarding the adoption of iron-fortified sorghum in Nigeria cited that environmentalists are strongly against the release because of the possible harmful effect on human health and the environment. Secondly, Nigerian consumers are very particular about their food choice like taste, aroma, and color. The vitamin A-enriched sorghum transgenics may change the color of the plant's parts making it less preferable by consumers. Moreover, seeds need to be bought every year and cannot be reused again and again breaking the seed saving culture and tradition of Nigerian farmers. Thirdly, the lack of funding can lead to the withdrawal of the transgenic sorghum project for which the government should support the scientist in this regard to continue their research program. Finally, the biggest challenge for the transgenic biofortification sorghum project in Nigeria is the lack of knowledge and negative perception regarding biotechnology by the stakeholders (Obi et al. 2017).

For successful and early adoption of ABS by SSA farmers, active involvement of farmers is recommended throughout the process of product development. Further, the government should create awareness about the nutritional benefits of the new product, subsidize the product, and participate in seed distribution of the transgenics. The scientists of ABS project are quite confident about the safety of introduced ABS genes hoping to see the light of the day.

10.9 Economical and Social Constraints for the Biofortified Sorghum

According to the UN Sustainable Development Goal 2 (SDG2), by 2030 all forms of hunger end, which is quite a daunting task. Malnutrition or hidden hunger due to micronutrient deficiencies has affected about one-third of the world population and severely endangered economic development. One of the workable strategies to diminish micronutrient malnutrition (MNM) is the biofortification of sorghum and increasing the intake of sorghum. Biofortification is the most promising option to improve the nutrition security of the poor. Biofortified foods can increase the levels of vitamins and minerals in our daily needs, and the nutrition status of vulnerable groups can be raised both at a national and worldwide scale, thus improving human nutrition (Lividini and Fiedler 2015; Meenakshi et al. 2010; Trijatmiko et al. 2016; Zhao and Shewry 2011; Saltzman et al. 2013; Waters and Sankaran 2011).

Biofortification is cost-effective as it ensures a nourishing future for all humans including the rural population (Grootboom et al. 2014; Saltzman et al. 2013; Zhao and Shewry 2011). With a one-time investment in biotechnology, farmers can sustain it for many years (Saltzman et al. 2013; Meenakshi et al. 2010). Finally, transgenic biofortification is a viable method of reaching micronutrient-deficient populations in the rural area who often have limited access to diverse, fortified, or supplemented meals. The success of using biofortified sorghum varieties in lowering down the problem of micronutrient malnutrition (MNM) has attracted the attention of communities, but most importantly it depends upon the attention from poor rural population. This can only be possible if the discouraging factors are eliminated. The government and development agencies are following many possible ways to mitigate MNM by increasing dietary diversification, supplementation of minerals, fortification, and enhancing the concentration and bioavailability of plant-based foods (White and Broadley 2009). It is a really very hazardous task to change people's diet. However genetic modification of food (biofortification), in general, has attracted high-rated controversies among scientists and policymakers.

There are four key issues that are expected to play a role in farmers' perception and attitude toward the biofortified sorghum; these include (1) source of seed, (2) market consideration, (3) experience with Bt cotton, and (4) external influence. Chinedu et al. (2018) found that many farmers still practice seed saving, but about 60% indicated that the source of their seed did not matter to them. To be able to strive in the presence of these challenges, the new biofortified sorghum seeds need to possess desirable agronomic attributes that could make it competitive with the local cultivars to encourage farmer's adoption; the biofortified alternative should possess a complete package of attributes including the addition of extra nutrient; it should be early maturing, low-cost, and high-yielding; and it should be provided through government institution to be more accessible to farmers.

Henceforth, there is a need for an extra effort by the government and NGOs to reach the rural people. By providing biofortified seeds, many benefits can be achieved. First, by growing the transgenic seed, it becomes a strategic means of developing micronutrient-dense trait on the major staple food (sorghum) which is widely consumed, from the rural area (Meenakshi et al. 2010). This is unlike food supplementation and fortification which often start from the urban areas. Secondly, during the laboratory production of biofortified crops, higher-yielding and disease resistance attributes can equally be added to biofortified crops, thereby improving their production. Finally, the production surplus generated from the biofortified crops can be marketed in the urban areas, providing more income for the farmers (Miller and Welch 2013; Saltzman et al. 2013).

10.10 Genome Editing Approaches for Biofortification of Sorghum

Although sorghum is recalcitrant to gene transformation as compared to other species (Raghuwanshi and Birch 2010), some genotypes have been successfully transformed with both particle bombardment (Casas et al. 1993) and *Agrobacterium* (Zhao et al. 2000) methods. CRISPR/Cas9 system-mediated targeted gene modification was reported for the first time in sorghum in 2013 (Jiang et al. 2013). Subsequently, various protocols were reported for *Agrobacterium* (Sander 2019; Char et al. 2020) and particle bombardment-mediated CRISPR/Cas-based gene editing (Liu et al. 2019). The practical application of CRISPR-based gene editing in sorghum was reported for improving protein quality and digestibility. Li et al. (2018) have successfully demonstrated editing of an alpha-kafirin gene family which increased the protein quality and digestibility in sorghum. Recently, Meng et al. (2020) have demonstrated an efficient protoplast assay in sorghum that can be used for transient gene expression and editing studies by CRISPR/Cas system. CRISPR-based gene editing holds a huge potential to expedite the goal of multi-nutrition enrichment in staple food like sorghum, but the product commercialization is highly dependent on the public perception and definition of natural products (Fig. 10.2).

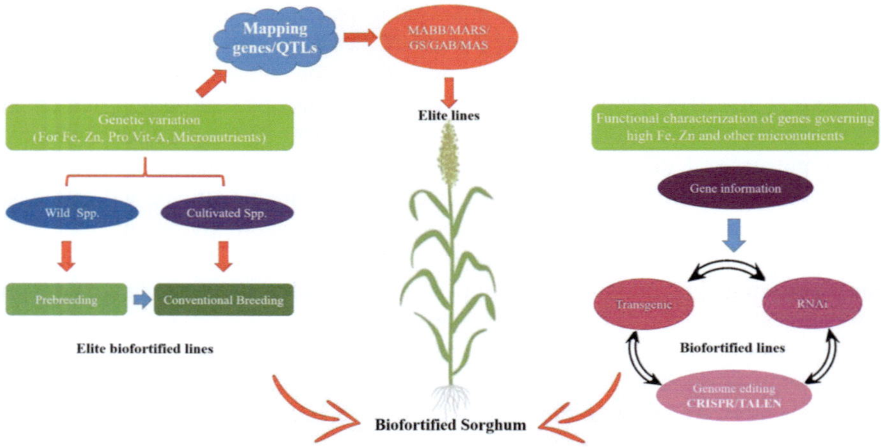

Fig. 10.2 Different approaches for improvement of nutraceutical properties in sorghum

References

Abbas SM (2012) Effects of low temperature and selenium application on growth and the physiological changes in sorghum seedlings. J Stress Physiol Biochem 8(1):268–286

Ahmad W, Tahir MUHAMMAD, Ahmad RIAZ, Ahmad RASHID (2018) Agronomic biofortification of fodder sorghum with zinc under different levels of nitrogen. Sains Malaysiana 47(6):1269–1276

Anuradha K, Prakash B, Ramu P, Shah T, Ashok Kumar A, Deshpande SP (2013) In silico identification of candidate genes involved for grain Fe and Zn concentration in sorghum using reported cereals gene homologs. In: Rakshit S et al (eds) Compendium of papers and abstracts: Global consultation on millets promotion for health & nutritional security. Society for Millets Research, Directorate of Sorghum Research, Hyderabad, pp 10–12

Aruna C, Cheruku D (2019) Genetic improvement of grain sorghum. In: Aruna C, Visarada KBRS, Bhat BV, Tonapi VA (eds) Breeding sorghum for diverse end uses. Woodhead Publishing, Cambridge, pp 157–173

Asensi-Fabado MA, Munné-Bosch S (2010) Vitamins in plants: occurrence, biosynthesis and antioxidant function. Trends Plant Sci 15(10):582–592

Bailey RL, West KP Jr, Black RE (2015) The epidemiology of global micronutrient deficiencies. Ann Nutr Metab 66(2):22–33

Bains K, Kaur H, Bajwa N, Kaur G, Kapoor S, Singh A (2015) Iron and zinc status of 6-month to 5-year-old children from low-income rural families of Punjab, India. Food Nutr Bull 36(3): 254–263

Barbosa GL, Gadelha FDA, Kublik N, Proctor A, Reichelm L, Weissinger E, Wohlleb GM, Halden RU (2015) Comparison of land, water, and energy requirements of lettuce grown using hydroponic vs. conventional agricultural methods. Int J Environ Res Public Health 12(6): 6879–6891

Batey I (2017) The diversity of uses for cereal grains. In: Wrigley C, Batey I, Miskelly D (eds) Cereal grains. Elsevier, Amsterdam, pp 41–53

Bauer P, Ling HQ, Guerinot ML (2007) FIT, the FER-like iron deficiency induced transcription factor in Arabidopsis. Plant Physiol Biochem 45(5):260–261

Bhardwaj D, Ansari MW, Sahoo RK, Tuteja N (2014) Biofertilizers function as key player in sustainable agriculture by improving soil fertility, plant tolerance and crop productivity. Microb Cell Factories 13(1):1–10

Black RE, Allen LH, Bhutta ZA, Caulfield LE, De Onis M, Ezzati M, Mathers F, Rivera J (2008) Maternal and child undernutrition: global and regional exposures and health consequences. Lancet 371(9608):243–260

Cakmak (2014) Agronomic biofortification. Harvest Plus Better Crops:31–32

Cakmak I, Kalaycı M, Ekiz H, Braun HJ, Kılınç Y, Yılmaz A (1999) Zinc deficiency as a practical problem in plant and human nutrition in Turkey: a NATO-science for stability project. Field Crop Res 60(1–2):175–188

Cardona, T. J., Speelman, S., & Sanou, E. (2018). Farmers attitudes towards GMO crops: comparison of attitudes towards first and second generation crops in Burkina Faso

Casas AM, Kononowicz AK, Zehr UB, Tomes DT, Axtell JD, Butler LG, Bressan RA, Hasegawa PM (1993) Transgenic sorghum plants via microprojectile bombardment. Proc Natl Acad Sci 90(23):11212–11216

Char SN, Lee H, Yang B (2020) Use of CRISPR/Cas9 for Targeted Mutagenesis in Sorghum. Current Protocols in Plant Biology 5(2):e20112

Chavan UD, Patil JV (2010) Grain sorghum processing. IBDC, Publishers, Lucknow, p 440

Che P, Zhao ZY, Glassman K, Dolde D, Hu TX, Jones TJ, Gruis DF, Obukosia S, Wambugu F, Albertsen MC (2016) Elevated vitamin E content improves all-trans β-carotene accumulation and stability in biofortified sorghum. Proc Natl Acad Sci 113(39):11040–11045

Chinedu O, Sanou E, Tur-Cardona J, Bartolini F, Gheysen G, Speelman S (2018) Farmers' valuation of transgenic biofortified sorghum for nutritional improvement in Burkina Faso: A latent class approach. Food Policy 79:132–140

Ciacci C, Maiuri L, Caporaso N, Bucci C, Del Giudice L, Rita Massardo D, Pontieri P, Di Fonzo N, Bean SR, Ioerger B, Londei M (2007) Celiac disease: *in vitro* and *in vivo* safety and palatability of wheat-free sorghum food products. Clin Nutr 26(6):799–805

Colangelo EP, Guerinot ML (2004) The essential basic helix-loop-helix protein FIT1 is required for the iron deficiency response. Plant Cell 16(12):3400–3412

Cremer JE, Bean SR, Tilley MM, Ioerger BP, Ohm JB, Kaufman RC, Wilson JD, Innes DJ, Gilding EK, Godwin ID (2014) Grain sorghum proteomics: integrated approach toward characterization of endosperm storage proteins in kafirin allelic variants. J Agric Food Chem 62(40):9819–9831

Cruet-Burgos C, Cox S, Ioerger BP, Perumal R, Hu Z, Herald TJ et al (2020) Advancing provitamin A biofortification in sorghum: Genome-wide association studies of grain carotenoids in global germplasm. The Plant Genome 13(1):e20013

Curie C, Panaviene Z, Loulergue C, Dellaporta SL, Briat J-F, Walker EL (2001) Maize yellow stripe1 encodes a membrane protein directly involved in Fe(III) uptake. Nature 409(6818): 346–349

Distelfeld A, Cakmak I, Peleg Z, Ozturk L, Yazici MA, Budak H, Saranga Y, Fahima T (2007) Multiple QTL-effects of wheat Gpc-B1 locus on grain protein and micronutrient concentrations. Physiol Plant 129:635–643

Eide D, Broderius M, Fett J, Guerinot ML (1996) A novel iron-regulated metal transporter from plants identified by functional expression in yeast. Proc Natl Acad Sci 93(11):5624–5628

Elkonin L, Italyanskaya J, Panin V (2018) Genetic modification of sorghum for improved nutritional value: state of the problem and current approaches. J Investig Genom 5:39–48

Elkonin LA, Italianskaya JV, Domanina IV, Selivanov NY, Rakitin AL, Ravin NV (2016) Transgenic sorghum with improved digestibility of storage proteins obtained by Agrobacterium-mediated transformation. Russ J Plant Physiol 63(5):678–689

Epstein, E. and Bloom, A. J. 2005. Mineral Nutrition of Plants: Principles and Perspectives, ed., 2nd, Sunderland, MA: Sinauer Associates

FAO (2021) World Food and Agriculture - Statistical Yearbook 2021. FAO, Rome. https://doi.org/10.4060/cb4477en

FAOSTAT. 2020. The State of Food Security and Nutrition in the World. http://www.fao.org/faost

Farooq M, ur-Rehman A, Aziz T, Habib M (2011) Boron nutripriming improves the germination and early seedling growth of rice (*Oryza sativa* L.). J Plant Nutr 34(10):1507–1515

Farooq M, Usman M, Nadeem F, ur Rehman H, Wahid A, Basra SM, Siddique KH (2019) Seed priming in field crops: potential benefits, adoption and challenges. Crop and pasture science 70(9):731–771

Grand Challenges in Global Health (2021). Create a full range of optimal, bioavailable nutrients in a single staple plant species. https://gcgh.grandchallenges.org/challenge/create-full-range-optimal-bioavailable-nutrients-single-staple-plant-species. Accessed 25 May 2021

Gregorio GB, Senadhira D, Htut H, Graham RD (2000) Breeding for trace mineral density in rice. Food Nutr Bull 21:382–386

Grootboom AW, Mkhonza NL, Mbambo Z, O'Kennedy MM, da Silva LS, Taylor J, Taylor JR, Chikwamba R, Mehlo L (2014) Co-suppression of synthesis of major α-kafirin sub-class together with γ-kafirin-1 and γ-kafirin-2 required for substantially improved protein digestibility in transgenic sorghum. Plant Cell Rep 33(3):521–537

Haussmann BI, Fred Rattunde H, Weltzien-Rattunde E, Traoré PS, Vom Brocke K, Parzies HK (2012) Breeding strategies for adaptation of pearl millet and sorghum to climate variability and change in West Africa. J Agron Crop Sci 198(5):327–339

Hefferon KL (2016) Can biofortified crops help attain food security? Curr Mol Biol Rep 2(4): 180–185

Henley EC, Taylor JRN, Obukosia SD (2010) The importance of dietary protein in human health: Combating protein deficiency in sub-Saharan Africa through transgenic biofortified sorghum. Adv Food Nutr Res 60:21–52

Hernández-Calderón E, Aviles-Garcia ME, Castulo-Rubio DY, Macías-Rodríguez L, Ramírez VM, Santoyo G, López-Bucio J, Valencia-Cantero E (2018) Volatile compounds from beneficial or pathogenic bacteria differentially regulate root exudation, transcription of iron transporters, and defense signaling pathways in *Sorghum bicolor*. Plant Mol Biol 96(3):291–304

Hodge J (2016) Hidden hunger: approaches to tackling micronutrient deficiencies. In: Gillespie S, Hodge J, Yosef S, Pandya-Lorch R (eds) Nourishing Millions: Stories of Change in Nutrition. International Food Policy Research Institute (IFPRI), Washington, pp 35–43

Jeong J, Guerinot ML (2009) Homing in on iron homeostasis in plants. Trends Plant Sci 14(5): 280–285

Jiang W, Zhou H, Bi H, Fromm M, Yang B, Weeks DP (2013) Demonstration of CRISPR/Cas9/ sgRNA-mediated targeted gene modification in Arabidopsis, tobacco, sorghum and rice. Nucleic Acids Res 41(20):e188–e188

Kayodé AP, Linnemann AR, Hounhouigan JD, Nout MJ, van Boekel MA (2006) Genetic and environmental impact on iron, zinc, and phytate in food sorghum grown in Benin. J Agric Food Chem 54(1):256–262

Kennedy G, Nantel G, Shetty P (2003) The scourge of" hidden hunger": global dimensions of micronutrient deficiencies. Food Nutr Agric 32:8–16

Khush GS, Lee S, Cho JI, Jeon JS (2012) Biofortification of crops for reducing malnutrition. Plant Biotechnol Rep 6(3):195–202

Kong L, Wang M, Bi D (2005) Selenium modulates the activities of antioxidant enzymes, osmotic homeostasis and promotes the growth of sorrel seedlings under salt stress. Plant Growth Regul 45(2):155–163

Kotla A, Phuke R, Hariprasanna K, Mehtre SP, Rathore A, Gorthy S et al (2019) Identification of QTLs and candidate genes for high grain Fe and Zn concentration in sorghum [*Sorghum bicolor* (L.) Moench]. J Cereal Sci 90:102850

Kruger J, Taylor JR, Du X, De Moura FF, Lönnerdal B, Oelofse A (2013) Effect of phytate reduction of sorghum, through genetic modification, on iron and zinc availability as assessed by an in vitro dialysability bioaccessibility assay, Caco-2 cell uptake assay, and suckling rat pup absorption model. Food Chem 141(2):1019–1025

Kumar A, Anuradha K, Ramaiah B, Grando S, Frederick H, Rattunde W, Virk P, Pfeiffer WH (2015) Recent advances in sorghum biofortification research. Plant Breeding Rev 39:89–124

Kumar A, Mehtre SA, Anuradha K, Jaganathan J, Prasanna H, Gorthy S, Gadakh SR, Chavan U, Kalpande HV, Tonapi VA (2018) Delivering bioavailable micronutrients through biofortifying sorghum and seed chain innovations. In: *Science Forum 2018*. South Africa, Stellenbosch

Kumar A, Reddy BV, Ramaiah B (2013b) Biofortification for combating micronutrient malnutrition: Identification of commercial sorghum cultivars with high grain iron and zinc concentrations. Indian J Dryland Agric Res Develop 28(1):89–94

Kumar A, Reddy BVS, Ramaiah B, Sahrawat KL, Pfeiffer WH (2012) Genetic variability and character association for grain iron and zinc contents in sorghum germplasm accessions and commercial cultivars. Eur J Plant Sci Biotechnol 6(1):66–70

Kumar A, Reddy BVS, Ramaiah B, Sanjana Reddy P, Sahrawat KL, Upadhyaya HD (2009) Genetic variability and plant character association of grain Fe and Zn in selected core collections of sorghum germplasm and breeding lines. *e-J*. SAT Agr Res 7:1–4

Kumar A, Sharma HC, Sharma R, Blummel M, Sanjana Reddy P, Belum Reddy VS (2013a) Phenotyping in sorghum [Sorghum bicolor (L.) Moench]. In: Panguluri SK, Kumar AA (eds) Phenotyping for plant breeding: applications of phenotyping methods for crop improvement. Springer Science+Business Media, New York, pp 73–109

Kumar AH, Kubsad VS (2017) Effect of fortification of organics with iron and zinc on growth, yield and economics of rabi sorghum [*Sorghum bicolor* (L.) Moench]. J Farm Sci 30(4):547–549

Laidlaw HKC, Mace ES, Williams SB, Sakrewski K, Mudge AM, Prentis PJ, Jordan DR, Godwin ID (2010) Allelic variation of the β-, γ-and δ-kafirin genes in diverse Sorghum genotypes. Theor Appl Genet 121(7):1227–1237

Lawson PG, Daum D, Czauderna R, Meuser H, Härtling JW (2015) Soil versus foliar iodine fertilization as a biofortification strategy for field-grown vegetables. Front Plant Sci 6:450

Lazarus JH (2015) The importance of iodine in public health. Environ Geochem Health 37(4):605–618

Leiser WL, Rattunde HFW, Piepho HP, Weltzien E, Diallo A, Melchinger AE, Parzies HK, Haussmann BI (2012) Selection strategy for sorghum targeting phosphorus-limited environments in West Africa: Analysis of multi-environment experiments. Crop Sci 52(6): 2517–2527

Li A, Jia S, Yobi A, Ge Z, Sato SJ, Zhang C, Angelovici R, Clemente TE, Holding DR (2018) Editing of an alpha-kafirin gene family increases, digestibility and protein quality in sorghum. Plant Physiol 177(4):1425–1438

Lipkie TE, De Moura FF, Zhao ZY, Albertsen MC, Che P, Glassman K, Ferruzzi MG (2013) Bioaccessibility of carotenoids from transgenic provitamin A biofortified sorghum. J Agric Food Chem 61(24):5764–5771

Liu G, Li J, Godwin ID (2019) Genome editing by CRISPR/Cas9 in sorghum through biolistic bombardment. In: Sorghum. Humana Press, New York, NY, pp 169–183

Liu ZQ, Li HL, Zeng XJ, Lu C, Fu JY, Guo LJ, Kimani WM, Yan HL, He ZY, Hao HQ, Jing HC (2020) Coupling phytoremediation of cadmium-contaminated soil with safe crop production based on a sorghum farming system. J Clean Prod 275:123002

Lividini K, Fiedler JL (2015) Assessing the promise of biofortification: a case study of high provitamin A maize in Zambia. Food Policy 54:65–77

Lonergan PF, Pallotta MA, Lorimer M, Paull JG, Barker SJ, Graham RD (2009) Multiple genetic loci for zinc uptake and distribution in barley (*Hordeum vulgare*). New Phytol 184:168–179

Long TA, Tsukagoshi H, Busch W, Lahner B, Salt DE, Benfey PN (2010) The bHLH transcription factor POPEYE regulates response to iron deficiency in Arabidopsis roots. Plant Cell 22(7): 2219–2236

Lung'aho MG, Mwaniki AM, Szalma SJ, Hart JJ, Rutzke MA, Kochian LV, Glahn RP, Hoekenga OA (2011) Genetic and physiological analysis of iron biofortification in maize kernels. PLoS One 6(6):e20429

Lyons G, Genc Y, Soole K, Stangoulis J, Liu F, Graham R (2009) Selenium increases seed production in Brassica. Plant Soil 318:73–80. https://doi.org/10.1007/s11104-008-9818-7

Maganur VR, Kubsad V (2020) Effect of zinc and iron enriched organics on yield and economics of kharif sorghum. J Farm Sci 33(2):188–191

Maleki FS, Chaichi MR, Mazaheri D, Tavakkol AR, Savaghebi G (2011) Barley grain mineral analysis as affected by different fertilizing systems and by drought stress. J Agric Sci Technol 13(3):315–326

Markole SK, Suresh K, Madhavi A, Shanti M (2020) Agronomic biofortification and nutrient uptake of sorghum (*Sorghum bicolor* L.) as influenced by fertilization strategies. Int J Chem Stud 8(3):2107–2110

Meenakshi JV, Johnson NL, Manyong VM, DeGroote H, Javelosa J, Yanggen DR, Naher F, Gonzalez C, Garcia J, Meng E (2010) How cost-effective is biofortification in combating micronutrient malnutrition? An ex ante assessment. World Develop 38(1):64–75

Mehlo L, Mbambo Z, Bado S, Lin J, Moagi SM, Buthelezi S, Stoychev S, Chikwamba R (2013) Induced protein polymorphisms and nutritional quality of gamma irradiation mutants of sorghum. Mutat Res/Fundam Mol Mech Mutagen 749(1–2):66–72

Meng R, Wang C, Wang L, Liu Y, Zhan Q, Zheng J, Li J (2020) An efficient sorghum protoplast assay for transient gene expression and gene editing by CRISPR/Cas9. PeerJ 8:e10077

Menguer PK, Vincent T, Miller AJ, Brown JK, Vincze E, Borg S, Holm PB, Sanders D, Podar D (2018) Improving zinc accumulation in cereal endosperm using Hv MTP 1, a transition metal transporter. Plant Biotechnol J 16(1):63–71

Miller DD, Welch RM (2013) Food system strategies for preventing micronutrient malnutrition. Food Policy 42:115–128

Mishra JS, Hariprasanna K, Rao SS, Patil JV (2015) Biofortification of post-rainy sorghum (*Sorghum bicolor*) with zinc and iron through fertilization strategy. Indian J Agric Sci 85(5): 721–724

Mudge SR, Campbell BC, Mustapha NB, Godwin ID (2016) Genomic approaches for improving grain quality of sorghum. In: The sorghum genome. Springer, Cham, pp 189–205

Obi CE, Ejiogu AO, Sanou EI (2017) The role of transgenic biofortified food in the reduction of hidden hunger in Nigeria. In: First International Conference of Food Security and Hidden Hunger. Federal University Ndufu-Aliku, Eboyi State, Nigeria, pp 8–11

Obukosia SD (2014) Biotechnology of nutritionally enhanced food crops targeting malnutrition in rural agricultural populations: the case study of Africa biofortified sorghum. In: Biotechnology in Africa. Springer, Cham, pp 157–177

Pearce EN, Andersson M, Zimmermann MB (2013) Global iodine nutrition: where do we stand in 2013? Thyroid 23(5):523–528

Pfeiffer WH, McClafferty B (2007) HarvestPlus: breeding crops for better nutrition. Crop Sci 47: S88–S105

Poblaciones MJ, Rengel Z (2016) Soil and foliar zinc biofortification in field pea (*Pisum sativum* L.): Grain accumulation and bioavailability in raw and cooked grains. Food Chem 212:427–433

Podar D, Scherer J, Noordally Z, Herzyk P, Nies D, Sanders D (2012) Metal selectivity determinants in a family of transition metal transporters. J Biol Chem 287(5):3185–3196

Prasanna R, Bidyarani N, Babu S, Hossain F, Shivay YS, Nain L (2015) Cyanobacterial inoculation elicits plant defense response and enhanced Zn mobilization in maize hybrids. Cogent Food Agric 1(1):998507

Prity SA, Sajib SA, Das U, Rahman MM, Haider SA, Kabir AH (2020) Arbuscular mycorrhizal fungi mitigate Fe deficiency symptoms in sorghum through phytosiderophore-mediated Fe mobilization and restoration of redox status. Protoplasma 257(5):1373–1385

Raghuwanshi A, Birch RG (2010) Genetic transformation of sweet sorghum. Plant Cell Rep 29(9): 997–1005

Ramzani PMA, Iqbal M, Kausar S, Ali S, Rizwan M, Virk ZA (2016) Effect of different amendments on rice (*Oryza sativa* L.) growth, yield, nutrient uptake and grain quality in Ni-contaminated soil. Environ Sci Pollut Res 23(18):18585–18595

Ramzani PMA, Shan L, Anjum S, Ronggui H, Iqbal M, Virk ZA, Kausar S (2017) Improved quinoa growth, physiological response, and seed nutritional quality in three soils having different stresses by the application of acidified biochar and compost. Plant Physiol Biochem 116:127–138

Rao PP, Basavaraj G, Ahmed W, Bhagavatula S (2010) An analysis of availability and utilization of sorghum grain in India. SAT eJournal 8

Rao PP, Birthal PS, Reddy BV, Rai KN, Ramesh S (2006) Diagnostics of sorghum and pearl millet grains-based nutrition in India. Int Sorghum Millets newsletter 47:93–96

Reddy BV, Ramesh S, Longvah T (2005) Prospects of breeding for micronutrients and b-carotene-dense sorghums. Int Sorghum Millets Newsletter 46:10–14

Rehman AU, Masood S, Khan NU, Abbasi ME, Hussain Z, Ali I (2021) Molecular basis of Iron Biofortification in crop plants, A step towards sustainability. *Plant Breeding* 140(1):12–22

Robinson NJ, Procter CM, Connolly EL, Guerinot ML (1999) A ferric-chelate reductase for iron uptake from soils. Nature 397(6721):694

Ruel-Bergeron JC, Stevens GA, Sugimoto JD, Roos FF, Ezzati M, Black RE, Kraemer K (2015) Global update and trends of hidden hunger, 1995-2011: the hidden hunger index. PLoS One 10(12):e0143497

Saltzman A, Birol E, Bouis HE, Boy E, De Moura FF, Islam Y, Pfeiffer WH (2013) Biofortification: progress toward a more nourishing future. Glob Food Sec 2(1):9–17

Sanchez PA, Swaminathan MS (2005) Hunger in Africa: the link between unhealthy people and unhealthy soils. Lancet 365(9457):442–444

Sander JD (2019) Gene editing in sorghum through Agrobacterium. In: Sorghum. Humana Press, New York, NY, pp 155–168

Serna-Saldivar SO, Espinosa-Ramírez J (2019) Grain structure and grain chemical composition. In: Sorghum and millets. AACC International Press, pp 85–129

Sinclair SA, Senger T, Talke IN, Cobbett CS, Haydon MJ, Krämer U (2018) Systemic upregulation of MTP2-and HMA2-mediated Zn partitioning to the shoot supplements local Zn deficiency responses. Plant Cell 30(10):2463–2479

Smoleń S, Skoczylas Ł, Ledwożyw-Smoleń I, Rakoczy R, Kopeć A, Piątkowska E et al (2016) Biofortification of carrot (*Daucus carota* L.) with iodine and selenium in a field experiment. Front Plant Sci 7:730

Stevens GA, Finucane MM, De-Regil LM, Paciorek CJ, Flaxman SR, Branca F, Peña-Rosas JP, Bhutta ZA, Ezzati M, Nutrition Impact Model Study Group (2013) Global, regional, and national trends in haemoglobin concentration and prevalence of total and severe anaemia in children and pregnant and non-pregnant women for 1995–2011: a systematic analysis of population-representative data. Lancet Glob Health 1(1):e16–e25

Sumithra M, Jee Hyun R, Jonathan DS, Franz FR, Klaus K, Robert EB (2013) The global hidden hunger indices and maps: an advocacy tool for action. PLoS One 8(6):e67860

Tauris B, Borg S, Gregersen PL, Holm PB (2009) A roadmap for zinc trafficking in the developing barley grain based on laser capture microdissection and gene expression profiling. J Exp Bot 60(4):1333–1347

Taylor J, Taylor JR (2011) Protein biofortified sorghum: effect of processing into traditional African foods on their protein quality. J Agric Food Chem 59(6):2386–2392

Teferra TF, Amoako DB, Rooney WL, Awika JM (2019) Qualitative assessment of 'highly digestible' protein mutation in hard endosperm sorghum and its functional properties. Food Chem 271:561–569

Tong J, Sun M, Wang Y, Zhang Y, Rasheed A, Li M, Xia X, He Z, Hao Y (2020) Dissection of molecular processes and genetic architecture underlying iron and zinc homeostasis for biofortification: from model plants to common wheat. Int J Mol Sci 21(23):9280

Trijatmiko KR, Dueñas C, Tsakirpaloglou N, Torrizo L, Arines FM, Adeva C, Balindong J, Oliva N, Sapasap MV, Borrero J, Rey J (2016) Biofortified indica rice attains iron and zinc nutrition dietary targets in the field. Sci Rep 6(1):1–13

Wairich A, de Oliveira BHN, Arend EB, Duarte GL, Ponte LR, Sperotto RA, Ricachenevsky FK, Fett JP (2019) The Combined Strategy for iron uptake is not exclusive to domesticated rice (*Oryza sativa*). Sci Rep 9(1):16144

Wakeel A, Farooq M, Bashir K, Ozturk L (2018) Micronutrient malnutrition and biofortification: recent advances and future perspectives. In: Plant micronutrient use efficiency, pp 225–243

Walker CF, Ezzati M, Black RE (2009) Global and regional child mortality and burden of disease attributable to zinc deficiency. Eur J Clin Nutr 63(5):591–597

Waters BM, Sankaran RP (2011) Moving micronutrients from the soil to the seeds: genes and physiological processes from a biofortification perspective. Plant Sci 180(4):562–574

Welch RM, Graham RD (2004) Breeding for micronutrients in staple food crops from a human nutrition perspective. J Exp Bot 55(396):353–364

White PJ, Broadley MR (2009) Biofortification of crops with seven mineral elements often lacking in human diets- Iron, zinc, copper, calcium, magnesium, selenium and iodine. New Phytol 182: 49–84

WHO. 2017. Malnutrition. www.who.int/mediacentre/factsheets/malnutrition/en/. Last accessed 4 February 2019.

Zhang H, Sun Y, Xie X, Kim MS, Dowd SE, Paré PW (2009) A soil bacterium regulates plant acquisition of iron via deficiency-inducible mechanisms. Plant J 58(4):568–577

Zhang X, Zhang D, Sun W, Wang T (2019) The Adaptive mechanism of plants to iron deficiency via iron uptake, transport, and homeostasis. Int J Mol Sci 20(10):2424

Zhao FJ, Shewry PR (2011) Recent developments in modifying crops and agronomic practice to improve human health. Food Policy 36:S94–S101

Zhao ZY, Cai T, Tagliani L, Miller M, Wang N, Pang H, Rudert M, Schroeder S, Hondred D, Seltzer J, Pierce D (2000) Agrobacterium-mediated sorghum transformation. Plant Mol Biol 44(6):789–798

Zhao ZY, Che P, Glassman K, Albertsen M (2019) Nutritionally enhanced sorghum for the arid and semiarid tropical areas of Africa. In: Sorghum. Humana Press, New York, NY, pp 197–207

Zhao ZY, Glassman K, Sewalt V, Wang N, Miller M, Chang S, Thompson T, Catron S, Wu E, Bidney D, Jung R (2003) Nutritionally improved transgenic sorghum. In: Plant biotechnology 2002 and beyond. Springer, Dordrecht, pp 413–416

Zuo Y, Zhang F (2009) Iron and zinc biofortification strategies in dicot plants by intercropping with gramineous species. A review. Agron Sustain Develop 29(1):63–71

Biofortification of Oats (*Avena sativa*) 11

Maneet Rana, Parichita Priyadarshini, Gunjan Guleria, Rahul Kaldate, Neeraj Kumar, Rahul Gajghate, K. K. Dwivedi, and Shahid Ahmed

Abstract

Oat (*Avena sativa* L.) is a multipurpose crop belonging to the family Poaceae and known for its grain and fodder nutritional qualities. Biofortification of oat is a sustainable solution to minimize micronutrient deficiency in the human and livestock population. The continuous growth of human population, together with the modifying eating habits, calls for an intensification of agricultural production. In around 66% of the world's population diet, there is a lack of more than one microelement. Inadequate concentrations of microelements necessitate the development of sustainable biofortified crops. This chapter deals with all aspects of oat biofortification including the role of essential and basic micronutrients in human and animal nutrition along with a strategy to develop biofortified oats. The molecular mechanisms underlying uptake, deposition, and mobilization of micronutrients in oat grain have also been discussed. The various strategies being used for the biofortification of oat have been discussed in great depth including agronomic, breeding, transgenic, and genome editing approaches. The areas requiring adequate attention, viz., challenges, limitations and success of breeding approaches, public release, and economic and social constraints, are also highlighted.

M. Rana (✉) · P. Priyadarshini · N. Kumar · R. Gajghate · K. K. Dwivedi · S. Ahmed
Division of Crop Improvement, ICAR-Indian Grassland and Fodder Research Institute, Jhansi, UP, India

G. Guleria
Department of Agronomy and Agrometeorology, Rani Lakshmi Bai Central Agricultural University, Jhansi, UP, India

R. Kaldate
Department of Agricultural Biotechnology, Assam Agricultural University, Jorhat, India

R. Deshmukh et al. (eds.), *Biofortification in Cereals*,
https://doi.org/10.1007/978-981-19-4308-9_11

Keywords

Agronomic · Breeding · Biofortification · Fodder · Grain · Oat · Micronutrients · Transgenic

11.1 Introduction

Biofortification directs the nutritional up-gradation of crops for higher bioavailability of micronutrients especially protein, vitamins, and minerals using genetic, agronomical, or any other intervention. Humans and livestock both require nutritionally enhanced food for effective growth and development (Welch and Graham 2005). Inadequate consumption of essential micronutrients leads to micronutrient deficiencies or malnutrition ("hidden hunger") (Muthayya et al. 2013).

As per FAO estimates, about 79.25 crore people around the globe are suffering from malnourishment, of which about 98% people are from developing countries only. About two billion people suffer from hidden hunger (McGuire 2015). In India, this situation is more serious, whereby one out of five children below 5 years of age is wasted (IFPRI 2019). Apart from humans, micronutrient deficiency is also rampant in the Indian livestock population. This sector contributes 4.5% and 17.5% of GDP and total agriculture GDP (National Accounts Statistics-2018; Central Statistical Organisation, GoI) and is economically established as fastest-growing sector. However, individual productivity of Indian livestock is very low, and this is due to severe nutritional deficiencies in animals (Ranjhan 1994).

In the developing world, iron (Fe), zinc (Zn), and selenium (Se) deficiencies are related with major health problems (Lyons et al. 2004; Cakmak et al. 2010; Bilski et al. 2012). Micronutrient deficiencies such as Fe, Ca, Zn, Se, vitamin A, and vitamin C lead to anemia, osteoporosis, dwarfism and hypogonadism, weak joints, night blindness, and scurvy, respectively (de Benoist et al. 2008; Heaney 2000; Prasad 2013; Sunde 2010; Dowling and Wald 1958; Agarwal et al. 2015), and can be overcome by adopting proper nutrition. Further as per WHO, a healthy or proper diet protects from all types of malnutrition as well as from noncommunicable diseases. In livestock, mineral deficiencies like Zn decrease feed ingestion and testicular growth and cause listlessness and skin lesions and alter prostaglandin synthesis and luteal function (Graham 1991). Cu deficiency causes low synthesis of thyroid hormone, progesterone production, and other reproductive deficiencies (Abdollahi et al. 2013). The requirement of Zn and Cu for an animal is 30–75 and 12–16 ppm, respectively, but the availability in major fodder crops is 15–30 and 2–11 ppm, respectively (Nocek et al. 2006). In spite of meeting the above mineral deficit through a commercial chemical in animal feed, the identification and development of high mineral content lines through biofortification in major fodder crops will be the permanent solution for the improvement of livestock health and productivity.

Oat (*Avena sativa* L.) being a multipurpose crop is grown for its grain as well as fodder value in various regions of the world (Ruwali et al. 2013). It is the main ingredient of muesli and granola (Gauldie 1981) and has also been used as a high energy feed due to its high-quality fodder value (Ruwali et al. 2013; Heuzé et al. 2016). Oat belongs to the family Poaceae and is known for lowering cholesterol and blood glucose level because of its high β-glucan content (Anderson and Chen 1986; Maier et al. 2000). In 2018, worldwide production of oats was about 23 million tonnes, led by Russia with 20% and Canada with 15% of world total production. Canada (379 M USD) is the top exporter of oats followed by Finland, Sweden, Poland, and Australia, while the top importer of oat is the USA (323 M USD) followed by Germany, Netherlands, Mexico, and Belgium-Luxembourg. India accounts for 7.5% of the total share for imports (FAOSTAT 2019). It occupied about 2% of the cultivated area which accounts for 11.7 million ha (Szatanik-Kloc et al. 2019). Table 11.1 represents the details of economic status, hunger index, child mortality, human population, and oat production of major oats-growing countries worldwide. If we look at the child mortality status (deaths per 1000 live births) of the top 20 oats-growing countries, then almost all countries except Romania, Poland, and Finland have less than 1% child mortality rate. This signifies the role of oats in eradicating malnutrition among children. Earlier, 95% of the world's oat production is used as animal fodder however keeping in view of its nutritional benefits, its consumption is increased by human population. In addition to this, oat extracts having natural skin conditioners and anti-inflammatory properties are also used in the pharmaceutical and cosmetics industry (Butt et al. 2008).

To combat micronutrient deficiency in humans and livestock together, oat can serve as the best dual-purpose crop. Biofortification of oat can be the best possible sustainable solution for improving micronutrient content in crops using various strategies like classical plant breeding, agronomic biofortification, and genetic engineering or with the latest genome editing tools. For biofortification to occur, there are various physiological processes that are altered in the plant system through intake of micronutrients by root, remobilization of micronutrients, and then sink to source translocation (Palmgren et al. 2008; Erenoglu et al. 2011). This chapter describes the role of various essential micronutrients for humans and animals along with various approaches of biofortification with the latest advancements in oat crops.

11.2 Top Priorities for Oat Biofortification

India, being one of the most populous countries, suffers from "hidden hunger" and lots of health problems. Among them, cardiovascular diseases (CVDs) and diabetes mellitus occupy the top position, about 31% (WHO 2017a) and 5% (Dattani and Jiang 2009) of global deaths with a projection of more than 187 and 578 million cases of CVDs and diabetes, respectively, by 2030 (WHO 2017a; Saeedi et al. 2019). Of late, functional foods have been widely utilized to reduce all these risks of diabetes and associated complications.

Nowadays oat has received great attention due to its enriched phytochemicals, dietary fibers, and nutritional values (Fig. 11.1). It has been considered as "supergrain" as it offers potential health benefits due to the presence of excellent

Table 11.1 Details of economic status and hunger index of major oats-growing countries worldwide

Country	Population (thousands)[a]	Oat production (tonne)[b]	GDP per capita (current USD)[c]	Hunger index[d]	Child mortality[e]
Russian Federation	145,934	4,424,433	10,792.47	5.2	5.87
Canada	37,742	4,237,300	45,870.64	–	5.01
Poland	37,847	1,209,580	16,739.73	–	79.38
Finland	5541	1,187,480	54,816.90	–	40.03
Australia	25,500	1,134,619	57,210.82	–	3.64
United Kingdom of Great Britain and Northern Ireland	67,886	1,076,000	42,235.94	–	4.31
Brazil	212,559	920,439	6728.17	<5	12.26
Spain	46,755	841,200	31,178.18	–	3.27
United States of America	331,003	771,440	66,144.14		6.67
Sweden	10,099	671,200	57,659.73	–	2.51
Argentina	45,196	571,630	9095.10	5.3	9.68
Germany	83,784	519,300	51,967.33	–	3.82
China, mainland	1,439,324	495,466	11,712.85	<5	8.40
Ukraine	43,734	422,000	3614.67	<5	7.95
France	65,274	402,420	44,769.80	–	4.41
Chile	19,116	384,922	14,208.69	<5	7.74
Belarus	9449	368,252	6513.12	<5	3.25
Romania	19,238	361,570	14,916.09	<5	83.76
Kazakhstan	18,777	267,006	9454.35	5.4	10.80
Turkey	84,339	265,000	7658.85	<5	10.91

[a] Data was obtained from World Population Prospects 2019
[b] Data was obtained from FAOSTAT 2021
[c] Data was obtained from World Economic Outlook Database 2021
[d] Data was obtained from Global Hunger Index 2020
[e] Data was obtained from UN IGME database 2020

amount of β-glucan content which is believed to reduce plasma cholesterol and blood glucose level and prevent coronary heart disease (Rasane et al. 2015). β-Glucan is found mainly in yeast, mushroom, oat, and barley. However, oat beta-glucans have been approved by FDA for multiple health benefits especially their ability in lowering cholesterol levels (Ho et al. 2016).

Further, oat being a dual-purpose crop is used as fodder for livestock and dietary supplement for both humans and livestock. Oat β-glucan enhances gut microbiome and immune responses in ruminants, while it lowers cholesterol levels and slows down glucose absorption in humans. Being an important dietary cereal with multiple health benefits, oat deserves attention for fortification of its grains with beneficial nutrients (Flander et al. 2007). Further, its demand is increasing day by day with its

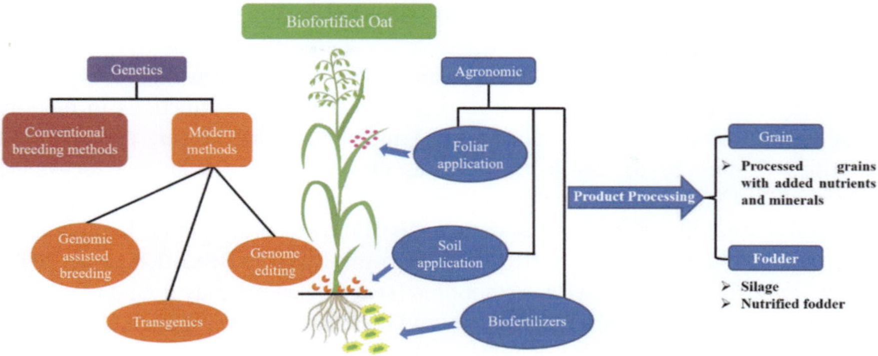

Fig. 11.1 Different approaches for improvement of nutraceutical properties in oat grain

global contribution of about 782 M USD in the international market. Looking at this huge market and high potential of oat both being as food and fodder crop, biofortification of Fe, Zn, Se, I, provitamins, and proteins is the need of the hour in oats. This will not only help in eradicating malnutrition in humans but also improve animals' health.

11.2.1 Essential Micronutrients/Metals: Zn and Fe

More than two billion people, mostly from low-income groups in developing countries, are affected due to micronutrient deficiencies in which Fe, Zn, and vitamins A and B9 deficiencies are most common (Kennedy et al. 2003; Bailey et al. 2015). As more than 50% of the human population in the world may be at threat of Zn deficiency due to insufficient Zn intake (Bell and Dell 2008). It has been reported that during 2008, more than 4.5 lakh children below the age of 5 years died because of diarrhea caused by Zn deficiency (Black et al. 2008). Apart from this, Zn deficiency leads to pneumonia and dwarfism in children (Cakmak et al. 1999; Walker et al. 2009). Zinc is an important micronutrient for plant's growth and development having a role in vital functions (photosynthesis and respiration), and it is also important for the nutritional value of feed and food-based plant products (Epstein and Bloom 2005).

Fe deficiency is the more prominent and prevalent nutrient deficiency. WHO reported that iron deficiency affected 38%, 29%, and 43% of pregnant, nonpregnant women, and preschool children, respectively (WHO 2017b). It was observed that approximately half of the cases of anemia are caused due to iron deficiency and are responsible for poor immunity and lower pregnancy outcomes (Stevens et al. 2013; WHO 2017b). It resulted in impaired cognitive development, low productivity, and growth retardation, with complications in health and economic systems (Bailey et al. 2015).

11.2.2 Basic Micronutrients: Selenium and Iodine

Se is an important micronutrient for human health, as it performs a key role in the brain, thyroid, gonads, and heart. It also performs important functions in binding with heavy metals and has antibacterial, antioxidant, anticancer, and antiviral activities (Lyons 2018). Se is identified as the 21st amino acid, selenocysteine (Rayman 2002). It is reported that around a billion people are Se deficient, which causes diseases, like Kashin-Beck disease (KBD), osteoarthropathy, hypothyroidism, rheumatoid arthritis, or heart failure (Wu et al. 2015; Bissardon et al. 2017; D'Amato et al. 2018). Likewise, both iodine (I) and selenium play a vital role in the regulation of thyroid functioning. The thyroid hormone turnover is controlled by the selenoenzymes, iodothyronine deiodinases D1, D2, and D3, which are important in thyroid gland metabolism (Köhrle 2013).

Iodine deficiency is highly common among people from both developing and developed nations (Cakmak et al. 2017). The recommended dietary allowance (RDA) of I for adults is 150–200 μg per day and for lactating or pregnant females is 230–260 μg per day (Lawson et al. 2015). About 75% of the school-aged children have inadequate iodine intake (Zimmermann and Andersson 2012) which leads to different health-related problems like goiter, mental disability, growth retardation, and increased miscarriage and infant mortality (Pearce et al. 2013; Lazarus 2015; Lv et al. 2017). Recent reports showed that even mild iodine deficiency affects pregnant women, which is associated with cognitive impairment in their children (Pearce et al. 2016).

11.2.3 Provitamins

Crop plants are the main source of energy and several vitamins. However, many crops are deficient in certain vitamins and shortage of which leads to malnutrition (Bailey et al. 2015). Vitamins play a vital role in human health and nutrition. Their deficiencies lead to blindness (vitamin A), beriberi (vitamin B1), pellagra (vitamin B3), anemia (vitamin B6), scurvy (vitamin C), and rickets (vitamin D) (Asensi-Fabado and Munné-Bosch 2010). Due to anemia, around two million children under 5 years die every year (Scott et al. 2014), whereas more than two billion people were affected globally. Similarly, vitamin A deficiency leads to morbidity and infectious disease mortality in around 125–130 million children (Kraemer et al. 2008).

Vitamin composition and level in crop plants vary significantly. However, in oat grains, an ample amount of vitamins like thiamine (B1), riboflavin (B1), niacin (B3), pantothenic acid (B5), pyridoxine (B6), and folate (B9) are present (Table 11.2). These vitamins are present in a lesser amount as compared to RDA. So, there is a need to enhance their concentrations using conventional or classic plant breeding, various agronomic, and genetic engineering approaches. Further, some important vitamins like vitamins A, B12, and C, D, and E are not present in oat grains; thus these can be enriched through genetic engineering or transgenic approaches.

Table 11.2 Comparative profiling of oat grains with other cereal grains

	Oats	Rice	Wheat	Maize	Barley	RDA
Minerals						
Iron (mg)	4.72	0.20	3.71	1.74	2.68	7–18
Zinc (mg)	3.97	0.49	2.96	2.24	2.0	3–11
Calcium (mg)	54.0	10.0	33.0	5.0	32.0	700–1300
Selenium (μg)	102	7.5	12.7	2.2	37.7	20–55
Iodine (μg)	16	0	0	0	0	90–250
Provitamins						
Vitamin A (IU) (μg/day)	0	0	0	214	0	300–900
Thiamin; B1 (mg)	0.763	0.02	0.297	0.16	0.37	0.5–1.2
Riboflavin; B2 (mg)	0.139	0.013	0.188	0.23	0.114	0.5–1.3
Niacin; B3 (mg)	0.961	0.4	5.35	2.6	6.27	6.0–16
Pantothenic acid; B5 (mg)	1.35	0.39	1.01	0.55	0.145	6.0–16
Pyridoxine; B6 (mg)	0.119	0.093	0.191	0.47	0.396	0.5–1.7
Folate; B9 (μg)	56.0	3.0	28.0	19.0	8.0	150–400
Vitamin B12	0	0	0	0	0	0.9–2.4
Vitamin C	0	0	0	0	0	15–90
Vitamin D	0	0	0	0	0	15–20
Vitamin E, alpha-tocopherol (mg)	0	0.04	0.53	0.49	0.57	6.0–15
Vitamin K1 (μg)	0	0	1.9	0.3	2.2	30–120
Beta-carotene (μg)	0	0	5	97	0	–
Lutein + zeaxanthin (μg)	0	0	220	1355	160	–
Protein						
Protein (g)	16.9	2.69	9.61	8.75	10.5	13–56

Data was obtained from USDA database; Acar et al. (2020); Dietary Reference Intakes 2011, 2019 [children (1–8 years), male and female (8 to >70)]

Several studies have been conducted to biofortified crop plants with several vitamins using conventional or classic plant breeding, agronomic, and genetic engineering approaches in various crops. Using genetic engineering strategy, vitamin content was increased in several studies such as vitamin B1 in rice (Dong et al. 2016); vitamin B6 in cassava (Li et al. 2015); and vitamin B9 in tomato and rice (de la Garza et al. 2007; Storozhenko et al. 2007). Likewise genetically modified vitamin-biofortified golden rice enriched with provitamin A (β-carotene) (Paine et al. 2005; Ye et al. 2000) and vitamin B9 (folate)-enhanced rice (Blancquaert et al. 2014; Storozhenko et al. 2007) were developed. Similarly, with help of breeding techniques, several varieties/accessions were identified with several-fold more vitamin B1 concertation in crops like wild potato (Goyer and Sweek 2011), cassava (Mangel et al. 2017), and rice (Kennedy and Burlingame 2003). With help of GWAS, Li et al. (2018) detected multiple QTLs, responsible for B1 content in common wheat. As such no such content enhancement of these vitamins has been reported in oats.

11.2.4 Protein

In poor developing countries, people depend mostly on low-protein staple crops for food. The biofortification of protein in staple food plant is important to fight childhood protein-energy malnutrition in these countries and ensure the sustainability of the crop (Taylor and Taylor 2011). Amino acids are the building blocks of proteins; out of the total 20 amino acids, 10 are essential; therefore these must be supplied in the food daily as excess amino acids cannot be stored in the body. In developing countries, the day-to-day ingestion of essential amino acids is often inadequate because of the shortage of protein-rich sources like meat, fish, or soybean as low protein content food sources such as rice, potato, and cassava are the only source of their meal. As per data obtained from USDA database, oat grains contain 16.9 g proteins. However, RDA for children (1–8 years) was 13–19 g; for male (9 to >70 years), it is 34–56 g; for females (9 to >70 years), it is 34–56, and during pregnancy 71 g per day is recommended.

11.3 Agronomic Biofortification of Oats

Agronomic biofortification is found to be a complementary approach to breeding strategy. Agronomic practices are timely and straightforward approaches as they can be easily utilized globally. Agronomic practices are also known as pre-harvest practices which enhance the nutrient content in crops. Food is categorized as biofortified if practices are pre-harvest and otherwise categorized as fortified if practices are postharvest. Some important agronomic biofortification approaches are soil/foliar application of organic and inorganic fertilizers and biofertilizers and nutrient priming (WHO 2021). Soil or foliar application method of biofortification is very prominent. It can be used in the case of inorganic fertilizers, organic manures, or biofertilizers. Organic manures, which contain animal or plant sources like vermicompost, farmyard, poultry manure, etc., are considered as a cost-effective, environmentally friendly alternative approach for inorganic synthetic fertilizers. Organic manures help in maintaining soil fertility. Maleki et al. (2011) observed that vermicompost application leads to an increase in the Zn and Fe content by 4% and 7%, respectively, in a barley crop. Another study increases iron content in rice and wheat by 10% and 15%, respectively, using the application of poultry manure (Ramzani et al. 2016, 2017). In addition to these, another source is biofertilizers; these are consisting of microorganisms that enhance the productivity and growth of plants. These include mycorrhizal fungi, blue-green algae, and cyanobacteria. These microorganisms are generally considered as plant growth-promoting bacteria. These bacteria showed enhanced growth along with increased nutrient content by modifying the supply or availability of nutrients in crops (Barbosa et al. 2015; Bhardwaj et al. 2014). Some biofertilizers like cyanobacteria were used in the zinc biofortification of wheat (Prasanna et al. 2015). Another approach is nutri-priming, which means using micro- and macronutrients for seed treatment before sowing (Farooq et al. 2011). It is the method used for soaking the seed in solutions that

Table 11.3 List of significant studies performed for the oat biofortification

Biofortification trait	Agronomic practices	Success/level of improvement	Other note	References
High zinc	Fertilization	Improved zinc concentration	Fertilization of coal fly ash	Bilski et al. (2012)
	Fertilization	Improved Zn concentration up to 32.3 and 48.95 mg/kg in oat grains and straw	Coating oat seeds with $ZnSO_4$ or ZnO pre-sowing	Shivay et al. (2013)
	Foliar application	Improved zinc concentration and fodder quality parameters of oats	Apply Zn @ 0.5%	Dhaliwal et al. (2020)
High copper	Foliar application	Improved copper content (28.9%), yield, and fodder quality	Apply Cu @ 0.2%	Sandhu et al. (2020)
High iron	Fertilization	Good accumulation of Jerry Oats	Fertilization of coal fly ash	Bilski et al. (2012)
High selenium	Fertilization	Improved selenium concentration	Fertilization of coal fly ash	Bilski et al. (2012)

contain micronutrients or macronutrients (Farooq et al. 2019; Raj and Raj 2019). It is a type of seed priming used for speedy germination and constant establishment of the crop in adverse conditions (Paparella et al. 2015). There are some examples of nutri-priming like grain zinc content which was found to be increased by 29% in chickpeas (Farooq et al. 2019). The nutri-priming method is considered as low-cost for nutrient enrichment (Poblaciones and Rengel 2016).

As far as oat is concerned, not much information is available on agronomic biofortification in oats. There are few studies that show that agronomic biofortification is very useful in increasing the concentration of micronutrients (Table 11.3). These include coating studies of Shivay et al. (2013), where they coat oat grains with zinc as zinc oxide (ZnO) or zinc sulfate ($ZnSO_4$) at the rate of 2 kg per 100 kg and found an increase in Zn concentration of about 32 mg/kg as compared to 25 mg/kg obtained with soil application at the same rate of application. They also observed that $ZnSO_4$ was better than ZnO in soil application. Both soil and foliar applications of fertilizers on oats were found to have a dominant effect on yield and concentration. Recently, Dhaliwal et al. (2020) observed that soil application of Zn at 25 kg/ha was better as compared to 0.5% foliar application at 60 DAS and 90 DAS. Further, both methods in combination are better for Zn enrichment and higher green and dry fodder yield in oat fodder as compared to other methods of Zn application.

In developing countries, Fe, Zn, and Se deficiencies become a serious problem to human health and crop production as well (Cakmak et al. 2010; Lyons et al. 2004). Bilski et al. (2012) give emphasis to these micronutrients and proved that nutrient concentration of these nutrients can be better using some agronomic biofortification. It is difficult to afford costly chemical fertilizers to apply in the field for biofortification. Low-priced material can be affordable for farmers to apply in large amounts, and in this direction, coal combustion by-products help in accumulating Fe, Zn, and Se levels in Jerry Oats and also increasing yields of oats (Bilski et al. 1995, 2011). More recently, Sandhu et al. (2020) concluded that foliar application of Cu @ 0.2% at 60 and 90 DAS increased oat yield and quality. However, these days, agronomic biofortification through chemical fertilizers has gained great attention all over the world as it is a quicker and faster approach (Cakmak 2008; De Valença et al. 2017).

11.4 Breeding Efforts for Oat Biofortification

The success of any crop breeding program lies in the availability of genetic material in that crop species. In oat being a self-pollinated and polypoid crop, lots of genetic variability exists in nature. Since oat has a higher level of Fe, Zn, Se, I, provitamins, and proteins than other cereal grains (Table 11.2), so as such no breeding work has been carried out earlier for its biofortification. Thus, to start breeding programs for oat biofortification, the breeder should first identify parents contrasting for the trait of interest say lines with high and low Fe or Zn content. Then breeder should develop a mapping population segregating for these traits. The advantage of developing mapping populations is that breeders have the chance to get transgressive segregants which means a chance to get lines with higher Fe or Zn content than the parents, and the other is the development of homozygous lines after selfing for over seven to eight generations. Once mapping populations will be developed, these will be used for mapping of Fe or Zn content. A similar strategy will be followed for Se, I, provitamin, and protein content in oats.

In contrast to this, breeders can directly use natural population or oat germplasm for mapping of these genes (Fe, Zn, Se, I, etc.) using an association mapping panel. Association mapping panel or core collection can be developed after studying genetic diversity, population structure, and removal of relative kinship between individuals. This will not only save time but also help in the identification of tightly linked markers to the targeted trait. Such type of study is known as genome-wide association studies (GWAS). Thus, using QTL mapping or GWAS, breeders can identify tightly associated markers to the trait of interest. Such markers can be used for introgression of biofortified traits to commercial variety using molecular breeding approaches. For more detail on these approaches, readers can go through excellent review on gene pyramiding and multiple character breeding by Rana et al. (2019).

11.5 Challenges, Limitations, and Success of Breeding Approaches for Oat Biofortification

To date, not even a single biofortified oat variety is developed throughout the world, so this will be an opportunity for the oat breeders to run their breeding programs in this direction, as oat is a dual-purpose crop that has tremendous potential. Since oat has a higher level of Fe, Zn, Se, I, provitamins, and proteins than other cereal grains, so no such work has been done for its enhancement in the past. However, nowadays looking at the higher recommended dietary allowances (RDA), the work on its enhancement can be initiated. Being a self-pollinated polyploid crop with a lot of genetic variability, these traits can be easily enhanced using conventional breeding approaches. To fasten this process, newer technologies like molecular breeding, genetic engineering, or more specifically genome editing tools using CRISPR-Cas technology can be used. This will not only help in eradicating malnutrition in the human population but also be beneficial for livestock population.

11.6 Molecular Understanding of Essential Micronutrient Uptake and Deposition in Oat Grain

11.6.1 Iron (Fe)

11.6.1.1 Iron Uptake and Transport

Plants opt for two types of approaches for Fe uptake from the soil. The first one (reduction-based) is common in dicotyledons and non-Poaceae monocotyledons. Under this strategy, the plant inaccessible Fe^{3+}, the predominant ionic form of Fe in the soil, is reduced at root surface to plant-accessible Fe^{2+} form (Zhang et al. 2019). Under iron deficiency, the uptake of chelated Fe^{3+} is facilitated by H+-ATPases (AHAs) localized in the plasma membrane which aid in the reduction of rhizospheric pH and thereby enhancing the solubility of Fe^{3+}. The FRO2 (ferric chelate reductase oxidase) gene catalyzes the reduction of Fe^{3+} to Fe^{2+} which is then imported into the root cells by high-affinity iron transporters, iron-regulated transporter (IRT1). Both FRO2 and IRT1 genes were first isolated and cloned from *Arabidopsis thaliana* (Eide et al. 1996; Robinson et al. 1999). The second strategy, also known as chelation-based, is mainly observed in graminaceous species. Plants in this category secrete phytosiderophores (PS), organic compounds belonging to the family of mugineic acids, acting as Fe^{3+} chelators (Rehman et al. 2021). The chelated Fe^{3+} is then transported to roots by yellow stripe-like (YSL) transporters (Curie et al. 2001).

Oat is a type 2 strategy plant (Jeong and Connolly 2009), and plant siderophore-mediated Fe chelation was first time reported in oats and rice (Takagi 1976). There is also evidence that oat cultivars exhibit microbial siderophore-mediated iron transport system which was also reported in maize suggesting strategy 3 mechanism for Fe acquisition in plants (Marschner and Römheld 1994). Of late, it has also been observed that rice plants use a combined strategy of iron uptake comprising the components and strategies of both strategies 1 and 2 (Wairich et al. 2019). These iron-related genes are governed by several transcription factors, for example, in *Arabidopsis* bHLH (basic helix loop helix) and FIT (FER-like iron deficiency-induced transcription factor) were detected to regulate *FRO2* and *IRT1* genes for iron acquisition under iron deficiency condition (Bauer et al. 2007). Similarly, another bHLH transcription factor, POPEYE (PYE), regulates growth and development under iron deficiency (Long et al. 2010). After Fe acquisition, the ions get transported and translocated to different organs which are facilitated by two iron efflux transporters (IRON REGULATED1/Ferroportin 1 (IREG1/FPN1) and IREG2/FPN2) identified in *Arabidopsis* (Colangelo and Guerinot 2004). However, the molecular mechanism behind the long-distance iron is still under gray area. The iron ion is highly reactive and less soluble inside the plant environment, so to avoid precipitation and toxic effect, the ions are translocated inside the plant as complexes with citrate, mugineic acid, nicotinamine, and phenolic compounds. Inside xylem, iron complexes with citrate at pH 5.5 and transmembrane protein ferric reductase defective 3 (*AtFRD3*) in *Arabidopsis* and the rice orthologue *OsFRDL1* (FRD-Like) help the transport of these complexes from root to shoot (Rehman et al. 2021). Iron translocation into actively growing plant sites such as shoot apex, root tips, and seeds and remobilization of iron from old parts to new ones occur via the phloem. Inside phloem tissue, the iron complexes with the nicotinamine at pH 7.5 and the resulting complex transport in phloem with the help of yellow spike-like transporter (YSL) family. This YSL transporter gene family is known to aid in unloading iron from xylem to phloem and loading it into developing seeds (Jeong and Guerinot 2009).

11.6.1.2 Fe Deposition in Grains

The distribution of iron in grains has been reported to be heterogeneous in nature. In rice, iron is mainly present in the aleurone layer, scutellum, and integument, whereas in peas the site of accumulation is mainly in the inner and outer epidermal layers of the embryo. Deposition of iron in the developing seed is mainly facilitated by *osYSL2* in rice and *YSL1* and *YSL3* in *Arabidopsis* (Rehman et al. 2021; Tong et al. 2020).

11.6.2 Zinc (Zn)

11.6.2.1 Zn Uptake and Transport

Under high pH conditions, zinc is tightly bound to the soil making it inaccessible for plant uptake. The Zn^{2+} uptake is facilitated by acidification and production of organic chelators like citrate and malate in the plant rhizosphere. The transporter family that contributes to this process belongs to the zinc import protein (ZIP) family (Tong et al. 2020). The zinc ion inside the plant root then makes complexes with nicotinamine and gets radially transported across different root layers which is facilitated by *metal tolerance protein 2 (MTP2)* in *Arabidopsis* (Sinclair et al. 2018). Zinc transport inside the xylem is facilitated by members of *HMA* (*heavy metal ATPase*) family of P_{1B}-type ATPases, like *HMA2* and *HMA9* in rice. Once

Zn^{2+} gets loaded into the xylem, it then moves to phloem tissues for long-distance Zn transport into the sink. In rice, Zn transport through xylem to phloem, transport through phloem to organs, and remobilization activity were governed by *OsZIP3/ OsHMA2*, *YSL* family transporters, and OsHMA9 transporters, respectively (Tong et al. 2020). Recently, a report about the expression of *zmZIP1* in oat cultivar JHO 822 under zinc-treated and control conditions was studied, and it was found that the gene got upregulated on micronutrient-applied soil suggesting the key role of the ZIP family in zinc acquisition in oats (Ranjan et al. 2021).

11.6.2.2 Zn Deposition in Grains

Like iron, zinc is localized in small vacuoles in aleurone and sub-aleurone layers of the grain. Transcriptional microarray analysis of grain tissues in barley showed expression of heavy metal ATPases (HMAs), metal tolerance proteins (MTPs), and natural resistance-associated macrophage proteins (Nramps), hinting their role in Zn deposition within the grain (Tauris et al. 2009). Particularly, MTPs, a member of the cation diffusion facilitator (CDF) transporter family, are shown to localize in the vacuolar membrane and transport zinc ions to the vacuole (Podar et al. 2012). Recently, *HvMTP1* has been characterized by overexpression studies in the endosperm of barley grains using endosperm-specific promoters (Menguer et al. 2018). The upregulation of this transporter led to increased zinc concentration in endosperm which opened a new strategy for zinc enrichment in the endosperm of cereal grains.

11.7 Transgenic Efforts for the Development of Biofortified Oats

The most important factors for the successful development of transgenics in any crop include the choice of explants with high regeneration capability, robust transformation protocol, and specific expression of the desired gene(s). Several efforts have been taken in oats to optimize these aforementioned factors and few successful transgenics have been reported.

(a) **Choice of explants for regeneration in oats:** Regeneration from explants depends on the type of tissue used. Initial studies on oat regeneration systems relied on callus culture developed from mature embryos, immature embryos, and seedling epicotyls (Bregitzer et al. 1989). The callus generated from these studies was embryogenic and non-friable. However, genetic manipulation at the cell level requires a friable embryogenic culture which was subsequently developed from non-friable culture, immature embryos, and seedling mesocotyl by modifying the MS medium with different types and concentrations of the hormone. The plants derived from this callus type were regenerated for more than 78 weeks after callus establishment. The friable embryogenic (FE) callus derived from immature embryos possessed loosely associated suspensor-borne somatic embryos (SE) and was found to develop directly into plantlets hinting that these types of SE are good targets for gene(s) introduction. Subsequently, the regeneration protocol of several other explants like seedling mesocotyls, apical meristems, leaf segments, axillary tiller buds, and leaf base segments (Nuutila et al. 2002) was optimized and utilized in oat studies because of their respective advantages over immature embryo-derived regeneration system. Regeneration capacity is also dependent on the cultivar type. Among many cultivars tested for, GAF/Park oat genotype and its selection GP-1 have been reported to be highly regenerable (Kelley and Sakhanokho 2009).

(b) **Transformation method for gene transfer in oats:** Biolistic method is the most preferred mode for gene transformation in cereals and has successfully been utilized to transform different explants in oats like embryogenic calli or cell suspensions derived from immature embryos, mature embryos, leaf base segments, cultures from seeds, and shoot apical meristem (Gasparis et al. 2008). However, this type of direct gene transfer has demerits like gene rearrangement and high copy number leading to gene silencing. Indirect gene transfer methods like *Agrobacterium*-mediated transformation which was initially thought to be effective for dicots only are now successfully used in monocots. Particularly in oats, *Agrobacterium*-based transformation has been utilized reported for implant gene transfer, immature embryos, and leaf explants (Gasparis et al. 2008).

(c) **Expression of desired genes:** The successful expression of the transgene is the most important aspect in transgenic development and depends on many factors like the type of promoter used, integration site, copy number, homology to endogenous genes, in vitro culture conditions, etc. In oats, unstable gene expression and non-Mendelian inheritance have been observed in both *Agrobacterium*-mediated and particle bombardment methods of transformation (Pawlowski et al. 1998; Gasparis et al. 2008). The instability in transgene expression might be attributed to the hexaploid nature which might be preventing unnecessary gene expression and triggering epigenetic silencing leading to a non-Mendelian pattern of inheritance.

A very few efforts have been made in the area of oat transgenic development which might be due to the large and complex nature of the genome, lack of genomic resources, and its lesser importance as a cereal crop (Somers et al. 1992; Maqbool et al. 2002; Oraby et al. 2005; Carlson et al. 2006). Oat is cultivated as a dual-purpose crop; it is grown for both grain and fodder use. The grains are a good source of minerals, vitamin E, phytic acid, aventhramides, phenolic compounds, and dietary fibers like beta-glucan and also a great source of winter fodder for livestock consumption (Kelley and Sakhanokho 2009; Kapoor and Batra 2016). Oat has been approved by FDA for high beta-glucan content, dietary fiber, with multiple health benefits, and FDA recommends daily human consumption of 3 g (FDA 1997). However, the development of biofortified oat has taken a back seat with few reports where oat biofortification has been attempted by agronomic means (Bilski et al. 2012; Shivay et al. 2013; Sandhu et al. 2020). Biofortification by the transgenic approach has been successfully utilized for the development of biofortified rice, wheat, sorghum, and barley for different traits, but no such study has been reported

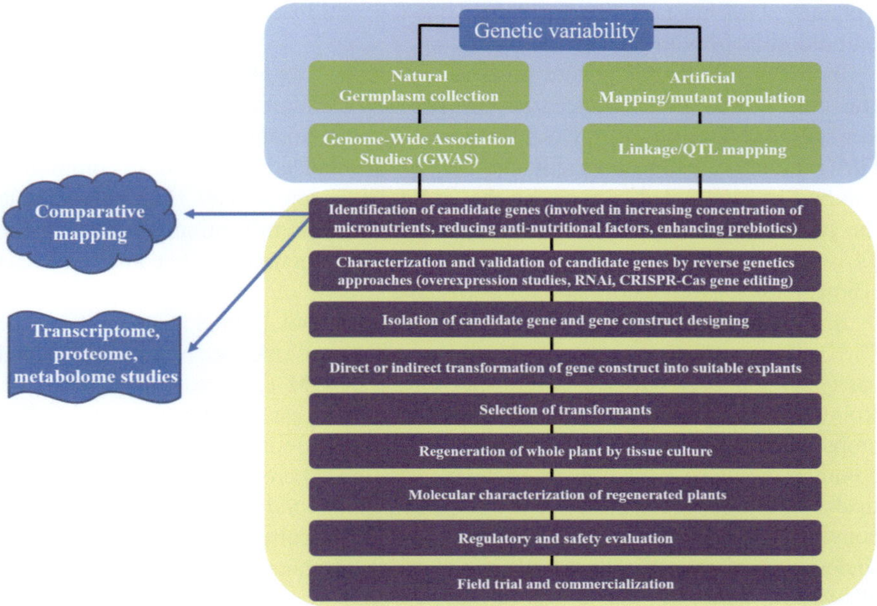

Fig. 11.2 Strategy used for the development of transgenic oats

in oat till date. Therefore, sincere efforts should be put in the area of oat biofortification with Zn, Fe, and beta-glucan with the help of a transgenic approach to expedite the goal of attaining nutritional security in the world (Fig. 11.2).

11.8 Challenges for the Public Release of Transgenic Oats

The success of developing transgenic crops solely depends on their commercialization and consumption. However, transgenic crop cultivation has raised numerous ethical issues like its impact on the environment, food safety, loss of biodiversity, resistant pest resurgence, gene flow to weedy species, etc. Several countries like the USA, Japan, Canada, Argentina, and Brazil are the largest producer of GM crops with positive acceptance by end users. In Europe, restricted use of GM crops for nonfood purposes is permitted in some countries (Shukla et al. 2018). But there still exist a lot of controversies related to human consumption of GM crops in countries like India. The negative response of the public regarding GMO is mainly due to ideological misbelief, political game, and lack of scientific knowledge. Although several scientific reports suggest the safety of GE food, in India, no GM food has been approved till date. Several efforts have been put for commercialization of the first GM food Bt brinjal way back in 2009 by the Genetic Engineering Appraisal Committee (GEAC), but the deregulation process was withheld in 2010 due to the concerns raised by anti-GM activists, farmer groups, NGOs, and scientists.

Similarly, GM mustard also faced the same fate and its commercialization was put on hold in 2014 with final call still pending.

In India, the central government has established regulatory bodies like IBSC (Institute Biosafety Committee), RCGM (Review Committee on Genetic Manipulation), RDAC (Recombinant DNA Advisory Committee), and GEAC for monitoring GMO-related research and facilitating its commercialization. But these committees and bodies have failed in handling the GMO issues completely. Thus to accelerate the process of GM product commercialization and its utilization by farmers, the Government of India is in the process of developing a new authority named as BRAI (Biotechnology Regulatory Authority of India).

Sincere efforts are needed to convince the general public for acceptance of GMOs. Issues related to ecological risk, environmental risk, and human and animal safety should be addressed before the release of GM crops for open-field trial. Robust biosafety guidelines should be framed to assess the performance of GM crops. Additionally, and most importantly, the scientific community should make the common public be aware regarding the benefits of GM crops through various social media platforms by citing success stories and also by conducting visits to villages to convince the farmers regarding the positive effects of cultivating GM crops on their socio-economic status. The actual impact of GMOs cannot be predicted until it is allowed for field trial; that is why recently the Indian government has approved the field trial of Bt brinjal variety that is in pipeline for commercial release for a decade ensuring that the current government is taking interest in transgenic commercialization. Hence, future research should be focused on transgenic biofortified oat development which if commercialized would definitely address the malnutrition issues in the world.

11.9 Economic and Social Constraints for the Biofortified Oats

Biofortification enhances food crops with added nutrition using conventional or classical plant breeding, various agronomic approaches, and genetic engineering (Talsma et al. 2017). Consumption of biofortified food grains will provide nutrients more easily to the human population. Even though a diet enriched with balanced food is recommended for daily micronutrient intake, the resources-poor population of the society depends on the cheap and reliable source of calories, i.e., rice and wheat. Biofortification promises to deliver micronutrients to the poor community as a cost-effective method since it involves only a one-time investment in breeding. The process of biofortification is mostly dependent on public funding (Bouis and Welch 2010). Therefore, a lot of factors will take into consideration including the research and development (R&D) resources dedicated for this effort. If private funding is involved, then returns would have to be generated either in form of hybrids or intellectual property rights that prohibit farmers from selling the seeds of their produce. Biofortification does not fall under this model as it has been generated for public use worldwide. Also, these processes have some challenging limitations such as high production costs required for equipment, technology, patenting, etc.

which are a cause of concern that can hamper this method to go global and the profit margin for private investors is very low in case of biofortified crops.

Participation of society during the designing of biofortification strategy plays a key role in making any biofortified crop a reality. The success of any biofortification program basically depends on the readiness of consumers and producers to accept the newly bred crop varieties (Saltzman et al. 2013). For producers, the adoption of nutrient-enriched crops will depend on various factors like yield, disease resistance, drought tolerance, and marketability, whereas for consumers, adoption can be influenced by the change in sensory traits of biofortified crops. Another major constraint is the lack of education which controls the socio-economic status. Consumers should be able to make an informed decision about acceptances of biofortified crops for which education in the necessary field is of utmost importance. Empowering women and children through education and capacity building would help make these crops more acceptable.

Biofortification can very well help in eradicating malnutrition. However, scientists need to move forward and disseminate the knowledge to the masses so that an informed decision can be taken. Finally, getting consumers' and farmers' awareness to accept biofortified crops will be a challenge, but advancements in good seed systems, expansion of markets and products, and demand-supply chain can become a reality (Nestel et al. 2006).

11.10 Genome Editing Approaches for Biofortification of Oats

There are mainly three types of genome editing (GE) tools that have been utilized successfully for mutating the gene of interest in plants. The tools reported till date vary based on the type of DNA-binding domain (DBD) and DNA cleavage domain, but they work on the same principle, that is, induction of double-stranded breaks (DSBs) followed by DNA repair by nonhomologous end joining (NHEJ) and homology-dependent recombination (HDR). Initially, tools like zinc finger nucleases (ZFNs) were introduced in 2002, and then transcription activator-like effector nucleases (TALENs) came into the picture in 2011. ZFNs consist of proteins having zinc finger motifs with each finger recognizing three base pairs (bp) of DNA. These fingers are customized according to the target sequence and are fused with endonucleases for cleavage of the target gene. TALENs, on the other hand, consist of transcription activator-like effector (TALE) protein (first reported to be secreted by *Xanthomonas* bacteria) and endonucleases. TALEs comprise of series of 33–35 amino acid repeat domains with 2 amino acid residues at 12th and 13th position of the domain (repeat variable diresidue) responsible for exploiting these proteins as DNA-binding domains. This residue recognizes a single DNA bp and is customized according to the target sequence. Although ZFNs and TALENs have comparable efficiencies when targeted to the same gene (Joung and Sander 2013), TALENs became the tool of choice subsequently because of the ease of designing and higher cleavage rate. Nevertheless, both these systems have their limitations: ZFNs show limited specificity and introduce off-target mutation (Puchta 2017), while vector construction in TALENs is a time- and labor-consuming process (Manghwar et al. 2019). Recently, clustered regularly interspaced palindromic repeats (CRISPR)-associated 9 (CRISPR-Cas9) system has captured a lot of attention because of its easy design and implementation. CRISPR-Cas9 is an RNA-guided endonuclease that cleaves the target DNA sequence at a precise location (3 bp upstream of protospacer-associated motif causing DSBs with the blunt end) (Jiang and Doudna 2017). This triggers a DNA repair mechanism leading to the introduction of indels/knockout of genes via NHEJ or knockin/gene replacement when template DNA is available via HDR. This CRISPR-based system is the tool of choice for genome editing in various crops because of its editing efficiency, simplicity in vector and guide RNA (gRNA) designing, easy use, and low cost compared to other aforementioned GE tools. However, this tool also has some limitations like the introduction of off-targets, limited protospacer-associated motif (PAM) site, and low HDR efficiency. Since the advent of CRISPR system in 2012, several researchers have taken an interest to enhance the efficiency and specificity of the system which led to the generation of many improved CRISPR-Cas systems. For example, Cas13 has been identified as a CRISPR effector which targets viral RNA and RNA endogenous to plants (Wolter and Puchta 2018). xCas9, a SpCas9 variant, recognizes a broader range of PAM with high editing efficiency and specificity (Rees and Liu 2018). CRISPR-Cas12a (Cpf1) isolated from Prevotella and Francisella (Cpf1) recognizes T-rich sequence at 5′ end of PAM sequence and not typical GG as in case of Cas9 (Zetsche et al. 2015). Recently, CRISPR-Cas14a has been isolated from non-culturable archaea that target single-stranded DNA (ssDNA) opening avenues for engineering plant resistance against ssDNA plant viruses (Khan et al. 2019). This system doesn't require flanking PAM sequence which is the major limitation in other Cas systems.

Although regeneration and transformation protocols have been standardized in many oat cultivars which are the prerequisites for GE in any crop, till date genome editing has not been attempted to develop GE oats. The reason might be due to lack of genome information of hexaploid oat, polyploidy, and the complex nature of the genome. However, available genomic resources from diploid *Avena* can be utilized to predict quality-related genes by a comparative genomics approach. This will facilitate in attempting gene/genome editing of desired genes in oats by the CRISPR-Cas system.

References

Abdollahi E, Kohram H, Shahir MH (2013) Plasma concentrations of essential trace microminerals and thyroid hormones during single or twin pregnancies in fat-tailed ewes. Small Rumin Res 113(2–3):360–364

Acar O, Izydorczyk MS, Kletke J, Atilla Yazici M, Imamoglu A, Cakmak I, Koksel H (2020) Comparison of short and long milling flows on yield and physicochemical properties of brans from biofortified and non-biofortified hull-less oats. Cereal Chem 97(4):859–867

Agarwal A, Shaharyar A, Kumar A, Bhat MS, Mishra M (2015) Scurvy in pediatric age group—a disease often forgotten? J Clin Orthop Trauma 6(2):101–107

Anderson JW, Chen WJL (1986) Cholesterol lowering properties of oat products. In: Webster EH (ed) Oats: chemistry and technology. AACC, St. Paul, pp 309–327

Asensi-Fabado MA, Munné-Bosch S (2010) Vitamins in plants: occurrence, biosynthesis and antioxidant function. Trends Plant Sci 15:582–592

Bailey RL, West KP Jr, Black RE (2015) The epidemiology of global micronutrient deficiencies. Ann Nutr Metab 66(2):22–33

Barbosa GL, Gadelha FDA, Kublik N, Proctor A, Reichelm L, Weissinger E, Wohlleb GM, Halden RU (2015) Comparison of land, water, and energy requirements of lettuce grown using hydroponic vs. conventional agricultural methods. Int J Environ Res Public Health 12(6): 6879–6891

Bauer P, Ling H-Q, Guerinot ML (2007) FIT, the FER-LIKE IRON DEFICIENCY INDUCED TRANSCRIPTION FACTOR in Arabidopsis. Plant Physiol Biochem 45(5):260–261. https://doi.org/10.1016/j.plaphy.2007.03.006

Bell RW, Dell B (2008) Micronutrients for sustainable food, feed, fibre and bioenergy production. International Fertilizer Industry Association (IFA), Paris, p 175

Bhardwaj D, Ansari MW, Sahoo RK, Tuteja N (2014) Biofertilizers function as key player in sustainable agriculture by improving soil fertility, plant tolerance and crop productivity. Microb Cell Factories 13(1):1–10

Bilski J, Alva AK, Sajwan OS (1995) Agricultural uses of coal fly ash. In: Rechcigl JE (ed) Environmental aspects of soil amendments, Inorganic fertilizers, vol 1. Lewis Publishers, Boca Raton, pp 255–291

Bilski J, Mclean K, McLean E, Soumaila F, Lander M (2011) Revegetation of coal ash by selected cereal crops and trace elements accumulation by plant seedlings. Int J Environ Sci 1(5): 1033–1046

Bilski J, Jacob D, Soumaila F, Kraft C, Farnsworth A (2012) Agronomic biofortification of cereal crop plants with Fe, Zn, and Se, by the utilization of coal fly ash as plant growth media. Adv Biores 3(4):130–136

Bissardon C, Bohic S, Francis L, Khan I, Charlet L (2017) Role of selenium in articular cartilage metabolism, growth, and maturation. In: Proceedings of the 11th international symposium on selenium in biology and medicine and the 5th international conference on selenium in the environment and human health, Stockholm

Black RE, Allen LH, Bhutta ZA, Caulfield LE, De Onis M, Ezzati M, Mathers F, Rivera J (2008) Maternal and child undernutrition: global and regional exposures and health consequences. Lancet 371(9608):243–260

Blancquaert D, De Steur H, Gellynck X, Van Der Straeten D (2014) Present and future of folate biofortification of crop plants. J Exp Bot 65(4):895–906

Bouis HE, Welch RM (2010) Biofortification—a sustainable agricultural strategy for reducing micronutrient malnutrition in the global south. Crop Sci 50:S20–S32

Bregitzer P, Somers DA, Rines HW (1989) Development and characterization of friable, embryogenic oat callus. Crop Sci 29(3):798–803

Butt MS, Tahir-Nadeem M, Khan MKI, Shabir R, Butt MS (2008) Oat: unique among the cereals. Eur J Nutr 47(2):68–79

Cakmak I (2008) Enrichment of cereal grains with zinc: agronomic or genetic biofortification? Plant Soil 302(1):1–17

Cakmak I, Kalaycı M, Ekiz H, Braun HJ, Kılınç Y, Yılmaz A (1999) Zinc deficiency as a practical problem in plant and human nutrition in Turkey: a NATO-science for stability project. Field Crop Res 60(1–2):175–188

Cakmak I, Pfeiffer WH, McClafferty B (2010) Biofortification of durum wheat with zinc and iron. Cereal Chem 87:10–20

Cakmak I, Guilherme LRG, Rashid A, Hora KH, Yazici A, Savasli E, Kalayci M, Tutus Y, Phuphong P, Rizwan M, Martins FAD, Dinali GS, Ozturk L (2017) Iodine biofortification of wheat, rice and maize through fertilizer strategy. Plant Soil 418(1):319–335

Carlson A, Skadsen R, Kaeppler HF (2006) Barley hordothionin accumulates in transgenic oat seeds and purified protein retains anti-fungal properties in vitro. In Vitro Cell Dev Biol Plant 42(4):318–323

Colangelo EP, Guerinot ML (2004) The essential basic helix-loop-helix protein FIT1 is required for the iron deficiency response. Plant Cell 16(12):3400–3412. https://doi.org/10.1105/tpc.104.024315

Curie C, Panaviene Z, Loulergue C, Dellaporta SL, Briat J-F, Walker EL (2001) Maize yellow stripe1 encodes a membrane protein directly involved in Fe(III) uptake. Nature 409(6818): 346–349. https://doi.org/10.1038/35053080

D'Amato R, Fontanella MC, Falcinelli B, Beone GM, Bravi E, Marconi O, Benincasa P, Businelli D (2018) Selenium biofortification in rice (Oryza sativa L.) sprouting: effects on Se yield and nutritional traits with focus on phenolic acid profile. J Agric Food Chem 66(16):4082–4090

Dattani N, Jiang A (2009) The diabetic pandemic: globalization, industrialization, and type 2 diabetes. The Meducator 1(15)

De Benoist B, McLean E, Egli I, Cogswell M (2008) Worldwide prevalence of anaemia 1993-2005. WHO Global Database of Anaemia, Geneva

de La Garza RID, Gregory JF, Hanson AD (2007) Folate biofortification of tomato fruit. Proc Natl Acad Sci 104(10):4218–4222

De Valença AW, Bake A, Brouwer ID, Giller KE (2017) Agronomic biofortification of crops to fight hidden hunger in sub-Saharan Africa. Glob Food Sec 12:8–14

Dhaliwal SS, Sandhu AS, Shukla AK, Sharma V, Kumar B, Singh R (2020) Bio-fortification of oats fodder through zinc enrichment to reduce animal malnutrition. J Agric Sci Technol 10:98–108

Dong W, Thomas N, Ronald PC, Goyer A (2016) Overexpression of thiamin biosynthesis genes in rice increases leaf and unpolished grain thiamin content but not resistance to Xanthomonas oryzae pv. oryzae. Front Plant Sci 7:616

Dowling JE, Wald G (1958) Vitamin A deficiency and night blindness. Proc Natl Acad Sci U S A 44(7):648–661

Eide D, Broderius M, Fett J, Guerinot ML (1996) A novel iron-regulated metal transporter from plants identified by functional expression in yeast. Proc Natl Acad Sci 93(11):5624–5628. https://doi.org/10.1073/pnas.93.11.5624

Epstein E, Bloom AJ (2005) Mineral nutrition of plants: principles and perspectives, 2nd edn. Sinauer Associates, Sunderland

Erenoglu EB, Kutman UB, Ceylan Y, Yildiz B, Cakmak I (2011) Improved nitrogen nutrition enhances root uptake, root-to-shoot translocation and remobilization of zinc (65Zn) in wheat. New Phytol 189(2):438–448

FAOSTAT (2019) Oats production in 2019, Crops/World Regions/Production Quantity from pick lists. Food and Agriculture Organization, Statistics Division. Accessed 12 Feb 2021

FAOSTAT (2021) Oats production in 2021, Crops/World Regions/Production Quantity from pick lists. Food and Agriculture Organization, Statistics Division. Accessed 12 Feb 2021

Farooq M, ur-Rehman A, Aziz T, Habib M (2011) Boron nutripriming improves the germination and early seedling growth of rice (Oryza sativa L.). J Plant Nutr 34(10):1507–1515

Farooq M, Usman M, Nadeem F, ur Rehman H, Wahid A, Basra SM, Siddique KH (2019) Seed priming in field crops: potential benefits, adoption and challenges. Crop Pasture Sci 70(9): 731–771

FDA (1997) FDA allows whole oat foods to make health claim on reducing the risk of heart disease. Food and Drug Administration. U.S. Department of Health and Human Services, Washington, DC

Flander L, Salmenkallio-Marttila M, Suortti T, Autio K (2007) Optimization of ingredients and baking process for improved wholemeal oat bread quality. LWT Food Sci Technol 40(5): 860–870

Gasparis S, Bregier C, Orczyk W, Nadolska-Orczyk A (2008) Agrobacterium-mediated transformation of oat (Avena sativa L.) cultivars via immature embryo and leaf explants. Plant Cell Rep 27(11):1721–1729

Gauldie E (1981) The Scottish Country Miller 1700-1900: a history of water-powered meal milling in Scotland. John Donald Publishers, Edinburgh

Goyer A, Sweek K (2011) Genetic diversity of thiamin and folate in primitive cultivated and wild potato (Solanum) species. J Agric Food Chem 59(24):13072–13080

Graham TW (1991) Trace element deficiencies in cattle. Vet Clin N Am Food Anim Pract 7(1): 153–215

Heaney RP (2000) Calcium, dairy products and osteoporosis. J Am Coll Nutr 19(2):83S–99S

Heuzé V, Tran G, Bound A, Lebas F (2016) Oat forage. In: Feedipedia, a programme by INRA. CIRAD, AFZ and FAO, Rome

Ho HV, Sievenpiper JL, Zurbau A, Mejia SB, Jovanovski E, Au-Yeung F et al (2016) The effect of oat β-glucan on LDL-cholesterol, non-HDL-cholesterol and apoB for CVD risk reduction: a systematic review and meta-analysis of randomised-controlled trials. Br J Nutr 116(8): 1369–1382

Jeong J, Connolly EL (2009) Iron uptake mechanisms in plants: functions of the FRO family of ferric reductases. Plant Sci 176(6):709–714

Jeong J, Guerinot ML (2009) Homing in on iron homeostasis in plants. Trends plant sci 14(5): 280–285

Jiang F, Doudna JA (2017) CRISPR–Cas9 structures and mechanisms. Annu Rev Biophys 46:505–529

Joung JK, Sander JD (2013) TALENs: a widely applicable technology for targeted genome editing. Nat Rev Mol Cell Biol 14(1):49–55

Kapoor R, Batra C (2016) Oats. In: Broadening the genetic base of grain cereals. Springer, New Delhi, pp 127–162

Kelley RY, Sakhanokho HF (2009) Oat. In: Compendium of transgenic crop plants. Wiley, New York, pp 139–156

Kennedy G, Burlingame B (2003) Analysis of food composition data on rice from a plant genetic resources perspective. Food Chem 80(4):589–596

Kennedy G, Nantel G, Shetty P (2003) The scourge of "hidden hunger": global dimensions of micronutrient deficiencies. Food Nutr Agric 32:8–16

Khan MZ, Haider S, Mansoor S, Amin I (2019) Targeting plant ssDNA viruses with engineered miniature CRISPR-Cas14a. Trends Biotechnol 37(8):800–804

Köhrle J (2013) Selenium and the thyroid. Curr Opin Endocrinol Diabetes Obes 20(5):441–448

Kraemer K, Waelti M, De Pee S, Moench-Pfanner R, Hathcock JN, Bloem MW, Semba RD (2008) Are low tolerable upper intake levels for vitamin A undermining effective food fortification efforts? Nutr Rev 66(9):517–525

Lawson PG, Daum D, Czauderna R, Meuser H, Härtling JW (2015) Soil versus foliar iodine fertilization as a biofortification strategy for field-grown vegetables. Front Plant Sci 6:450

Lazarus JH (2015) The importance of iodine in public health. Environ Geochem Health 37(4): 605–618

Li KT, Moulin M, Mangel N, Albersen M, Verhoeven-Duif NM, Ma Q, Zhang P, Fitzpatrick TB, Gruissem W, Vanderschuren H (2015) Increased bioavailable vitamin B6 in field-grown transgenic cassava for dietary sufficiency. Nat Biotechnol 33(10):1029–1032

Li J, Liu J, Zhang P, Wan Y, Xia X, Zhang Y, He Z (2018) Genome-wide association mapping of vitamins B1 and B2 in common wheat. Crop J 6(3):263–270

Long TA, Tsukagoshi H, Busch W, Lahner B, Salt DE, Benfey PN (2010) The bHLH transcription factor POPEYE regulates response to iron deficiency in Arabidopsis roots. Plant Cell 22(7): 2219–2236

Lv C, Yang Y, Jiang L, Gao L, Rong S, Darko GM, Jiang W, Gao Y, Sun D (2017) Association between chronic exposure to different water iodine and thyroid cancer: a retrospective study from 1995 to 2014. Sci Total Environ 609:735–741

Lyons G (2018) Biofortification of cereals with foliar selenium and iodine could reduce hypothyroidism. Front Plant Sci 9:730

Lyons GH, Lewis J, Lorimer MF, Holloway RE, Brace DM, Stangoulis JC, Graham RD (2004) High-selenium wheat: agronomic biofortification strategies to improve human nutrition. Food Agric Environ 2(1):171–178

Maier SM, Turner ND, Lupton JR (2000) Serum lipids in hypercholesterolemic men and women consuming oat bran and amaranth products. Cereal Chem 77(3):297–302

Maleki FS, Chaichi MR, Mazaheri D, Tavakkol AR, Savaghebi G (2011) Barley grain mineral analysis as affected by different fertilizing systems and by drought stress. J Agric Sci Technol 13(3):315–326

Mangel N, Fudge JB, Fitzpatrick TB, Gruissem W, Vanderschuren H (2017) Vitamin B1 diversity and characterization of biosynthesis genes in cassava. J Exp Bot 68(13):3351–3363

Manghwar H, Lindsey K, Zhang X, Jin S (2019) CRISPR/Cas system: recent advances and future prospects for genome editing. Trends Plant Sci 24(12):1102–1125

Maqbool S, Zhong H, El-Maghraby Y, Ahmad A, Chai B, Wang W, Sabzikar R, Sticklen M (2002) Competence of oat (Avena sativa L.) shoot apical meristems for integrative transformation, inherited expression, and osmotic tolerance of transgenic lines containing hva1. Theor Appl Genet 105(2–3):201–208

Marschner H, Römheld V (1994) Strategies of plants for acquisition of iron. Plant Soil 165(2): 261–274

McGuire S (2015) FAO, IFAD, and WFP. The state of food insecurity in the world 2015: meeting the 2015 international hunger targets: taking stock of uneven progress. Rome: FAO, 2015. Adv Nutr 6(5):623–624

Menguer PK, Vincent T, Miller AJ, Brown JK, Vincze E, Borg S et al (2018) Improving zinc accumulation in cereal endosperm using Hv MTP 1, a transition metal transporter. Plant Biotechnol J 16(1):63–71

Muthayya S, Rah JH, Sugimoto JD, Roos FF, Kraemer K, Black RE (2013) The global hidden hunger indices and maps: an advocacy tool for action. PLoS One 8(6):e67860

Nestel P, Bouis HE, Meenakshi JV, Pfeiffer W (2006) Biofortification of staple food crops. J Nutr 136(4):1064–1067

Nocek JE, Socha MT, Tomlinson DJ (2006) The effect of trace mineral fortification level and source on performance of dairy cattle. J Dairy Sci 89(7):2679–2693

Nuutila AM, Villiger C, Oksman-Caldentey KM (2002) Embryogenesis and regeneration of green plantlets from oat (Avena sativa L.) leaf-base segments: influence of nitrogen balance, sugar and auxin. Plant Cell Rep 20(12):1156–1161

Oraby HF, Ransom CB, Kravchenko AN, Sticklen MB (2005) Barley HVA1 gene confers salt tolerance in R3 transgenic oat. Crop Sci 45(6):2218–2227

Paine JA, Shipton CA, Chaggar S, Howells RM, Kennedy MJ, Vernon G et al (2005) Improving the nutritional value of Golden Rice through increased pro-vitamin A content. Nat Biotechnol 23(4):482–487

Palmgren MG, Clemens S, Williams LE, Krämer U, Borg S, Schjørring JK, Sanders D (2008) Zinc biofortification of cereals: problems and solutions. Trends Plant Sci 13(9):464–473

Paparella S, Araújo SS, Rossi G, Wijayasinghe M, Carbonera D, Balestrazzi A (2015) Seed priming: state of the art and new perspectives. Plant Cell Rep 34(8):1281–1293

Pawlowski WP, Torbert KA, Rines HW, Somers DA (1998) Irregular patterns of transgene silencing in allohexaploid oat. Plant Mol Biol 38(4):597–607

Pearce EN, Andersson M, Zimmermann MB (2013) Global iodine nutrition: where do we stand in 2013? Thyroid 23(5):523–528

Pearce EN, Lazarus JH, Moreno-Reyes R, Zimmermann MB (2016) Consequences of iodine deficiency and excess in pregnant women: an overview of current knowns and unknowns. Am J Clin Nutr 104(3):918S–923S

Poblaciones MJ, Rengel Z (2016) Soil and foliar zinc biofortification in field pea (Pisum sativum L.): Grain accumulation and bioavailability in raw and cooked grains. Food Chem 212:427–433

Podar D, Scherer J, Noordally Z, Herzyk P, Nies D, Sanders D (2012) Metal selectivity determinants in a family of transition metal transporters. J Biol Chem 287(5):3185–3196

Prasad R (2013) Fertilizer nitrogen, food security, health and environment. Proc Indian Natl Sci Acad 79(4):997–110

Prasanna R, Bidyarani N, Babu S, Hossain F, Shivay YS, Nain L (2015) Cyanobacterial inoculation elicits plant defense response and enhanced Zn mobilization in maize hybrids. Cogent Food Agric 1(1):998507

Puchta H (2017) Applying CRISPR/Cas for genome engineering in plants: the best is yet to come. Curr Opin Plant Biol 36:1–8

Raj AB, Raj SK (2019) Seed priming: an approach towards agricultural sustainability. J Appl Nat Sci 11(1):227–234

Ramzani PMA, Khalid M, Naveed M, Ahmad R, Shahid M (2016) Iron biofortification of wheat grains through integrated use of organic and chemical fertilizers in pH affected calcareous soil. Plant Physiol Biochem 104:284–293

Ramzani PMA, Khalid M, Anjum S, Ali S, Hannan F, Iqbal M (2017) Cost-effective enhanced iron bioavailability in rice grain grown on calcareous soil by sulfur mediation and its effect on heavy metals mineralization. Environ Sci Pollut Res 24(2):1219–1228

Rana M, Sood A, Hussain W, Kaldate R, Sharma TR, Gill RK et al (2019) Gene pyramiding and multiple character breeding. In: Lentils. Academic, New York, pp 83–124

Ranjan R, Chauhan VS, Chaudhary M, Singh KK, Ahmed S, Dwivedi KK (2021) Expression of ZmZIP1 a gene involved in zinc transport after nutrient application in Oat (Avena Sativa L.). Adv Biores 12(2):23–29

Ranjhan SK (1994) Availability and requirements of feed and fodder for livestock and poultry. Report, Department of Animal Husbandry, Government of India, New Delhi

Rasane P, Jha A, Sabikhi L, Kumar A, Unnikrishnan VS (2015) Nutritional advantages of oats and opportunities for its processing as value added foods—a review. J Food Sci Technol 52(2): 662–675

Rayman MP (2002) The argument for increasing selenium intake. Proc Nutr Soc 61(2):203–215

Rees HA, Liu DR (2018) Base editing: precision chemistry on the genome and transcriptome of living cells. Nat Rev Genet 19(12):770–788

Rehman AU, Masood S, Khan NU, Abbasi ME, Hussain Z, Ali I (2021) Molecular basis of iron biofortification in crop plants; a step towards sustainability. Plant Breed 140(1):12–22

Robinson NJ, Procter CM, Connolly EL, Guerinot ML (1999) A ferric-chelate reductase for iron uptake from soils. Nature 397(6721):694. https://doi.org/10.1038/17800

Ruwali Y, Singh K, Kumar S, Kumar L (2013) Molecular diversity analysis in selected fodder and dual purpose oat (Avena sativa L.) genotypes by using random amplified polymorphic DNA (RAPD). Afr J Biotechnol 12(22)

Saeedi P, Petersohn I, Salpea P, Malanda B, Karuranga S, Unwin N et al (2019) Global and regional diabetes prevalence estimates for 2019 and projections for 2030 and 2045: results from the International Diabetes Federation Diabetes Atlas. Diabetes Res Clin Pract 157:107843

Saltzman A, Birol E, Bouis HE, Boy E, De Moura FF, Islam Y, Pfeiffer WH (2013) Biofortification: progress toward a more nourishing future. Glob Food Sec 2(1):9–17

Sandhu A, Dhaliwal SS, Shukla AK, Sharma V, Singh R (2020) Fodder quality improvement and enrichment of oats with Cu through biofortification: a technique to reduce animal malnutrition. J Plant Nutr 43(10):1378–1389

Scott SP, Chen-Edinboro LP, Caulfield LE, Murray-Kolb LE (2014) The impact of anemia on child mortality: an updated review. Nutrients 6:5915–5932. https://doi.org/10.3390/nu6125915

Shivay YS, Prasad R, Pal M (2013) Zinc fortification of oat grains through zinc fertilisation. Agric Res 2(4):375–381

Shukla M, Al-Busaidi KT, Trivedi M, Tiwari RK (2018) Status of research, regulations and challenges for genetically modified crops in India. GM Crops Food 9(4):173–188

Sinclair SA, Senger T, Talke IN, Cobbett CS, Haydon MJ, Krämer U (2018) Systemic upregulation of MTP2-and HMA2-mediated Zn partitioning to the shoot supplements local Zn deficiency responses. Plant Cell 30(10):2463–2479

Somers DA, Rines HW, Gu W, Kaeppler HF, Bushnell WR (1992) Fertile, transgenic oat plants. Biotechnology 10(12):1589–1594

Stevens GA, Finucane MM, De-Regil LM, Paciorek CJ, Flaxman SR, Branca F, Peña-Rosas JP, Bhutta ZA, Ezzati M, Nutrition Impact Model Study Group (2013) Global, regional, and national trends in haemoglobin concentration and prevalence of total and severe anaemia in children and pregnant and non-pregnant women for 1995–2011: a systematic analysis of population-representative data. Lancet Glob Health 1(1):e16–e25

Storozhenko S, De Brouwer V, Volckaert M, Navarrete O, Blancquaert D, Zhang GF, Lambert W, Van Der Straeten D (2007) Folate fortification of rice by metabolic engineering. Nat Biotechnol 25(11):1277–1279

Sunde RA (2010) Selenium. In: Coates PM, Bertz JM, Blackman MR (eds) Encyclopedia of dietary supplements. Informa Healthcare, London

Szatanik-Kloc A, Ambrozewicz-Nita A, Franus W, Józefaciuk G (2019) Early effect of clinoptilolite on yield and quality of oat (Avena sativa L.). Int Agrophys 33(1):107

Takagi SI (1976) Naturally occurring iron-chelating compounds in oat- and rice-root washings: I. Activity measurement and preliminary characterization. Soil Sci Plant Nutr 22(4):423–433

Talsma EF, Melse-Boonstra A, Brouwer ID (2017) Acceptance and adoption of biofortified crops in low- and middle-income countries: a systematic review. Nutr Rev 75(10):798–829

Tauris B, Borg S, Gregersen PL, Holm PB (2009) A roadmap for zinc trafficking in the developing barley grain based on laser capture microdissection and gene expression profiling. J Exp Bot 60(4):1333–1347

Taylor J, Taylor JR (2011) Protein biofortified sorghum: effect of processing into traditional African foods on their protein quality. J Agric Food Chem 59(6):2386–2392. https://doi.org/10.1021/jf104006v

Tong J, Sun M, Wang Y, Zhang Y, Rasheed A, Li M et al (2020) Dissection of molecular processes and genetic architecture underlying iron and zinc homeostasis for biofortification: from model plants to common wheat. Int J Mol Sci 21(23):9280

Wairich A, de Oliveira BHN, Arend EB, Duarte GL, Ponte LR, Sperotto RA, Ricachenevsky FK, Fett JP (2019) The combined strategy for iron uptake is not exclusive to domesticated rice (Oryza sativa). Sci Rep 9(1):16144. https://doi.org/10.1038/s41598-019-52502-0

Walker CF, Ezzati M, Black RE (2009) Global and regional child mortality and burden of disease attributable to zinc deficiency. Eur J Clin Nutr 63(5):591–597

Welch RM, Graham RD (2005) Agriculture: the real nexus for enhancing bioavailable micronutrients in food crops. J Trace Elem Med Biol 18(4):299–307

WHO (2017a) Fact sheets. https://www.who.int/news-room/fact-sheets/detail/cardiovascular-diseases-(cvds)

WHO (2017b) Malnutrition. www.who.int/mediacentre/factsheets/malnutrition/en/. Accessed 4 Feb 2019

Wolter F, Puchta H (2018) The CRISPR/Cas revolution reaches the RNA world: Cas13, a new Swiss Army knife for plant biologists. Plant J 94(5):767–775

World Health Organization (2021) Biofortification of staple crops. http://www.who.int/elena/titles/complementary_feeding/en/. Accessed 12 Jan 2021

Wu Q, Rayman MP, Lv H, Schomburg L, Cui B, Gao C, Chen P, Zhuang G, Zhang Z, Peng X, Li H, Zhao Y, He X, Zeng G, Qin F, Hou P, Shi B (2015) Low population selenium status is associated with increased prevalence of thyroid disease. J Clin Endocrinol Metabol 100(11):4037–4047

Ye X, Al-Babili S, Klöti A, Zhang J, Lucca P, Beyer P, Potrykus I (2000) Engineering the provitamin A (β-carotene) biosynthetic pathway into (carotenoid-free) rice endosperm. Science 287(5451):303–305

Zetsche B, Gootenberg JS, Abudayyeh OO, Slaymaker IM, Makarova KS, Essletzbichler P, Volz SE, Joung J, Van Der Oost J, Regev A, Koonin EV (2015) Cpf1 is a single RNA-guided endonuclease of a class 2 CRISPR-Cas system. Cell 163(3):759–771

Zhang X, Zhang D, Sun W, Wang T (2019) The adaptive mechanism of plants to iron deficiency via iron uptake, transport, and homeostasis. Int J Mol Sci 20(10):2424. https://doi.org/10.3390/ijms20102424

Zimmermann MB, Andersson M (2012) Assessment of iodine nutrition in populations: past, present, and future. Nutr Rev 70(10):553–570

Nutrigenomics in Cereals

12

Shashank Kumar Yadav, Pragya Yadav,
and Viswanathan Chinnusamy

Abstract

The sequencing of human genome has opened floodgates of immense knowledge and opportunities in the realm of personalized nutrition and health. Nutrigenomics is a rapidly emerging field that employs tools of bioinformatics, genomics, metabolomics, proteomics, epigenomics, and transcriptomics to bridge the existing gap and build up a holistic understanding of the interaction of dietary components and genes at the molecular level. Recent scientific evidence has fortified that genetic polymorphism plays a key role in daily nutritional requirements, metabolic response to food, and potency of dietary factors in response to diseases. Every individual has a categorical response to nutrients which results in nutrient impairment leading to alteration of gene expression. Several reports have highlighted that nutrients like carbohydrates, amino acids, fatty acids, vitamins, and minerals play a pivotal role in the regulation of gene expression. Cereal crops predominantly constitute 50% of daily dietary energy and protein source for majority of the worlds population. With the advent of molecular biology tools like genetic engineering, genome editing, and marker-assisted breeding, cereal crops are being enriched in order to maximize their nutritional potential while minimizing the anti-nutrient contents. The chapter highlights the potential role of cereals in nutrigenomics with emphasis on the current advances and challenges in the field.

Keywords

Gene-diet interactions · Nutritive value of cereals · Functional foods · Omics · Non-communicable diseases · Genome engineering

S. K. Yadav (✉) · P. Yadav · V. Chinnusamy
Division of Plant Physiology, ICAR-Indian Agricultural Research Institute, New Delhi, India

R. Deshmukh et al. (eds.), *Biofortification in Cereals*,
https://doi.org/10.1007/978-981-19-4308-9_12

12.1 Introduction

Nutrigenomics is "the study of bidirectional interactions between genes and diet". Deciphering of the human genome in 2001 marked the beginning of a revolutionary era in the realm of nutrigenomics, particularly its impact on human health. The central dogmatic flow of genetic information from gene expression to protein synthesis is largely influenced by lifestyle, and diet particularly nutrients and non-nutrient components present therein. The novel example for the crosstalk between the human genome and its environment includes genome-food interaction which eventually translates into a healthy or diseased state on the basis of a given genome. Phenotype, which represents the physical characteristics or observable traits of an organism, is governed by the spatio-temporal expression of a gene or multiple genes. Apart from genetic expression, phenotypic characteristics are also influenced by nutrition, e.g. cardiovascular diseases are mainly caused by cholesterol present in the food. Cereals have remained an indispensable part of the human diet since the dawn of agriculture, but the true nutritive potential of cereals has largely remained unexplored. In the present chapter, emerging paradigms of nutritional genomics with respect to cereals are discussed as they comprise the functional food market. Emphasis has been laid on leveraging the role of genomics tools to produce better food for improving human nutrition and health and thereby delivering societal and economic benefits.

12.2 Nutrition Value of Cereals

Cereals and cereal products have always been an indispensable part of most human diets since the dawn of agriculture till the present irrespective of socio-economic and geographical divergence across the world. The importance of cereals and cereal products can be accessed from the fact that cereal crops are cultivated across 50% of harvested area worldwide [FAO Food Outlook], contributing to the greatest degree to the global food security, which extends to approximately 2800 million tons annually [FAO 2020]. Cereals and cereal products alone contribute to 75% of the daily dietary energy and nutritional requirement of the global population as they are a mammoth source of macronutrients, primarily carbohydrates (75%), proteins (5–15%), and fat (1–5%) inclusive of other micronutrients such as vitamins and minerals (World Health Organization 2003). Nutritive significance of major cereal crops has been summarized in Table 12.1. Cereals solely impart 10,000–15,000 kJ/kg of energy, which is approximately 15–20 times more than fruits and vegetables. Moreover, cereals are the paramount source of dietary fibres and bioactive compounds, particularly as whole grains with augmented health benefits (Hall et al. 2017).

Globally, standard nutritional guidelines are being devised emphasizing upon the embodiment of a larger proportion of whole grains in diet for boosting health (EU Science Hub n.d.; U.S. Department of Health and Human Services and U.-S. Department of Agriculture 2015; Dietitians Association of Australia (DAA) n.d.).

Table 12.1 Nutritive content of major cereals

Cereal	Protein (g)	Fat (g)	Carbohydrate (g)	Minerals (g)	Calcium (g)	Fibre (g)	Energy (kcal)
Wheat (whole)	11.9	1.5	71.2	1.5	41	1.2	346
Rice (raw)	6.8	0.5	78.2	0.6	10	0.2	345
Finger millet	12.3	1.3	72	2.7	344	3.6	328
Foxtail millet	12.3	2.7	66	2.1	31	8	365
Proso millet	12.5	4.2	73	1.7	14	2.2	378
Pearl millet	11.6	5.0	67.5	2.3	42	1.2	361
Barley	13.6	1.2	74	1.5	26	2.4	352
Sorghum	10.4	1.9	72.6	1.6	25	1.6	349
Amaranth	14	7.0	65	1.5	37	2.7	371
Maize	9.2	4.7	72	1.9	1.2	7.3	365
Oat	16.9	6.9	66.3	1.8	4.0	10.6	389

Source: Gopalan C, Rama Sastri B.V., and Balasubramanian, S.C., 2004, Nutritive Value of Indian Foods, National Institute of Nutrition, ICMR, Hyderabad

One of the many pivotal functions of whole grains, recently discovered, highlights their prebiotic functionality for gut microflora, a principal component for the host's wellbeing (Dietitians Association of Australia (DAA) n.d.; Costabile et al. 2008). In native or unprocessed form, cereals are the key source of carbohydrates, proteins, fats, essential oils, fibre, minerals, and vitamins which upon post-harvest processing (dehusking and dehulling) for value enhancement and other commercial aspects like polishing results in loss of essential minerals and oils. For human consumption, cereals are routinely marketed in native form or as additives in processed food. As animal fodder, they are primarily consumed by poultry and livestock, which are ultimately consumed by humans as poultry products, dairy, and meat. Cereals also have multitudinal commercial objectives, such as in adhesives, oils, paper industry, textiles, laundering preparations, and cosmetics (Rosentrater and Evers 2018).

12.2.1 Composition and Nutritional Aspects of Cereals

12.2.1.1 Carbohydrates

Carbohydrates dominantly constitute approximately 80% of total dry matter in cereals, primarily as crude fibre and soluble carbohydrates. Cellulose, pentosans, and hemicelluloses are the key constituents of fibre, while starch forms the major portion of soluble carbohydrates across all cereals. Trace quantities of dextrin and other free sugars including glucose and disaccharides like sucrose and maltose are also present.

12.2.1.2 Protein

Protein accounts second after carbohydrate in terms of constituents, but its content varies across different species as well as lineage amongst the same species of cereals. Protein is omnipresent in different proportions across all tissues of cereal grain. Elevated concentration of protein occurs in embryo, scutellum, and aleurone layer compared to endosperm, pericarp, and testa. Categorically, protein in cereals belongs to albumin, globulin, prolamine (gliadins), and glutelin subclass which varies in different cereals.

12.2.1.3 Lipids

Lipids constitute 1–2% in major cereal crops like rice and wheat while 3% in maize with highest proportion present in germ and bran compared to other tissues of grain. Chemically lipids in cereals are mostly triglycerides of palmitic, oleic, and linoleic acid. Apart from these, cereals also are sources phospholipids and lecithin. Owing to the amount of cereals consumed, it is assessed that cereals solely can meet 50% of our essential fatty acid requirement while cereals in conjugation with pulses can suffice the essential fatty acid requirement.

12.2.1.4 Minerals

The majority of minerals (about 95%) naturally exist as sulphates and phosphates of potassium, magnesium, and calcium. Despite a substantial proportion of phosphorus and calcium present in phytin, these remain largely unabsorbed by the body. Phytates also block iron uptake by the body, but upon germination, phytate content declines sharply due to enzymatic breakdown leading to improved availability of iron to the body. Unprocessed cereals contain more phytates than refined or polished cereals. Moreover, cereals are also a rich source of essential trace elements like zinc, manganese, and copper.

12.2.1.5 Vitamins

Unprocessed and whole-grain cereals are a rich source of vitamins, particularly vitamin B. Since the majority of these vitamins are located on the bran, therefore polishing of grains considerably reduces the vitamin content of cereals. Apart from postharvest processing, parboiling of cereals leads to depletion of vitamin content present on the outer layer of grains. Except for maize, cereals lack either vitamin A or C. Cereal grains-derived oils are also a rich source of vitamin E.

12.2.1.6 Enzymes

Certain grains contain many enzymes such as proteases, amylases, oxio-reductase, and lipases which are of prime importance. Proteases are relatively more in the germ compared to other tissues, while amylase activity accentuates during germination. Lipase enzyme is primarily responsible for fatty acid metabolism during the storage of cereals.

12.3 Cereal and Cereal Product Contribution to the Diet

Being heterotrophs, humans depend upon different plants as a source of food and energy, where cereals occupy the paramount position amongst them [FAO 2002]. Cereals are the edible seeds or caryopsis, belonging to common grass members of the monocot family Poaceae, also known as Gramineae (Bender and Bender 2009). Most extensively consumed cereals include triticale, rice, oat, maize, sorghum, barley, millet, and amaranth. The massive use of cereals as functional food can be attributed to large-scale cultivation, higher harvest index, ease of storage, greater mean life, and substantial nutritional and calorific contribution worldwide.

12.3.1 Cereals and Cereal-Based Food

Cereal grains apart from being the primary source for obtaining daily energy requirements are also part of many cultures in many countries and serve as the raw material of many foods and beverages. Apart from lineage and genomic constituents, nutritional abundance of cereal crops depends on the degree of postharvest processing like hulling, husking, and milling resulting in the removal of outer bran layer which is a rich source of vitamins, fibre, and minerals. Cereal-derived/cereal-based foods are primarily made from diverse grains comprising wheat, maize, rice, oats, rye, barley, millet, and sorghum, with rice, maize, and wheat together accounting for two-thirds of human dietary consumption (Sarwar 2008). Examples of cereal-based products majorly include cornmeal, corn grits, wheat, oat, rice and rye flour-based dough and bread, snack foods, tortillas, cakes, dry mixes, etc. Moreover, cereal-based products are also used as basal material for coating, batter, sweeteners, thickeners, baby food, bakery products, and alcoholic beverages such as wine and beer. Cereal-derived food products can broadly be majorly grouped into four categories:

1. Flour-based baked products like bread, cakes, pastries, cookies, dough, and cakes.
2. Processed or value-enhanced grains (milled, hulled, and polished), cornmeal, pearled barley, hominy, farina, corn grits, soup, and thickening agents rich in starch.
3. Whole-grain products like black rice, rolled oats, puffed and shredded grains.
4. Fermented, roasted, and boiled grains-derived beverages.

12.3.1.1 Bread

Bread making can be traced back to 450 BC in Rome with the advent of water milling. Various types of bread are prepared using different cereals as basal material, and breads using a mixture of different cereal flours mixed in definite proportion to be used as a functional food with superior nutritive significance were used since ancient times in South Asian regions, particularly in the Indian sub-continent which is nowadays being marketed under multigrain flour-based bread. High consumption of bread across the world can also be credited to its high nutritive value with 40% carbohydrate, 8–9% protein, and fat <3 g/100 g.

12.3.1.2 Breakfast Cereals

Breakfast cereals are the key source of daily nutrition and calorific significance. Deficiency of nutrients like riboflavin, niacin, folate, and vitamin B12 was more profound in infants not consuming fortified breakfast cereals (Patient and Ainsworth 1994). Research on adults exhibited that fibre- and carbohydrate-rich breakfast was associated with the highest post-breakfast alertness (McNulty et al. 1996). In another study, positive correlation between cereal consumption in breakfast and effect on health showed that adults having cereals daily in breakfast reported superior physical and mental wellbeing compared to the ones consuming in lesser frequency (Holt et al. 1999).

12.3.1.3 Bakery Products

Cereal foods contribute 5% of daily fat intake in infants and 7% amongst adults, despite low account of fat in cereal-based foods (Smith 1999). Bakery products like biscuits, pancakes, cookies, and buns constitute foods comprising sugar and fats. Therefore, under the umbrella of "Balance of Good Health", they do not fall into the same category as the cereal products discussed above.

12.4 What Is Nutrigenomics?

Nutrigenomics is a rapidly evolving realm of science that bridges the understanding of the effect of genetic diversity in response to food and its constituents. It focuses on unravelling the underlying mechanism of signalling cascade triggered as a result of the interlinking of dietary bioactives with the genome at the molecular level. It also helps in understanding how a particular nutrient or an anti-nutrient regime present in our diet configures human health. Nutrigenomics teaches us what is the specific nutrient requirement of our body which ultimately determines the genetic messages our body receives.

12.4.1 Origin of Nutrigenomics

Nutrigenomics might sound like a new domain of science to some, but its advent can be dated back to the genesis of the role of genes in shaping the architecture of all the living beings on the planet. The term "nutrigenomics" was coined by Peregrin in the year 2001 (Gregory et al. 2000). In the mid-1970s, the field of nutritional research gained momentum with the advancements in the field of genetics where genes were characterized based on functionality, particularly their role in human health. Later nutrient components were considered as "signalling molecules" which relay the signals to the cells and ultimately to the nucleus to trigger the change in expression of a particular gene or sets of the gene leading to alteration in protein or a metabolite turnover (Public Health Nutrition n.d.).

12.4.2 Genesis and Components of Nutrigenomics

The postulation about the effect of food on human health has been ages old. Policies were formulated and amendments concerned were made perpetually worldwide to emphasize the prominence of diet and nutrients present therein in preventing and combating chronic diseases. But the major breakthrough came with the publishing of the draft of human genome that eventually led to the interlinking of nutrients and genome. Nutrigenomics is an amalgamation of multidisciplinary sciences under one umbrella, encompassing genetics, proteomics, metabolomics, transcriptomics, and epigenomics (Fig. 12.1) which helps in designing a "genome-tailored diet" for each individual leading to the stabilized genome. Genome stability implies to minimizing damage to DNA, epigenetic alterations (acetylation and methylation during chromatin remodelling), transcriptome (RNA and non-coding miRNAs during post-transcriptional gene regulation), and proteomics (alterations in turnover of a protein).

12.4.3 Modern Nutrigenomics: Personalized Nutrition

Scientists are deciphering diverse interplay between the genes, nutrition, and disease. Modern-day nutrigenomics takes advantage of advanced high-throughput techniques for the identification and prevention of nutritionally based diseases. The role of genes and its association with disease, primarily, the role of dietary and bioactive components affecting the regulation of these genes is also being explored. Different types of diets have been shown to alter body weight, metabolic rate, blood pressure, cholesterol, and other health variables distinctively relying on the genetic constituent of an individual (Bayless and Rosensweig 1966). These findings have put forth resolutely the concept of personalized nutrition, which can provide dietary propositions based on the health and genetic profile of a person.

12.5 Nutrigenomic Diseases and Molecular Diagnosis

Research pertaining to nutrigenomics has relied on the doctrine of nutrition-gene-disease interactions and thereby preventing mankind from so-called modern-day diseases like obesity, cardiovascular diseases, respiratory disorders, diabetes, and different forms of cancers. Collectively such diseases come under the umbrella of non-communicable diseases (NCDs) which are largely mediated by overexposure to a particular type of diet or precisely junk food. Upon overexposure to a certain type of unhealthy diet, our body responds to it by abnormal or disturbed metabolism leading to the synthesis of certain biomolecules which act as biomarker for the identification and diagnosis of diseases. On molecular basis damage or mutation in

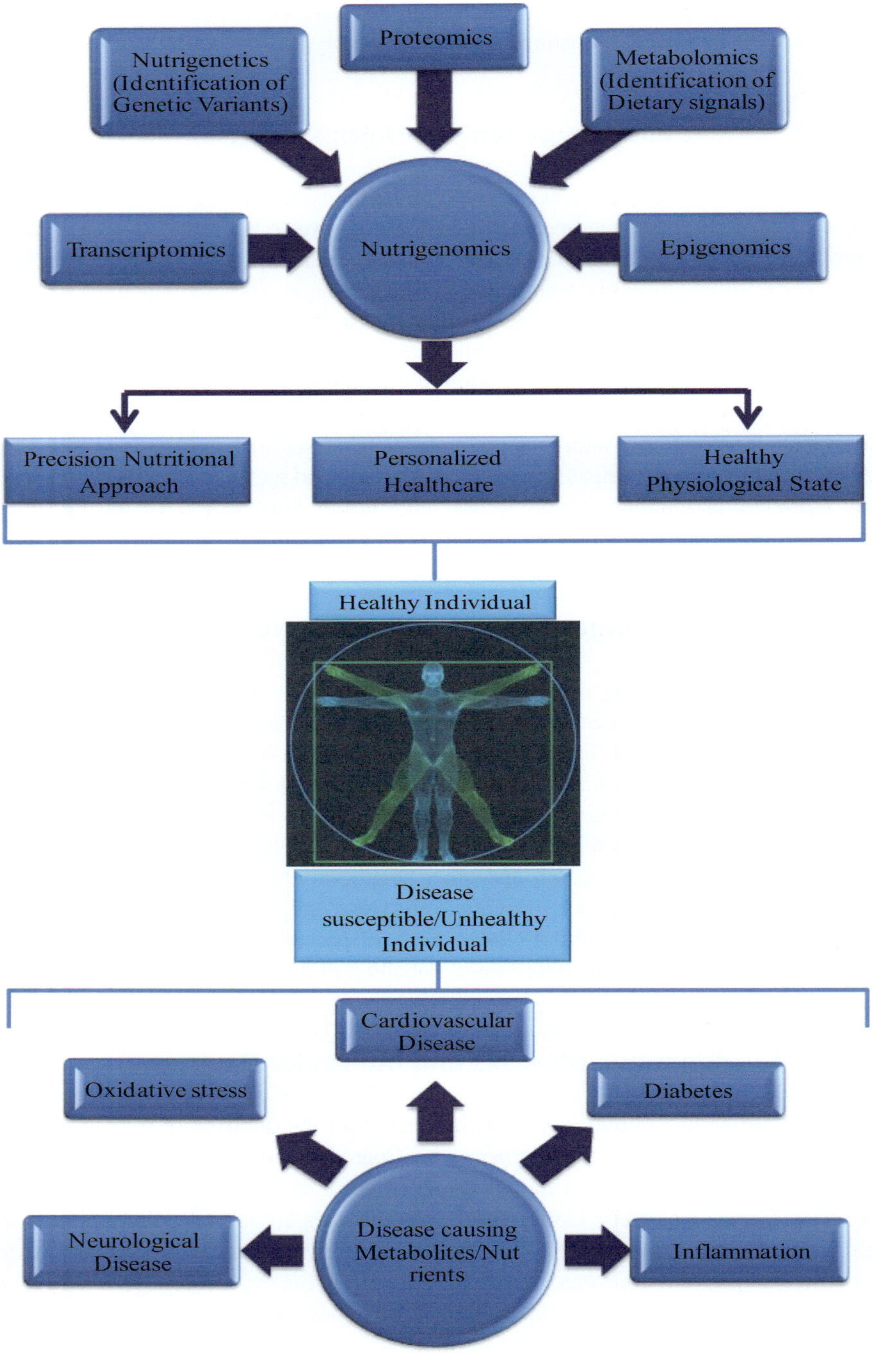

Fig. 12.1 Several components of nutrigenomics and its role in disease prevention

DNA or a portion of DNA fragment can lead to certain genetic or hereditary disorders which can be diagnosed by molecular techniques, for example, chip-based diagnosis, hybridization assays, PCR-based assays, SNP genotyping, etc. In addition, a widely exploited nutrigenomics tool is transcriptomics, which makes use of microarray assay to quantify mRNA copies of all transcribed genes in a spatio-temporal manner.

12.5.1 Phenylketonuria (PKU)

Phenylketonuria (PKU) is a metabolic genetic disorder characterized by a mutation in a single gene coding for hepatic enzyme phenylalanine hydroxylase (PAH) causing the fatal build-up of phenylalanine in the blood. PAH enzyme plays a key role in metabolizing aromatic amino acid phenylalanine, and affected individual is prohibited from consuming phenylalanine-rich foods like milk, cheese, chicken, fish, egg white, etc. (Enattah et al. 2002). Infants and young-aged children are more prone to phenylketonuria (PKU), but nowadays infants are routinely screened for PKU, and if found affected are recommended a special diet lacking phenylalanine.

12.5.2 Identification of the Gene(s) for Lactose Intolerance (LI)

Lactose intolerance (LI) or hypolactasia is a genetic disorder faced by the majority of adults worldwide with an estimate of 68% of people facing issues digesting lactose. In the early 1960s, researchers unravelled that inability to digest milk is primarily a genetic disorder (Moraes and Pereira 2009). Loktionov (2003) identified that mutation in *lactase* (*LCT*) gene, encoding for lactase enzyme, causes incompetence to break down lactose. Further, the advent of lactose intolerance was traced back to 10,000–12,000 years ago in Europe, where polymorphism in a single nucleotide (SNP) of a gene resulted in constitutive expression of lactase gene in adults. The person with lactose intolerance could utilize nutrition-rich dairy products much more efficiently than normal persons. The most commonly recommended treatment focuses on the avoidance of dairy products.

12.5.3 Galactosemia

Galactosemia is a rare inherited metabolic disorder in which the affected person is unable to utilize galactose. In other words, it's the inability of body to convert galactose into glucose for generating energy in the form of ATP. The molecular basis of galactosemia has been found to be a defect in *galactose 1-phosphate uridylyltransferase* (GALT) gene, the product of which is involved in the catabolism of galactose (Tucker et al. 2013). Alterations in dietary intake can prove to be fruitful; further in-depth knowledge is imperative where a single nutrient may affect our biological system.

12.6 Malnutrition: The Genesis of Chronic Diseases

A diet rich in all essential nutrients is a decisive factor for the promotion and maintenance of sound health throughout the life of an individual. Malnutrition is found to be the key factor for chronic-related disorders in approximately 40% of patients (Ilnytska and Argyropoulos 2008). Malnutrition can be correlated with complication rates, disability, need for care, as well as mortality. Malnutrition can be primarily attributed to two factors: inadequate feeding and anomalous nutrient uptake or metabolism by the body. Some of the nutrition-dependent chronic diseases that pose a serious threat to public health include cardiovascular diseases, obesity, cancer, diabetes, and osteoporosis.

12.6.1 Obesity

Obesity is the most prevalent multifactorial disease that forms the core of numerous metabolic disorders affected by variable environmental and genetic factors. Nearly one-third, i.e. 30% of the world population, is either obese or overweight, leading to a rise in chronic diseases [WHO 2018]. Excessive food intake has been found to be associated with polymorphism in genes encoding several sensory or taste receptors including leptin, insulin, ghrelin, and cholecystokinin (Ferguson 2006). Nutrigenomics assists in the formulation of novel functional diets for restraining obesity by exploiting scientific and molecular mechanisms of effect of bioactive components on basal metabolic rate (BMR) and body weight. Obesity has been found to drastically augment the peril of various chronic diseases including CVD and cancer (Kopelman 2000; Bianchini et al. 2002). Increasing physical activity is complemented with the shift from fat-rich foods and beverages to food rich in anti-inflammatory bioactive, such as caffeic acid, tyrosol, quercetin, and seeds of *Salvia hispanica*. These bioactive compounds have been found to modulate gene expression via activation/inactivation of transcription factors (Cozzolino and Cominetti 2013; Costa and Rosa 2011).

12.6.2 Cancers

Cancer can be defined as anomalous and unconstrained cellular growth with the potential to spread to distinct cells and tissues. Being a multifactorial disease, cancer is characterized by mutations in several oncogenes and alteration in gene expression, transcriptome, and eventually metabolic operations. Mutation in tumour suppressor gene p53 resulting in mutated or non-function p53 protein has been found to be the primary cause of cancer in >50% of different types of human tumours. Apart from being genetically governed, natural and dietary components heterogeneously manifest tumour stimulation, invasion, and progression by targeting and altering several pathways. The majority of cancers (35–40%) are primarily influenced by diet. However, numerous dietary components having nutraceutical and phytochemical significance help in the prevention of cancer through the genetic and epigenetic mechanism, for example, certain nutrients play a key role in the maintenance of normal cellular methylation. Numerous preclinical studies have stipulated the role of nutrients in cancer prevention. Deciphering the crosstalk between nutrigenomics in the regulation of cancer is already ascertained. Another prominent cause of cancer is irreversible damage to DNA and the inability of cellular machinery to repair the genetic material because of overexposure to certain genotoxins and deficiency of nutrients required as cofactors for DNA repair enzymes. Apart from macronutrients, micronutrient plays an indispensable role in shaping up individual's health depending upon genetic makeup, physical state, and age (Costa and Rosa 2011; Almendro and Gascan 2012). Studies have highlighted that deficiency of micronutrients, such as vitamins, selenium, niacin, and zinc, can lead to mutation in DNA in a similar manner as observed during radiation exposure. These mutations can lead to degeneration of the helical structure of DNA and oxidative lesions leading to cancer (Cozzolino and Cominetti 2013; Costa and Rosa 2011; Almendro and Gascan 2012). Unhealthy eating practices lead to the cellular build-up of obnoxious metabolites which can interact and alter DNA structure causing mutation at nucleotide residue. For example, aflatoxin B1 present in fungi (*Aspergillus flavus*)-contaminated food causes apurination of DNA causing acute damage to liver including cirrhosis, necrosis, and carcinoma (Moraes and Pereira 2009; Mahan and Scott Stump 2005). Methionine, an essential amino acid, is synthesized from its precursor 5-methyltetrahydrofolate which in turn is synthesized from folate present in food. Insufficient dietary uptake of folate can lead to hindrance of DNA mutilation leading to increased risk of cancer (Cozzolino and Cominetti 2013; Costa and Rosa 2011; Almendro and Gascan 2012; Cozzolino 2012). Several bioactives such as selenium, prostacyclins, zinc, ascorbic acid, etc. act as antioxidants for coping with imbalanced levels of reactive oxygen species (ROS), which cause irreversible oxidative damage to cell membrane and biomolecules such as DNA, lipid, and lipoproteins.

12.6.3 Type 2 Diabetes

Diabetes alone accounts for >90% of all diseases afflicting ~200 million people worldwide (Surh 2003). Rapid transition from traditional diets to excessive uptake of food rich in carbohydrates and saturated fats is one of the primary causative factors of type 2 diabetes. Mutation in the gene causes alteration in the metabolic function of insulin, the key hormone which regulates glucose and lipid metabolism which is rendered non-functional. Diabetes acts as a genesis for other life-threatening diseases such as cardiovascular disease, stroke, and renal dysfunction (Calder et al. 1998). On the genomic level, a total of 65 SNPs has been identified to be associated with the precarious level of developing diabetes. Reliable and economically cheap genome-based rapid detection has helped early detection of genetic predisposition of SNPs related to cause type 2 diabetes (Chang et al. 1998). Early detection of diabetes helps in tailoring lifestyle, particularly dietary intake to minimize any obnoxious increase in blood sugar without hampering the health of a person.

12.6.4 Cardiovascular Diseases

Cardiovascular diseases (CVDs) are alone the principal cause of a high mortality rate accounting for 17.9 million deaths (representing ~31% worldwide) in 2017 [WHO 2019]. CVDs are primarily a multifactorial condition involving other health complications such as thrombosis, obesity, and hypertension which can be attributed to environmental factors. The interrelation between dietary factors and peril for CVD can be rooted back to the 1950s, when heart diseases were linked to the consumption of saturated fatty acids. Through advanced technologies, relationship between nutrients and CVD risk has now been well established. Studies have focused on the identification of genes and genetic variation associated with CVDs and related health complications. Atherosclerosis is one of the substituent elements of CVD pathogenesis causing metabolic disorder and lipid transport with chronic inflammation (Cozzolino and Cominetti 2013). Studies have shown that polymorphism in E4 gene coding for apolipoprotein E displayed higher LDL cholesterol. Similarly, allelic variation in the gene encoding for angiotensinogen (AGT), angiotensin-converting enzyme (ACE), and aldosterone synthase (CYP11B2) has been accounted for blood pressure regulation and thus hypertension (Hooper et al. 2001; Schaefer 2002). Several loci have been identified which regulate the progression of CVD: e.g. FTO and 9p21 have been characterized to be associated with obesity and CVD, respectively, which in turn are regulated by gene-diet interaction (Corella and Ordovas 2009; Ye and Kwiterovich 2000). Several biomarkers have been characterized for the detection of CVD; one of such biomarkers is an elevated level of homocysteine which is regulated by dietary folate and alcohol consumption (Mahley and Rall 2000; Luft and Weinberger 1997). Furthermore, reports have fortified the role of macro- and micronutrients in epigenetic regulation of cardiometabolic risks. Some of primary nutrients include methionine, choline, vitamin B12, zinc, and folate which are required for methylation of histones in DNA. Thus, the role of personalized or precision dietary nutrients has become the need of the hour with accentuating CVD worldwide.

12.7 Global Status of Nutrigenomics Research

12.7.1 Global Health Scenario

Despite exceptional advancements being made in the realm of improving public health and disease prevention, a large proportion of the population worldwide suffers from chronic diseases and remains astoundingly overlooked in the global health agenda. Chronic diseases can largely be attributed to the transition in dietary and lifestyle habits which is a direct consequence of rapid globalization. Developing countries account for the largest proportion of people suffering from non-communicable and infectious diseases. Therefore, a lot of emphasis and efforts need to be made to combat chronic diseases with utmost priority at various international and national levels. Amongst non-communicable diseases (NCDs), cancer, respiratory disorders, cardiovascular diseases, and diabetes attribute 71% of death globally. Every year more than 15 million people die from NCDs in age group of 30–69 years. Such a large percentage lays a considerable strain on health budgets, particularly of developing and underdeveloped economies. Thus, the significance of nutrigenomics in the present era has grown by leaps and bounds to tackle global health issues, particularly NCDs. This is evident from the fact that the global nutrigenomics market is expected to grow by approximately 11% (USD 425.61 million) in the next 5 years from 2021 to 2026 [WHO 2020].

12.7.2 India's Health Scenario

Compared to western countries and other developed nations, the rate of fatal diseases in India is lower. However, India is witnessing an expeditious epidemiological, socio-economic, health, and nutritional transformation over the last two to three decades which can be accredited to the rapid rate of urbanization and changes in lifestyle, particularly eating habits (Franco and Reitsma 2001; Godard and Hurlimann 2009; Rao 2001). Transition in eating habits from consumption of coarse grains to processed rice and wheat has led to an increased rate of cardiovascular and cancers in the Indian population (Shetty 2002). Malnutrition, particularly undernutrition-related disorders, has a deleterious impact on major sections of the population. Contrarily, overnutrition has led to a steep increase in the number of people suffering from obesity and chronic lifestyle disorders, primarily CVD, diabetes, and cancer (Anderson et al. 2001; Sharma and Majumdar 2009; Sinha et al. 2003). Taking such conditions into account, a common consensus needs to be made pertaining to diet-related disorders leading to alteration at molecular and metabolic levels; therefore, nutrigenomics should be employed efficiently and effectively.

12.8 Role of Nutrigenomics for Better, Healthier, and Longer Life

Health is a barometer for measuring the quality of life and prosperity of an individual or a society. The health of a person is also reflected by the ability to adapt and self-acclimatize to adverse conditions. In recent times a radical change in the pattern of diseases has been observed which can be attributed largely to eating habits, rapid urbanization, environment, and lifestyle (Hossain et al. 2007). The interaction between genomic and environmental cues plays a decisive role in the buildout and progression of numerous life-threatening diseases. Diet is one such environmental factor that not only meets the daily metabolic energy requirement of the body but also regulates epigenetic changes associated with aging (Kaput et al. 2007; Neeha and Kinth 2013). Physical and mental stresses along with lifestyle considerably contribute to the regulation of gene expression in response to the dynamic environment. Bioactive and chemopreventive molecules present in food help in preventing the risk of chronic diseases through regulation at the molecular level (World Health Organization 1990; Bacalini et al. 2014; Jeffery and O'Toole 2013). Traditionally it was believed that nutritional requirements for all the beings are the same, but with the advancement of nutrigenomics, the role of personalized diet or functional foods gained recognition. Every age group and sex has a specific recommended dietary allowance (RDA), specifying the daily nutrient requirement for a healthy and disease-free life. Nutrigenomics is creating a power shift, from an unhealthy disease-prone lifestyle to a healthy disease-free lifestyle. Incurable genetic disorders like cardiovascular disease, obesity, diabetes, cancer, and Alzheimer's cannot be avoided, but following genome-tailored personalized functional foods, one can lead a healthy lifestyle with minimal complications and illness.

12.9 Nutrigenomics Research Tools

The remarkable advancements being made in the field of nutrigenomics can be attributed to various "omic" technologies such as genomics (e.g. SNP genotyping for polymorphism analysis), transcriptomics (e.g. gene expression arrays), metabolomics (e.g. assessment of active metabolites and bioactive compounds), and proteomics (e.g. abundance of a particular bioactive peptide and proteins). These technologies are substantially used to explore the molecular mechanism underlying the role of dietary components in diseases pertaining to nutrition such as cancer, diabetes, inflammatory disease, cardiovascular disease, and obesity (Chae et al. 2007). These technologies will pave way for genome-wide transcript and protein analyses, coupled with bioinformatic tools and databases which can prove to be handy in the identification of the novel gene, proteins, and other bioactive components for unravelling mechanisms underlying disease development and progression.

12.9.1 Genomics

The sequencing of human genome has opened floodgates of immense information and opportunities in the field of nutrigenomics and disease prevention. Novel genes and polymorphism amongst a susceptible and tolerant population in response to disease are now being functionally characterized. Reliable and rapid sequencing technologies have helped in the identification of mutation in a particular gene or several genes at early stages in life which can prove instrumental in minimizing the risk of disease progression.

12.9.1.1 Single Nucleotide Polymorphism

With the advancements in the realm of molecular biology, scientists have been able to identify genes coding for nutritionally imperative proteins primarily digestive enzymes, hormones regulating metabolic pathways, and molecules responsible for the intercellular and intracellular transport of nutrients and cofactors at the site of metabolism. Numerous common SNPs have been reported to alter nutrient requirements. One such example is SNP methylenetetrahydrofolate dehydrogenase (MTHFD1-G1958A) related to organ impairment when a person gets choline low or has deficient diet, most prominent in premenopausal women. Infected mother with such polymorphism has fourfold high risk of having a child born with neural tube dysfunction in comparison to the women on a choline-rich diet (Arts et al. 2001). A new preventive medicine approach is alternative therapies utilizing nutritional approach for disease prevention; one such technology is the development of SNP array which helps in the identification of haplotypes amongst a population. The role of SNPs in relation to diseases like cancer, diabetes, cardiovascular disorders, leukaemia, Down syndrome, and neural tube defect (NTD) has been investigated. SNP profiling for intracellular folate metabolic pathway has been investigated in the Indian population (Najafian et al. 2012). Similarly, the role of vitamin B12 (cobalamine), homocysteine, and folate in acute lymphoblastic leukaemia (ALL) fortifies the imperativeness of gene-environment-nutrition interaction in the development and progression of cancer (Trujillo et al. 2006).

12.9.2 Transcriptomics and RNA-Seq Technology

Transcriptomics is the most extensively used technique for real-time expression profiling of genes in a spatio-temporal manner as well as in response to the defined nutritional state of a being. Variation in intracellular RNA transcript abundance in response to dietary interventions forms the elemental basis for deciphering the complex interrelation between genes and nutrients (Wickramasinghe et al. 2012; O'Brien et al. 2012). In-depth diligent transcriptomic analysis pools all forms of RNA (coding and non-coding) for understanding the post-transcriptional modification, particularly mRNA splicing of a gene. Several types of chips or microarrays are now available which are extensively being used for gene expression and genotyping pertaining to SNPs, point mutations, and short tandem repeats (SSR). Compared to microarray-based transcriptional profiling, a much more efficient and less error-prone technique is RNA-Seq to study nutrient-gene interactions with high unparalleled accuracy and reliability. RNA-seq has led to the identification of novel genes that are upregulated in response to dietary factors, nutritional and physiological state of an individual, micro- and macronutrient deficiencies, and diseases (Wickramasinghe et al. 2012; Swanson et al. 2003; Külahoglu and Bräutigam 2014). High-throughput RNA-seq has become the most preferable choice for enabling inexpensive and routine comprehensive analysis of human transcriptomes or genomes (Shendure and Ji 2008; Chen et al. 2015). This technology is primarily used for quantitative expression analysis of known and unknown transcripts and mRNA splice variants and for the analysis of SNPs which can be used as potential biomarkers pertaining to a trait (Külahoglu and Bräutigam 2014; Han et al. 2015; Piskol et al. 2013).

12.9.3 Proteomics

Proteins are the key component of the central dogma of inevitably all biological processes; thus, proteomics acquires centre stage as one of the key tools in nutrigenomics research. The role of post-translational modification (e.g. acetylation, ubiquitylation, phosphorylation, etc.) of proteins will help in deciphering the regulation mechanism in response to nutrients and the environment. For example, protein phosphorylation of extracellular signal-regulated protein kinase (ERK) is altered upon exposure to diallyl disulphide, a compound present in processed garlic causing cell cycle arrest (Ghodke et al. 2011). Qualitative and quantitative profiling of proteins for the development of peptide and protein-based markers can be instrumental in addressing questions pertaining to nutritional competency of various bioactive dietary components (Adiga et al. 2008; Ordovas and Corella 2004). Proteomics, structural dynamism, and crystallographic studies of peptide and complex protein can help in the identification of aberrant protein structures and their effect in response to diet. For example, effects of dietary fish oil, trans fat elaidic acid, or conjugated linoleic acid on lipoprotein metabolism and insulin levels have been well characterized in animal models. Fish oil supplement and elaidic acid have been reported to lower plasma and liver cholesterol. Proteomics coupled with physiologic analysis has proved to be instrumental in unravelling the mechanism underlying the regulatory role of dietary fatty acids in lipid metabolism (Knowles and Milner 2003).

12.9.3.1 Applications of Proteomics in Nutrigenomics

Researchers world over are employing proteomics as a tool to decipher the kinetics of transcript/protein abundance in disease progression in response to nutritional intervention. Proteomics in nutritional studies helps in the identification and quantification of dietary proteins and peptides and assesses their potent nutritional bio-efficacy. The inhibitory effect of sodium butyrate on the succession of human HT-29 cancer cells was assessed in vitro using 2DMS-based proteomic tool. Butyrate has been found to be involved in regulating the expression of genes involved in ubiquitin-mediated proteosomal degradation of proteins involved in regulation of apoptosis and cell cycle and differentiation (Kussmann and Affolter 2006). Combinatorial use of DNA microarrays and proteomics has been exploited to characterize numerous bioactives for their anti-colorectal cancer properties and their role in the progression of cancer in vivo using colon epithelial cell lines (Kussmann et al. 2005).

12.9.4 Metabolomics

One of the contemporary tools of nutrition biology is metabolomics, which attributes to the quantification and extensive analysis of macro and micro biomolecules, within cells, tissues, or organism, and their temporal changes in response to the external environment. Metabolomics is a key tool for the identification of food-derived biomarkers and their variability for metabolizing the same in a healthy and diseased individual. For example, biochemical profiling occurring after soy-rich dietary intervention in humans led to the analysis of changes that occur in plasma components, like alterations in amino acid, plasma lipoprotein, and carbohydrate metabolism (Solanky et al. 2003). Foods contain myriad of non-nutrient molecules that are absorbed, metabolized, and released into body fluid which variably alters the metabolome of an individual. Nuclear magnetic resonance (NMR) spectroscopy pattern recognition and other spectroscopic technologies along with high-throughput chromatographic techniques are being used for metabolic profiling of blood resulting from different dietary treatments. The new technologies for the analysis of metabolites together with the bioinformatic tools and data processing are proving instrumental for researchers and healthcare professionals to detect predispositions to disease and its management to improve individual health. Metabolomics will help in the characterization of the population on the basis of metabolic activities and thereby designing personalized food for maximizing health benefits (Table 12.2).

12.10 Nutrigenomics and Public Awareness

Despite growing scientific evidences and expanding horizons in the realm of nutrigenomics as a panacea for improving individual and community health, debate pertaining to the spread and public acceptance of nutrigenomics still remains the primary daunting task. Some of the key issues that needs to be addressed from public health perspective are highlighted herein as follows: Firstly, availability of scientific evidences to fortify the advancements and co-relation of dietary components with genome and its impact on individual health. Secondly, privacy issue pertaining to sharing genetic information and health profile used to generate a nutritional prescription. It is imperative for the consumers to know who has the access to their personal information and up to what extent it will be used to make a public database. Thirdly, the regulation and presence of a concerned authority to supervise test, health claims, and how the services will be provided keeping the risk assessment into the equation. Much of the public concerns are associated with scientific know-how and techniques involved therein. Fourth and lastly, public fear related to genetically modified organisms (GMO) and genome editing (CRISPR technology) needs to be addressed keeping scientific and social, health, and ethical issues in mind. Bridging the trench between nutrition and genomics will help in overcoming these challenges, create more awareness, and build trust and confidence amongst the public.

Table 12.2 List of nutrients and related deficiency disorder along with preventive food sources

Nutrient	Deficiency/disease and symptoms	Preventive food sources
Vitamin A (retinol)	Poor vision/night blindness	Spinach, carrot, butter, mangoes, green leafy vegetables, apricot
Vitamin B1 (thiamine)	Extreme weakness, beriberi	Yeast, eggs, meat, whole grain, cereals, dried beans
Vitamin B2 (riboflavin)	Retarded growth, bad skin	Green leafy vegetables, beans, peas, milk
Vitamin B3 (niacin)	Pellagra, diarrhoea, dermatitis, and dementia	Mushroom, brown rice, green leafy vegetables, peanut, sweet potato
Vitamin B5 (pantothenic acid)	Fatigue; malaise; apathy; restlessness; insomnia; cramps; gastrointestinal disorders	Sunflower seed, milk, whole grain, meat, lentils, oats
Vitamin B6 (pyridoxine)	Seborrheic dermatitis, microcytic anaemia, epileptiform convulsions, weakened immune function	Poultry products, beans, oats, banana, potato, avocado, soya bean
Vitamin B9 (folic acid)	Megaloblastic anaemia, neural tube defects	Beans, peanuts, sunflower seeds, fruits, whole grains, seafood
Vitamin B12 (cyanocobalamin)	Anaemia	Pomegranate, beetroot, spinach, seafood
Vitamin C (ascorbic acid)	Scurvy, swollen gums, loose teeth	Lemon, oranges, Indian gooseberry
Vitamin D (calciferol)	Rickets, brittle bones in children	Milk, fish, liver oil, sun exposure
Vitamin E (tocopherol)	Chronic pancreatitis, cholestasis, cystic fibrosis, neuromuscular dysfunction	Wheat germ oil, nuts, sunflower, safflower, rapeseed, corn, and soybean and olive oils
Vitamin K (phylloquinone)	Excessive bleeding due to injury	Green leafy vegetables, dairy products
Biotin	Unhealthy skin, hair fall, neurological abnormalities	Egg yolk, beans, legumes, nuts, liver oil, linseed, mushroom, banana
Calcium	Brittle bones, excessive bleeding, bad muscular movement	Milk, green leafy vegetables
Phosphorus	Bad teeth and bones	Pulses, cereals, milk
Iron	Anaemia, lack of red blood cells	Green leafy vegetables, drumstick, pulses
Iodine	Goitre, enlarged thyroid gland	Iodine-fortified salt, fish
Copper	Low appetite, retarded growth	Pulses and leafy vegetables
Zinc	Undeveloped genitals, skin lesions, poor appetite, impotency	Seafood, nuts, whole grain, dairy products, fortified cereals
Magnesium	Hypocalcaemia; hypokalaemia; cardiac and neuromuscular manifestation; latent tetany; osteoporosis; insulin resistance and impaired insulin secretion	Pumpkin seeds, almond, legumes, whole grains, green leafy vegetables
Manganese	Dermatitis, hypocholesterolaemia	Whole grains, nuts, chickpea, clams, mussels, soybeans, brown

(continued)

Table 12.2 (continued)

Nutrient	Deficiency/disease and symptoms	Preventive food sources
		rice, sweet potato, leafy vegetables, coffee, tea
Protein	Kwashiorkor, extremely underweight	Sprouted beans, cereals, fish, milk, soya, milk, poultry
Carbohydrate	Non-insulin-dependent diabetes mellitus (NIDDM), ketosis, cardiovascular disease, stunted or poor growth in children	Whole grains, milk, potato, fruits, vegetables, legumes, cereals
Essential fatty acids	Improper fat absorption, scaly dermatitis, alopecia, thrombocytopenia (impaired wound healing), stunted growth in infants	Meat, oil-rich beans, linseed, walnut, almond

12.10.1 Public Awareness Events

The relationship between rapidly changing lifestyle and nutritional status quo of the Indian population has been investigated extensively for the formulation of policies and programme to spread knowledge and public awareness (Acharya et al. 2004). Community health programmes are being organized to make common people understand that the genetic makeup of an individual or community cannot be altered, but the effect of environment (dietary) component on genes can be regulated to attain desired health benefits. An executive course on genomic policies was conducted to provide a platform for stakeholders to deliberate the significance of nutrigenomics for health improvement in India (Krishnaswamy 2008). The course aimed to enlighten participants with the significance of genomics for improving health. A common multidisciplinary consensus is being devised for issues related to policy drafting and ethics. Therefore, food-based dietary guideline (FBDG) was developed by a multidisciplinary group comprising nutritionists, molecular biologists, agriculturists, statisticians, technologists, dieticians, etc. to overpower diet-related disorders and promote health has been emphasized (Fafournoux et al. 1990). The objective of FBDG is the prevention of both under- and overnutrition and to promote a healthier lifestyle. Routine conferences and workshops are being organized where people from different realms involved in public health spread awareness and address queries related to nutrigenomic-based personalized nutrition service amongst people.

12.11 Nutrition and Gene Interactions

Recent advancements in the realm of genomics, proteomics, and metabolomics have propelled genome-wide studies on the association of diseases and health in response to dietary nutrients. The nutrient-gene interaction at cellular, individual, and community extent is now being investigated comprehensively for the identification of novel genes and pathways. The response to a nutrient implies to be genotype-specific, where the nutrition status of a person alters gene expression. Bioactive molecules, carbohydrates, fatty acids, minerals, and vitamins, are invariably involved in the regulation of gene expression. In eukaryotes, the regulation of gene expression involves intricate interaction of neural, hormonal, and environmental factors. Although genetics is the central driving force governing phenotypic or visible traits of an organism, a smaller proportion of overall phenotype including health profile is a result of metabolic and environmental effect it exerts. In the era of post-genomics, technological advancements have led to the identification of several biomarkers related to nutrient-gene interactions for genomic and metabolic profiling in diseased and healthy individuals.

12.11.1 Nutrition and Gene Regulation

Gene regulation is a very complex mechanism where dietary intervention impacts successive steps of signal transduction from transcription, translation, to post-translational modifications (Trayhurn 2000). Using nutrigenomics tools such as genomics, proteomics, and metabolomics, researchers are able to elucidate the genotypic and phenotypic effect of dietary constituents on cellular and intermediary functioning simultaneously (Hinds et al. 2005). Transcriptional regulation via activation and deactivation of transcription factors is the most prominent mechanism for gene control. Nuclear receptors are one of the most abundant classes of transcriptional regulators in animals which act as sensors to nutritional, pathophysiological, developmental, and endocrine dynamism and thereby trigger adaptive response via gene regulation. Ligand binding to the receptor induces a conformational change, resulting in segregation of corepressor and recruitment of coactivator proteins, triggering downstream effector targets. Nuclear receptors in metabolically active organs like the intestine, liver, and adipose tissues are induced by the nutritional state of an individual, thereby regulating transcription factors to alter several nutrient-responsive genes. Nuclear receptors thus act as a sensory receptor in regulation of numerous nutrition and metabolic responsive genes to further regulate various cellular development and differentiation processes.

12.11.2 Synergism of Nutrients and Gene Expression

Predominantly, three types of nutrient-gene interaction can lead to altered phenotype, which are summarized as follows:

1. **Genetic variations**: These include polymorphism at genetic level, e.g. single nucleotide polymorphisms (SNPs) leading to altered functionality of genes. Degree of variability is high across human genome, accounting for 50,000 SNPs in genes (Gibney and Gibney 2004), which results in altered gene expression and ultimately change in protein which might be structurally and functionally distinct from the native one. A number of SNPs have been identified known

to affect nutrient requirement. For example, mutation in *MTHFR* gene encoding for methylenetetrahydrofolate reductase enzyme, required for metabolizing one form of vitamin B, and folate into another, results in elevated plasma homocysteine levels which may increase the risk of cardiovascular diseases, formation of blood clots (thrombosis), and stroke. Two primary MTHFR SNP variants called C677T and A1298C are the most prevalent forms, comprising 15–30% of the affected individuals (da Costa et al. 2006). Similarly, *phosphatidylethanolamine N-Methyltransferase (PEMT)* gene encodes for protein involved in the endogenous synthesis of choline, an imperative nutrient involved in cellular growth and metabolism, but is rendered non-functional because of SNP in the promoter of *PEMT* gene (rs12325817). Such variants are associated with susceptibility to choline deficiency, which causes fatty acid liver and organ dysfunction (Oommen et al. 2005).

2. **Direct interactions**: Sometimes nutrients interact with a receptor which acts as a transcription factor by binding to DNA and induces expression of a gene. For example, vitamin A (retinol) interacts with retinoic acid receptor which further activates or represses transcription by binding to DNA motifs (e.g. retinoic acid response elements) in promoter regions (Muller and Kersten 2003). Likewise, vitamin D interacts with vitamin D receptor, calcium with calcineurin, and zinc with metal-responsive transcription factor-1 (Hsu and Huang 2006). Dietary fatty acids interact with peroxisome proliferators-activated receptors (PPARs), which then bind to DNA and alter gene expression (Davis and Uthus 2004). Nutrients that are components in single-carbon metabolism provide the most substantial evidence of synergy between nutrients and DNA methylation because they alter the availability of methyl groups and therefore the biochemical pathways of methylation processes (Steinmetz et al. 1998). Such single-carbon metabolites include vitamin B6, folate, vitamin B12, methionine, and choline, deficiencies of which in combination cause global hypomethylation, cirrhosis, hepatic steatosis, and ultimately hepatic tumorigenesis in rodents in the absence of carcinogen treatment (Steinmetz et al. 1998; Christou and Twyman 2004).
3. **Epigenetic interactions**: Sometimes nutrients alter the structure of DNA through chromatin remodelling and alter gene regulation by acetylation, methylation, or biotinylation of histones. Such expression can occur in a spatio-temporal manner or can even persist through generations. Epigenetic interaction with histones leads to the uncoiling of chromatin structure, creating grooves through which transcription factors can intercalate and activate gene promoters (Lane and Bailey 2005).

12.12 The Possibilities of Transgenic Foods for Malnutrition Eradication Through Cereals

Plants are the paramount source of energy and nutrients for all humans and livestock. Despite occupying an indispensable role, the majority of our food crops lack certain essential nutrients, although a mixed diet assists in overcoming diet-based malnutrition. More than 50% of the world's population suffers from undernutrition or malnutrition which has become a global health concern (Zhu et al. 2007). The majority of global population notably across the developing nations relies on cereal crops primarily rice, wheat, and maize which lack the full complement of essential nutrients (Al-Babili and Beyer 2005). Great advancement has been made to address the issue through the application of plant biotechnology via engineering of cereal food crops for nutrition enhancement and improving public health. Nutritionally enhanced genetically modified cereal crops have consistently shown efficacy in providing safe and available nutrients to combat malnutrition. Research on up-scaling essential nutrients such as minerals, fats, vitamins, and amino acids with a simultaneous reduction in anti-nutrient factors in cereal crops has become a daunting task for researchers worldwide. Biofortification of key cereal food crops is now being achieved fundamentally through two main approaches: transgene introgression (genetic engineering) and conventional breeding approach. The selection of approach relies predominantly on whether the bioactive compound is synthesized de novo by the plant or acquired from the surrounding environment. In de novo approach, enhancing the nutritional content requires engineering of metabolites or the existing metabolic pathway with the intent to accentuate the basal level of the bioactive compound or downturning the antagonistic compound to generate a novel product with phyto-nutritive significance.

12.12.1 Genetically Engineered Rice (*Oryza sativa*)

Rice has been one of the most imperative food crops worldwide and has acquired central stage for enhancement of its nutritive content in a way to combat the global malnutrition challenge. In this realm provitamin A (beta-carotene)-fortified golden rice has been marked as a major breakthrough in confronting vitamin A deficiency (VAD) in underprivileged economically compromised populations, particularly infants and children less than 5 years old. Genes encoding maize *phytoene synthase* (*ZmPSY*) gene and *carotene desaturase* were overexpressed in rice resulting in enhancement of beta-carotene precursor, i.e. phytoene by up to 23-fold in transgenic rice (Paine et al. 2005; Tang et al. 2009, 2012; Shumskaya and Wurtzel 2013; Tanumihardjo et al. 2010). In a similar approach, enhanced folic acid (vitamin B9) content was achieved in genetically engineered rice through overexpression of *Arabidopsis GTP-cyclohydrolase I* (*AtGTPCHI*) and *aminodeoxy-chorismate synthase* gene, thereby increasing folate content (up to 150-fold) (Haskell 2012; Moghissi et al. 2015). Rice has also been targeted to enhance iron content in order to combat anaemia. Several reports have shown that overexpression of genes encoding iron transporter *OsIRT1* (Xudong et al. 2000), *nicotianamine aminotransferase* (*NAAT*) (Zheng et al. 2010), *nicotianamine synthase* 1 (*OsNAS1*) and 2 (*OsNAS2*) (Trijatmiko et al. 2016; Lee et al. 2009), soybean *ferritin* (Goto et al. 1999; Vasconcelos et al. 2003; Lucca et al. 2002), and common bean ferritin (Hurrell and Egli 2010) resulted in increased iron content. Silencing genes responsible for the synthesis of anti-nutrient compounds like phytic acid have shown to increase iron bioavailability (Lee and An 2009). Similarly, overexpression of metal homeostasis gene *OsIRT1* (Masuda et al. 2008) and *mugineic acid synthesis* genes from barley (*HvNAS1, HvNAS1, HvNAAT-A, HvNAAT-B, IDS3* (Anai et al. 2003)) leads to increased iron and zinc content in genetically modified rice. Overexpression of soybean *omega-3 fatty acid desaturase* (*GmFAD3*) gene (Crawford et al. 2000) in rice leads to enhanced polyunsaturated fatty acid (PUFA) content in seed which can help in reduction of bad cholesterol levels in the body (Zheng et al. 1995). Seed-specific overexpression of bean *β-phaseolin* (Lee et al. 2003); sesame 2S albumin (Katsube et al. 1999); soybean glycinin (Sindhu et al. 1997); pea legumin (Yang et al. 2016); *dihydrodipicolinate synthase* (*DHPS*) (Lee et al. 2001); maize *DHPS* (Wakasa et al. 2006); rice *anthranilate synthase* α-subunit (Zhou et al. 2009); and *E. coli aspartate aminotransferase* (Shin et al. 2006) has been reported to enrich protein quality in transgenic rice. Antioxidant activity has been increased in rice by overexpressing maize C1 and R-S regulatory genes (*ZmMYB*-HLH transcription factor) leading to increased flavonoid content (Wang et al. 2014).

12.12.2 Genetically Engineered Wheat (*Triticum aestivum*)

Wheat is one of the key staple food crops with an annual consumption of 733.2 million metric tons globally [FAOStat 2020]. In terms of nutritive significance, wheat is a rich source of energy in the form of carbohydrates and starch and a substantial amount of protein, dietary fibres, and phytochemicals. Through genetic engineering, overexpression of bacterial *phytoene synthase* (*PSY*) and *carotene desaturase* genes (*CrtB*, *CrtI*) (Cong et al. 2009; Xiaoyan et al. 2012) and provitamin A content has been increased in transgenic wheat. Overexpressing the ferritin gene from soybean and wheat (TaFer1-A) has been reported to increase iron content in transgenic wheat (Borg et al. 2012; Brinch-Pederson et al. 2000). The bioavailability of iron has been increased either by overexpressing the phytochrome gene (*phyA*) (Bhati et al. 2016) to accentuate phytase activity or by silencing ABCC13 transporter gene to decrease phytic acid content (Tamas et al. 2009). Protein content and primarily essential amino acids, cysteine, lysine, methionine, and tyrosine, content in wheat grain have been enhanced by overexpressing *Amaranthus albumin* gene (*ama1*) (Doshi et al. 2006). Further overexpression of maize regulatory genes (C1, B-peru) involved in anthocyanin production has been reported to enhance antioxidant activity in transgenic wheat (Ramesh et al. 2004).

12.12.3 Genetically Engineered Barley (*Hordeum vulgare*)

Despite nutrimental significance, barley lacks certain micronutrients. Therefore, zinc content has been improved by overexpression of zinc transporters in transgenic barley (Holme et al. 2012). Similarly, the bioavailability of iron and zinc has been fortified by enhancing phytase activity by overexpressing the barley *phytase* gene (*HvPAPhy*_a) (Ohnoutkova et al. 2012). Lysine content has been increased by overexpressing the DHPS gene (*dapA*) (Dikeman and Fahey 2006) in transgenic barley. Overexpression of *cellulose synthase-like* gene (*HvCslF*) (Carciofi et al. 2012) has been reported to increase β-glucan in transgenic barley which plays a key role in combating cardiovascular disease and type 2 diabetes in humans (Burton et al. 2011). Overexpression of *delta-6-desaturase* (*D6D*) gene has been reported to enhance polyunsaturated fatty acids (PUFA), γ-linolenic acid, and stearidonic acid (STA) in transgenic barley (Mihalik et al. 2014).

12.12.4 Genetically Engineered Maize (*Zea mays*)

Maize is one of the globally imperative cereal crops that has been addressed to increase its vitamin, protein, mineral, and fibre content with simultaneous lowering of its anti-nutrient components by means of genetic modification. Enrichment of endosperm-specific provitamin A and other carotenoids was accomplished by overexpressing bacterial *crtB* (Decourcelle et al. 2015) and other carotenoid-encoding genes (Zhu et al. 2008; Cahoon et al. 2003). Overexpression of *homogentisic acid geranylgeranyl transferase* (*HGGT*) gene has been reported to upscale tocotrienol and tocopherol content in maize (Chen et al. 2003). Vitamin C (L-ascorbic acid) content has been increased by nearly 100-fold in transgenic maize by expression of *dehydroascorbate reductase* (DHAR gene) (Naqvi et al. 2009). Multivitamin corn, containing a higher percentage of beta-carotene, double-fold folate, and sixfold the normal ascorbate content, has been developed by engineering three distinct metabolic pathways (Drakakaki et al. 2005). Overexpressing soybean ferritin (Chen et al. 2008) and *phyA2* gene from *Aspergillus niger* (Shi et al. 2007) has been successfully reported to increase iron bioavailability in transgenic maize. Maize kernels have destitute nutritional quality particularly essential amino acids like lysine and tryptophan. Several molecular approaches have been devised for improving protein profiles. Lysine-rich protein has been increased by overexpressing the *sb401* gene from potato (Tang et al. 2013; Huang et al. 2006), while lysine and tryptophan content has been increased by silencing alpha-zeins in transgenic maize (Yang et al. 2002). Methionine content has been enhanced in maize via alteration in *cis*-acting site for *Dzs10* gene (Lipkie et al. 2013). Further, overexpression of milk α-lactalbumin has been reported to balance amino acid content in maize (Lai and Messing 2002).

12.12.5 Genetically Engineered Sorghum (*Sorghum Bicolor*)

Sorghum bicolor is a multipurpose cereal crop is also popularly known as "food of the poor" feeding millions of poverty-ridden populations around the world as it can be grown anywhere and withstand harsh environmental conditions. Provitamin A (beta-carotene) content in transgenic sorghum has been enhanced by overexpressing *Homo188-A* (Zhao et al. 2002). Overexpressing the high-lysine protein (*HT12*) gene has been reported to enhance lysine content in transgenic sorghum (Elkonin et al.

2016). Sorghum is difficult to digest when compared to other cereal crops because of the presence of seed storage protein γ-kafirin, which is inert to protease digestion. Therefore, RNAi-mediated silencing of gene encoding *γ-kafirin* along with combined suppression of its isoform genes (*γ-kafirin-1*, *γ-kafirin-2*, and *α-kafirin A1*) has been reported to improve the digestibility index of transgenic sorghum (Grootboom et al. 2014; Malzahn et al. 2017).

12.13 Genome Editing Approaches for Nutrient Enrichment in Cereals

With the advent of high-throughput genome sequencing, numerous crop species have been sequenced, enabling genome editing approaches to tailor genomes for optimizing desirable traits irrespective of genetic barriers across species. Several genome editing approaches such as zinc finger nucleases (ZFNs), transcription activator-like effector nucleases (TALENs), and clustered regularly interspaced short palindromic repeats (CRISPR-Cas) system are being extensively used to precisely target and edit any gene of interest in several crops such as wheat, rice, maize, and barley. Genome editing involves precise and regulated changes in the genome employing site-specific nucleases (SSNs), generating double-stranded breaks (DSBs) at specific loci. Contrary to the transgenic approach which leads to the generation of random insertion events and off-target mutations resulting in undesirable and random phenotype, genome editing methods generate mutants with minimal chances of having off-target phenotype. Using chimeric DNA/RNA oligonucleotides (ONDs), *acetohydroxyacid synthase* genes (*AHAS108* and *AHAS109*) were edited in maize for generating herbicide-tolerant lines (Gao et al. 2010). Likewise, heritable targeted mutagenesis was induced in maize using engineered meganuclease I-CreI to target specific mutation in *LIGULELESS1* (LG1) gene promoter (Kim et al. 1996). These molecular approaches will prove to be instrumental in functional genomics studies and crop improvement breeding programmes resulting in varieties with mass acceptability and relatively lesser regulatory concerns in comparison to the conventional genetically modified (GM) crops (Zhu et al. 1999).

12.13.1 Zinc Finger Nucleases (ZFNs)

In recent years several novel approaches and techniques have enabled precision modification of plant genomes. Amongst these zinc finger nucleases (ZFNs) are engineered multidomain endonucleases, harbouring zinc finger DNA-binding domain fused with a nuclease, primarily a *Fok*I endonuclease. These nuclease functions as a dimer, where pairs of zinc finger domain binds to upstream and downstream target sites, thereby generating double-stranded nicks in the targeted DNA. Zinc finger nucleases are assembled from C2H2 zinc finger domains where each finger recognizes three nucleotides; thereby three-finger adheres to nine nucleotides of a DNA sequence. One of the major negative aspects in employing ZFNs is low specificity resulting from off-target dimerization leading to genotoxicity (Porteus and Baltimore 2003; Szczepek et al. 2007; Porteus 2006; Ramirez et al. 2008). Despite certain drawbacks, ZFNs have been applied across many cereal crops. *Inositol-pentakisphosphate 2-kinase* (*IPK1*) gene catalysing terminal step in phytic acid biosynthesis pathway has been edited in maize by generating 66 ZFNs against 5 intragenic positions (Cantos et al. 2014). In a recent study, non-coding genomic regions were identified in the rice genome for site-specific integration and higher gene expression using ZFNs which resulted in localization of 28 genomic regions including only 1 non-coding discovered for the safe integration of ZFN constructs carrying a *β-glucuronidase* gene (Christian et al. 2010).

12.13.2 Transcription Activator-Like Effector Nucleases (TALENs)

Transcription activator-like effector nucleases (TALENs) are protein-based DNA targeting systems that leverage engineered or synthetic restriction enzymes generated through the fusion of TAL effector DNA-binding domain with DNA-cleaving nuclease (*Fok*I) subunit. TALENs introduce highly specific targeted mutations via repair of double-stranded breaks (DSBs) either by homology-directed repair (HDR) or by non-homologous end joining (NHEJ). The TAL DNA-binding domain is composed of repeats which may vary in number from 16 to 30 constituting a protein encoded by ~3.8 kb CDS. TALENs have evolved from the *Xanthomonas* type III AvrBs3 superfamily effectors which function as transcription factors *in planta* (Boch and Bonas 2010; Li et al. 2012). Each protein constituting AvrBs3 superfamily has different DNA-binding motif repeats that regulate host-pathogen diversity. A higher frequency of repeats enables TALENs to have higher target specificity and affinity resulting in escalated genome editing rates with low genotoxicity. TALEN-based genome editing proves to be one of the most accurate systems with a high mutability rate, but the system possesses certain drawbacks. Similar to ZFNs, TALENs also employ an engineered *Fok*I nuclease subunit and thereby require designing two monomers for individual genome target resulting in large ORF length which cannot be used in viral-based vectors. Moreover, length and the need to synthesize novel pairs of enzymes for each target may constrain the ability to edit multiple targets.

12.13.3 Clustered Regularly Interspaced Short Palindromic Repeats (CRISPR-CAS System)

CRISPR or clustered regularly interspaced short palindromic repeats were discovered in bacteria *E. coli* as a host defence mechanism against foreign genetic elements such as those present within plasmids and phage DNA by inducing RNA-guided DNA cleavage. The CRISPR-CAS system incorporates fragments of foreign DNA known as spacers into CRISPR cassettes followed by transcription of CRISPR arrays and spacers to make a guide crRNA (CRISPR RNA) which then specifically cleaves the targeted genome (Wiedenheft et al. 2012; Van der Oost et al. 2009; Makarova et al. 2011; Barrangou 2013). The key factor for high specificity and efficiency is (protospacer adjacent motif) PAM recognition that distinguishes between bacterial encoding RNA from the bacteriophage target sequence. Cas9 binds to the PAM sequence, resulting in the unfolding of DNA, allowing RNA/DNA hybridization or R-loop formation followed by cleavage of both DNA/RNA and ssDNA strands. CRISPR/Cas9 employs binding of sgRNA at conserved 20 bp complementary sequence and cleaving the target sequence through the action of Cas9 nuclease protein (HNH domain cleaves the complementary DNA strand, while RuvC domain cleaves the non-complementary DNA strand), thereby generating double-stranded break (DSB). Plants' endogenous repair mechanism then repairs DSB generated by NHEJ (non-homologous end joining) or by HR (homologous recombination) which introduces small indels resulting in frame-shift mutations or premature stop codon (Szczelkun et al. 2014; Tsui and Li 2015; Shan et al. 2013). The CRISPR-CAS9 system has been successfully used for genome editing in rice using codon-optimized *sp*Cas9 for targeting the *phytoene desaturase* (*OsPDS*) gene causing non-functionality of carotenoid biosynthetic pathway resulting in white kernels and albino seedlings (Sun et al. 2017). Furthermore, amylose content in rice has been successfully increased by targeting two starch branching enzymes (SBE) SBEIIb (Tang et al. 2017). CAS9-based editing has also been used to decrease heavy metal toxicity in the engineered rice. For example, cadmium (Cd) is a highly toxic heavy metal causing negative health effects upon consuming high-Cd content rice. CRISPR/Cas9 editing system has been employed to knock down metal transporter gene *OsNramp5* to generate mutant *indica* rice lines having low cadmium content (Connorton et al. 2017). Biofortification of wheat to alleviate the intrinsic iron content has been achieved by genome engineering of vacuolar iron transporter (Liang et al. 2014). Similarly, anti-nutritional phytic acid content has been decreased in maize by knocking down genes encoding for phytic acid synthesis (*ZmIPK1A*, *ZmIPK*, and *ZmMRP4*) in maize protoplast by designing two gRNAs targeting respective genes at frequencies of 16.4% and 19.1%, respectively (Zhu et al. 2016).

12.14 Nutrigenomics Future

The imperativeness of nutrition in disease prevention and treatment has gained a central stage with the emergence of high-precision next-generation sequence (NGS) technologies allowing full-genome sequencing in a time- and cost-effective manner. NGS has revolutionized the way we think about daily nutrition, health, illness, and disease prevention and has exhibited that the present modern-day diet and eating habits of human beings are different from its original food niche. Genome-tailored or personalized nutritional counselling can be explored not just for changing eating habits for improving lifestyle but also assisting in the early and precise diagnosis of diseases, thereby retarding the progression of chronic illness and also assisting in the treatment of others. Proponents believe that health care can be improved if nutritionists and practitioners recommend and promote personalized rather community guidelines based upon genetic profile, health status, diet preference, phenotype, and environmental factors.

The realm of nutrigenomics is still in its infancy, and some uncertainties related to its public acceptance and technological advancements still pose a bigger challenge. Ethical issues pertaining to consumer confidentiality and rights need to be thoroughly addressed. Instilling public awareness and trust in terms of knowledge, on how the genes interact with their environment and dietary and lifestyle choices that affect their health status, needs to be prioritized across different public health policies and nutrigenomic based programs at community as well as national levels. Identification of genes and subsidiary pathways underlying combating and progression of disease will help in using "food as medicine". Furthermore, it will also help in escalating levels of essential micronutrients in staple food crops and thus will have a significant impact on the improvement of the nutritional status of populations worldwide. Diseases related to heredity and malnutrition, which are prevalent across various ethnic population groups throughout the world, could be reduced by a personalized and nutrition-rich food supplementation, particularly if efforts are channelized on staple food crops including rice, wheat, beans, maize, and cassava. Synergism between healthcare practice and nutritionally enhanced food product development will enable a swift uptake and translation of novel knowledge in the realm of nutrigenomics; however steps in this direction must be taken with utter discretion and foresightedness. In terms of scientific perspective, focus on cheap and easily accessible knowledge and technological advancement that can largely benefit population particularly socio-economic weaker sections of the world need to occupy centre stage. Novel and affordable technologies can effectively coordinate alongside clinical practices; the information generated thereby must be overlaid onto the integrated metabolic pathway matrix that health professionals already understand. An in-depth understanding of human intermediary metabolism by nutrition scientists is a valuable asset required for the advancement of nutrigenomics which further could fortify a place for nutrition clinicians. Future development in the field of nutrigenomics undoubtedly will place its seemingly huge potential in a better perspective.

References

Acharya T, Kumar NK, Muthuswamy V et al (2004) Harnessing genomics to improve health in India—an executive course to support genomics policy. Health Res Policy Syst 2:1–13

Adiga MNS, Chandy S, Ramaswamy G et al (2008) Homocysteine, vitamin B12 and folate status in pediatric acute lymphoblastic leukemia. Indian J Pediatr 7:235–238

Al-Babili S, Beyer P (2005) Golden Rice—five years on the road—five years to go? Trends Plant Sci 10:565–573

Almendro V, Gascan P (2012) Nutrigenomics and cancer. http://www.fundacionmhm.org/pdf/Mono9/Articulos/articulo9

Anai T, Koga M, Tanaka H et al (2003) Improvement of rice (Oryza sativa L.) seed oil quality through introduction of a soybean microsomal omega-3 fatty acid desaturase gene. Plant Cell Rep 21(10):988–992

Anderson RA, Jones CJ, Goodfellow J (2001) Is the fatty meal a trigger for acute coronary syndromes. Atherosclerosis 159:9–15

Arts IC, Hollman PC, Feskens EJ et al (2001) Catechin intake might explain the inverse relation between tea consumption and ischemic heart disease: the Zutphen elderly study. Am J Clin Nutr 74:227–232

Bacalini MG, Friso S, Olivieri F et al (2014) Present and future of anti-ageing epigenetic diets. Mech Ageing Dev 136–137:101–115, 138

Barrangou R (2013) CRISPR-Cas systems and RNA-guided interference. Wiley Interdiscip Rev RNA 4:267–278

Bayless TM, Rosensweig NS (1966) A racial difference in incidence of lactase deficiency. A survey of milk intolerance and lactase deficiency in healthy adult males. JAMA 197(12):968–972

Bender DA, Bender AE (2009) Benders' dictionary of nutrition and food technology, 7th edn. Woodhead Publishing, Abington

Bhati KK, Alok A, Kumar A et al (2016) Silencing of ABCC13 transporter in wheat reveals its involvement in grain development, phytic acid accumulation and lateral root formation. J Exp Bot 67(14):4379–4389

Bianchini F, Kaaks R, Vainio H (2002) Overweight, obesity, and cancer risk. Lancet Oncol 3:565–574

Boch J, Bonas U (2010) *Xanthomonas* AvrBs3 family-type III effectors: discovery and function. Annu Rev Phytopathol 48:419–436

Borg S, Brinch-Pedersen H, Tauris B et al (2012) Wheat ferritins: improving the iron content of the wheat grain. J Cereal Sci 56:204–213

Brinch-Pederson H, Olesen A, Rasmussen SK et al (2000) Generation of transgenic wheat (*Triticum aestivum* L.) for constitutive accumulation of an *Aspergillus phytase*. Mol Breed 6:195–206

Burton RA, Collins HM, Kibble NA et al (2011) Over-expression of specific *HvCslF* cellulose synthase-like genes in transgenic barley increases the levels of cell wall (1,3;1,4)-β-d-glucans and alters their fine structure. Plant Biotechnol J 9(2):117–135

Cahoon EB, Hall SE, Ripp KG et al (2003) Metabolic redesign of vitamin E biosynthesis in plants for tocotrienol production and increased antioxidant content. Nat Biotechnol 21:1082–1087

Calder PC, Davis J, Yaqoob P et al (1998) Dietary fish oil suppresses human colon tumour growth in athymic mice. Clin Sci (London) 94:303–311

Cantos C, Francisco P, Trijatmiko KR et al (2014) Identification of "safe harbor" loci in indica rice genome by harnessing the property of zinc-finger nucleases to induce DNA damage and repair. Front Plant Sci 5:302

Carciofi M, Blennow A, Jensen SL et al (2012) Concerted suppression of all starch branching enzyme genes in barley produces amylose-only starch granules. BMC Plant Biol 12(1):223

Chae YJ, Kim CH, Ha TS, Hescheler J, Ahn HY, Sachinidis A (2007) Epigallocatechin-3-O-gallate inhibits the angiotensin-II induced adhesion molecule expression in human umbilical vein endothelial cell via inhibition of MAPK pathways. Cell Physiol Biochem 20:859–866

Chang WL, Chapkin RS, Lupton JR (1998) Fish oil blocks azoxymethane-induced rat colon tumorigenesis by increasing cell differentiation and apoptosis rather than decreasing cell proliferation. J Nutr 128:491–497

Chen Z, Young TE, Ling J et al (2003) Increasing vitamin C content of plants through enhanced ascorbate recycling. Proc Natl Acad Sci U S A 100:3525–3530

Chen R, Xue G, Chen P et al (2008) Transgenic maize plants expressing a fungal phytase gene. Transgenic Res 17(4):633–643

Chen D, Li W, Du M et al (2015) Sequencing and characterization of divergent marbling levels in the beef cattle (longissimus dorsi muscle) transcriptome. Asian Australas J Anim Sci 28:158–165

Christian M, Cermak T, Doyle EL et al (2010) Targeting DNA double-strand breaks with TAL effector nucleases. Genetics 186(2):757–761

Christou P, Twyman RM (2004) The potential of genetically enhanced plants to address food insecurity. Nutr Res Rev 17:23–42

Cong L, Wang C, Chen L et al (2009) Expression of phytoene synthase1 and carotene desaturase crtI genes result in an increase in the total carotenoids content in transgenic elite wheat (*Triticum aestivum* L.). J Agric Food Chem 57(18):8652–8660

Connorton JM, Jones ER, Rodríguez-Ramiro I et al (2017) Wheat vacuolar iron transporter TaVIT2 transports Fe and Mn and is effective for biofortification. Plant Physiol 174(4):2434–2444

Corella D, Ordovas JM (2009) Advances in genetics. Nutrigenomics in cardiovascular medicine. Nutrition and Genomics Laboratory, JM-USDA Human Nutrition Research Center on aging at Tufts University, Boston

Costa NMB, Rosa COB (2011) Functional foods: bioactive components and physiological effects. 1 Reprint, R'ubio, Rio de Janeiro

Costabile A, Klinder A, Fava F, Napolitano A et al (2008) Whole-grain wheat breakfast cereal has a prebiotic effect on the human gut microbiota: a double-blind, placebo-controlled, crossover study. Br J Nutr 99:110–120

Cozzolino SMF (2012) The bioavailability of nutrients, 4th edn. Monole, São Paulo

Cozzolino SMF, Cominetti C (2013) Biochemical and physiological bases of nutrition in different stages of life in health and disease, 1st edn. Monole, São Paulo

Crawford M, Galli C, Visioli F et al (2000) Role of plant-derived omega-3 fatty acids in human nutrition. Ann Nutr Metab 44:263–265

da Costa K, Kozyreva OG, Song J et al (2006) Common genetic polymorphisms have major effects on the human requirement for the nutrient choline. FASEB J 20:1336–1344

Davis CD, Uthus EO (2004) DNA methylation, cancer susceptibility and nutrient interactions. Exp Biol Med 229:988–995

Decourcelle M, Perez-Fons L, Baulande S et al (2015) Combined transcript, proteome, and metabolite analysis of transgenic maize seeds engineered for enhanced carotenoid synthesis reveals pleotropic effects in core metabolism. J Exp Bot 66(11):3141–3150

Dietitians Association of Australia (DAA) (n.d.) The Australian Dietary Guidelines. https://daa.asn.au/smart-eating-for-you/smart-eating-fast-facts/healthy-eating/the-australian-dietary-guidelines

Dikeman CL, Fahey GC (2006) Viscosity as related to dietary fiber: a review. Crit Rev Food Sci Nutr 46:649–663

Doshi KM, Eudes F, Laroche A et al (2006) Transient embryo specific expression of anthocyanin in wheat. In Vitro Cell Dev Biol Plant 42:432–438

Drakakaki G, Marcel S, Glahn RP et al (2005) Endosperm-specific co-expression of recombinant soybean ferritin and *Aspergillus phytase* in maize results in significant increases in the levels of bioavailable iron. Plant Mol 59(6):869–880

Elkonin LA, ItalianskayaI JV, Domanina VN et al (2016) Transgenic sorghum with improved digestibility of storage proteins obtained by Agrobacterium-mediated transformation. Russ J Plant Physiol 63:678–689

Enattah NS, Sahi T, Savilahti E et al (2002) Identification of a variant associated with adult-type hypolactasia. Nat Genet 30(2):233–237

EU Science Hub (n.d.) Food-based Dietary Guidelines in Europe. https://ec.europa.eu/jrc/en/health-knowledge-gateway/promotion-prevention/nutrition/food-based-dietary-guidelines

Fafournoux P, Remesy C, Demigne C (1990) Fluxes and membrane transport of amino acids in rat liver under different protein diets. Am J Physiol Endocrinol Metab 259:E614–E625

FAO (2002) Food outlook, no 4, october 2002 (fao.org)

FAO (2020) World food and agriculture – statistical yearbook 2020 (fao.org)

FAOStat (2020) https://www.fao.org/3/cb1329en/CB1329EN.pdf

Ferguson LR (2006) Nutrigenomics: integrating genomic approaches into nutrition research. Mol Diagn Ther 10(2):101–108

Franco RF, Reitsma PH (2001) Gene polymorphisms of the haemostatic system and the risk of arterial thrombotic disease. Br J Haematol 115:491–506

Gao H, Smith J, Yang M et al (2010) Heritable targeted mutagenesis in maize using a designed endonuclease. Plant J 61(1):176–187

Ghodke Y, Chopra A, Shintre P et al (2011) Profiling single nucleotide polymorphisms (SNPs) across intracellular folate metabolic pathway in healthy Indians. Indian J Med Res 133:274–279

Gibney MJ, Gibney ER (2004) Diet, genes and disease: implications for nutrition policy. Proc Nutr Soc 63(3):491–500

Godard B, Hurlimann T (2009) Nutrigenomics for global health: ethical challenges for underserved populations. Curr Pharmacogenomics Person Med 7:205

Goto F, Yoshihara T, Shigemoto N et al (1999) Iron fortification of rice seed by the soybean ferritin gene. Nat Biotechnol 17:282–286

Gregory J, Lowe S, Bates CJ et al (2000) National Diet and nutrition survey: young people aged 4 to 18 years, Report of the diet and nutrition survey, vol 1. The Stationary Office, London

Grootboom AW, Mkhonza NL, Mbambo Z et al (2014) Co-suppression of synthesis of major α-kafirin sub-class together with γ-kafirin-1 and γ-kafirin-2 required for substantially improved protein digestibility in transgenic sorghum. Plant Cell Rep 33(3):521–537

Hall C, Hillen C, Garden Robinson J (2017) Composition, nutritional value and health benefits of pulses. Cereal Chem 94:11–31

Han Y, Gao S, Muegge K et al (2015) Advanced applications of RNA sequencing and challenges. Bioinform Biol Insights 9(Suppl 1):29–46

Haskell MJ (2012) The challenge to reach nutritional adequacy for vitamin A: β-carotene bioavailability and conversion—evidence in humans. Am J Clin Nutr 96:1193S–1203S

Hinds DA, Stuve LL, Nilsen GB et al (2005) Whole-genome patterns of common DNA variation in three human populations. Science 307:1072–1079

Holme IB, Dionisio G, Brinch-Pedersen H et al (2012) Cisgenic barley with improved phytase activity. Plant Biotechnol J 10(2):237–247

Holt SH, Delargy HJ, Lawton CL et al (1999) The effects of high-carbohydrate vs high-fat breakfasts on feelings of fullness and alertness, and subsequent food intake. Int J Food Sci Nutr 50(1):13–28

Hooper L, Summerbell CD, Higgins JPT et al (2001) Dietary fat intake and prevention of cardiovascular disease: systemic review. Br Med J 322:757–763

Hossain P, Kawar B, Nahas ME et al (2007) Obesity and diabetics in the developing world—a growing challenge. N Engl J Med 356:213–215

Hsu SC, Huang CJ (2006) Reduced fat mass in rats fed a high oleic acid-rich safflower oil diet is associated with changes in expression of hepatic PPARa and adipose SREBP-1c-regulated genes. J Nutr 136:1779–1785

Huang S, Frizzi A, Florida CA et al (2006) High lysine and high tryptophan transgenic maize resulting from the reduction of both 19- and 22-kD alpha-zeins. Plant Mol Biol 61(3):525–535

Hurrell R, Egli I (2010) Iron bioavailability and dietary reference values. Am J Clin Nutr 91:1461S–1467S

Ilnytska O, Argyropoulos G (2008) The role of the agouti-related protein in energy balance regulation. Cell Mol Life Sci 65(17):2721–2731

Jeffery IB, O'Toole PW (2013) Diet microbiota interactions and their implications for healthy living. Nutrients 5:234–252

Kaput J, Noble J, Hatipoglu B et al (2007) Application of nutrigenomic concepts to type 2 diabetes mellitus. Nutr Metab Cardiovasc 17:89–103

Katsube T, Kurisaka N, Ogawa M et al (1999) Accumulation of soybean glycinin and its assembly with the glutelins in rice. Plant Physiol 120:1063–1073

Kim YG, Cha J, Chandrasegaran S (1996) Hybrid restriction enzymes: zinc finger fusions to Fok I cleavage domain. Proc Natl Acad Sci U S A 93(3):1156–1160

Knowles LM, Milner JA (2003) Diallyl disulfide induces ERK phosphorylation and alters gene expression profiles in human colon tumor cells. J Nutr 133:2901–2906

Kopelman PG (2000) Obesity as a medical problem. Nature 404:635–643
Krishnaswamy K (2008) Developing and implementing dietary guidelines in India. Asia Pac J Clin Nutr 17(Suppl 1):66–69
Külahoglu C, Bräutigam A (2014) Quantitative transcriptome analysis using RNA-seq. In: Staiger D (ed) Plant circadian networks. Humana, New York, pp 71–91
Kussmann M, Affolter M (2006) Proteomic methods in nutrition. Curr Opin Clin Nutr Metab Care 9(5):575–583
Kussmann M, Affolter M, Fay LB (2005) Proteomics in nutrition and health. Comb Chem High Throughput Screen 8(8):679–696
Lai JS, Messing J (2002) Increasing maize seed methionine by mRNA stability. Plant J 30:395–402
Lane MA, Bailey SJ (2005) Role of retinoid signalling in the adult brain. Prog Neurobiol 75:275–293
Lee S, An G (2009) Over-expression of OsIRT1 leads to increased iron and zinc accumulations in rice. Plant Cell Environ 32:408–416
Lee SI, Kim HU, Lee YH et al (2001) Constitutive and seed-specific expression of a maize lysine-feedback-insensitive dihydrodipicolinate synthase gene leads to increased free lysine levels in rice seeds. Mol Breed 8:75–84
Lee TTT, Wang MMC, Hou RCW et al (2003) Enhanced methionine and cysteine levels in transgenic rice seeds by the accumulation of sesame 2S albumin. Biosci Biotechnol Biochem 67:1699–1705
Lee S, Jeon US, Lee SJ et al (2009) Iron fortification of rice seeds through activation of the nicotianamine synthase gene. Proc Natl Acad Sci U S A 106:22014–22019
Li T, Liu B, Spalding MH et al (2012) High-efficiency TALEN-based gene editing produces disease-resistant rice. Nat Biotechnol 30(5):390–392
Liang Z, Zhang K, Chen K et al (2014) Targeted mutagenesis in *Zea mays* using TALENs and the CRISPR/Cas system. J Genet Genomics 41(2):63–68
Lipkie TE, De Moura FF, Zhao ZY et al (2013) Bioaccessibility of carotenoids from transgenic provitamin a biofortified sorghum. J Agric Food Chem 61(24):5764–5771
Loktionov A (2003) Common gene polymorphisms and nutrition: emerging links with pathogenesis of multifactorial chronic diseases. J Nutr Biochem 14:426–451
Lucca P, Hurrell R, Potrykus I (2002) Fighting iron deficiency anemia with iron-rich rice. J Am Coll Nutr 21:184S–190S
Luft FC, Weinberger MH (1997) Heterogeneous responses to changes in dietary salt intake: the salt-sensitivity paradigm. Am J Clin Nutr 65(Suppl 2):612S–617S
Mahan LK, Scott Stump S (2005) Food, nutrition & diet therapy, 6th edn. Roca, Sao Paulo
Mahley RW, Rall SC (2000) Apolipoprotein E: far more than a lipid transport protein. Ann Rev Genomics Hum Genet 1:507–537
Makarova KS, Haft DH, Barrangou R et al (2011) Evolution and classification of the CRISPR-Cas systems. Nat Rev Microbiol 9:467–477
Malzahn A, Lowder L, Qi Y (2017) Plant genome editing with TALEN and CRISPR. Cell Biosci 24(7):21
Masuda H, Suzuki M, Morikawa KC et al (2008) Increase in iron and zinc concentrations in rice grains via the introduction of barley genes involved in phytosiderophore synthesis. Rice 1:100–108
McNulty H, Eaton-Evans J, Cran G et al (1996) Nutrient intakes and impact of fortified breakfast cereals in schoolchildren. Arch Dis Child 75(6):474–481
Mihalik D, Gubisova M, Klempova T et al (2014) Transgenic barley producing essential polyunsaturated fatty acids. Biol Plant 58(2):348–354
Moghissi AA, Pei S, Liu Y (2015) Golden rice: scientific, regulatory and public information processes of a genetically modified organism. Crit Rev Biotechnol 21:1–7
Moraes AAC, Pereira FEL (2009) Nutrigenomics of the soy. In: Waitzberg D, Enteral L (eds) Parenteral nutrition in clinical practice, 4th edn. Atheneu, São Paulo, pp 2039–2047
Muller M, Kersten S (2003) Nutrigenomics: goals and strategies. Nat Rev Genet 4:315–322

Najafian M, Jahromi MZ, Nowroznejhad MJ et al (2012) Phloridzin reduces blood glucose levels and improves lipids metabolism in streptozotocin induced diabetic rats. Mol Biol Rep 39:5299–5306

Naqvi S, Zhu C, Farre G et al (2009) Transgenic multivitamin corn through biofortification of endosperm with three vitamins representing three distinct metabolic pathways. Proc Natl Acad Sci U S A 106(19):7762–7767

Neeha VS, Kinth P (2013) Nutrigenomics research: a review. J Food Sci Technol 50:415–428

O'Brien MA, Costin BN, Miles MF (2012) Using genome-wide expression profiling to define gene networks relevant to the study of complex traits: from RNA integrity to network topology. In: Chesler EJ, Haendel MA (eds) International review of neurobiology. Academic, London, pp 91–133

Ohnoutkova L, Zitka O, Mrizova K et al (2012) Electrophoretic and chromatographic evaluation of transgenic barley expressing a bacterial dihydrodipicolinate synthase. Electrophoresis 33(15): 2365–2373

Oommen AM, Griffin JB, Sarath G et al (2005) Roles for nutrients in epigenetic events. J Nutr Biochem 16:74–77

Ordovas JM, Corella D (2004) Nutritional genomics. Annu Rev Genomics Hum Genet 5:71–118. https://doi.org/10.1146/annurev.genom.5.061903.180008

Paine JA, Shipton CA, Chaggar S et al (2005) Improving the nutritional value of Golden Rice through increased pro-vitamin A content. Nat Biotechnol 23:482–487

Patient D, Ainsworth P (1994) The chemistry of flour and bread. Nutr Food Sci 94(3):22–24

Piskol R, Ramaswami G, Li JB (2013) Reliable identification of genomic variants from RNA-seq data. Am J Hum Genet 93:641–651

Porteus MH (2006) Mammalian gene targeting with designed zinc finger nucleases. Mol Ther 13(2):438–446

Porteus MH, Baltimore D (2003) Chimeric nucleases stimulate gene targeting in human cells. Science 300(5620):763

Public Health Nutrition (n.d.) ISSN: 1475-2727 (electronic), 1368–9800 (paper)

Ramesh SA, Choimes S, Schachtman DP (2004) Over-expression of an *Arabidopsis* zinc transporter in *Hordeum vulgare* increases short-term zinc uptake after zinc deprivation and seed zinc content. Plant Mol Biol 54(3):373–385

Ramirez CL, Foley JE, Wright DA et al (2008) Unexpected failure rates for modular assembly of engineered zinc fingers. Nat Methods 5(5):374–375

Rao S (2001) Nutritional status of Indian population. J Biosci 26(4):481–489

Rosentrater KA, Evers AD (2018) Feed and industrial uses for cereals. In: Kent's technology of cereals. Woodhead Publishing, Abington, pp 785–837

Sarwar M (2008) Laboratory studies on different wheat genotypes for their resistance against Khapra Beetle Trogoderma granarium everts (Coleoptera: Dermestidae). Pak J Seed Technol 2(11&12):46–53

Schaefer EJ (2002) Lipoproteins, nutrition, and heart disease. Am J Clin Nutr 75:191–212

Shan Q, Wang Y, Li J et al (2013) Targeted genome modification of crop plants using a CRISPR-Cas system. Nat Biotechnol 31(8):686–688

Sharma M, Majumdar PK (2009) Occupational life style diseases: an emerging issue. Indian J Occup Environ Med 13(3):109–112

Shendure J, Ji H (2008) Next-generation DNA sequencing. Nat Biotechnol 26:1135–1145

Shetty PS (2002) Nutrition transition in India. Public Health Nutr 5(1A):175–172

Shi J, Wang H, Schellin K et al (2007) Embryo-specific silencing of a transporter reduces phytic acid content of maize and soybean seeds. Nat Biotechnol 25(8):930–937

Shin YM, Park HJ, Yim SD et al (2006) Transgenic rice lines expressing maize C1 and R-S regulatory genes produce various flavonoids in the endosperm. Plant Biotechnol J 4:303–315

Shumskaya M, Wurtzel ET (2013) The carotenoid biosynthetic pathway: thinking in all dimensions. Plant Sci 208:58–63

Sindhu AS, Zheng Z, Murai N (1997) The pea seed storage protein legumin was synthesized, processed, and accumulated stably in transgenic rice endosperm. Plant Sci 130:189–196
Sinha R, Anderson DE, McDonal SS et al (2003) Cancer risk and diet in India. J Postgrad Med 49: 222–228
Smith JP (1999) Healthy bodies and thick wallets: the dual relation between health and economic status. J Econ Perspect 13(2):144–166
Solanky KS, Bailey NJ, Beckwith-Hall BM et al (2003) Application of biofluid 1H nuclear magnetic resonance based metabonomic techniques for the analysis of the biochemical effects of dietary isoflavones on human plasma profile. Anal Biochem 323:197–204
Steinmetz KL, Pogribny IP, James SJ et al (1998) Hypomethylation of the rat glutathione S-transferase p (GSTP) promoter region isolated from methyl-deficient livers and GSTP positive liver neoplasms. Carcinogenesis 19(8):1487–1494
Sun Y, Jiao G, Liu Z et al (2017) Generation of high-amylose rice through CRISPR/Cas9-mediated targeted mutagenesis of starch branching enzymes. Front Plant Sci 8:298
Surh YJ (2003) Cancer chemoprevention with dietary phytochemicals. Nat Rev Cancer 3:768–780
Swanson KS, Schook LB, Fahey GC Jr (2003) Nutritional genomics: implications for companion animals. J Nutr 133:3033–3040
Szczelkun MD, Tikhomirova MS, Sinkunas T et al (2014) Direct observation of R-loop formation by single RNA-guided Cas9 and Cascade effector complexes. Proc Natl Acad Sci U S A 111(27):9798–9803
Szczepek M, Brondani V, Büchel J et al (2007) Structure-based redesign of the dimerization interface reduces the toxicity of zinc-finger nucleases. Nat Biotechnol 25(7):786–793
Tamas C, Kisgyorgy BN, Rakszegi M et al (2009) Transgenic approach to improve wheat (*Triticum aestivum* L.) nutritional quality. Plant Cell Rep 28(7):1085–1094
Tang G, Qin J, Dolnikowski GG et al (2009) Golden Rice is an effective source of vitamin A. Am J Clin Nutr 89:1776–1783
Tang G, Hu Y, Yin SA et al (2012) β-carotene in Golden Rice is as good as β-carotene in oil at providing vitamin A to children. Am J Clin Nutr 96:658–664
Tang M, He X, Luo Y et al (2013) Nutritional assessment of transgenic lysine-rich maize compared with conventional quality protein maize. J Sci Food Agric 93:1049–1054
Tang L, Mao B, Li Y et al (2017) Knockout of *OsNramp5* using the CRISPR/Cas9 system produces low Cd-accumulating indica rice without compromising yield. Sci Rep 7(1):14438
Tanumihardjo SA, Palacios N, Pixley KV (2010) Provitamin A carotenoid bioavailability: what really matters? Int J Vitam Nutr Res 80:336–350
Trayhurn P (2000) Proteomics and nutrition—a science for the first decade of the new millennium. Br J Nutr 83:1–2
Trijatmiko K, Duenas C, Tsakirpaloglou N et al (2016) Biofortified indica rice attains iron and zinc nutrition dietary targets in the field. Sci Rep 6:19792
Trujillo E, Davis C, Milner J (2006) Nutrigenomics, proteomics, metabolomics, and the practice of dietetics. J Am Diet Assoc 106:403–413
Tsui TK, Li H (2015) Structure principles of CRISPR-Cas surveillance and effector complexes. Annu Rev Biophys 44:229–255
Tucker KL, Smith CE, Lai CQ et al (2013) Quantifying diet for nutrigenomic studies. Annu Rev Nutr 33:349–371
U.S. Department of Health and Human Services and U.S. Department of Agriculture (2015) 2015–2020 Dietary Guidelines for Americans, 8th Edition. https://health.gov/sites/default/files/2019-09/2015-2020_Dietary_Guidelines.pdf
Van der Oost J, Jore MM, Westra ER et al (2009) CRISPR-based adaptive and heritable immunity in prokaryotes. Trends Biochem Sci 34:401–407
Vasconcelos M, Datta K, Oliva N et al (2003) Enhanced iron and zinc accumulation in transgenic rice with the ferritin gene. Plant Sci 164:371–378

Wakasa K, Hasegawa H, Nemoto H et al (2006) High-level tryptophan accumulation in seeds of transgenic rice and its limited effects on agronomic traits and seed metabolite profile. J Exp Bot 57:3069–3078

Wang C, Zeng J, Li Y et al (2014) Enrichment of provitamin A content in wheat (*Triticum aestivum* L.) by introduction of the bacterial carotenoid biosynthetic genes CrtB and CrtI. J Exp Bot 65(9):2545–2556

WHO (2018) https://apps.who.int/iris/rest/bitstreams/1137482/retrieve

WHO (2019) https://apps.who.int/iris/bitstream/handle/10665/311696/WHO-DAD-2019.1-eng.pdf

WHO (2020) https://www.who.int/data/gho/publications/world-health-statistics

Wickramasinghe S, Rincon G, Islas-Trejo A et al (2012) Transcriptional profiling of bovine milk using RNA sequencing. BMC Genomics 13:45

Wiedenheft B, Sternberg SH, Doudna JA (2012) RNA-guided genetic silencing systems in bacteria and archaea. Nature 482:331–338

World Health Organization (1990) Diet, nutrition and the prevention of chronic diseases. Report of a WHO Study Group (WHO technical report series, no. 797), Geneva

World Health Organization (2003) Diet, nutrition and the prevention of chronic diseases. WHO, Geneva. ISSN 0512-3054

Xiaoyan S, Yan Z, Shubin W (2012) Improvement Fe content of wheat (*Triticum aestivum*) grain by soybean ferritin expression cassette without vector backbone sequence. J Agric Biotechnol 20: 766–773

Xudong Y, al-Babili S, Klöti A et al (2000) Engineering the provitamin A (b-carotene) biosynthetic pathway into (carotenoid-free) rice endosperm. Science 287:303–305

Yang SH, Moran DL, Jia HW et al (2002) Expression of a synthetic porcine alpha-lactalbumin gene in the kernels of transgenic maize. Transgenic Res 11:11–20

Yang QQ, Zhang CQ, Chan ML et al (2016) Biofortification of rice with the essential amino acid lysine: molecular characterization, nutritional evaluation, and field performance. J Exp Bot 67(14):4285–4296

Ye SQ, Kwiterovich PO (2000) Influence of genetic polymorphisms on responsiveness to dietary fat and cholesterol. Am J Clin Nutr 52(Suppl 5):1275S–1284S

Zhao ZY, Glassman K, Sewalt V et al (2002) Nutritionally improved transgenic sorghum. In: Plant biotechnology 2002 and beyond. Springer, Dordrecht, pp 413–416

Zheng A, Sumi K, Tanaka K et al (1995) The bean seed storage protein β-phaseolin is synthesized, processed and accumulated in the vacuolar type-II protein bodies of transgenic rice endosperm. Plant Physiol 109:777–786

Zheng L, Cheng Z, Ai C et al (2010) Nicotianamine, a novel enhancer of rice iron bioavailability to humans. PLoS One 5(4):e10190

Zhou Y, Cai H, Xiao J et al (2009) Over-expression of aspartate aminotransferase genes in rice resulted in altered nitrogen metabolism and increased amino acid content in seeds. Theor Appl Genet 118:1381–1390

Zhu T, Peterson DJ, Tagliani L et al (1999) Targeted manipulation of maize genes *in vivo* using chimeric RNA/DNA oligonucleotides. Proc Natl Acad Sci U S A 96(15):8768–8773

Zhu C, Naqvi S, Gomez-Galera S et al (2007) Transgenic strategies for the nutritional enhancement of plants. Trends Plant Sci 12:548–555

Zhu C, Naqvi S, Breitenbach J et al (2008) Combinatorial genetic transformation generates a library of metabolic phenotypes for the carotenoid pathway in maize. Proc Natl Acad Sci U S A 105(47):18232–18237

Zhu J, Song N, Sun S et al (2016) Efficiency and inheritance of targeted mutagenesis in maize using CRISPR-Cas9. J Genet Genom 43:5–36

Genetically Modified Cereal Crops Regulation Policies

13

Ram Krishna, P. S. Soumia, Waquar Akhter Ansari, Kiran Khandagale, and Major Singh

Abstract

The history of crop genetic manipulation through conventional breeding (artificial selection and selective breeding) dates back to more than 10,000 years. To feed the intense growing population, conventional breeding is unsuitable due to time, money consumption, and lack of desirable traits in plant genetic pool. The introduction of biotechnology in the late twentieth century and the start of the twenty-first century revolutionized modern agriculture by introducing the unavailable desired traits from other sources. The adaptation of genetically modified (GM) crops may create many socio-economic, food, and sustainability opportunities for both farmer ecosystem and farmers. In the last two decades GM crop adaptation increased due to its ability to multiply the quality agricultural productivity. Worldwide during 2017, 30% of canola, 80% of cotton, 32% of maize, and 77% of GM soybean were cultivated. Globally, 26 countries (21 developing and 5 industrialized countries) planted 191.7 million hectares of biotech crops. Furthermore, 43 other countries have formally cultivated GM crops to measure the utilization of GM crops. Despite the above facts a huge gap exists in both rapid acceptance of GM crops by farmers in many countries and for food, feeds, and limited acceptance by consumers in global market. These facts also characterized the various opinions of consumers. The significant factors influencing consumer's attitudes are the awareness of benefit and risk, knowledge and trust, and personal values. GM crops have sparked tremendous public outrage, particularly on the rising concerns over GM food labelling, prompting the government to withdraw Bt brinjal from India. The increasing GM crop

R. Krishna (✉) · P. S. Soumia · M. Singh
ICAR-Directorate of Onion and Garlic Research, Pune, India

W. A. Ansari · K. Khandagale
Department of Botany, Savitribai Phule Pune University, Pune, India

R. Deshmukh et al. (eds.), *Biofortification in Cereals*,
https://doi.org/10.1007/978-981-19-4308-9_13

cultivation has augmented a wide range of distresses with respect to environmental, socio-economic, and food safety issues. In this chapter, we explained the present status of GM crops research, regulatory framework, and challenges involved in GM research globally.

Keywords

Cereal crops · Genetic modification · Food policy · Transgene

13.1 Introduction

Plants are the most important resource for life as they provide about 90 and 80 percent of calories and protein, respectively, to the global human population. Furthermore, the plants also provide foods directly or indirectly to animals. About 3000 plant species were being cultivated for food purposes by human, but presently total global population mainly relies upon 20 species of crop for nutritional requirements of which 50 percent is shared by eight crop species (Krishna et al. 2019). Minerals and vitamins are obtained from 30 species (fruit and vegetables). As per estimation, the earth can feed 15 billion strict vegetarians or 5 billion mixed diet population but the world total population by 2050 will reach nearly 10 billion. Hence global agriculture is a major challenge to feed and nourish the increasing global population. The world food security status, i.e., the equilibrium between increasing food requirement of the world population and worldwide agricultural output, associated with inconsistencies between supply and demand at the regional, national, and local scales is disturbing (Ingram 2011). It has perceptibly compounded during the ongoing decades, finishing as of late in the 2008 nourishment emergency. It is basic to take note that in mid-2011, nourishment costs were back to their statures of the center of the 2008 emergency (FAO 2011). This is because of abiotic (drought, heat, salt, water logging, etc.) and biotic (virus, bacteria, fungi, insect, and weeds) stresses which potentially hampers the agricultural productivity and quality in natural ecosystem. As conventional breeding procedures were unable to overcome these biotic and abiotic stresses, genetic modifications of crops were initiated (Krishna et al. 2019). Twenty-first century is the era of biotechnology, which deals with the genetic modification (GM) of genetic materials in living organisms, thereby achieving specific functions (Raman 2017; Zhang et al. 2016). Roughly 10,000 years ago, most basic theory of adaptation for domestication and consumption of plants was reported, where our predecessors often selected superior parents to manipulate the genetic material in living organisms, enabling them to perform specific functions which were collectively termed as "selective breeding" and "artificial selection" by Darwin. Although recombinant DNA technology first emerged in the 1960s, the basic principle of recombination was discovered many years earlier. After this discovery of transferable nature of the genetic material between different species in 1946, double helical structure of DNA and the concept of central dogma by Watson and Crick were reported in 1954. Consequently, Boyer and Cohen in 1973 made the world's first genetically modified organism using restriction endonucleases and DNA ligase, commonly referred to as "molecular scissors and glue" that allowed the direct modification of the genome. These advances allowed scientist to manipulate the genetic material of the organism and induce different effects. Rudolf Jaenisch in 1974 created the first genetically modified animal (mouse), while in 1983 first genetically modified plant (antibiotic resistant tobacco) was produced. In 1992, transgenic tobacco for virus resistance was first commercialized in China. Later in 1994, first genetically modified food, Flavr Savr tomato (Calgene, USA) was approved for human consumption. Antisense technology was used in the modification allowing the tomato to delay ripening after picking as the polygalacturonase enzyme production got hampered. Subsequently, few transgenic crops like canola with modified oil composition (Calgene), *Bacillus thuringiensis* (*Bt*) corn (Ciba-Geigy), bromoxynil resistant cotton (Calgene), *Bt* cotton (Monsanto), *Bt* potatoes (Monsanto), glyphosate resistant soybeans (Monsanto), virus-resistant squash (Asgrow), and delayed ripening tomatoes (DNAP, Zeneca/ Peto, and Monsanto) received marketing approval in 1995 (James 2011). Up till 1996, nearly 35 approvals were granted for commercial production of 8 transgenic crops and one flower crop (carnation) with 8 different traits in 6 countries (James 1996). After two decades of commercialization of biotech crops, nearly 70 countries covering 191.7 million hectares area have adopted this technology by 2018, thereby making it fastest adopted crop technology in the history of modern agriculture (ISAAA 2018). As of 2019, the USA leads the list of countries for commercial production of genetically modified crops. Presently, GM crops like canola, corn, carrots, cantaloupe, cotton, tomatoes, potatoes, brinjal, soybean, strawberries, lettuce, etc. are easily available in the market. Furthermore, the GM products like medicines, vaccines, foods, feeds, and fibers are currently in the pipeline (Bawa and Anilakumar 2013; Zhang et al. 2016). With the advent of biotech crops, global food crop production has increased by >370 million tonnes from a relatively smaller acreage (Zhang et al. 2016). Furthermore, GM crops have been beneficial to both economy and the environment. As referenced before, these biotech crops pose less impact on the environment, bringing about expansion in species diversity. Therefore, it is obvious that GM crops have been recommended by various agricultural scientists, growers, and most environmentalists worldwide. However, questions related to safety and efficacy have been raised during their advancements. More precisely, the GM seed industry has been plagued with several issues related to human health and insect resistance which truly dilute their beneficial effects. Besides, lack of clear understanding and knowledge of GM technologies, safety studies, and mistrust regarding GMOs have only aggravated the problems. As the result, many countries, particularly the European Union and Middle East have either imposed partial or full restrictions on GM crops. Hence, GM crops are still being one of the hottest topics of debate at public and policymaking levels. Despite the mistrust regarding GMOs still prevails in society, why do scientists often recommend incorporation of transgenic crops into conventional agriculture?

13.2 Need of Transgenic Food Crops

Agriculture sector alone contributes a large share of the GDP which is estimated at US$ 3.2 trillion worldwide and also generates employment in both developing and underdeveloped nations (World Bank 2017). For example: agriculture contributes only 1.4% to the GDP and engages nearly 1.62% of the workforce in USA, whereas it is 18.6% of the GDP involving 50% of the workforce in the developing countries of South Asia (Nayar 2011). Although agriculture industry has contributed much towards the GDP and employment generation with 19% of the world's population, it is projected to suffer significant setbacks by 2050 due to the burgeoning population, pest resistance, and depletion of natural resources. The details of which are elaborated further in this section.

13.2.1 Population Explosion

According to United Nations report, the current world population of 7.6 billion is expected to reach 9.8 billion in 2050 and further to an estimated 11.2 billion in 2100 (www.un.org). In comparison to 2013, nearly 50% increase in the population is expected by 2050; henceforth, the present agricultural practices alone cannot sustain this burgeoning population and eradicate malnutrition on a global scale. In a recent report by FAO, nearly 653 million people will remain undernourished in 2030, regardless of the significant reduction in global hunger (FAO 2017). Besides, previous studies revealed that the top four global crops (soybean, maize, wheat, and rice) are increasing at 1.0%, 0.9%, 1.6%, and 1.3% each year, respectively, which is less than the required growth rate of 2.4%/year needed to sustain the global population by 2050 (Ray et al. 2013). Further, problems like improved nutritional standards of lower-middle class population and estimated decline in arable land (from 0.242 ha/person in 2016 to 0.18 ha/person in 2050) owing to degradation and accelerated urbanization, rapid population explosion will increase the demand for food resources.

13.2.2 Biotic Stresses in Plants (Pests and Diseases)

Biotic stresses pose major economic losses in agriculture every year. Annually about 20–40% of global crop loss is due to pest alone. In order to combat these crop pest and diseases, an expenditure of approximately $290 m annually is incurred by the agriculture industry (FAO 2017). It is estimated that disease and pest of crop occurrence become more frequent and are expanding 2.7 km per year towards the poles (Bebber et al. 2014), which is noted in Central America as wheat rust and coffee leaf rust outbreaks. This phenomenon is attributed to globalization which has tremendously increased the movement of plant materials, associated pest and disease, vectors, and climate change (FAO 2017). However, integrated pest and disease management techniques had tried to manage crop losses due to these biotic stresses to some extent but are incapable to solve the transboundary crop-demics. For example, Tropical Race-4 (TR4) strains of *Fusarium oxysporum* f.sp. *cubense* (Foc) have significantly crippled the global banana industry by causing Panama disease (or Panama wilt) during early-mid 1990s (Ordonez et al. 2015). Later in 2013 nearly 5900 hectares of bananas in Philippines and >20% of total banana plantations from Mozambique in 2015 were abandoned due to TR4 infestation. Moreover, in terms of economic value, this strain had also caused nearly US$ 388.4 m loss in countries like Taiwan, Malaysia, and Indonesia (ProMusa Organization 2017). Hence, increasing transboundary crop and pest diseases movement has environmental, social, and economic impacts on farmers and threatens food security.

13.2.3 Burden on Natural Resources

The FAO's 2050 estimations propose an estimated shortage of natural resources crop care (FAO 2017). In spite of full agricultural efficiency, unsustainable competition has strengthened because of population growth, industrialization, urbanization, and climate change. Agriculture alone accounts for 80% of all global deforestation. Deforestation is still common for agriculture in tropical and subtropical regions and responsible for seven million hectares loss of natural forests per year during 2000–2010 (FAO 2017). Furthermore, excessive groundwater exploitation for agricultural practices alone accounted for 70% of total water exploitation, which severely depletes naturally occurring water resources in many countries. This has been especially reported in the region of low rainfall, like Central Asia, North Africa, and Middle East where 80–90% of total water exploitation is used for agriculture (FAO 2017). The same trends are estimated to be continuing for the twenty-first century and therefore increase the pressure on natural resources globally.

13.3 GM Crops: The Way out

Globally genetically modified crops provide numerous benefits to the farmers and also are potential enough to cope with major challenges faced by agriculture. Benefit from the global farm income alone is estimated to be $117.6 billion from 1996 to 2013. Wherein the yearly global net income has increased by 34.3% in 2010–2012 (Zhang et al. 2016; Chen and Lin 2013; Brookes and Barfoot 2012). Although GM crops increase the global yield by 22%, it has also drastically reduced the usage of pesticide by 37% and its impact on the environment by 18% (Sibhatu and Qaim 2018; Klümper and Qaim 2014). In order to attain the same yield standards through growing conventional crops, >300 million acres of arable land need to be engaged in the cultivation process which may add to the current environmental and socio-economic problems in agriculture (Zhang et al. 2016). Further the impact of GM crops on the economy can be better understood through the success stories from Australia (GM canola) and India (GM cotton) (Brookes and Barfoot 2014).

13.4 GM Cereal Crops and Food Security

The world's population is expected to increase from 7.7 billion to 9.7 billion in 2050, owing to 34% hike during the next 30 years as per the United Nations report 2019. Due to urbanization, population explosion will occur mainly in developing countries, wherein 70% of the world's population will be urban as compared to present day urban population of 49%. Against the background of diminishing natural resource, the immediate priority in global agriculture is to increase the productivity to ensure sufficient availability of food and other raw materials for a growing population (Von Braun 2007). Though burgeoning population has always instilled pressure on food production, our agricultural systems have been strengthened to mitigate food insufficiency through various technological interventions. Cereals are the basic source of food energy (56%) and protein consumed (50%) on earth (Krishna et al. 2019; Zhang et al. 2016). In order to meet the requirements of massive population, global food production must increase by 70% indicating two-fold increase in cereal production from the same available resources (Raman 2017). In modern-day agriculture, transgenic plants play an integral part in ensuing difficulty in differentiating the transgenics from its counterparts in some regions. The genetic transformation of crop plants based on recombinant DNA technology during the early 1980s has enabled breeders to transfer novel gene(s) across species boundaries, unlike conventional breeding. Genetically modified (GM) traits can be distinguished into three categories: (1) First-generation GM crops involve improved agronomic traits (resistance to pests and diseases); (2) Second-generation GM crops involve enhanced quality traits (higher nutrient contents of food products); and (3) Third-generation GM crops involve plants designed to produce special substances for pharmaceutical or industrial purposes. At present, only a few first-generation technologies have been commercialized, of which the dominant being herbicide tolerance (HT) in soybeans, which made up 81% of the global GM crop area in 2010. Since the inception biotech crops, about 148 million ha of GM crops have been grown in 29 countries, signifying 10% of 1.5 billion hectares of cropland in the world (https://www.isaaa.org/resources/publications/briefs/55/default.asp). As far as the area of GM crops is concerned, there is an unprecedented 100-fold increase from 1.7 million hectares in 1996 to 170 million hectares in 2012 (Osmond and Colombo 2019; Bawa and Anilakumar 2013), thus making it the fastest adopted agricultural technology of the recent past.

The concept of integrated pest management (IPM) appeared in the 1970s, when the negative impact on the environment and human health was evident due to injudicious use of chemical pesticides. The indirect (preventive) crop protection practices act as the basis of IPM module, which mainly rely on understanding the ecosystem including the crop, pest, and natural enemy biology and use of optimized farming practices to manage pests. Host plant resistance, either developed through conventional breeding or genetic engineering is the keystone of IPM and also complements the other pest management practices. Most GM crops provide tolerance to herbicides (like glyphosate, dicamba, or 2–4 D), insect pests (like moths, flies, or beetles), or a combination of both traits. *Bt* crop cultivation reduces the use of chemical insecticides and thus provides environmental and economic benefits leading to sustainable agricultural production. The concept of using *Bt* genes was not novel as *Bt* formulations (like Dipel, Foil) were already been commercially exploited for more than four decades to control insect pest in particular Lepidoptera (Cannon 1996). *Bt* toxins exhibit high level of species-specificity against insect pests belonging to the order Lepidoptera, Diptera, and Coleoptera, without affecting predators and other beneficial insects (World Bank 2017; Nayar et al. 2012).

Genetically engineered crops against insect pests were first commercialized during mid-1990s with the introduction of GM maize, potato, and cotton plants expressing genes encoding the entomocidal δ-endotoxin (including Cry and Cyt toxins) from a Gram positive, spore-forming soil bacterium, *Bacillus thuringiensis* (*Bt*). Bt Cry and Cyt toxins belong to a class of bacterial toxins known as pore-forming toxins (PFT) that are secreted as water-soluble proteins which undergo conformational changes in order to insert into or to translocate across the cell membranes of their host. These PFTs are broadly classified into two main groups: (i) α-helical toxins that include the Cry proteins containing three domains (forms the trans-membrane pore) and (ii) β-barrel toxins that include Cyt proteins (aid in insertion into the membrane) (Parker and Feil 2005; Bravo et al. 2007). The Cry genes are located on plasmids of large molecular weight. Currently, more than 70 classes of Cry genes are described (cry1 the cry70). These endotoxins have been classified as Cry1-Cry69 and Cytl-Cyt3 and different subgroups depending on their amino acid sequence (http://www.lifesci.sussex.ac.uk/home/Neil_Crickmore/Bt/). In most commercial crop varieties, these Cry proteins are usually expressed in their active forms, whereas in biopesticide formulations these Cry proteins are present as protoxins. The relevance of Cry proteins is due to their toxic properties produced after ingestion by insects, which clearly indicated that the plants are to be fed by the insects to get the desired control and their spray forms cannot kill the insects.

13.4.1 Genetically Engineered Cereal Crops against Biotic Stress

Biotic stress is one of the major constraints for plants to release their potential yield. One way to increase the crop yield is to reduce damages caused by biotic stresses such as insects, diseases, and weeds. Pathogens can cause about 10–16% loss of the global harvest (Chakraborty and Newton 2011), whereas insect pest can cause about 14–25% of the total production (DeVilliers and Hoisington 2011). Naturally available gene pool lacks resistance source to biotic stress which limits the plant breeders either to create resistance or introgress this trait into new varieties. Therefore, it is necessary to search for alternative sources of genes in other completely unrelated species of plants or in microbial organisms. Besides, traditional methods are resource- and time-consuming and germplasm dependent (Bidhan et al. 2011). Genetic engineering has transformed plants with foreign genes to enhance their resistance or tolerance against different biotic stresses.

13.4.2 GM Cereals against Insect Pests

Globally, there are very few commercially released GM cereals, including maize and rice being particularly effective against insect pests. The first transgenic cereal crop released commercially was *Bt* maize during 1996 in the USA. Thereafter several countries like Canada, Argentina, Philippines, South Africa, Spain, and France have adopted its commercial cultivation. The area under *Bt* maize has extended to 60.9 million hectares globally, which is 31% of the global maize production in 2019 (https://www.isaaa.org/resources/publications/briefs/55/default.asp). Besides cultivation of *Bt* maize provides both economic and environmental benefits as it decreases the load of active ingredient (*a.i.*) of insecticides by 35% globally (Brooke sand Barfoot, 2010). Reduction in the pesticide load is attributed to coleopteran active *Bt* maize against *Diabroticaspp* which otherwise would have contributed to 25–30% of the global total in maize (James 2003). *Bt* maize has been transformed with either cry1Ab, cry1Ac, or cry9C against *Ostrinia nubilalis* and *Sesamia nonagrioides*, or with cry1F against *Spodoptera frugiperda*, and with cry3Bb, cry34Ab, and cry35Ab against rootworms of the genus *Diabrotica* (James 2012). Similarly, rice (*Oryza sativa* L.) is the staple food crop in several countries all over the world including India, which feeds more than half of the global population. The crop suffers severe yield loss mainly due to the infestation of stem borers and estimated to be 5–10% (Hammond et al. 2004). Use of chemical pesticide is the major method to control insect pests in rice crop. The excessive use of these insecticides not only increased production cost but also pollutes environment and threatens human health. Developing resistant varieties through conventional breeding approaches were not found successful due to the non-availability of resistant source against the pests like striped stem borer (*Chilo suppressalis*), Yellow stem borer (*Scirpophaga incertulas*), and leaf folder (*Cnaphalocrocis medinalis*). However, transgenic crops expressing *Bt* toxins were found to be effective in controlling the pests and have shown some yield advantage too.

Early commercial varieties of insect tolerant GM crops expressed single Cry proteins against lepidopteran pests, for example, *Bt* cotton expressing Cry1Ac (Bollgard I; developed by Monsanto) and *Bt* maize expressing Cry1Ab (developed by Syngenta). Later on, other lepidopteran-active Bt toxins, such as Cry1F and Cry2Ab2, were also introduced and pyramided into a single variety. For instance, Widestrike cotton expresses both Cry1F + Cry1Ac (developed by Dow Agrosciences) and Bollgard II cotton expressing Cry1Ac + Cry2Ab2 (developed by Monsanto). Likewise, Yieldgard maize expressing Cry3Bb1 (developed by Monsanto) was used against coleopteran pests (chrysomelid rootworms). With regard to GM crops, the success story of GM cereals is less perceptible than other economically important crops. Development of transgenics in cereals took a longer period due to lack of techniques for stable transgene production, horizontal gene transfer, and issues regarding its acceptability.

13.4.3 GM Cereals Against Plant Diseases

At the early stages of infection, fungal pathogens usually secrete polygalacturonases (PGs) to degrade pectin, while during the course of evolution, plants have developed strategies to combat it through the production of polygalacturonase-inhibiting proteins (PGIPs) (Oelfose et al. 2006). In cereal crops like wheat, diseases in particular fusarium head blight (FHB) caused by *Fusarium graminearum* result in significant yield loss and mycotoxin (trichothecene and deoxynivalenol-DON) contamination worldwide. Food contamination with DON is a risk for human and animal health. Recently, transgenic wheat expressing a L3 gene (N-terminal fragment of yeast ribosomal protein) showed resistance to Fusarium disease and improved level of DON in transgenic wheat kernel (Di et al. 2010). Likewise, GM wheat with bean PvPGIP2 in their flowers also showed reduced *F. graminearum* infection (Ferrari et al. 2012). Moreover, transgenic wheat and Barley plants expressing bovine lactoferrin gene (a broad-spectrum antimicrobial gene) conferred resistance to head blight (Han et al. 2012). Likewise, in rice diseases such as blast (*Magnaporthe grisea*), bacterial leaf blight (*Xanthomonas oryzae pv. oryzae*) and sheath blight (*Rhizoctonia solani*) are some major constraints for high productivity. GM rice plants expressing wheat puroindoline genes PinA and/or PinB produce puroindolines which reduced the growth of *M. grisea* and *R. solani* by 35–50% in vitro conditions, thereby conferring resistance (Krishnamurthy et al. 2001). Likewise, genes encoding chitinase or 1, 3-glucanase from plants and microbes have been used in developing transgenic rice resistant to fungal pathogens (Fujikawa et al. 2012). In other study, GM rice expressing AtNPR1 showed increased disease resistance against *M. grisea* and *Xanthomonas oryzae pv. oryzae* by priming the expression of salicylic acid-responsive endogenous genes PR1b, PR5, PR10, and PBZ1 (Li et al. 2020; Fitzgerald et al. 2004). Genome sequencing of rice has revealed five NR1-like genes of which three genes, namely OsNPR1, OsNPR2, and OsNPR3 were induced by the infection of *Xanthomonas oryzae* pv. *oryzae* and *M. grisea*. OsNPR1 is the rice orthologue of *Arabidopsis* NPR1 gene; whose overexpression conferred disease resistance to bacterial blight, however enhanced herbivore susceptibility (Chern et al. 2005; Yuan et al. 2007). Another strategy to confer resistance to plants against disease is through activating phytoalexins (part of plant defense mechanisms in some species). Stilbene synthase gene (STS) of Vst1 (a key enzyme phytoalexin biosynthesis in grape) could improve resistance in rice against *Pyricularia oryzae* (Coutos-Thévenot et al. 2001) and in barley against powdery mildew (Liang et al. 2000). More recently, mitogen-activated protein kinase (MAPK) cascade (especially OsMKK6) regulates genes responsible for phytoalexin synthesis in rice in response to UV and blast infestation (Wankhede et al. 2013). Moreover, transgenic rice lines containing OsMKK6 gene showed overexpression of phytoalexins under UV stress.

13.4.4 Genetically Engineered Cereal Crops against Herbicide

In the agroecosystem, weeds reduce crop yield because they compete with the crop for nutrients, water, and light. They occasionally produce allelopathic substances that are toxic to plants and also act as reservoirs for disease inoculum and insect pests during the off-seasons. Yield losses in crops due to weeds were estimated to be approximately AUD 3.3 billion in Australia (Llewellyn et al. 2016), whereas in India it costs over USD 11 billion annually (Gharde et al. 2018). When left unattended, weeds can cause up to 100% yield loss. Several herbicides are available in the market for weed management; however, its efficacy depends on selective or nonselective mode of action. Globally, two nonselective herbicides glyphosate and glufosinate are most widely used. Glyphosate is the nonselective post-emergence herbicide which acts as an analog of enolpyruvate that binds and inhibits the enzyme 5-enolpyruvylshikimate-3-phosphate synthase (EPSPS) involved in shikimate pathway leading to the synthesis of chorismate-derived metabolites including the aromatic amino acids. Inactivating this enzyme by glyphosate would interfere with the growth and kill the weedy plants due to the absence of aromatic amino acids such as tryptophan, tyrosine, and phenylalanine (Steinrücken and Amrhein 1980). Roundup ready was the first transgenic glyphosate resistant corn developed by Monsanto in 1998 (USDA 1997). Subsequently, many commercial cultivars with tolerance to other herbicides were developed such as Liberty Link Corn against glufosinate. Likewise, GM maize against dicamba at pre- and post-emergence crop stages showed tolerance due to dicamba monooxygenase (DMO) enzyme which is linked with chloroplast peptide (CTP) (Cao et al. 2011). Recently, an imidazolinone resistance (IR) XA17 gene was introduced into maize which showed resistance to imazaquin and nicosulfuron herbicides (Menkir et al. 2010). Another mechanism that deactivates glyphosate into a non-toxic N-acetyl glyphosate is by introducing the glyphosate N-acetyltransferase (GAT gene) from *Bacillus licheniformis* to maize (Castle et al. 2004).

Furthermore, stacking of genes in a single cultivar was preferred over GM crops with a single gene for improved insect pest and weed management. For example, GM maize developed by pioneer expressing two Cry genes (Cry34Ab1, Cry35Ab1) pyramided with PAT (phosphinothricin acetyl transferase) genes was found tolerant to insect pests as well as herbicides (Cao et al. 2011). Commercially herbicide tolerant rice plants were developed by targeting either of these three pathways, such as (1) shikimate pathway (Roundup Ready® rice), (2) glutamine biosynthesis pathway (Liberty Link®), and (3) branched chain amino acid synthesis (Clearfield®). Clearfield rice is non-transgenic, whereas Roundup ready and Liberty Link rice are transgenic (Rodenburg and Demont 2009). Likewise, transgenic rice plants with enhanced melatonin levels were developed recently to provide protection against oxidative stress due to herbicide application (Park et al. 2013).

13.4.5 Genetically Engineered Cereal Crops Against Abiotic Stress

Transgenic cereal crops potentially improved the yield under abiotic stresses like drought, salt, cold, and heat. The transgenes from the different sources are transferred to cereal crops aiming to regulate different molecular pathways. Genes responsible for regulating signaling cascade and transcription like *ABF/ABRE* (ABA-responsive element binding factor/ABA-responsive element) *CBF/DREB* (C-repeat-binding factor/dehydration responsive element binding protein), *HSF* (heat shock factor), *MAP* (mitogen-activated protein), phospholipases, and salt oversensitive kinases (Hussain et al. 2011; Shou et al. 2004; Thiery et al. 2004; Qiu et al. 2002) have been transferred in cereals crop and studied thoroughly. DREB genes have been used in transformation of cereal crops especially rice and wheat to increase drought tolerance (Chen et al. 2008). Recently, overexpression of OsDREB2A significantly enhanced drought and salt tolerance of transgenic rice plants (Cui et al. 2011) and overexpression of *ZmDREB2A* with *CaMV35S* or *rd28A* promoter resulted in better tolerance to drought in maize (Qin et al. 2007). The *WRKY* superfamily of plant transcription factors (TFs) has a conserved sequence (*WRKYGQK*) at their N-terminal end (Wu et al. 2008). Transgenic rice expressing OsWRKY11 under control of heat shock protein promoter (HSP101) was shown to survive longer and retain water under a short severe drought treatment than wild type plants (Wu et al. 2009). Regardless of the concerns raised above, the area under commercial cultivation of GM cereal crops is expanding year by year. Seeing the GM cereals production pattern, it may be expected for commercialization of abiotic stress tolerance GM cereal crops in near future.

13.5 Regulation of GM Cereal Crops

The introduction of GM cereal crops sparked debate and piqued public interest in agriculture. As GM cereal crops are consumed as food, feed, and fodder in many countries, multiple regulatory approaches to regulate GM crops have been devised and implemented. However, the key scientific risk element remains same for all regulatory approaches, but the risks and advantage vary significantly by the policy decisions that are influenced by the political and cultural scenario (Smyth and Phillips 2014; McHughen and Smyth 2012). The decision of policymakers is influenced by different factors like tradition of the culture, condition of environment and society, and risk tolerance (Shukla et al. 2018). The policymakers may face pressure from food safety and environmentalist groups, natural crop producers, farmers (large scale), animal husbandry group, animal consumers, global agricultural companies, and other things engage in the chain of complex global food production and distribution (Hicks 2017).

There are many countries which have approved a "process-based" method to regulate GM crops in which modified crops through specific genetic engineering approach are subjected to premarket safety review for environmental and food safety. Some regulation systems for GM crops are beyond the safety of food and environmental protection to tackle economical and social issues, like protection of non-GM crop production, labeling products for consumer information and considering the concern of society and economy. In GM crop regulatory system, a committee first examines the international agreements which have importance to GM crop regulation and then gives illustrations of three countries and European Union (EU) to reveal various methods which may consider for the commercialization of GM crops by national or regional governments (Morris and Spillane 2010).

13.5.1 International Cereal Crop Regulation Frameworks

Internationally, there are limits on international trade agreements due to national product regulation policies of the countries which are parties of the agreements. The WTO (World Trade Organization) agreements and Cartagena Protocol for biosafety protocol are especially followed for the GM crop and food regulation. The GM crop and food safety assessment regulation system of the member countries must be uniform to the WTO principles set in the WTO Agreement on the Application of SPS (Sanitary and Phytosanitary Measures) Agreement (National Academies of Sciences, Engineering, and Medicine 2016). The SPS Agreement regulates measurement of GM crops to protect animals, human health, plant life as well as food safety. The SPS measures scientific fact based evidence except those for which scientific information is not available, in such cases, country may regulate by resolving scientific uncertainty. To encourage similarity in measurement, the SPS Agreement accepts global standards and guidelines set up by CAC (Codex Alimentarius Commission) and other different international organizations. Generally, the guidelines and principles of Codex direct GM foods developer to give information which facilitates regulators to evaluate various risks related to food safety:

- GM plant description (involved crop and genetic modification nature).
- Host plant description and its utilization as food along with cultivation, breeding, and known allergenicity or toxicity problems.
- Gene donor organism's description including allergenicity or toxicity problems related to them.
- Genetic modifications description consisting of transformation method details, utilized DNA and vector, and any other intermediate host utilized in the process.
- Genetic modification characterization, including inserted DNA copy number, left and right regions of border, DNA sequences expression and impact on host gene expression.
- Assessment of safety, consisting:
  - Substances expressed: Toxicity analysis expressed products from individual genetic events and an ensuring evaluation for toxic compound from donor organisms for accidental transformation. In case of protein, the allergenicity should be analyzed for amino acid sequences.

- Key components composition analysis: An analysis of the host plant key component with GM plants under field trial and natural conditions is closely resembled for large-scale production.
- Metabolite analysis: GM plants metabolite analysis is dissimilar to the original host. If any metabolite is identified, its potential impact on human health must be evaluated.
- Processing of food: Analysis of food processing treatment impacts on metabolites of GM crops. It is needed to assess the potential toxicity of a modified metabolite or protein expressed in GM crops vs non-GM crops.
- Analysis of nutrients: Similar to the compositional evaluation, except that when DNA is inserted, the key nutritional compound is expected to change. In such circumstances, more testing may be required to determine the level of the questioned nutrient and its effects on human health, taking into account typical consumption trends and trait stability in variable environments.

13.5.1.1 USDA Regulation of Pharma Crops

The U.S. Department of Agriculture's Animal and Plant Health Inspection Services (APHIS) is the regulating authority of GM crops established under Plant Protection Act of 2000 (PPA). According to the act "plant pests" are the organisms which cause disease, damage, or injury to plant parts or products, including viruses, bacteria, fungi, and parasitic plants. The generated GM plants are legalized under the (PPA) if they were generated by gene transfer using *Agrobacterium tumefaciens*, which is supposed to be a pest of plant, or DNA transfer from a pest of plant (like terminator gene). USDA controls GM plants either by permission or a notification procedure. Like for the regular Bt crop field trials, viz. Bt cotton and Bt corn, notification procedure is utilized, which are normal formalities. The institution, organization, company, or universities give a notice of APHIS trial and give consent to follow specific rules and regulation set by USDA, and USDA normally signs off. In case of field trial of GM crops having higher risk, like those which are extremely outcrossing or which persist in ground or water for a long period need a permit. The GM crops field trial which produces industrial or pharmaceutical chemicals, a permit is for all time needed. The process of permission may be more or less extensive, needing either an Environmental Impact Statement or an Environmental Assessment.

When an institution, organization, company, or universities decides its desire to commercialize a GM crop and seed of the same for the farmer's cultivation purpose, it can appeal APHIS for deregulated class. This procedure needs submission of risk-assessment details (data) for demonstrating that the crop does not have a plant-pest risk. The appropriate data must be disclosed in public and contain disease susceptibility and insect pests, effects on non-target organisms and beneficial organisms, weediness, and the gene flow risk to wild or weedy relatives. After the incident of ProdiGene 2000, USDA implemented a higher level of scrutiny for the GM crops having higher risk of inherent. As a consequence, GM crops for industrial and pharmaceutical purposes are not suitable for deregulation and must remain under permit even after commercialization. Nevertheless, numerous gaps continue. The present USDA regulatory system does not ensure an in-depth assessment of the environmental impact prior to the planting of pharmaceutical crops. As an alternative, USDA's policy of gene-confinement measures is planned to "minimize" rather than prevention of non-GM crops contamination. In general, USDA is too short-handed to work out sufficient supervision and mostly leaves biotech companies to control themselves. Furthermore, USDA holds the locations of all test fields secret from neighboring farmers and the public, without disclosing the drug or chemical identity being produced, and overlooks biotech companies' pharma crop plantation practices anonymously, without identification.

13.5.1.2 U.S. Regulation of Genetically Modified Crops

The regulation of genetically modified crops in USA is regulated by three different regulatory agencies: viz. EPA (Environmental Protection Agency), FDA (Food and Drug Administration), and USDA (U.S. Department of Agriculture). These three agencies regulate the genetically modified crops from a different point of view with each other (Smyth and Phillips 2014; McHughen and Smyth 2012). EPA is responsible for the regulation of biopesticides like Bt toxins under the FIFRA (Federal Insecticide, Fungicide, and Rodenticide Act) (Ledford 2013). In case of crop developed against insect pest with foreign gene, EPA needs the developing organization to verify the toxin to be expressed in crop for environmental safety and also food safety to insure non-allergic nature of expressed protein1. FDA regulates the safety of GM crops consumed by humans or animals as food and feed. As per the policy in 1992, most GM crops were treated as "substantially equivalent" to non-GM crops by FDA; wherein these GM crops were generally recognized as "Safe" under the FFDCA (Federal Food, Drug, and Cosmetic Act) and does not need prior-market approval. If the expressed protein in edible transgenic crops differs significantly from natural plant proteins in terms of structure, function, or quality and is harmful to humans, the FDA has the authority to impose more stringent standards of Federal Food, Drug, and Cosmetic Act (FFDCA) mandating the premarket approval of biotechnological products.

13.5.1.3 The FDA Consultative Process for GM Crops

FDA set up a willful consultation process in 1997 in collaboration with the developer of GM crops for reviewing the purpose of "substantial equivalence" prior to crop marketing, like assessment of transgene product and plant toxicity and allergenicity. If the results in the food-safety assessment are satisfactory, the FDA notifies the developer that the crop can be marketed (Bonetta 2001).

13.5.2 Regulation of GM Crops in India

In India, Ministry of Environment, Forest and Climate Change (MoEFCC) regulates the GMO experiments, trials, and release under the environment protection act (EPA) 1986. This act has made several rules to solve the environmental issues arising due to hazardous chemicals, hazardous wastes, solid wastes, biomedical wastes, etc. To address the problems associated with microbes and genetic engineering MoEFCC notified the “Rules for manufacture, use/import/export and storage of hazardous microorganisms/genetically engineered organisms or cells, 1989” as per Sections 8 and 25 of EPA, 1986 (Shukla et al. 2018). Sections 8 and 25 deal with the regulation of genetic engineering and gene technology in India (http://geacindia.gov.in/acts-and-rules.aspx). These rules are referred as Rules 1989, which covers all the activities involving GMOs and products thereof including new gene technologies (Kandasamy and Padmavati 2014; Chimata and Bharti 2019). Rules, 1989 defined the term gene technology and genetic engineering as follows: “Gene Technology” means the application of the gene technique called genetic engineering, including self-cloning and deletion as well as cell hybridization. “Genetic engineering” means the technique by which heritable material, which does not usually occur or will not occur naturally in the organism or cell concerned, generated outside the organism or the cell is inserted into said cell or organism. It shall also mean the formation of new combinations of genetic material by incorporation of a cell into a host cell, where they occur naturally (self-cloning) as well as modification of an organism or in a cell by deletion and removal of parts of the heritable material.

These rules were enforced by MoEFCC, Department of Biotechnology (DBT) and state governments through six competent authorities: rDNA Advisory Committee (RDAC), Institutional Biosafety Committee (IBSC), Review Committee on Genetic Manipulation (RCGM), Genetic Engineering Appraisal Committee (GEAC), State Biotechnology Coordination Committee (SBCC), and District Level Committee (DLC). RDAC is constituted by DBT and acts as advisory body on emerging issues on DNA technology. IBSC is set up in each institute included in recombinant DNA research and responsible for following RDNA guidelines in transgenic experiments. RCGM is regulatory body under DBT which involves scientific risk assessment and development of guidelines for GMO research. GEAC is the apex regulatory committee under the MoEFCC and is responsible for final approval for environmental release of GMOs. SBCC and DLC are for monitoring purpose and act as nodal point at state and district level for coordinating GMO related activities (Fig. 13.1). Apart from the Rules of 1989, the following acts are engaged in the regulation of GMOs in India: Plant Quarantine Order, 2003, Biological Diversity Act, 2002, and Food Safety and Standards Act, 2006 (http://in.biosafetyclearinghouse.net/phase2/publications.shtml).

13.5.2.1 Biosafety Assessment Guidelines

GMO regulatory authorities periodically established several guidelines to evaluate the impact of recombinant DNA technology development in the nation (http://geacindia.gov.in/guidelines-and-protocols.aspx). Recombinant DNA Safety Guidelines and Regulations, 1990, categorized recombinant DNA (RDNA) into three categories based on risks and provided guidelines for the measurement of containment in accordance with each risk category. Later, as DNA technology developed in the country, the rules were periodically updated to address the issues that emerged. For instance, updated guidelines for transgenic plant research from 1998, guidelines and standard operating procedures for conducting confined field

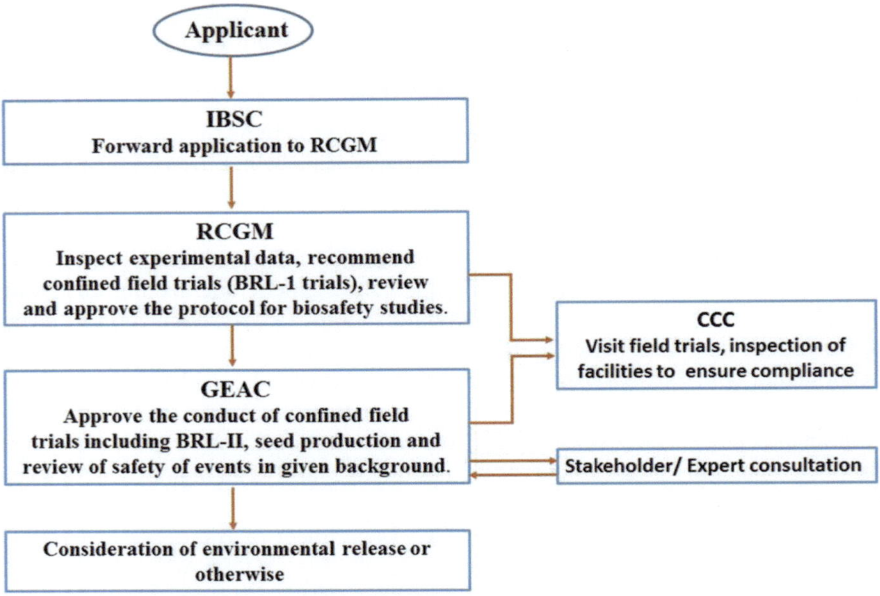

Fig. 13.1 Procedure for approval of confined field trials and environmental release of genetically engineered plants

trials of regulated GE plants from 2008, guidelines for the safety assessment of foods derived from GE plants from 2008, guidelines and a handbook for IBSCs from 2011, and guidelines for the environmental risk of GE plants from 2016 are just a few examples.

Biosafety assessment data has to be generated at various stages of transgenic plant development such as laboratory research, greenhouse studies, field testing, and at environmental release. These data broadly include effect of genetic modification and protein characterization, food and feed safety, environmental safety (http://geacindia.gov.in/resource-documents/biosafety-regulations/guidelinesandprotocols/RiskAnalysisFrameworkforWeb).

13.6 Conclusion

Cereal crops (wheat, paddy, maize, etc.) are globally considered as staple crop. Urbanization and the demand to feed the ever-increasing global population are exerting pressure on the agricultural resources (land, water, and soil nutrients), which have become increasingly scarce. The excessive exploitation of groundwater for irrigation resulted in depletion of groundwater which is also a threat for drinking water in some parts of the world. In contrast, overuse of pesticides and herbicides not only disrupts the agricultural ecosystem but also increases the input required for agriculture, which eventually lowers profits and has a negative impact on public health. In the twenty-first century, biotechnology made it possible to genetically modify cereal crops to have specific traits such biotic and abiotic stress tolerance, and herbicide resistance. In cereal crops many genetically modified plants have been developed and released for commercial cultivation in some countries. Future food security and sustainable agriculture will require the adaption of GM crops with various traits. The acceptance of GM cereal crops around the world is in doubt due to genetic manipulation and the issues it raises. In spite of the fact that most regulatory systems around the world are comparable to one another, synchronising them is still necessary to enable the commercial cultivation and trading of genetically modified crops for the benefit of sustainable agriculture and global food security.

References

Bawa AS, Anilakumar KR (2013) Genetically modified foods: safety, risks and public concerns—a review. J Food Sci Technol 50(6):1035–1046

Bebber DP, Holmes T, Gurr SJ (2014) The global spread of crop pests and pathogens. Glob Ecol Biogeogr 23(12):1398–1407

Bidhan R, Noren SK, Mandal AB, Basu AK (2011) Genetic engineering for abiotic stress tolerance in agricultural crops. Biotechnology 10(1):1–22

Bonetta L (2001) GM crops under new US scrutiny. Curr Biol 11(6):R201

Bravo A, Gill SS, Soberon M (2007) Mode of action of bacillus thuringiensis cry and Cyt toxins and their potential for insect control. Toxicon 49(4):423–435

Brookes G, Barfoot P (2012) Global impact of biotech crops: environmental effects, 1996–2010. GM Crops Food 3(2):129–137

Brookes G, Barfoot P (2014) Economic impact of GM crops: the global income and production effects 1996–2012. GM Crops Food 5(1):65–75

Cannon RJC (1996) Bacillus thuringiensis use in agriculture: a molecular perspective. Biol Rev 71(4):561–636

Cao M, Sato SJ, Behrens M, Jiang WZ, Clemente TE, Weeks DP (2011) Genetic engineering of maize (Zea mays) for high-level tolerance to treatment with the herbicide dicamba. J Agric Food Chem 59(11):5830–5834

Castle LA, Siehl DL, Gorton R, Patten PA, Chen YH, Bertain S et al (2004) Discovery and directed evolution of a glyphosate tolerance gene. Science 304(5674):1151–1154

Chakraborty S, Newton AC (2011) Climate change, plant diseases and food security: an overview. Plant Pathol 60(1):2–14

Chen H, Lin Y (2013) Promise and issues of genetically modified crops. Curr Opin Plant Biol 16(2): 255–260

Chen JQ, Meng XP, Zhang Y, Xia M, Wang XP (2008) Over-expression of OsDREB genes lead to enhanced drought tolerance in rice. Biotechnol Lett 30(12):2191–2198

Chern M, Fitzgerald HA, Canlas PE, Navarre DA, Ronald PC (2005) Overexpression of a rice NPR1 homolog leads to constitutive activation of defense response and hypersensitivity to light. Mol Plant-Microbe Interact 18(6):511–520

Chimata MK, Bharti G (2019, August) Regulation of genome edited technologies in India. In: *Transgenic research*, vol 28(2), Springer International Publishing, pp 175–181

Coutos-Thévenot P, Poinssot B, Bonomelli A, Yean H, Breda C, Buffard D et al (2001) In vitro tolerance to Botrytis cinerea of grapevine 41B rootstock in transgenic plants expressing the stilbene synthase Vst 1 gene under the control of a pathogen-inducible PR 10 promoter. J Exp Bot 52(358):901–910

Cui M, Zhang W, Zhang Q, Xu Z, Zhu Z, Duan F, Wu R (2011) Induced over-expression of the transcription factor OsDREB2A improves drought tolerance in rice. Plant Physiol Biochem 49(12):1384–1391

DeVilliers SM, Hoisington DA (2011) The trends and future of biotechnology crops for insect pest control. Afr J Biotechnol 10(23):4677–4681

Di R, Blechl A, Dill-Macky R, Tortora A, Tumer NE (2010) Expression of a truncated form of yeast ribosomal protein L3 in transgenic wheat improves resistance to fusarium head blight. Plant Sci 178(4):374–380

FAO (2011). http://faostat.fao.org/. Accessed 3 June 2021.

FAO (2017). The future of food and agriculture (FAO) Food and Agriculture Organization of the United Nations. 2017. Available at: http://www.fao.org/publi cations/fofa/en/. Accessed 3 Jun. 2021.

Ferrari S, Sella L, Janni M, De Lorenzo G, Favaron F, D'ovidio R (2012) Transgenic expression of polygalacturonase-inhibiting proteins in Arabidopsis and wheat increases resistance to the flower pathogen fusarium graminearum. Plant Biol 14:31–38

Fitzgerald HA, Chern MS, Navarre R, Ronald PC (2004) Overexpression of (at) NPR1 in rice leads to a BTH-and environment-induced lesion-mimic/cell death phenotype. Mol Plant-Microbe Interact 17(2):140–151

Fujikawa T, Sakaguchi A, Nishizawa Y, Kouzai Y, Minami E, Yano S et al (2012) Surface α-1, 3-glucan facilitates fungal stealth infection by interfering with innate immunity in plants. PLoS Pathog 8(8):e1002882

Gharde Y, Singh PK, Dubey RP, Gupta PK (2018) Assessment of yield and economic losses in agriculture due to weeds in India. Crop Prot 107:12–18

Hammond B, Dudek R, Lemen J, Nemeth M (2004) Results of a 13 week safety assurance study with rats fed grain from glyphosate tolerant corn. Food Chem Toxicol 42(6):1003–1014

Han J, Lakshman DK, Galvez LC, Mitra S, Baenziger PS, Mitra A (2012) Transgenic expression of lactoferrin imparts enhanced resistance to head blight of wheat caused by fusarium graminearum. BMC Plant Biol 12(1):1–9

Hicks DJ (2017) Genetically modified crops, inclusion, and democracy. Perspect Sci 25(4): 488–520

Hussain SS, Iqbal MT, Arif MA, Amjad M (2011) Beyond osmolytes and transcription factors: drought tolerance in plants via protective proteins and aquaporins. Biol Plant 55(3):401–413

Ingram J (2011) A food systems approach to researching food security and its interactions with global environmental change. Food security 3(4):417–431

ISAAA (2018). https://www.isaaa.org/resources/publications/briefs/55/default.asp

James C (2003) Global status of commercialized biotech/GM crops: 2003. ISAAA brief no. 31. ISAAA, Ithaca, New York

James C (2012) Global status of commercialized biotech/GM crops. Brief no. 44. ISAAA, Ithaca, NY

James, C. (1996) Global review of the field testing and commercialization of transgenic plants: 1986 to 1995. ISAAA Brief No.1

James C (2011) Global status of commercialized biotech/GM crops, vol 44. Isaaa, Ithaca, NY

Kandasamy M, Padmavati M (2014) Transgenic crop research and regulation in India: whether legislation rightly drives the motion? J Commer Biotechnol 20(4)

Klümper W, Qaim M (2014) A meta-analysis of the impacts of genetically modified crops. PLoS One 9(11):e111629

Krishna R, Karkute SG, Ansari WA, Jaiswal DK, Verma JP, Singh M (2019) Transgenic tomatoes for abiotic stress tolerance: status and way ahead. *3*. Biotech 9(4):1–14

Krishnamurthy K, Balconi C, Sherwood JE, Giroux MJ (2001) Wheat puroindolines enhance fungal disease resistance in transgenic rice. Mol Plant-Microbe Interact 14(10):1255–1260

Ledford H (2013) US regulation misses some GM crops. Nature News 500(7463):389

Li W, Deng Y, Ning Y, He Z, Wang GL (2020) Exploiting broad-spectrum disease resistance in crops: from molecular dissection to breeding. Annu Rev Plant Biol 71:575–603

Liang H, Zheng J, Duan X, Sheng B, Jia S, Wang D et al (2000) A transgenic wheat with a stilbene synthase gene resistant to powdery mildew obtained by biolistic method. Chin Sci Bull 45(7): 634–638

Llewellyn R, Ronning D, Clarke M, Mayfield A, Walker S, Ouzman J (2016) Impact of weeds in Australian grain production. Grains Research and Development Corporation, Canberra, ACT

McHughen A, Smyth SJ (2012) Regulation of genetically modified crops in USA and Canada: American overview. In: *Regulation of agricultural biotechnology: the United States and Canada*. Springer, Dordrecht, pp 35–56

Menkir A, Chikoye D, Lum F (2010) Incorporating an herbicide resistance gene into tropical maize with inherent polygenic resistance to control Striga hermonthica (Del.) Benth. Plant Breeding 129(4):385–392

Morris SH, Spillane C (2010) EU GM crop regulation: a road to resolution or a regulatory roundabout? Eur J Risk Regul 1(4):359–369

National Academies of Sciences, Engineering, and Medicine (2016) Genetically engineered crops: experiences and prospects. National Academies Press

Nayar R (2011) More and better jobs in South Asia. World Bank Publications

Nayar R, Gottret P, Mitra P, Betcherman G, Lee YM, Santos I, Dahal M, Shrestha M (2012) More and better jobs in South Asia. South Asia Development Matters, World Bank. https://openknowledge.worldbank.org/handle/10986/2391

Oelfose D, Dubery IA, Meyer R, Arendse MS, Gazendam I, Berger DK (2006) Apple polygalacturonase inhibition potential expressed in transgenic tobacco inhibits polygalacturonases from fungal pathogens of apple and anthracnose of lupins. Phytochemistry 67:255–263

Ordonez N, Seidl MF, Waalwijk C, Drenth A, Kilian A, Thomma BP et al (2015) Worse comes to worst: bananas and Panama disease—when plant and pathogen clones meet. PLoS Pathog 11(11):e1005197

Osmond AT, Colombo SM (2019) The future of genetic engineering to provide essential dietary nutrients and improve growth performance in aquaculture: advantages and challenges. J World Aquacult Soc 50(3):490–509

Park S, Lee DE, Jang H, Byeon Y, Kim YS, Back K (2013) Melatonin-rich transgenic rice plants exhibit resistance to herbicide-induced oxidative stress. J Pineal Res 54(3):258–263

Parker MW, Feil SC (2005) Pore-forming protein toxins: from structure to function. Prog Biophys Mol Biol 88(1):91–142

ProMusa Organization (2017). Tropical race 4 – TR4 j News, knowledge and information on bananas. Available at: http://www.promusa.org/. TropicalCraceC4C–CTR4#Impact. Accessed 11 June 2021.

Qin F, Kakimoto M, Sakuma Y, Maruyama K, Osakabe Y, Tran LSP et al (2007) Regulation and functional analysis of ZmDREB2A in response to drought and heat stresses in Zea mays L. Plant J 50(1):54–69

Qiu QS, Guo Y, Dietrich MA, Schumaker KS, Zhu JK (2002) Regulation of SOS1, a plasma membrane Na+/H+ exchanger in Arabidopsis thaliana, by SOS2 and SOS3. Proc Natl Acad Sci 99(12):8436–8441

Raman R (2017) The impact of genetically modified (GM) crops in modern agriculture: a review. GM Crops Food 8(4):195–208

Ray DK, Mueller ND, West PC, Foley JA (2013) Yield trends are insufficient to double global crop production by 2050. PLoS One 8(6):e66428

Rodenburg, J., & Demont, M. (2009). Potential of herbicide-resistant rice technologies for sub-Saharan Africa

Shou H, Bordallo P, Fan JB, Yeakley JM, Bibikova M, Sheen J, Wang K (2004) Expression of an active tobacco mitogen-activated protein kinase kinase kinase enhances freezing tolerance in transgenic maize. Proc Natl Acad Sci 101(9):3298–3303

Shukla M, Al-Busaidi KT, Trivedi M, Tiwari RK (2018) Status of research, regulations and challenges for genetically modified crops in India. GM Crops Food 9(4):173–188

Sibhatu KT, Qaim M (2018) Meta-analysis of the association between production diversity, diets, and nutrition in smallholder farm households. Food Policy 77:1–18

Smyth SJ, Phillips PW (2014) Risk, regulation and biotechnology: the case of GM crops. GM crops & food 5(3):170–177

Source: ISAAA GM Approval Database. http://www.isaaa.org/gmapprovaldatabase/

Steinrücken HC, Amrhein N (1980) The herbicide glyphosate is a potent inhibitor of 5-enolpyruvylshikimic acid-3-phosphate synthase. Biochem Biophys Res Commun 94(4): 1207–1212

Thiery L, Leprince AS, Lefebvre D, Ghars MA, Debarbieux E, Savouré A (2004) Phospholipase D is a negative regulator of proline biosynthesis in Arabidopsis thaliana. J Biol Chem 279(15): 14812–14818

USDA (1997) Environmental assessment and finding of no significant impact for Monsanto/Dekalb petition 97–099-01p for determination of nonregulated status for transgenic glyphosate tolerant corn line GA21: 1–14

Von Braun J (2007) The world food situation: new driving forces and required actions, Food policy rep. 18, Int. Food Policy Res Inst, Washington, DC

Wankhede DP, Kumar K, Singh P, Sinha AK (2013) Involvement of mitogen activated protein kinase kinase 6 in UV induced transcripts accumulation of genes in phytoalexin biosynthesis in rice. Rice 6(1):1–8

World Bank. Agriculture, value added (current US$). 2017. Available at: http://data.worldbank.org/indicator/NV.AGR.TOTL.CD. Accessed 5 June 2021.

Wu W, Su Q, Xia XY, Wang Y, Luan YS, An LJ (2008) The Suaeda liaotungensis kitag betaine aldehyde dehydrogenase gene improves salt tolerance of transgenic maize mediated with minimum linear length of DNA fragment. Euphytica 159(1):17–25

Wu X, Shiroto Y, Kishitani S, Ito Y, Toriyama K (2009) Enhanced heat and drought tolerance in transgenic rice seedlings overexpressing OsWRKY11 under the control of HSP101 promoter. Plant Cell Rep 28(1):21–30

Yuan Y, Zhong S, Li Q, Zhu Z, Lou Y, Wang L et al (2007) Functional analysis of rice NPR1-like genes reveals that OsNPR1/NH1 is the rice orthologue conferring disease resistance with enhanced herbivore susceptibility. Plant Biotechnol J 5(2):313–324

Zhang C, Wohlhueter R, Zhang H (2016) Genetically modified foods: a critical review of their promise and problems. Food Sci Human Wellness 5(3):116–123

Nanotechnological Approaches for Biofortification Concept and Concern in Cereal Crops

14

Jyoti Prakash Sahoo, Upasana Mohapatra, Sushil Kumar Singh, Kailash Chandra Samal, Vinod Kumar Yadav, Ankit Moharana, Ambika Prasad Mishra, and Ashish Kumar Dash

Abstract

In the generation of frequent climatic conditions, the international agricultural scenario is going through numerous and remarkable demanding situations. To obtain food security, nano-engineering and nanotechnology are a practical biotechnological application on agriculture for increasing the productivity of food crop plants, particularly cereals. Nanotechnology, a recent scientific tool, helps increase agricultural productivity by enhancing the efficiency of agronomic inputs and minimizing the relevant losses due to the drastic climate change. Human health in rural areas is highly affected by consuming nutrient-deficient

J. P. Sahoo
Department of Agricultural Biotechnology, Odisha University of Agriculture and Technology, Bhubaneswar, India

U. Mohapatra
Department of Plant Biotechnology, University of Agricultural Science, GKVK, Bengaluru, India

S. K. Singh (✉)
DBT North East Centre for Agricultural Biotechnology, Assam Agricultural University, Jorhat, India

K. C. Samal · V. K. Yadav
University Department of Botany, Ranch University, Ranchi, India

A. Moharana
Department of Seed Science and Technology, Odisha University of Agriculture and Technology, Bhubaneswar, India

A. P. Mishra
Faculty of Agriculture, Sri Sri University, Cuttack, India

A. K. Dash
Department of Soil Science and Agricultural Chemistry, Odisha University of Agriculture and Technology, Bhubaneswar, India

R. Deshmukh et al. (eds.), *Biofortification in Cereals*,
https://doi.org/10.1007/978-981-19-4308-9_14

food crops, and nanotechnology might be the sustainable crop biotechnology approach to accept this challenge. There are divergent and contrasting strategies of fortifying the food crops, especially cereals, with the valuable vitamins and minerals, which constitute the concept of nutritional diversification. However, the sustainability and affordability of these strategies have now no longer been achieved. Biofortification through the application of nanotechnology is a current, impending, hopeful, economical and durable agricultural method of providing the micronutrients which are essential to a diversified population of human beings that has narrow access to healthy diets. This chapter concerns about all the relevant factors of crop biofortification in cereals. It attempts to encapsulate and outline all of the biofortification studies through nanotechnology that has been performed on important cereal crop plants. Besides the challenges in biofortified cereal crops, nanotechnology and nano-engineering strategies have a great prospect to convene the challenges of malnutrition across the globe.

Keywords

Biofortification · Cereals · Nanotechnology · Nutritional security · Crop improvement

14.1 Introduction

The basic need to use the concept of nanotechnology in modern agriculture emerges from the reality that, the population is rising continuously, and it essentially creates the necessity to feed the growing population of the human race. Due to the activity carried out by human beings for agricultural practices, the lands gradually lose their fertile status. The unfertile quality of lands used for crop production affects crop productivity drastically. Modern agriculturists need some advanced plan of action to improve the crop plants to encourage and enhance crop productivity. If precision agriculture is a debate, nanotechnology is a recently evolved biotechnology in modern agricultural development to formulate strategies for mitigating the vast need of food demand for the exponentially growing human population. The space between the crop nutrient loss, and the crops with the essential nutrients is now filled by the recent valuable developments and applications of nanobiotechnology in agriculture for the biofortification of major food crops like cereals. A survey of FAO (Food and Agriculture Organization) on human health has evaluated that across the world, around 792.5 million human population are suffering from malnutrition, out of which 780 million human population are from the emerging nations (McGuire 2015). Urban agriculture and farming have enriched the novel science of nanotechnology and can contribute to food security and minimize malnutrition. Biofortification in cereals is defined as the nutritional enhancements of the cereal food crops with increased bioavailability of essential nutrients and vitamins to the exponentially growing human population. The biofortified crops are developed and recommended for cultivation using modern and improved biotechnology tools, plant breeding policies and the required agronomic practices. Hence, cereal biofortification in different cereal crop varieties can provide long-term sustainability of micronutrient-available crops to the people for consumption. The essential international organizations like WHO (World Health Organization) and the CGIAR (Consultative Group on International Agricultural Research) have now involved them willingly in the development of nutritionally rich biofortified cereal crops (Bouis 2000), which are high-yielding.

Nanotechnology appears to be the opportunity to revolutionize this subject of agriculture, because the whole nanobiotechnology enterprise, including its applications in the field of modern agriculture, has grown to 1 trillion US dollars (Harper 2015). As a result, there is an excessive correlation among staple foods, particularly cereals and the nutrients of its consumers, specifically some of the rural groups, who do not often produce other assets of nutrient supplementation. So, biofortification using the recent nanotechnological tools in cereals is an idea of growing the nutrient content material of cereal crop plants for the duration of their cultivation and other agronomical purposes (Bouis et al. 2011). India's food grain production increased significantly from 50.82 million tonnes in 1950–1951 to 252.22 million tonnes during 2015–2016, and a close drift has been seen in the food grain production since the last 10 years (Neeraja et al. 2017). Agronomic and breeding procedures, and also current nanotechnological interventions are used in the concept of biofortification to improve the nutritional content of cereal crops (Stein et al. 2007). The affordability and accessibility of cereal crop biofortification for addressing vitamin deficiencies are unrivalled. This is because fortified cereal grains are commonplace in many people's diets and constitute a key source of nutrients.

14.2 Essentials of Biofortification Research in Agriculture

Humans require micronutrients like microminerals, vitamins, essential amino acids, and fatty acids in relatively adequate amounts to maintain a healthy and better lifestyle. The microminerals like zinc, iron, copper, manganese, iodine, molybdenum, etc. and the macrominerals like potassium, calcium, magnesium, phosphorus, sulphur, chlorine and sodium are essentially desired in the regularly consumed food of the human population. Aside from all these, the food should be fortified with vitamins and fatty acids, which are essential amino acids for crucial mental and physical health development in a human being (Garg et al. 2018). If human nutrition is the concern, consumable agricultural commodities must be the preliminary source of nutrients (Graham et al. 2001). The nutrient deficiency of consumables is not supporting the health of the human population, and leads to poor health status, mental and physical illness, enhanced disability, and reduced national socio-economic development in the underdeveloped countries (Chizuru et al. 2003). The deficiency of micronutrients in cereals genuinely influences 43% of school going children and 38% of pregnant women globally (Stevens et al. 2013). Cereal biofortification is directed towards enriching the percentage of essential and desired micronutrients in cereal crops. Biofortification of agricultural plants, particularly cereals, using modern nanotechnological technologies can give good calorific values to fulfil the required energy demands while also supplementing all critical micro and macronutrients for optimal human health.

14.3 Nano-Farming: A New Era in Biofortified Agriculture

Agro-nanotechnological interventions and developments commonly provide short-term technical solutions to current agricultural operations. Nanoparticle engineering and nanotechnology are two of the most recent modern biotechnological revolutions, both of which have shown new properties with enhanced strength. Some of the current agricultural applications of nanotechnology include the controlled release of nano-fertilizers; increased crop growth, yield and productivity; nano-based target gene delivery methods; and the use of nano-pesticides. There are two primary ways for crop improvement using nanotechnology. The materials used in the "bottom-up" approach are molecular components that assemble chemically utilizing molecular recognition principles. The "top-down" technique creates nano-objects from more important things without atomic-level control. The continued development of nano-technology for recognizing problems and launching collaborative approaches for sustainable agricultural growth, with a particular preference for staple cereal food crops, has incredible potential to deliver social and equitable benefits. Recent advances in tissue engineering strategies and bioengineered nanomaterials-based CRISPR (Clustered Regularly Interspaced Short Palindromic Repeats) and CRISPR-associated (Cas) protein mRNA and sg (single-guided) RNA for cereal crop genetic modification are a significant scientific achievement (Miller et al. 2017). As a result, nano-farming might herald a new era of fortification for crucial agricultural goods such as cereals, pulses and oilseeds and play a vital role in the development of sustainable biofortified agriculture. The nanotechnology applications in agriculture are illustrated in Fig. 14.1, including nano-fertilizers, precision farming, plant stress management, crop improvement, crop protection, etc.

Human civilization gets the essential nutrients from their diet. Unfortunately, the daily diets of human being frequently lack sufficient amounts of vital nutrients, resulting in malnutrition. Heart disease, blindness, cancer and mortality are reduced by eating biofortified foods (Hossain and Mohiuddin 2012). Crop fortification is required to address this worldwide issue. Supplementation, dietary diversity, industrial biofortification and nanotechnology-based biofortification are some of the unique solutions available to solve this problem (Smith 2000). Diversification of diets may be defined as a community-wide intake of various food sources aimed at treating micronutrient deficits (Gibson and Anderson 2009). Dietary diversity is that foods with high micronutrient content are difficult to come by in rural areas. It becomes an untraced cycle of nutritional insufficiency in people if food plants, particularly grains, are not fortified with micronutrients (Joy et al. 2015). Micronutrients are induced during food processing employing nano-farming techniques in commercial food fortification, providing acceptable nutritious levels to humans, when the processed product is ingested.

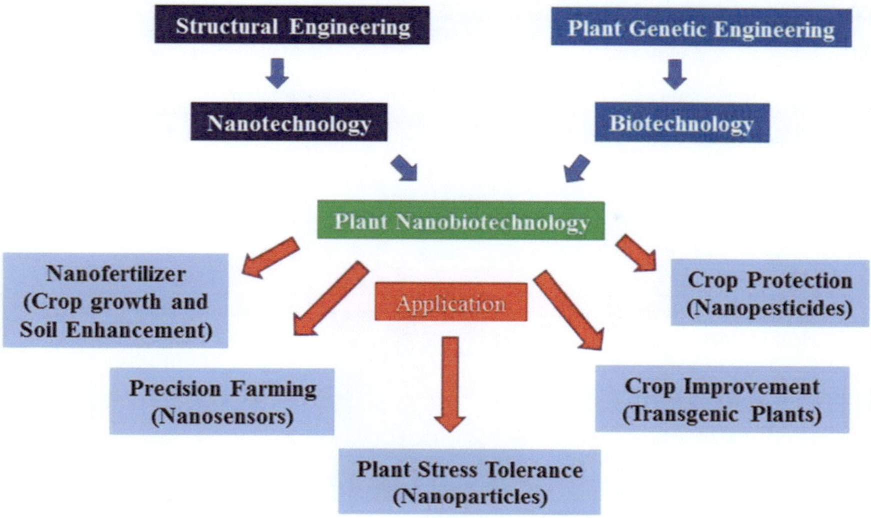

Fig. 14.1 Application of nanotechnology in agriculture

14.4 Use of Nanoparticles in Cereal Crop Improvement

A heavy number of agrochemicals, including fertilizers for yield improvement, affect soil fertility by increasing the remnants from these fertilizers. Conventional fertilizers through leaching sometimes get accumulated in the soil, underground water and other water reservoirs, which have environmental consequences like pollution and greenhouse gas emission. The best alternative to conventional fertilizer is nano-fertilizers. They are elegant and effective in delivering nutrients because their high surface area to volume ratio causes a slow release of fertilizers and increases the nutrient uptake by crops. The application of nano-fertilizers also makes the crop drought and disease resistant and helps in reducing environmental hazards. The outcome from usage of nano-fertilizers varies according to the species of the plant and physical properties like size, concentration and chemical properties like the composition of nanomaterial (Thakur et al. 2018). A list of approved nano-fertilizers used in the world is indexed in Table 14.1. These are extensively used for the fortification of cereal crops. Some other nanomaterials and their relevance in agriculture are discussed below.

14.4.1 Zeolite-Based Nano-Fertilizer

Recently the zeolites, alkali and alkaline earth aluminosilicates are used to develop nano-fertilizers because of their availability, less or inexpensive nature and safety (Morales-Díaz et al. 2017). The high cationic exchange activity, excellent porosity,

Table 14.1 Nano-fertilizers approved for use across the globe

Sl. no.	Approved Nano-fertilizers	Market logo	Manufacturer for marketing	Country
1	JU-11 NanoMax-NPK (Nano max NPK fertilizer)		JU Agri sciences Pvt. ltd., new Delhi	India
2	Magic green Nano calcium leaf fertilizer		AC international network co., ltd.	Germany
3	NanoGreen AGRI-NANO foliar fertilizer & agricultural soap		Nano green sciences, Inc.	India
4	Microtrace liquid Nano encapsulated micro nutrient fertilizers	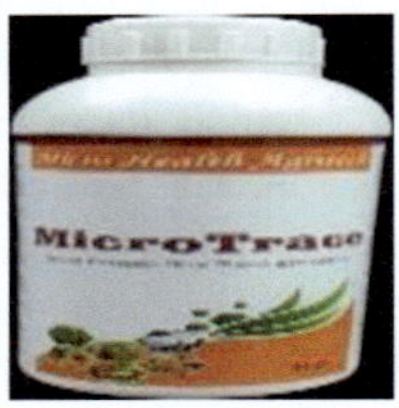	Jhosna corporation, Raichur, Karnataka	India
5	Nanoplant ultra		Nanoplants innovative micronutrients	Trakai
6	TAG Nano NPK - 4G Nano FERTILISER		Tropical Agrosystem India (P) ltd.	India

(continued)

Table 14.1 (continued)

Sl. no.	Approved Nano-fertilizers	Market logo	Manufacturer for marketing	Country
7	IFFCO Nano nitrogen		Indian farmers Fertiliser cooperative (IFFCO)	India
8	IFFCO Nano urea liquid		Indian farmers Fertiliser cooperative (IFFCO)	India

extensive surface area and capability to retain both negative and positive nutrient ions for a long time make the zeolite suitable to be used in nano-fertilizers. On account of the effectiveness of contained hydrated ions, the exposed surface area of zeolites has excellent interaction with cations and polar molecules. The zeolites are converted into nano-form by the top-down approach of ball milling, decreasing the size and increasing the surface area for adsorption or desorption of ions or proper interaction. This helps in holding few ions and releasing few ions in an adequate way required for the crop. Therefore, nano-zeolites are efficient and hazardless and should be promoted to deliver both macro and micronutrients.

Along with the nutrient enhancement, these are also used to regulate soil acidity, quality seed germination, pesticides, wound healing (Eroglu et al. 2017). When urea was applied with the nano-zeolites, a low fertilizer concentration was found in the soil and plants, as the slow release of fertilizers increases nitrogen uptake (Manikandan and Subramanian 2016). The lower concentration of fertilizer in the rhizosphere region also reduces leaching losses and eutrophication, and the greenhouse effect.

14.4.2 Zinc-Based Nanoparticles

The Zn is considered an essential micronutrient for the crop because of its role in activating some enzymes, hormonal regulation and chlorophyll synthesis and carbohydrate metabolism. Zinc oxides and sulphates are the primary sources of zinc micronutrients. Zinc oxides are first absorbed and then metabolized and at the end get accumulated in the plant system. In nanoparticle form, the zinc oxide can be applied in lower concentrations as the higher concentration is toxic to the plants. As the oxide is not completely soluble, zinc oxide is used along with sulphate. Many soil characteristics like soil type, soil pH, mineral composition, percentage of organic matter and ionic strength affect the plant nutrient uptake (Milani et al. 2015).

14.4.3 Iron Oxide-Based Nanoparticles

One of the most abundant elements in the Earth's crust is iron which is primarily available in phases: magnetite, maghemite and hematite. Though iron is pyrophoric with magnetic properties and highly reactive, the low toxicity nature, and excess availability properties make iron be used as a nanoparticle in agriculture. The iron-based nano-fertilizer γ-Fe_20_3 nanoparticles are recently used for crop fortification as it increases seed germination percentage, root growth and the water content of leaf chlorophyll. After the application of γ-Fe_20_3 nanoparticles in soil, the iron gets released from the nanoparticle because of its dynamic nature and migrated to other plant parts from the root. An increase in iron concentration in the shoots of the *C. maxima* was observed after the plants were applied with the γ-Fe_20_3 nanoparticles and Fe^{3+}. The insignificance in the concentration of iron level of both control and treated plants recommended that the iron is transported to the other plant parts through roots. Also, in some cases, the nutrients are translocated from the leaves to the different plant parts (Hu et al. 2017).

14.4.4 Copper-Based Nanoparticles

Being an essential micronutrient, copper should be applied in lower doses. The insolubility nature of copper oxide limits its use because of its toxicity. Also, it gets accumulated in the leafy vegetables when applied as fertilizer or plant growth regulator, pesticide, herbicide, fungicide and additive. This accumulation depends on the soil and environmental factors and plant types. So, the reported recommended dose for copper oxide nanoparticles is 0.3 mg/L Cu^{2+} released from the 1000 mg/L of copper nanoparticles which is nontoxic and increases plant growth. Copper nanoparticles are safer than copper oxide nanoparticles. The release of Cu2+ ions from copper oxide makes the plants resistant to the microbes.

14.4.5 Titanium Dioxide-Based Nanoparticles

Titanium dioxide nanoparticles can be applied in less than 4% concentrations to increase the nitrogen fixation, photosynthesis and growth of the plant in general. The nanoparticle size is a hindrance in its usage for crop fortification as the size of 30 nm nanoparticle cannot pass through the root cells of maize plant causing accumulation in the soil leading to the soil toxicity. Also, some nanoparticles can be translocated through the root cells in wheat, and some cannot. The particle size of less the 20 nm makes the nanoparticles to be penetrated through the root cells and reduces the soil toxicity (Du et al., 2011).

14.4.6 Cerium Oxide-Based Nanoparticles

Cerium oxide nanoparticles are used for crop improvement and nutritional benefits. But the success of the application of nanoparticles varies according to the concentration, soil composition and type of plant species. At higher concentrations, the plant's growth is reduced, and at a lower concentration, the enhanced growth rate is observed. In the case of lettuce, CeO_2 nanoparticles at 100 mg kg-1 amplified the plant growth, but at 1000 mg kg-1, the plant growth was hindered. The moisture content of soil plays a more significant role in delivering fertilizer through CeO_2-based nanoparticles. An improved photosynthetic rate of soybean plants was observed under high moisture content and less in lower moisture content when CeO_2 nanoparticle was applied to the plants. The effect of CeO_2 nanoparticles on the crops in the laboratory varies from the significant field condition, which should be minimized to use CeO_2 as a nano-fertilizer (Cao et al. 2018).

14.4.7 Noble Metal-Based Nanoparticles

The primary noble metals used to develop nano-fertilizers are silver, gold, platinum and some nanocrystalline metals like Fe, Co and Cu. Silver nanoparticles are used to promote the plant's growth, reducing the unwanted microbial load in hydroponic culture, enhancing seed germination (Duhan et al. 2017). Still, the negative effect of silver nanoparticles on nitrification was observed due to oxidative dissolution of silver polyvinyl pyrrolidone (Masrahi et al. 2014). Gold nanoparticle at minute concentration enhances the ratio of the shoot and root length of *Lactuca sativa* seeds. In some crops, increased seed germination, growth rate, yield, pod length, chlorophyll content, sugar concentration, free radical scavenging and early flowering are observed (Masrahi et al. 2014). As gold is not a micronutrient required for plant growth, the utilization of gold nanoparticles as fertilizer is limited in agriculture. Platinum nanoparticles are reported as growth enhancers as they increase the length and weight of the plant root system (Astafurova et al. 2015). When *Sinapis alba* and *Lepidium sativum* were applied with the platinum nanoparticle, nutrient uptake and translocation were observed in shoot and roots of both the species without any toxicity effect (Astafurova et al. 2015). Few nanocrystalline metals such as iron, cobalt and copper were applied to soybean seeds, and improved chlorophyll index, nodule number and number of crops were observed (Ngo et al. 2014).

14.4.8 Selenium-Based Nanoparticles

More than one billion humans who suffer from selenium malnutrition require selenium as a dietary supplement. The one way to get selenium from food is through enriching the selenium level in agricultural produce by applying selenium fertilizer. As higher concentration causes toxicity, selenium nanoparticles are the best alternative to satisfy the requirement. Selenium nanoparticles are used as a detoxifying agent for heavy metals and free radicals. They also have anticancer, antioxidant and antimicrobial properties. Fortification with selenium nanoparticles enhances the biochemical properties by increasing the amino acids, proteins and secondary metabolites, including phenolics, flavonoids and glucosinolates.

14.4.9 Carbon- and Silicon Dioxide-Based Nanoparticles

Fullerenes, fullers, single-walled carbon nanotubes and multiwalled carbon nanotubes are the carbon nanomaterials used in nanotechnology. Multiwalled carbon nanotubes enhance seed germination in tomatoes without causing toxicity. Also, in the long term, they don't have any harmful effect on the crop and positively affect plant growth (Lahiani et al. 2017). Single-walled nanotube also increases the seed germination percentage in tomatoes at a lower concentration of 25 μg/mL (Lahiani et al. 2017). The essential micronutrient silicon is needed for plant growth, high yield and biotic and abiotic stresses resistance. Nanosilicon dioxide nanoparticles are used as fortifying agents due to their purity, ultrathin particle size, high surface adsorption, energy and thermal resistance. Improved seed germination, reduced transpiration rate, improved green colouration and shoot expansion were observed in tomatoes using 8 g/L of 12 nm nanosilicon dioxide (Avestan et al. 2016).

14.4.10 Biofortification in Cereal Crops

To achieve the country's national security after understanding the importance of crop biofortification, the Indian Council of Agricultural Research (ICAR) has started a Consortia Research Platform (CRP) and sanctioned it to the Indian Institute of Rice Research (IIRR) for enhancement of the nutritional status of major food crops like rice, maize, sorghum, pearl millet and minor millets. ICAR institutes have carried out this, with the Indian Council of Medical Research, state agricultural universities and the traditional universities in India, to develop biofortified cereals with increased β-carotene, quality protein, iron and zinc. Many biofortified rice varieties are released through conventional breeding for high protein (Distelfeld et al. 2007), zinc (Velu et al. 2014) and iron (Velu et al. 2014) content by screening a vast number of genotypes and identifying the donor and the QTL associated with it. Also, transgenic rice has been developed using the iron transporter gene with three to four times higher concentrations of iron than the wild types.

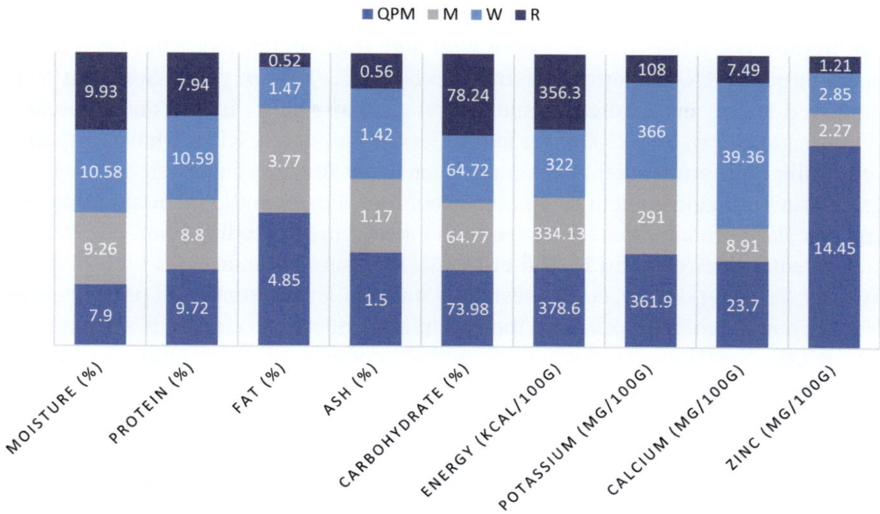

(QPM – Quality Protein Maize, M – Maize, W – Wheat and R – Rice)

Fig. 14.2 Nutritional comparison of QPM with other important staple cereal crop grains. *QPM* quality protein maize, *M* maize, *W* wheat, *R* rice)

Maize with low phytate content has been released (Pixley et al. 2011). The nutritional profile of QPM (quality protein maize) has been improved with the help of conventional breeding techniques. Therefore, it is said to be a biofortified food and its comparison with other nutrients can be possible. A comparison of the nutritional composition of quality protein maize with other staple cereal grains is indexed in Fig. 14.2 (Sumbo and Victor 2014; Longvah et al. 2017). The maize, wheat and rice crops are compared with the QPM concerning the moisture percentage, protein percentage, fat percentage, ash percentage, carbohydrate percentage, energy-providing capacity, potassium, calcium and zinc content. The fat content, zinc content and ash content of QPM are relatively low, whereas the carbohydrate content (Singla and Grover 2017) is relatively high compared to maize, wheat and rice (Fig. 14.2).

Golden rice with 23 times increased β-carotene concentration has been developed by transferring the genes required for the synthesis of the β-carotene (Ram et al. 2010). The development of biofortified wheat with higher iron content, provitamin and proteins through transgenesis is still under research (Yadav et al. 2015). Several biofortified maize verities have been developed by the introgression of the allele into the inbreds and marker-assisted selection. Multivitamin corn with an increased amount of beta-carotene, folate and ascorbate by engineering three distinct metabolic pathways (Wang et al. 2010).

In barley, the essential amino acid lysine content is enhanced by expressing the DHPS gene, and beta-glucan concentration is increased by overexpressing the cellulose synthase gene. Research for transgenic barley variety for biofortification is still in progress. Transgenic sorghum with high digestibility index has been developed through RNA interference by silencing the *kafirin* gene. Sorghum with improved lysine and carotene content through transgenesis is still under research (Parthasarathy Rao et al. 2006). In India, HarvestPlus and ICRISAT have supported pearl millet biofortification programme intensively for the genetic improvement of grain iron with zinc content as an associated trait. For both the micronutrients iron and zinc, large variabilities have been found in advanced breeding lines, populations and parental lines of hybrids (Rai et al. 2012). Using parental lines with high Fe, hybrids are being developed and tested by national and international programmers. New Fe and Zn content sources in the germplasm collections are also being explored for genetic diversification for high Fe and Zn content. Biofortification for various traits in cereal crops is indexed in Table 14.2.

14.5 Biosafety of Nanomaterials in Sustainable Agriculture

Nanomaterials depend on several factors such as substances used, process method, biological substrate, size and structure and reactions in the application medium. So, the assessment of safety is complex in the case of nano-fertilizers. Depending on the synthesis method, nanoparticles may not be considered entirely safe without considering the synthesis material (Lee et al. 2003). In general, biologically engineered nanomaterials are less toxic than chemically synthesized, and are more biocompatible and safer. The behaviour, properties and decomposition of the nanomaterial should be studied to minimize the toxicity and make it environmentally safe (Astafurova et al. 2015). According to the group of researchers from the School of Agricultural, Food and Biosystems Engineering (ETSIAAB) under Universidad Politécnica de Madrid (UPM), zinc oxide nanoparticles can be used as a source of nano-fertilizer without any toxicity (Chemgroup 2017). Zinc should not be used in excess amounts, either in ionic form or oxide form, to avoid toxicity. Generally, metal and metal oxide nanoparticles are more detrimental to soil microbes and affect the nitrogen-fixing bacteria and phosphorus- and potassium-solubilizing bacteria. But SiO_2 and TiO_2 nanoparticles enhance the activities of nitrifying bacteria and the uptake of nutrients (Changmei et al. 2002). Biologically synthesized silver nanoparticles are less toxic and positively affect protein and carbohydrate synthesis (Krishnaraj et al. 2012). Along with the development of nano-fertilizers and usage in agriculture, the safety and toxicity of nanoparticles must be investigated by evaluating the thin line between deficiency and toxicity of the nutrients (Chen et al. 2010).

Table 14.2 Biofortification for various traits in cereal crops

Crop name	Types of biofortification	Genes	References
Rice	Beta-carotene and phytoene	Phytoene synthase and carotene 7,8-desaturase	Ye et al. 2000
	Folate (vitamin B9)	*GTPCHI* – GTP cyclohydrolase I *ADCS* – Aminodeoxychorismate synthase	Blancquaert et al. 2015
	Iron	*OsIRT1* (nicotianamine aminotransferase, iron transporter), *OsNAS1*, *OsNAS2* (nicotianamine synthase 1 and 2) Soybean ferritin, standard bean ferritin	Masuda et al. 2008
	Zinc	Overexpression of *OsIRT1* and *HvNAS family*, *HvNAAT-B*, *IDS3* (mugineic acid synthesis gene from barley)	Lee and An, 2009
	Resistant starch	Antisense RNA inhibition of SEB (starch-branching enzymes)	Wei et al. 2010
	Protein and amino acid (content high)	Pea legumin, rice anthranilate synthase α-subunit, sesame 2S albumin	Zhouet al. 2009
	Flavonoids and antioxidants	Maize C1 and R-S regulatory genes and phenylalanine ammonia-lyase and chalcone synthase (*CHS*) genes	Ogo et al. 2013
	Alpha-linolenic acid	Soybean omega-3 fatty acid desaturase (*FAD3*) gene (*GmFAD3*)	Anai et al. 2003
Wheat	Provitamin A and carotenoids	CrtB, CrtI – Bacterial phytoene synthase and carotene 7,8-desaturase gene	Wang et al., 2014
	Iron	*TaFer1-A* – Soybean ferritin gene	Xiaoyan et al. 2012
	Phytase or phytic acid	Wheat ABCC13 transporter gene silencing, *phyA* (phytochrome gene)	Bhati et al. 2016
	Amino acid composition	*ama1* – Albumin gene of *Amaranthus*	Tamás et al. 2009
	Anthocyanin	C1, B-Peru (maize regulatory genes)	Doshi et al. 2006
	Amylose content	SBE (SBEIIa) gene silencing	Francesco et al. 2010
Maize	Provitamin A and carotenoids	By expressing bacterial *crtB*, carotenogenic genes	Decourcelle et al. 2015
	Tocopherol	Overexpression of homogentisic acid, *HGGT* gene, geranylgeranyl transferase gene	Cahoon et al. 2003
	Ascorbic acid	DHAR – Dehydroascorbate reductase	Chen et al. 2003
	Phytase, ferritin (iron Bioavailability)	Soybean ferritin and *aspergillus* phytase, silencing the expression of ATP-binding transporter gene	Aluru et al. 2011
	Lysine, tryptophan and methionine	*sb401* from potato, antisense dsRNA targeting alpha-zeins, *cis*-acting site for *Dzs10*	Frizzi et al. 2008

(continued)

Table 14.2 (continued)

Crop name	Types of biofortification	Genes	References
	Human lactoferrin	α-Lactalbumin	Yang et al. 2002
Barley	Zinc	Zinc transporters	Ramesh et al. 2004
	Phytase	Phytase gene	Holme et al. 2012
	Lysine	Deoxyhypusine synthase gene	Ohnoutkova et al. 2012
	Beta-glucan	*HvCslF* – Cellulose synthase-like gene	Burton et al. 2011
	Resistant starch	SBE gene family suppression	Carciofi et al. 2012
	Polyunsaturated fatty acids	Δ6-desaturase (*D6D*)	Mihálik et al. 2014
	Human lactoferrin	Human lactoferrin (*HLF*) gene	Kamenarova et al. 2007
Sorghum	Provitamin A	*Homo188-A*	Lipkie et al. 2013
	Lysine	HT12 – High lysine protein	Zhao et al., 2003
	Improved protein digestibility	*γ-Kafirin gene* suppression by RNAi technology	Elkonin et al. 2016

14.6 Conclusion

As demonstrated schematically, nanotechnology has been used in agriculture to enhance crop output, including cereal crops, while also enhancing quality through updating farming infrastructure. Engineered nanomaterials and their application in sustainable agriculture have radically altered the world agricultural landscape due to their novelty, rapid expansion and enormous potential for fulfilling anticipated world food demand. Biofortification is a potentially cost-effective agricultural method for improving the nutritional quality of crops especially in underprivileged populations worldwide, as described above. Traditional crop breeding, genetic change of a specific trait and nanoparticle application are all promising biofortification approaches for treating human nutritional deficiency. The era of biofortified cereal crops with increased nutritional content provides appropriate amounts of micronutrients, that are typically lacking in diets in both the developing and developed countries across the globe. Many people throughout the world consume fortified cereal grains, which are an essential source of nutrients. For treating micronutrient deficiencies, cereal crop biofortification is economical and accessible to everyone. It can be stated that, nanobiotechnology and nanomaterials are the most current sources of cereal crop biofortification.

Conflict of Interest The authors declare that they have no conflict of interest.

References

Chemgroup AG (2017) Study finds potential use for zinc oxide nanoparticles for fertilizer feedstock. AG Chemgroup, Praha

Aluru MR, Rodermel SR, Reddy MB (2011) Genetic modification of low phytic acid 1-1 maize to enhance iron content and bioavailability. J Agric Food Chem 59(24):12954–12962

Anai T, Koga M, Tanaka H, Kinoshita T, Rahman SM, Takagi Y (2003) Improvement of rice (Oryza sativa L.) seed oil quality through introduction of a soybean microsomal omega-3 fatty acid desaturase gene. Plant Cell Rep 21(10):988–992

Astafurova T, Zotikova A, Morgalev Y, Verkhoturova G, Postovalova V, Kulizhskiy S, Mikhailova S (2015) Effect of platinum nanoparticles on morphological parameters of spring wheat seedlings in a substrate-plant system. In: IOP Conference Series: Materials Science and Engineering, vol 98, No. 1. IOP Publishing, p 012004

Avestan S, Naseri LA, Hassanzade A, Sokri SM, Barker AV (2016) Effects of nanosilicon dioxide application on in vitro proliferation of apple rootstock. J Plant Nutr 39(6):850–855

Bhati KK, Alok A, Kumar A, Kaur J, Tiwari S, Pandey AK (2016) Silencing of ABCC13 transporter in wheat reveals its involvement in brain development, phytic acid accumulation and lateral root formation. J Exp Bot 67(14):4379–4389

Blancquaert D, Van Daele J, Strobbe S, Kiekens F, Storozhenko S, De Steur H et al (2015) Improving folate (vitamin B 9) stability in biofortified rice through metabolic engineering. Nat Biotechnol 33(10):1076–1078

Bouis HE (2000) Enrichment of food staples through plant breeding: a new strategy for fighting micronutrient malnutrition. Nutrition 16:701–704

Bouis HE, Hotz C, McClafferty B, Meenakshi JV, Pfeiffer WH (2011) Biofortification: a new tool to reduce micronutrient malnutrition. Food Nutr Bull 32(1_suppl1):S31–S40

Burton RA, Collins HM, Kibble NA, Smith JA, Shirley NJ, Jobling SA et al (2011) Over-expression of specific HvCslF cellulose synthase-like genes in transgenic barley increases the levels of cell wall (1, 3; 1, 4)-β-d-glucans and alters their fine structure. Plant Biotechnol J 9(2): 117–135

Cahoon EB, Hall SE, Ripp KG, Ganzke TS, Hitz WD, Coughlan SJ (2003) Metabolic redesign of vitamin E biosynthesis in plants for tocotrienol production and increased antioxidant content. Nat Biotechnol 21(9):1082–1087

Cao Z, Rossi L, Stowers C, Zhang W, Lombardini L, Ma X (2018) The impact of cerium oxide nanoparticles on the physiology of soybean (Glycine max (L.) Merr.) under different soil moisture conditions. Environ Sci Pollut Res 25(1):930–939

Carciofi M, Blennow A, Jensen SL, Shaik SS, Henriksen A, Buléon A et al (2012) Concerted suppression of all starch branching enzyme genes in barley produces amylose-only starch granules. BMC Plant Biol 12(1):1–16

Changmei L, Chaoying Z, Junqiang W, Guorong W, Mingxuan T (2002) Research of the effect of nanometer materials on germination and growth enhancement of Glycine max and its mechanism. Soybean Sci 21(3):168–171

Chen R, Ratnikova TA, Stone MB, Lin S, Lard M, Huang G et al (2010) Differential uptake of carbon nanoparticles by plant and mammalian cells. Small 6(5):612–617

Chen Z, Young TE, Ling J, Chang SC, Gallie DR (2003) Increasing vitamin C content of plants through enhanced ascorbate recycling. Proc Natl Acad Sci 100(6):3525–3530

Chizuru N, Ricardo U, Shiriki K, Prakash S (2003) The joint WHO/FAO expert consultation on diet, nutrition and the prevention of chronic diseases: process, product and policy implications. Public Health Nutr 7(1a):245–250

Decourcelle M, Perez-Fons L, Baulande S, Steiger S, Couvelard L, Hem S et al (2015) Combined transcript, proteome, and metabolite analysis of transgenic maize seeds engineered for enhanced carotenoid synthesis reveals pleiotropic effects in core metabolism. J Exp Bot 66(11): 3141–3150

Distelfeld A, Cakmak I, Peleg Z, Ozturk L, Yazici AM, Budak H (2007) Multiple QTL-effects of wheat Gpc-B1 locus on grain protein and micronutrient concentrations. Physiol Plant 129:635–643

Doshi KM, Eudes F, Laroche A, Gaudet D (2006) Transient embryo-specific expression of anthocyanin in wheat. In Vitro Cell Develop Biol Plant 42(5):432–438

Duhan JS, Kumar R, Kumar N, Kaur P, Nehra K, Duhan S (2017) Nanotechnology: the new perspective in precision agriculture. Biotechnol Rep 15:11–23

Elkonin LA, Italianskaya JV, Domanina IV, Selivanov NY, Rakitin AL, Ravin NV (2016) Transgenic sorghum with improved digestibility of storage proteins obtained by agrobacterium-mediated transformation. Russ J Plant Physiol 63(5):678–689

Eroglu N, Emekci M, Athanassiou CG (2017) Applications of natural zeolites on agriculture and food production. J Sci Food Agric 97(11):3487–3499

Francesco S, Michela J, Angela D (2010) Increasing the amylose content of durum wheat through silencing of the SBEIIa genes. BMC Plant Biol 144(10):1–12

Frizzi A, Huang S, Gilbertson LA, Armstrong TA, Luethy MH, Malvar TM (2008) Modifying lysine biosynthesis and catabolism in corn with a single bifunctional expression/silencing transgene cassette. Plant Biotechnol J 6(1):13–21

Garg M, Sharma N, Sharma S, Kapoor P, Kumar A, Chunduri V, Arora P (2018) Biofortified crops generated by breeding, agronomy, and transgenic approaches are improving the lives of millions of people around the world. Front Nutr 5:12

Gibson RS, Anderson VP (2009) A review of interventions based on dietary diversification or modification strategies with the potential to enhance intakes of total and absorbable zinc. Food Nutr Bull 30:108–143

Graham RD, Welch RM, Bouis HE (2001) Addressing micronutrient malnutrition through enhancing the nutritional quality of staple foods: principles, perspectives and knowledge gaps. Adv Agron 70:77–142

Harper T (2015) The year of the trillion Dollar nanotechnology market. AZoNetwork UK Ltd., Manchester

Holme IB, Dionisio G, Brinch-Pedersen H, Wendt T, Madsen CK, Vincze E, Holm PB (2012) Cisgenic barley with improved phytase activity. Plant Biotechnol J 10(2):237–247

Hossain SM, Mohiuddin AKM (2012) Study on biofortification of Rice by targeted genetic engineering. Int J Agric Res Innov Technol 2:25–35

Hu J, Guo H, Li J, Wang Y, Xiao L, Xing B (2017) Interaction of γ-Fe 2 O 3 nanoparticles with Citrus maxima leaves and the corresponding physiological effects via foliar application. J Nanobiotechnol 15(1):51

Joy EJM, Kumssa DB, Broadley MR, Watts MJ, Young SD, Chilimba ADC, Ander EL (2015) Dietary mineral supplies in Malawi: spatial and socioeconomic assessment. BMC Nutr 1:42

Kamenarova K, Gecheff K, Stoyanova M, Muhovski Y, Anzai H, Atanassov A (2007) Production of recombinant human lactoferrin in transgenic barley. Biotechnol Biotechnol Equip 21(1): 18–27

Krishnaraj C, Ramachandran R, Mohan K, Kalaichelvan PT (2012) Optimization for rapid synthesis of silver nanoparticles and its effect on phytopathogenic fungi. Spectrochim Acta A Mol Biomol Spectrosc 93:95–99

Lahiani MH, Nima ZA, Villagarcia H, Biris AS, Khodakovskaya MV (2017) Assessment of effects of the long-term exposure of agricultural crops to carbon nanotubes. J Agric Food Chem 66(26): 6654–6662

Lee TT, Wang MM, Hou RC, Chen LJ, Su RC, Wang CS, Tzen JT (2003) Enhanced methionine and cysteine levels in transgenic rice seeds by the accumulation of sesame 2S albumin. Biosci Biotechnol Biochem 67(8):1699–1705

Lipkie TE, De Moura FF, Zhao ZY, Albertsen MC, Che P, Glassman K, Ferruzzi MG (2013) Bioaccessibility of carotenoids from transgenic provitamin a biofortified sorghum. J Agric Food Chem 61(24):5764–5771

Longvah T, Ananthan R, Bhakarachary K, Venkaiah K (2017) Indian food composition tables. National Institute of Nutrition, Indian Council of Medical Research, Hyderabad

Manikandan A, Subramanian KS (2016) Evaluation of zeolite-based nitrogen nano-fertilizers on maize growth, yield and quality on inceptisols and alfisols. Int J Plant Soil Sci 9(4):1–9

Masrahi A, VandeVoort AR, Arai Y (2014) Effects of silver nanoparticle on soil-nitrification processes. Arch Environ Contam Toxicol 66(4):504–513

Masuda H, Suzuki M, Morikawa KC, Kobayashi T, Nakanishi H, Takahashi M, Nishizawa NK (2008) Increase in iron and zinc concentrations in rice grains via the introduction of barley genes involved in phytosiderophore synthesis. Rice 1(1):100–108

McGuire S (2015) FAO, IFAD, and WFP. In: The state of food insecurity in the world 2015: Meeting the 2015 international hunger targets: taking stock of uneven progress. FAO, Rome, p 2015

Mihálik D, Gubišová M, Klempová T, Čertík M, Ondreičková K, Hudcovicová M et al (2014) Transgenic barley producing essential polyunsaturated fatty acids. Biol Plant 58(2):348–354

Milani N, Hettiarachchi GM, Kirby JK, Beak DG, Stacey SP, McLaughlin MJ (2015) Fate of zinc oxide nanoparticles coated onto micronutrient fertilizers in an alkaline calcareous soil. PLoS One 10(5):e0126275

Miller JB, Zhang S, Kos P, Xiong H, Zhou K, Perelman SS, Zhu H, Siegwart DJ (2017) Non-viral CRISPR/Cas gene editing in vitro and in vivo enabled by synthetic nanoparticle co-delivery of Cas9 mRNA and sgRNA. Angew Chem Int Ed 56:1059–1063

Morales-Díaz AB, Ortega-Ortíz H, Juárez-Maldonado A, Cadenas-Pliego G, González-Morales S, Benavides-Mendoza A (2017) Application of nanoelements in plant nutrition and its impact in ecosystems. Adv Nat Sci Nanosci Nanotechnol 8(1):013001

Neeraja CN, Babu VR, Ram S, Hossain F, Hariprasanna K, Rajpurohit BS, Datta SK (2017) Biofortification in cereals: progress and prospects. Curr Sci 113(6):1050–1057

Ngo QB, Dao TH, Nguyen HC, Tran XT, Van Nguyen T, Khuu TD, Huynh TH (2014) Effects of nanocrystalline powders (Fe, co and cu) on the germination, growth, crop yield and product quality of soybean (Vietnamese species DT-51). Adv Nat Sci Nanosci Nanotechnol 5(1):015016

Ogo Y, Ozawa K, Ishimaru T, Murayama T, Takaiwa F (2013) Transgenic rice seed synthesizing diverse flavonoids at high levels: a new platform for flavonoid production with associated health benefits. Plant Biotechnol J 11(6):734–746

Ohnoutkova L, Zitka O, Mrizova K, Vaskova J, Galuszka P, Cernei N et al (2012) Electrophoretic and chromatographic evaluation of transgenic barley expressing a bacterial dihydrodipicolinate synthase. Electrophoresis 33(15):2365–2373

Parthasarathy Rao P, Birth PS, Reddy BVS, Rai KN, Ramesh S (2006) Diagnostics of sorghum and pearl millet grain-based nutrition in India. Int Sorghum Millets Newsl 47:93–96

Pixley KV, Palacios N, Glahn RP (2011) The usefulness of iron bioavailability as a target trait for breeding maize (Zea mays L.) with enhanced nutritional value. Field Crops Res 123:153–160

Rai KN, Govindaraj M, Rao AS (2012) Genetic enhancement of grain iron and zinc content in pearl millet. Qual Assur Saf Crop 4:119–125

Ram S, Verma A, Sharma S (2010) Large variability exists in phytase levels among Indian wheat varieties and synthetic hexaploids. J Cereal Sci 52:486–490

Ramesh SA, Choimes S, Schachtman DP (2004) Over-expression of an Arabidopsis zinc transporter in Hordeum vulgare increases short-term zinc uptake after zinc deprivation and seed zinc content. Plant Mol Biol 54(3):373–385

Singla P, Grover K (2017) Biofortified cereal crops: a sustainable approach for food and nutritional security. Curr J Appl Sci Technol 24:1–13

Smith IF (2000) Micronutrient interventions: options for Africa. Food Nutr Bull 21:532–537

Stein AJ, Nestel P, Meenakshi JV, Qaim M, Sachdev HPS, Bhutta ZA (2007) Plant breeding to control zinc deficiency in India: how cost-effective is biofortification? *Public Health Nutr* 10(5): 492–501

Stevens GA, Finucane MM, De-Regil L, Paciorek CJ, Flaxman SR, Branca F et al (2013) Global, regional, and national trends in haemoglobin concentration and prevalence of total and severe anaemia in children and pregnant and non-pregnant women for 1995–2011: a systematic analysis of population-representative data. Lancet Glob Health 1(1):e16–e25

Sumbo HA, Victor IA (2014) Comparison of chemical composition, functional properties and amino acids composition of quality protein maize and common maize (Zea mays L). Afr J Food Sci Technol 5(3):81–89

Tamás C, Kisgyörgy BN, Rakszegi M, Wilkinson MD, Yang MS, Láng L, Bedő Z (2009) Transgenic approach to improve wheat (Triticum aestivum L.) nutritional quality. Plant Cell Rep 28(7):1085–1094

Thakur S, Thakur S, Kumar R (2018) Bio-nanotechnology and its role in agriculture and food industry. J Mol Genet Med 12(324):1747–0862

Velu G, Ortiz-Monasterio I, Cakmak I, Hao Y, Singh RP (2014) Biofortification strategies to increase grain zinc and iron concentrations in wheat. J Cereal Sci 59:365–372

Wang Z, Libault M, Joshi T, Valliyodan B, Nguyen HT, Xu D et al (2010) SoyDB: a knowledge database of soybean transcription factors. BMC Plant Biol 10(1):14

Wei C, Zhang J, Chen Y, Zhou W, Xu B, Wang Y, Chen J (2010) Physicochemical properties and development of wheat large and small starch granules during endosperm development. Acta Physiol Plant 32(5):905–916

Xiaoyan S, Yan Z, Shubin W (2012) Improvement Fe content of wheat (Triticum aestivum) grain by soybean ferritin expression cassette without vector backbone sequence. J Agric Biotechnol 20(7):766–773

Yadav OP et al (2015) Genetic improvement of maize in India: retrospect and prospects. Agric Res 4. https://doi.org/10.1007/s40003-015-0180-8

Yang SH, Moran DL, Jia HW, Bicar EH, Lee M, Scott MP (2002) Expression of a synthetic porcine α-lactalbumin gene in the kernels of transgenic maize. Transgenic Res 11(1):11–20

Ye X, Al-Babili S, Kloti A, Zhang J, Lucca P, Beyer P, Potrykus I (2000) Engineering the provitamin A (β-carotene) biosynthetic pathway into (carotenoid-free) rice endosperm. *Science 287*(5451):303–305

Zhou Y, Cai H, Xiao J, Li X, Zhang Q, Lian X (2009) Over-expression of aspartate aminotransferase genes in rice resulted in altered nitrogen metabolism and increased amino acid content in seeds. Theor Appl Genet 118(7):1381–1390

Correction to: Biofortification in Cereals

Rupesh Deshmukh, Altafhusain Nadaf, Waquar Akhter Ansari,
Kashmir Singh, and Humira Sonah

Correction to:
R. Deshmukh et al. (eds.), Biofortification in Cereals,
https://doi.org/10.1007/978-981-19-4308-9

The original version of the book was published with errors. The following corrections have been made after publication.

1. The first editor Dr. Rupesh Deshmukh's affiliation has been changed to read as "Central University of Haryana, Mahendragarh, Haryana, India".

2. The title of Chapter 3 has been updated to read as "Molecular Breeding Approaches for Biofortification of Cereal Crops".

The updated original version for this book and chapter 3 can be found at
https://doi.org/10.1007/978-981-19-4308-9
https://doi.org/10.1007/978-981-19-4308-9_3

R. Deshmukh et al. (eds.), *Biofortification in Cereals*,
https://doi.org/10.1007/978-981-19-4308-9_15

MIX
Papier aus verantwortungsvollen Quellen
Paper from responsible sources
FSC® C105338

If you have any concerns about our products,
you can contact us on
ProductSafety@springernature.com

In case Publisher is established outside the EU,
the EU authorized representative is:
Springer Nature Customer Service Center GmbH
Europaplatz 3, 69115 Heidelberg, Germany

Printed by Libri Plureos GmbH
in Hamburg, Germany